ELECTRIC MACHINES

STEADY-STATE THEORY AND DYNAMIC PERFORMANCE

ELECTRIC MACHINES

STEADY-STATE THEORY AND DYNAMIC PERFORMANCE

MULUKUTLA S. SARMA
NORTHEASTERN UNIVERSITY

wcb

WM. C. BROWN PUBLISHERS
DUBUQUE, IOWA

wcb group

WM. C. BROWN   CHAIRMAN OF THE BOARD
MARK C. FALB   PRESIDENT AND CHIEF EXECUTIVE OFFICER

wcb

WM. C. BROWN PUBLISHERS
COLLEGE DIVISION

LAWRENCE E. CREMER   PRESIDENT
JAMES L. ROMIG   VICE-PRESIDENT, PRODUCT DEVELOPMENT
DAVID A. CORONA   VICE-PRESIDENT, PRODUCTION AND DESIGN
E. F. JOGERST   VICE-PRESIDENT, COST ANALYST
MARCIA H. STOUT   MARKETING MANAGER
MARILYN A. PHELPS   MANAGER OF DESIGN
WILLIAM A. MOSS   PRODUCTION EDITORIAL MANAGER
MARY M. HELLER   PHOTO RESEARCH MANAGER

BOOK TEAM

ROBERT B. STERN   SENIOR EDITOR
NOVA A. MAACK   ASSISTANT EDITOR
MARY BETH TAUKE   DESIGNER
BARBARA ROWE DAY   PRODUCTION EDITOR
CAROL M. SCHIESSL   PHOTO RESEARCH EDITOR
MAVIS M. OETH   PERMISSIONS EDITOR

TO MY CHILDREN,
SATISH, SANTI, AND SURESH
AND MY WIFE,
SAVITRI

CONTENTS

PREFACE

I have always enjoyed engineering (teaching, research, and consultation), particularly because the subject area involves devices that move rather than static pieces of equipment such as electronic amplifiers. To some of our young would-be scientists and engineers who are inclined to believe that electrical machines is a difficult or dull subject and one in which the fundamental research has all been done, I have tried to convey some of the thrills of exploring the field of electromechanical energy conversion and, in most cases, succeeded in rapidly dispelling their misconceived ideas. The urgent need for updating of power engineering programs and production of highly competent engineers capable of working on the frontiers of fast-developing, competitive technology has been recognized.

Electrical machinery is one of the subject areas that has been most affected by changing curricula patterns. Whereas it once formed the core of the electrical engineering syllabus, with as many as three or four courses devoted to the subject, it must now constitute a more modest portion of the curriculum so that more time may be devoted to covering the vast range of developments in other fields. Nowadays there is barely a single course assigned for electrical machines, and in some rare instances a two-course sequence appears. Coupled with this significant reduction in time, however, is the increased scope of the subject to include steady-state performance as well as dynamic behavior of electrical machines, which is essential for analyzing the overall stability of a system of which the machine happens to be an integral part. The study of electrical machines continues to be one of the most fundamental subjects in an engineering curriculum because it provides a basic physical and mathematical understanding of electromechanical energy conversion.

This book focuses on electromechanical energy conversion, and it is aimed at the typical undergraduate course of one or two semesters in duration (or of one semester followed by an elective course in a second semester). Because fundamental physical concepts underlie creative engineering and become the most valuable as well as permanent part of a student's background, they have been highlighted while giving due attention to mathematical techniques. The theory has been developed from simple beginnings in such a manner that it can readily be extended to new and complex situations. The art of reducing a practical device to an appropriate mathematical model and recognizing its limitations has been adequately presented. Sufficient motivation is provided for the student to develop interest in the analytical procedures to be applied. To accomplish all this in a shorter time than formerly available, much thought has gone into rationalizing the theory and conveying in a concise manner the essential details concerning the physical nature of machines.

Providing a match with the modern engineering curricula, comprehensively under one cover, the full text adequately meets the needs of those whose training requires a deep understanding of various aspects of machine behavior for further study of power systems, control systems, machine design, or general industrial applications. Latest developments such as the practical utilization of superconducting materials in electrical machines have been indicated in the text wherever appropriate, thereby making the book up-to-date from the viewpoints of both subject matter and analytical procedures. Most space is devoted to a careful explanation of four main electromagnetic devices: the transformer, the induction motor, the synchronous machine, and the direct-current machine. A sound knowledge of these should permit the ready understanding of all other electrical machines, which for the most part are modifications of these four types. Similarities among the main kinds of machines are brought out wherever possible and similar lines of treatment are followed for each machine where this is practicable, in spite of the distinguishing features peculiar to each class of machines. The theory has been extended to introduce different modes of operation in a manner adequate for general understanding, but without detailed explanation in all cases in view of the scope of the text and

PART 1

INTRODUCTION

The theme of this book is electromechanical energy conversion. Electromechanical devices are involved in every industrial and manufacturing process of a technological society. The importance of these devices in almost every aspect of life does not need to be emphasized. An understanding of the principles of electromechanics is quite important for all those who desire to extend the usefulness of electrical technology in order to ameliorate the problems of energy, pollution, and poverty that presently face mankind.

The main purpose of this book is to present the fundamentals of electromechanics, apply these to some of the basic configurations of electromechanical devices, and stimulate the reader to further investigation of new and complex situations with exciting possibilities for advanced development. Most space is devoted to a careful explanation of four main electromagnetic devices: the transformer, the induction motor, the synchronous machine, and the direct-current machine. The book is up-to-date from the viewpoints of both subject matter and analytical procedures to meet the needs of those whose training requires a deep understanding of various aspects of machine behavior for further study of either power systems, control systems, machine design, or general industrial applications.

The transformer, although not an electromechanical energy-conversion device, is an important auxiliary in the overall problem of energy conversion for the transformation of electrical power between different voltage-current levels. The concepts of the transformer behavior and the analysis of mutually coupled circuits serve as a useful adjunct to the study of electromagnetic rotating machines. The generators, converting mechanical energy to electrical energy, as well as the motors, converting electrical energy to mechanical energy, provide the power on which modern industrialized societies depend so much. Electromechanical energy conversion is effected by using the magnetic field as a coupling medium between a stationary member and a moving member. The great ease with which energy may be stored in magnetic fields accounts largely for the wide use of electromagnetic devices for the interconversion of electrical and mechanical energy.

Electromagnetic devices utilize magnetic materials for shaping the magnetic fields that act as the medium for transferring and converting energy. The interaction between the electric circuits and the magnetic fields plays an important part in the operation of various types of equipment discussed in this book. That is why a chapter on magnetic circuits, including those involving permanent magnets, is presented. Broader aspects of electromechanical energy conversion are given due attention in order to present a proper perspective and motivate the reader to think beyond the devices that are treated here.

Machine windings are the means by which the theory is translated into practice, and they form a unifying link between different machines; hence, a chapter is devoted for presenting the essentials of these.

The modern interest in electronics has influenced circuit texts and courses to the detriment of the study of polyphase systems. Since this topic is quite fundamental to the study of commercial and industrial power applications that utilize polyphase devices, a chapter is devoted for the analysis of balanced three-phase circuits.

Analysis by means of phasor diagrams is a very useful approach for the understanding of mathematical equivalent-circuit models of practical devices. As such a review of phasor diagrams is considered to be appropriate at the very beginning.

1

A REVIEW OF PHASOR DIAGRAMS

1.1 PHASORS

Let us consider a complex function of time,

$$f(t) = re^{j\omega t} \tag{1.1.1}$$

whose magnitude, given by r, is a constant and whose angle, given by ωt radians at time t, changes with time. Such a function can be graphically interpreted in terms of a rotating *phasor* whose magnitude is given by r units and direction of rotation is counterclockwise (considered positive for positive ω and positive t). Further, if ω is a constant, the frequency of rotation becomes a constant equal to ω rad/s. $e^{-j\omega t}$ would imply clockwise (or negative) rotation.

By the use of Euler's identity, Equation 1.1.1 may be written as

$$f(t) = re^{j\omega t} = r(\cos \omega t + j \sin \omega t) \tag{1.1.2}$$

in which the real part (or the projection on the real axis) varies as the cosine of ωt, while the imaginary part (or the projection on the imaginary axis) varies as the sine of ωt. This concept is illustrated in Figure 1.1.1. The variation of the exponential function with time is sinusoidal in nature and corresponds to the case of the sinusoidal steady state.

The sum of rotating phasors of the same frequency ω is a rotating phasor of the same frequency. This property becomes a key to the use of rotating phasors for sinusoidal steady-state analysis. Also, it can be shown that if a rotating-phasor input signal is applied to a linear time-invariant network or system, the steady-state output is a rotating phasor of the same frequency.

Consider a general sinusoidal voltage of the form

$$v(t) = V_m \cos(\omega t + \theta) \tag{1.1.3}$$

Once the frequency ω is known, then v can be completely specified by its amplitude V_m and its phase θ. Based on the following relationship

$$V_m \cos(\omega t + \theta) = \text{Real part of } [V_m e^{j\theta} e^{j\omega t}] \tag{1.1.4}$$

a related *phasor* can be defined such that

$$\overline{V} = V_m e^{j\theta} = V_m \angle \theta \tag{1.1.5}$$

and

$$v(t) = \text{Real part of } [\overline{V} e^{j\omega t}] \tag{1.1.6}$$

FIGURE 1.1.1 A rotating phasor with its projection on the imaginary axis varying as the sine and its projection on the real axis varying as the cosine.

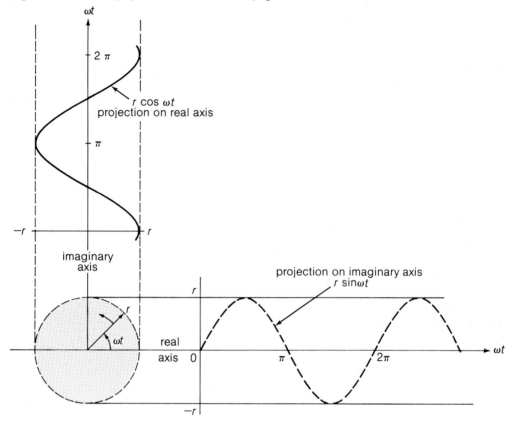

It should be pointed out that although we have chosen to represent sinusoids and their related phasors on the basis of cosine functions, we could have chosen sine functions just as easily. Useful equations in changing sine functions to cosines are given below:

$$\sin(\omega t + \alpha) = \cos\left(\omega t + \alpha - \frac{\pi}{2}\right) \qquad (1.1.7)$$

and

$$\cos(\omega t + \beta) = \sin\left(\omega t + \beta + \frac{\pi}{2}\right) \qquad (1.1.8)$$

As an example, let us consider an ac voltage of constant amplitude and constant frequency given by the following in the time domain:

$$v(t) = 100 \sqrt{2} \cos(\omega t + 30°) \text{ volts} \qquad (1.1.9)$$

in which ω is given by $(2 \times \pi \times 60)$ rad/s corresponding to a frequency f of 60 Hz. In the case of ac voltages and currents, it is more convenient to choose the rms values rather than the amplitudes for the magnitudes of the phasors. Suppressing the explicit time variation, the phasor representation in the frequency domain then becomes

$$\overline{V} = 100 \angle 30° \text{ volts} \qquad (1.1.10)$$

Starting from Equation 1.1.10, it is easy to go back to Equation 1.1.9 in the time domain by making use of Equation 1.1.6 and remembering that the amplitude is given by $\sqrt{2}$ times the rms value for sinusoidal variation. The phasor representation of the time-domain current such as

$$i(t) = 10\sqrt{2} \sin(\omega t + 30°) \text{ A} \qquad (1.1.11)$$

will be given by

$$\overline{I} = 10 \angle -60° \text{ A} \qquad (1.1.12)$$

since by Equation 1.1.7

$$\begin{aligned} \sin(\omega t + 30°) &= \cos(\omega t + 30° - 90°) \\ &= \cos(\omega t - 60°) \end{aligned} \qquad (1.1.13)$$

Note that in Equation 1.1.12 rms value of the current has been chosen for the magnitude of the phasor.

Sinusoidal waveforms of the *same* wavelength or frequency can be represented by phasors that can be added or subtracted like coplanar vectors. Such waveforms may be functions of time as in the case of ac voltages and currents of constant amplitude and constant frequency; they can also be space waves of flux density and magnetomotive force of constant amplitude and constant wavelength. Sinusoidal distributions of flux density and magnetomotive force in the air gaps of electric machines are treated as phasors. However, the magnitude of a space phasor is generally made equal to the amplitude of the wave, and the angles represent space angles rather than time angles.

For the three linear time-invariant passive elements, R (the pure resistance), L (the pure inductance), and C (the pure capacitance), the relationships between voltage and current in the time domain as well as in the frequency domain are shown in Figure 1.1.2.

FIGURE 1.1.2 Voltage and current relationships in time domain and frequency domain for the elements *R, L,* and *C. (a)* The current is in phase with voltage in a purely resistive circuit (unity power factor), *(b)* the current lags the voltage by 90° in the case of a pure inductor (zero power factor lagging), *(c)* the current leads the voltage by 90° in the case of a pure capacitor (zero power factor leading).

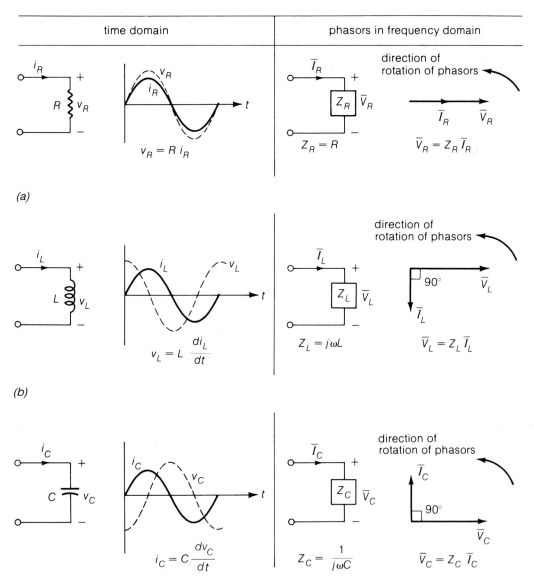

FIGURE 1.1.3   Sinusoidal steady-state analysis method by the use of phasors.

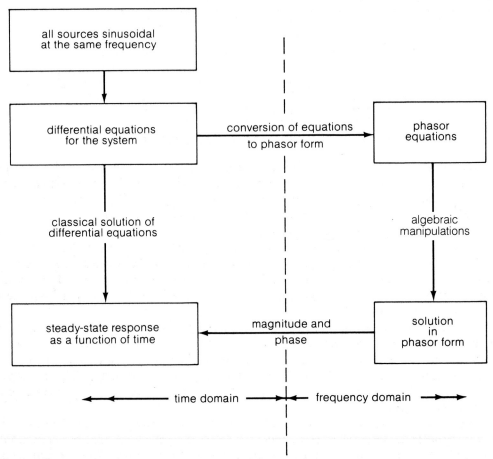

In general, the real solutions are time-domain functions, and their phasors are frequency-domain functions, being the functions of the frequency ω. Thus, to solve the time-domain problem, one may convert to phasors and solve the corresponding frequency-domain problem, which is generally much easier. Finally, one can convert back to the time domain by finding the corresponding time function from its phasor representation. Thus, when all sources are sinusoidal, operating at the same frequency, and only the steady-state response is desired, the method of solution using phasors is shown in Figure 1.1.3.

1.2   ANALYSIS WITH PHASOR DIAGRAMS

We have seen that phasors, being complex numbers, can be represented in the complex plane in the conventional polar form as an arrow, with a length corresponding to the magnitude of the phasor, and an angle (with respect to the positive real axis) that is the phase of the phasor. In a

FIGURE 1.2.1 Series-parallel circuit.

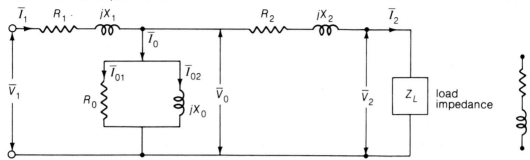

phasor diagram, the various phasor quantities corresponding to a given network may be combined in such a way that one or both of Kirchhoff's laws are satisfied. The phasor method of analyzing circuits is credited generally to Charles Proteus Steinmetz (1865–1923), a famous electrical engineer with the General Electric Company in the early part of the 20th century.

The phasor diagram may be drawn in a number of ways such as in the polar (or ray) form, with all phasors originating at the origin, or in the polygonal form, with one phasor located at the end of another, or in a combination of the above depending on the convenience and the point to be made. Such diagrams provide geometrical insight into the voltage and current relationships in a network; they are particularly helpful in visualizing steady-state phenomena in the analysis of networks with sinusoidal signals.

In constructing a phasor diagram, each sinusoidal voltage and current is represented by a phasor of length equal to the rms value of the sinusoid, and with an angular displacement from the positive real axis, which is the angle of the equivalent cosine function at t equal to zero. This use of the cosine is arbitrary, and so also the use of rms value. Formulation in terms of the sine function could just as well have been chosen, and the amplitude rather than the rms value could have been chosen for the magnitude.

The *reference* of a phasor diagram is the line $\theta = 0$. It is not really necessary that any voltage or current phasor coincide with the reference, even though it is often more convenient for the analysis when a given voltage or current phasor is taken as the reference. Since a phasor diagram is a frozen picture at one instant of time giving the relative locations of various phasors involved and the whole diagram of phasors is assumed to be rotating at a constant frequency, different phasors may be made to be the reference simply by rotating the entire phasor diagram in either the clockwise or counterclockwise direction.

From the viewpoints of ease and convenience, it would be a matter of common sense to choose the current phasor as the reference in the case of series circuits, in which all the elements are connected in series and the common quantity for all the elements involved is the current. Similarly for the case of parallel circuits, in which all the elements are connected in parallel and the common quantity for all the elements involved is the voltage, a good choice for the reference is the voltage phasor. As for the series-parallel circuits, no firm rule applies to all situations.

We shall now consider the series-parallel circuit shown in Figure 1.2.1 and develop the corresponding phasor diagram. We shall assume the load to be inductive in nature (a combination of

FIGURE 1.2.2 Development of the phasor diagram for the circuit shown in Figure 1.2.1 for the case of an inductive load.

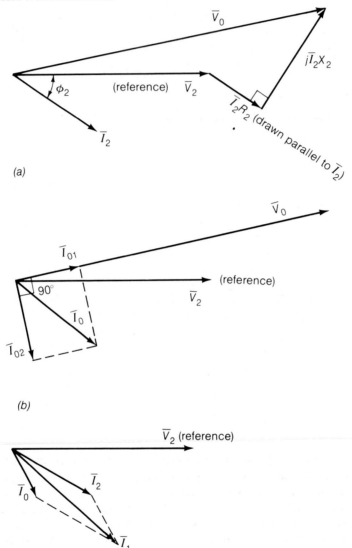

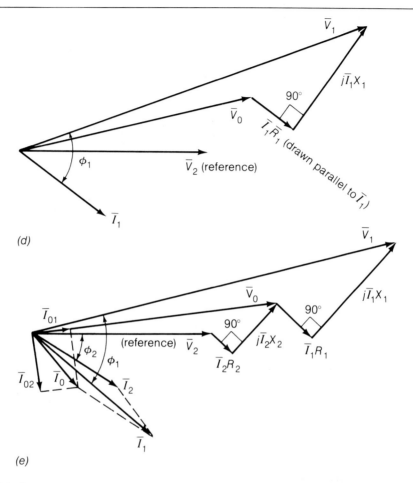

(d)

(e)

resistance and inductance) and the corresponding load power-factor angle to be ϕ_2 lagging the voltage $\overline{V}_2$. That is to say the current $\overline{I}_2$ lags the voltage $\overline{V}_2$ by an angle of ϕ_2 as shown in Figure 1.2.2(a). Phasor voltages and currents are indicated in Figure 1.2.1 and their magnitudes, unless otherwise mentioned, will be assumed to be their corresponding rms values, which would be the same as the meter readings if voltmeters and ammeters were to be connected appropriately.

For the sake of developing the phasor diagram, we shall also assume that R_1 is approximately equal to R_2, X_1 is approximately equal to X_2, R_1 is less than X_1, R_0 is greater than R_1, and X_0 is greater than X_1.

$$R_1 \simeq R_2; X_1 \simeq X_2; R_1 < X_1$$
$$R_0 > R_1; X_0 > X_1 \tag{1.2.1}$$

Given $\overline{V}_2$ and $\overline{I}_2$ at the load as well as the parameters of the circuit of Figure 1.2.1, we shall try to determine $\overline{V}_0$, $\overline{I}_0$, $\overline{I}_1$, and $\overline{V}_1$ by means of a phasor diagram. As seen in Figure 1.2.2, the voltage $\overline{V}_2$ is chosen as the *reference*. Starting with the reference voltage $\overline{V}_2$, $\overline{I}_2$ is drawn at an angle of ϕ_2 lagging the voltage $\overline{V}_2$ in Figure 1.2.2*(a)*, because the load is assumed to be inductive in its nature. Looking at Figure 1.2.1, the relationship to be satisfied by $\overline{V}_0$ and $\overline{V}_2$ is given by the following according to Kirchhoff's loop equation:

$$\overline{V}_0 = \overline{V}_2 + \overline{I}_2 R_2 + j\overline{I}_2 X_2 \qquad (1.2.2)$$

Figure 1.2.2*(a)* is drawn such that Equation 1.2.2 is satisfied; the voltage drop $\overline{I}_2 R_2$ is drawn parallel to $\overline{I}_2$, while adding $\overline{I}_2 R_2$ to $\overline{V}_2$ (by locating $\overline{I}_2 R_2$ at the end of the phasor $\overline{V}_2$); $j\overline{I}_2 X_2$ is drawn perpendicular to $\overline{I}_2$, while adding $j\overline{I}_2 X_2$ to $(\overline{V}_2 + \overline{I}_2 R_2)$ yielding $\overline{V}_0$.

   Next, looking at the shunt portion of the series-parallel circuit of Figure 1.2.1, knowing $\overline{V}_0$, R_0, and X_0, it is easy to find $\overline{I}_{01}$ and $\overline{I}_{02}$, which when added will yield $\overline{I}_0$.

$$\overline{I}_0 = \overline{I}_{01} + \overline{I}_{02} \qquad (1.2.3)$$

This is what is shown in Figure 1.2.2*(b)*; $\overline{I}_{01}$, being the current in a pure resistor, is drawn in line with (i.e., parallel to) $\overline{V}_0$; $\overline{I}_{02}$, being the current in a pure inductor, is drawn perpendicular to $\overline{V}_0$, lagging $\overline{V}_0$ by $90°$; $\overline{I}_{01}$ and $\overline{I}_{02}$ are then added, using the parallelogram law, to give $\overline{I}_0$.
   The relationship that has to be satisfied by the currents $\overline{I}_1$, $\overline{I}_0$, and $\overline{I}_2$ is given by

$$\overline{I}_1 = \overline{I}_2 + \overline{I}_0 \qquad (1.2.4)$$

consistent with the notation shown in Figure 1.2.1. Figure 1.2.2*(c)* corresponds to Equation 1.2.4.
   Then $\overline{V}_1$ can be found by solving the loop equation

$$\overline{V}_1 = \overline{V}_0 + \overline{I}_1 R_1 + j\overline{I}_1 X_1 \qquad (1.2.5)$$

This part is depicted in Figure 1.2.2*(d)*. The power-factor angle ϕ_1 between $\overline{V}_1$ and $\overline{I}_1$ can also be found.
   Thus, the entire phasor diagram is put together by superposition as in Figure 1.2.2*(e)*. The mathematical steps associated with the steady-state solution of the network are omitted here, as the student should be capable of doing those on his or her own from the background received in the basic circuit-theory courses.

BIBLIOGRAPHY

Budak, A. *Circuit Theory Fundamentals and Applications*. Englewood Cliffs, N.J.: Prentice–Hall, 1978.

Cruz, J. B., Jr., and Van Valkenburg, M. E. *Signals in Linear Circuits*. Boston: Houghton Mifflin, 1974.

Johnson, D. E.; Hilburn, J. L.; and Johnson, J. R. *Basic Electric Circuit Analysis*. Englewood Cliffs, N.J.: Prentice–Hall, 1978.

Van Valkenburg, M. E. *Network Analysis*. 3d ed., Englewood Cliffs, N.J.: Prentice–Hall, 1974.

PROBLEMS

1-1.

Considering $\overline{A}$ to be an arbitrary phasor, relative to $\overline{A}$ sketch the following:

    a. $-j\overline{A}$
    b. $2j\overline{A}$
    c. $(1 - 2j)\overline{A}$
    d. conjugate of $\overline{A}$, denoted by $\overline{A}^*$.

1-2.

With the use of phasors, add $10 \sin(377t + 30°)$ and $5 \cos(377t - 45°)$.

1-3.

For the *RC* network shown, let $v_2(t) = \sqrt{2} \cos t$ volts

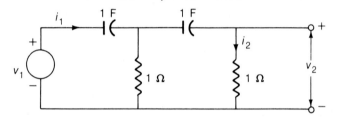

    a. Draw a phasor diagram (approximately to scale, preferably on graph paper) showing all voltages and currents.
    b. Based on the results of the phasor diagram, find $v_1(t)$.
    c. What values will an rms meter record for the voltages $v_1(t)$ and $v_2(t)$?

1-4.

Consider the series-parallel circuit of Figure 1.2.1. Draw the corresponding phasor diagrams [similar to that of Figure 1.2.2(e)] for the following two cases:

    a. Let the load be purely resistive, in which case $\overline{I}_2$ will be in phase with $\overline{V}_2$, with the load power factor being unity.
    b. Let the load be capacitive in nature (a combination of resistance and capacitance), for which case $\overline{I}_2$ will be leading $\overline{V}_2$ by an angle ϕ_2.
    c. It appears from Figure 1.2.2 that the magnitude of $\overline{V}_1$ is bigger than that of $\overline{V}_2$. Comment on whether this would be true for all types of loads under all conditions. If not, try to identify such a case and show the corresponding phasor diagram.

1-5.

Consider the circuit shown in the figure.

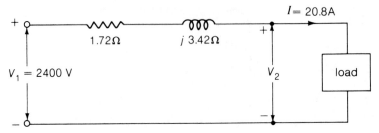

Sketch the corresponding phasor diagram, and find the corresponding V_2 for the following cases:

    a. A load power factor of 0.8 lagging.
    b. A load power factor of 0.8 leading.
    c. Unity load power factor.

2

THREE-PHASE CIRCUITS

FIGURE 2.1.1   Elementary three-phase, 2-pole ac generator.

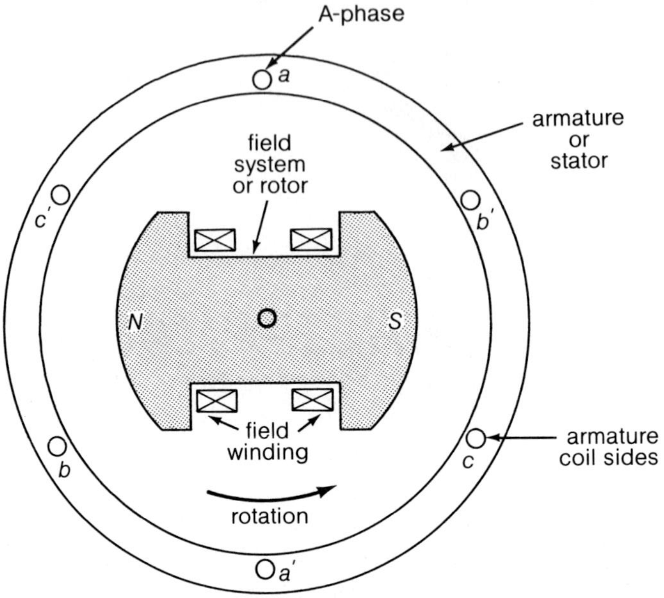

2.1 THREE-PHASE SOURCE VOLTAGES AND PHASE SEQUENCE

The three-phase system is by far the most common polyphase system used for generation, transmission, and heavy power utilization of ac electric energy, because of its definite economic and operating advantages. An ideal three-phase source generates three sinusoidal voltages of equal amplitudes displaced from each other by an angle of 120° in time phase. The voltages generated by the giant synchronous generators in power stations are practically sinusoidal with a frequency of 60 Hz in the United States, or 50 Hz in the United Kingdom and many other countries. Even though voltages and currents are sinusoidal, the power delivered to a balanced load is constant for a three-phase system, while it is sinusoidal and pulsating for a single-phase system. The three-phase scheme of power transmission is the simplest of the polyphase methods to offer the advantages of using the ac mode, constant power flow, and high power transfer capability.

The elementary three-phase, two-pole generator shown in Figure 2.1.1 has three identical stator coils (*aa'*, *bb'*, and *cc'*), of one or more turns, displaced by 120° in space from each other. The rotor carries a field winding excited by the dc supply through brushes and slip rings. When the rotor is driven counterclockwise at a constant speed, voltages will be generated in the three phases in accordance with Faraday's law. Each of the three stator coils constitutes one phase in this single

FIGURE 2.1.2
(a) Waveforms in time domain, *(b)* phasors of the generated voltages.

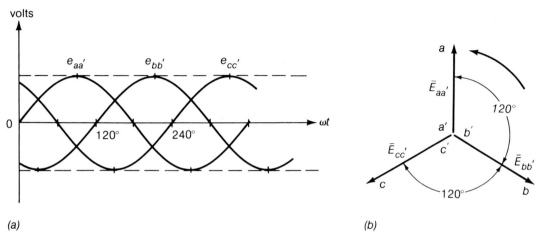

(a)   *(b)*

generator. If the field structure is so designed that the flux is distributed sinusoidally over the poles, the flux linking any phase will vary sinusoidally with time, and sinusoidal voltages will be induced in the three phases. These three induced voltage waves will be displaced 120 electrical degrees in time because the stator phases are displaced by 120° in space. Figure 2.1.2 shows the wave forms and the corresponding phasors of the three voltages. The time origin and the reference axis are chosen on the basis of analytical convenience.

The stator phase windings may be connected in either wye (known also as star, or symbolically represented as Y) or delta (known also as mesh, or symbolically represented as Δ) as shown schematically in Figure 2.1.3. Almost all alternating current generators (otherwise known as alternators) have their stator phase windings connected in wye. By connecting together either all three primed terminals or all three unprimed terminals to form the neutral of the wye, a wye connection results. If a neutral conductor is brought out, the system is known as a four-wire, three-phase system; otherwise, it is a three-wire, three-phase system. A delta connection is effected for the armature of the generator by connecting terminals a' to b, b' to c, and c' to a. The generator terminals A, B, C (and sometimes N for a wye connection) are brought out as shown in Figure 2.1.3. In the Δ-connection, no neutral exists and hence only a three-wire, three-phase system can be formed.

From the nature of the connections shown in Figure 2.1.3, it can be seen that the line-to-line voltage (V_{L-L} or V_L) is equal to the phase voltage (V_{ph}) for the Δ-connection, and the line current is equal to the phase current for the Y-connection. A balanced wye-connected three-phase source

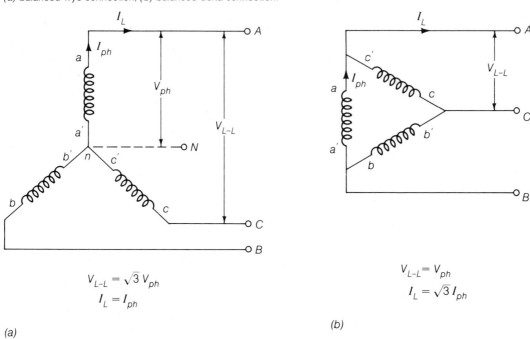

FIGURE 2.1.3 Schematic representation of generator windings: *(a)* balanced wye connection, *(b)* balanced delta connection.

$$V_{L-L} = \sqrt{3}\, V_{ph}$$
$$I_L = I_{ph}$$

(a)

$$V_{L-L} = V_{ph}$$
$$I_L = \sqrt{3}\, I_{ph}$$

(b)

and its associated phasor diagram are shown in **Figure 2.1.4**, from which it can be seen that the line-to-line voltage is equal to $\sqrt{3}$ times the phase voltage (or the line-to-neutral voltage). The student should be able to reason on similar lines and conclude that, for the balanced delta-connected three-phase source, the line current will be equal to $\sqrt{3}$ times the phase current.

The notation for subscripting is such that V_{AB} is the potential at point A with respect to point B, I_{AB} is a current with positive flow from point A to point B, and I_A, I_B, and I_C are line currents with positive flow from source to the load, as shown in Figure 2.1.5. The notation used is rather arbitrary; in some textbooks a different notation for the voltage is adopted such that the order of subscripts indicates the direction in which the voltage rise is taken. The student should be careful not to get confused, but try to be consistent with any chosen conventions. The rms values are usually chosen as magnitudes of the phasors for convenience. It is customary to use the letter symbol E for generated emf and V for terminal voltage. Sometimes the two are equal, but sometimes not. If we should neglect the existence of the generator winding impedance, the generated emf will be equal to the terminal voltage of the generator. Although the three-phase voltages are generated in one three-phase alternator, for analytical purposes it is modeled by three identical, interconnected, single-phase sources.

FIGURE 2.1.4 *(a)* Balanced wye-connected, three-phase source, *(b)* phasor diagram for the three-phase source (sequence *ABC*), *(c)* single-line equivalent circuit.

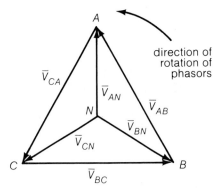

direction of rotation of phasors

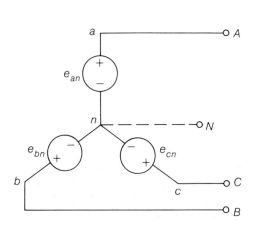

(a)

$$\bar{V}_{BC} = V_L \angle 0° \text{ (Ref.)}; \qquad \bar{V}_{AN} = \frac{V_L}{\sqrt{3}} \angle 90°$$

$$\bar{V}_{AB} = V_L \angle 120°; \qquad \bar{V}_{BN} = \frac{V_L}{\sqrt{3}} \angle -30°$$

$$\bar{V}_{CA} = V_L \angle 240°; \qquad \bar{V}_{CN} = \frac{V_L}{\sqrt{3}} \angle -150°$$

Note that such relations as $\bar{V}_{AB} = \bar{V}_{AN} + \bar{V}_{NB}$ are satisfied; also $\bar{V}_{AN} + \bar{V}_{BN} + \bar{V}_{CN} = 0$; $\bar{V}_{AB} + \bar{V}_{BC} + \bar{V}_{CA} = 0.$

(b)

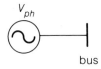

bus

or

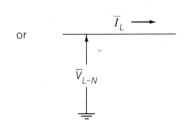

(c)

FIGURE 2.1.5   Notation for subscripting.

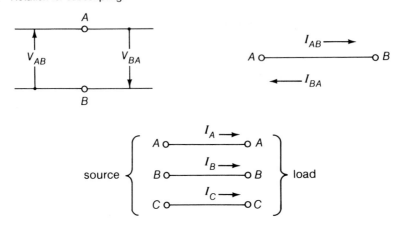

The one-line equivalent circuit of the balanced wye-connected three-phase source is shown in Figure 2.1.4*(c)*. The line-to-neutral (otherwise known as phase) voltage is used; it may be taken as a reference with a phase angle of zero for convenience. This procedure yields the equivalent single-phase circuit, in which all quantities correspond to those of one phase in the three-phase circuit. Except for the 120° phase displacements in the currents and voltages, the conditions in the other two phases are the same, and there is no need to investigate them individually. Line currents in the three-phase system are the same as in the single-phase circuit, and total three-phase real power, reactive power, and volt-amperes are three times the corresponding quantities in the single-phase circuit. Line-to-line voltages, if desired, can be obtained by multiplying voltages in the single-phase circuit by $\sqrt{3}$.

When a system of sources is so large that its voltage and frequency remain constant regardless of the power delivered or absorbed, it is known as an *infinite bus*. Such a bus has a voltage and frequency that are unaffected by external disturbances. The infinite bus is treated as an ideal voltage source.

PHASE SEQUENCE   It is a standard practice in the United States to designate the phases *A-B-C* such that under balanced conditions the voltage and current in *A*-phase lead in time the voltage and current in *B*-phase by 120° and in *C*-phase by 240°. This is known as *positive phase sequence A-B-C*. The phase sequence should be observed either from the waveforms in the time domain shown in Figure 2.1.2*(a)* or from the phasor diagram shown in Figure 2.1.2*(b)* or 2.1.4*(b)*, and not from the space or schematic diagrams such as Figures 2.1.3 and 2.1.4*(a)*. If the rotation of the generator in Figure 2.1.1 is reversed, or if any two of the three leads from the armature (not counting the neutral) to the generator terminals are reversed, the phase sequence becomes *A-C-B* (or *CBA* or *BAC*), which is known as *negative phase sequence*.

Only the balanced three-phase sources are considered in this chapter. Selection of one voltage as the reference with a phase angle of zero determines the phase angles of all the other voltages

FIGURE 2.2.1 Wye-delta and delta-wye transformations.

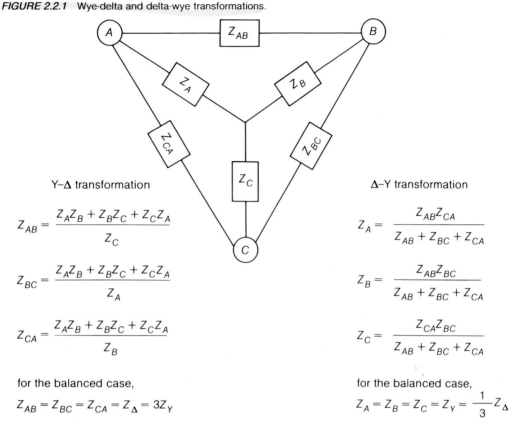

Y–Δ transformation

$$Z_{AB} = \frac{Z_A Z_B + Z_B Z_C + Z_C Z_A}{Z_C}$$

$$Z_{BC} = \frac{Z_A Z_B + Z_B Z_C + Z_C Z_A}{Z_A}$$

$$Z_{CA} = \frac{Z_A Z_B + Z_B Z_C + Z_C Z_A}{Z_B}$$

for the balanced case,

$$Z_{AB} = Z_{BC} = Z_{CA} = Z_\Delta = 3Z_Y$$

Δ–Y transformation

$$Z_A = \frac{Z_{AB} Z_{CA}}{Z_{AB} + Z_{BC} + Z_{CA}}$$

$$Z_B = \frac{Z_{AB} Z_{BC}}{Z_{AB} + Z_{BC} + Z_{CA}}$$

$$Z_C = \frac{Z_{CA} Z_{BC}}{Z_{AB} + Z_{BC} + Z_{CA}}$$

for the balanced case,

$$Z_A = Z_B = Z_C = Z_Y = \frac{1}{3} Z_\Delta$$

in the system for a given phase sequence. As indicated before, the reference phasor is arbitrarily chosen for convenience. In Figure 2.1.4(b), $\overline{V}_{BC}$ is the reference phasor, and with the counterclockwise rotation (assumed positive) of all the phasors at the same frequency, the sequence can be seen to be *A-B-C*. Unless otherwise mentioned, the positive phase sequence is to be assumed.

2.2 BALANCED THREE-PHASE LOADS

Three-phase loads can be connected in either wye (also known as star or Y) or delta (also known as mesh or Δ). If the load impedance in each of the three phases is the same, both in magnitude and phase angle, the load is said to be balanced.

For the analysis of network problems, transformations for converting a delta-connected network to an equivalent wye-connected network and vice versa will be found to be useful. The relationships for interconversion of wye and delta networks are given in Figure 2.2.1. These can be obtained by imposing the condition of equivalence that the impedance between any two terminals for one network be equal to the corresponding impedance between the same terminals for the other network.

FIGURE 2.2.2   Balanced wye-connected load: *(a)* connection diagram,
(b) phasor diagram, *(c)* single-line equivalent circuit.

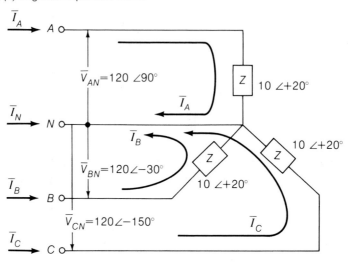

(a)

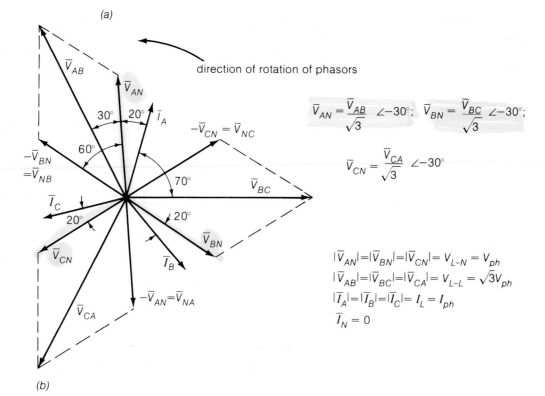

direction of rotation of phasors

$$\overline{V}_{AN} = \frac{\overline{V}_{AB}}{\sqrt{3}} \angle -30°; \quad \overline{V}_{BN} = \frac{\overline{V}_{BC}}{\sqrt{3}} \angle -30°;$$

$$\overline{V}_{CN} = \frac{\overline{V}_{CA}}{\sqrt{3}} \angle -30°$$

$$|\overline{V}_{AN}| = |\overline{V}_{BN}| = |\overline{V}_{CN}| = V_{L-N} = V_{ph}$$
$$|\overline{V}_{AB}| = |\overline{V}_{BC}| = |\overline{V}_{CA}| = V_{L-L} = \sqrt{3}V_{ph}$$
$$|\overline{I}_{A}| = |\overline{I}_{B}| = |\overline{I}_{C}| = I_{L} = I_{ph}$$
$$\overline{I}_{N} = 0$$

(b)

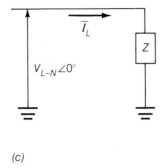

(c)

The details are left as a desirable exercise to the student. For the balanced case, each Y-impedance is one-third of each Δ-impedance; conversely, each Δ-impedance is three times each Y-impedance.

BALANCED WYE-CONNECTED LOAD   Let us consider a three-phase, four-wire, 208-volt supply system connected to a balanced wye-connected load with impedances of 10 ∠20° ohms, as shown in Figure 2.2.2(a). We shall solve for the line currents and draw the corresponding phasor diagram.

Conventionally it is assumed that 208 volts is the rms value of the line-to-line voltage of the supply system, and the phase sequence is positive or *A-B-C*, unless otherwise mentioned. The magnitude of the line-to-neutral (or phase) voltages is given by $208/\sqrt{3}$ or 120 volts. Selecting the line currents returning through the neutral conductor as shown in Figure 2.2.2(a), we have

$$\overline{I}_A = \frac{\overline{V}_{AN}}{Z} = \frac{\frac{208}{\sqrt{3}} \angle 90°}{10 \angle 20°} = 12 \angle 70° \qquad (2.2.1)$$

$$\overline{I}_B = \frac{\overline{V}_{BN}}{Z} = \frac{\frac{208}{\sqrt{3}} \angle -30°}{10 \angle 20°} = 12 \angle -50° \qquad (2.2.2)$$

$$\overline{I}_C = \frac{\overline{V}_{CN}}{Z} = \frac{\frac{208}{\sqrt{3}} \angle -150°}{10 \angle +20°} = 12 \angle -170° \qquad (2.2.3)$$

Note that $\overline{V}_{BC}$ has been arbitrarily chosen as the reference phasor as in Figure 2.1.4(b). Assuming the direction of the neutral current toward the load as positive, we obtain

$$\overline{I}_N = -(\overline{I}_A + \overline{I}_B + \overline{I}_C)$$
$$= -(12 \angle 70° + 12 \angle -50° + 12 \angle -170°) = 0 \qquad (2.2.4)$$

That is to say that the system neutral and the star point of the wye-connected load are at the same potential, even if they are not connected together electrically. It makes no difference whether they are interconnected or not. *Thus for a <u>balanced</u> wye-connected load, the neutral current is always zero; the line currents and phase currents are equal (in magnitude); and the line-to-line voltage (in magnitude) is* $\sqrt{3}$ *times the phase voltage.* The phasor diagram is drawn in Figure 2.2.2(b) from which it can be observed that the balanced line (or phase) currents lag the corresponding line-to neutral voltages by the impedance angle (20° in our example). The load power factor is given by cos 20° for our problem, and it is said to be *lagging* in this case, as the impedance angle is positive and the phase current lags the corresponding phase voltage by that angle.

The problem can also be solved in a simpler way by making use of a single-line equivalent circuit shown in Figure 2.2.2(c)

$$\overline{I}_L = \frac{\overline{V}_{L-N}}{Z} = \frac{\frac{208}{\sqrt{3}}\angle 0°}{10\angle 20°} = 12\angle -20° \tag{2.2.5}$$

in which $\overline{V}_{L-N}$ is chosen as the reference for convenience. The magnitude of the line current and the power-factor angle are known; the negative sign associated with the angle indicates that the power factor is lagging. By knowing that the line (or phase) currents $\overline{I}_A$, $\overline{I}_B$, and $\overline{I}_C$ lag their respective voltages $\overline{V}_{AN}$, $\overline{V}_{BN}$, and $\overline{V}_{CN}$ by 20°, the phase angles of various voltages and currents, if desired, can be obtained with respect to any chosen reference such as $\overline{V}_{BC}$.

BALANCED DELTA-CONNECTED LOAD   Next let us consider the case of a balanced delta-connected load with impedances of 5 ∠45° ohms supplied by a three-phase, three-wire, 100-volt system as shown in Figure 2.2.3(a). We shall determine the line currents and draw the corresponding phasor diagram.

With the assumed positive phase sequence *(A-B-C)* and with $\overline{V}_{BC}$ as the reference phasor, the line-to-line voltages $\overline{V}_{AB}$, $\overline{V}_{BC}$, and $\overline{V}_{CA}$ are shown in Figure 2.2.3. The rms value of the line-to-line voltage is 100 volts for our example. Choosing the positive directions of the line and phase currents as in Figure 2.2.3(a), we have

$$\overline{I}_{AB} = \frac{\overline{V}_{AB}}{Z} = \frac{100\angle 120°}{5\angle 45°} = 20\angle 75° \tag{2.2.6}$$

$$\overline{I}_{BC} = \frac{\overline{V}_{BC}}{Z} = \frac{100\angle 0°}{5\angle 45°} = 20\angle -45° \tag{2.2.7}$$

$$\overline{I}_{CA} = \frac{\overline{V}_{CA}}{Z} = \frac{100\angle 240°}{5\angle 45°} = 20\angle 195° \tag{2.2.8}$$

FIGURE 2.2.3 Balanced delta-connected load: *(a)* connection diagram, *(b)* phasor diagram, *(c)* single-line equivalent circuit.

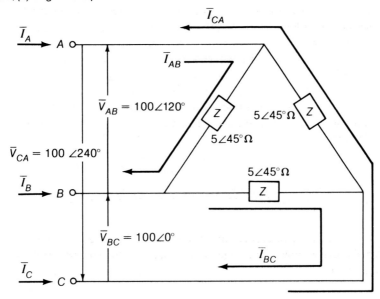

(a)

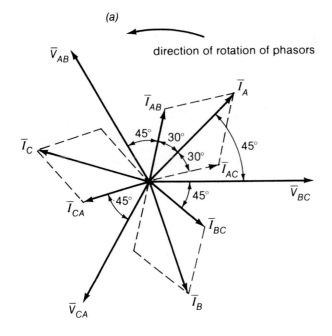

direction of rotation of phasors

(b)

$$|\overline{V}_{AB}|=|\overline{V}_{BC}|=|\overline{V}_{CA}|= V_{L-L} = V_{ph}$$
$$|\overline{I}_{AB}|=|\overline{I}_{BC}|=|\overline{I}_{CA}|= I_{ph}$$
$$|\overline{I}_{A}|=|\overline{I}_{B}|=|\overline{I}_{C}|= I_{L} = \sqrt{3}\, I_{ph}$$
$$\overline{I}_{A} = \sqrt{3}\, \overline{I}_{AB}\, \angle{-30°};\ \overline{I}_{B} = \sqrt{3}\, \overline{I}_{BC}\, \angle{-30°};$$
$$\overline{I}_{C} = \sqrt{3}\, \overline{I}_{CA}\, \angle{-30°}$$

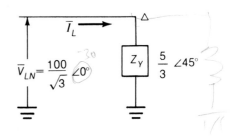

(c)

By the application of Kirchhoff's current law at each of the vertices of the delta-connected load, we obtain

$$\bar{I}_A = \bar{I}_{AB} + \bar{I}_{AC} = 20 \angle 75° - 20 \angle 195° = 34.64 \angle 45° \quad (2.2.9)$$

$$\bar{I}_B = \bar{I}_{BA} + \bar{I}_{BC} = -20 \angle 75° + 20 \angle -45°$$
$$= 34.64 \angle -75° \quad (2.2.10)$$

$$\bar{I}_C = \bar{I}_{CA} + \bar{I}_{CB} = 20 \angle 195° - 20 \angle -45°$$
$$= 34.64 \angle 165° \quad (2.2.11)$$

The phasor diagram showing the line-to-line voltages, phase currents, and line currents is drawn in Figure 2.2.3(b). The load power factor is lagging, given by cos 45°.

For a balanced delta-connected load, the phase voltages and the line-to-line voltages are equal (in magnitude); and the line current (in magnitude) is $\sqrt{3}$ times the phase current.

The example worked above can also be solved by the one-line equivalent method for which the delta-connected load is replaced by its equivalent star-connected load. The single-line equivalent circuit is shown in Figure 2.2.3(c). The details are left as an exercise for the student.

POWER IN BALANCED THREE-PHASE CIRCUITS   The total power delivered by a three-phase source or consumed by a three-phase load is found simply by adding the power in each of the three phases. In a balanced circuit, however, this is the same as multiplying the average power in any one phase by 3, since the average power is the same for all phases. Thus one has

$$P = 3 V_{ph} I_{ph} \cos \phi \quad (2.2.12)$$

where V_{ph} and I_{ph} are the magnitudes of any phase voltage and phase current, cos ϕ is the load power factor, and ϕ is the power-factor angle between the phase voltage, $\bar{V}_{ph}$, and the phase current, $\bar{I}_{ph}$, corresponding to any phase. In view of the relationships between the line and phase quantities for balanced wye- or delta-connected loads, Equation 2.2.12 can be rewritten in terms of the line-to-line voltage and the line current for either wye- or delta-connected balanced loading as follows:

$$P = \sqrt{3} V_L I_L \cos \phi \quad (2.2.13)$$

V_L and I_L are the magnitudes of the line-to-line voltage and the line current; ϕ is still the load power-factor angle as in Equation 2.2.12, being the angle between the phase voltage and the corresponding phase current.

In a balanced three-phase system, the sum of the three individually pulsating phase powers adds up to a constant, nonpulsating total power of magnitude three times the average real power in each phase. That is, in spite of the sinusoidal nature of the voltages and currents, the total instantaneous power delivered into the three-phase load is a constant, equal to the total average power. The real power P is expressed in watts when the voltage and the current are expressed in volts and amperes, respectively.

The total reactive power Q (expressed as reactive voltamperes or vars) and the voltamperes for either wye- or delta-connected balanced loadings are given by

$$Q = 3 V_{ph} I_{ph} \sin \phi \qquad (2.2.14)$$

or

$$Q = \sqrt{3} \, V_L I_L \sin \phi \qquad (2.2.15)$$

and

$$|S| = \sqrt{P^2 + Q^2} = 3V_{ph} I_{ph} = \sqrt{3} \, V_L I_L \qquad (2.2.16)$$

The complex power, S, is given by

$$S = P + jQ \qquad (2.2.17)$$

In speaking of a three-phase system, unless otherwise specified, balanced conditions are assumed; and the terms voltage, current, and power, unless otherwise identified, are conventionally understood to imply the line-to-line voltage (rms value), the line current (rms value), and the total power of all three phases.

2.3 MEASUREMENT OF POWER

A wattmeter is an instrument with a potential coil and a current coil so arranged that its deflection is proportional to $VI \cos \theta$, where V is the voltage (rms value) applied across the potential coil, I is the current (rms value) passing through the current coil, and θ is the angle between $\overline{V}$ and $\overline{I}$. By inserting a wattmeter to measure the average real power in each phase (with its current coil in series with one phase of the load and its potential coil across the phase of the load), the total real power in a three-phase system can be determined by the sum of the wattmeter readings. However, in practice, this may not be possible due to the nonaccessibility of either the neutral of the wye connection or the individual phases of the delta connection. Hence it is more desirable to have a method for measuring the total real power drawn by a three-phase load, while we have access to only three line terminals.

The three-phase power can be measured by three wattmeters having current coils in each line and potential coils connected across the given line and any common junction. Since this common junction is completely arbitrary, it may be placed on any one of the three lines; in which case the wattmeter connected in that line will indicate zero power because its potential coil has no voltage across it. Hence that wattmeter may be dispensed with, and three-phase power can be measured by means of only two wattmeters having a common potential junction on any one of the three lines in which there is no current coil. This is known as the two-wattmeter method of measuring three-

FIGURE 2.3.1 Connection diagram for two-wattmeter method of measuring three-phase power.

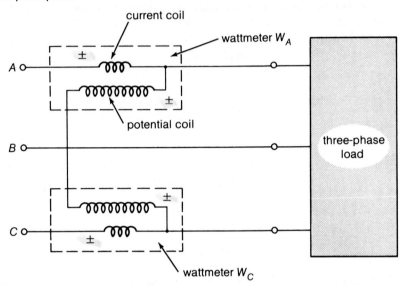

phase power. In general, m-phase power can be measured by means of $m-1$ wattmeters; the method is valid for both balanced and unbalanced circuits with either the load or the source unbalanced.

Figure 2.3.1 illustrates the connection diagram for the two-wattmeter method of measuring three-phase power. The total real power delivered to the load is given by the *algebraic sum* of the two wattmeter readings:

$$P = W_A + W_C \tag{2.3.1}$$

The significance of the algebraic sum will be realized in the paragraphs to follow. Two wattmeters can be connected with their current coils in any two lines, while their potential coils are connected to the third line, as shown in Figure 2.3.1. The wattmeter readings are given by

$$W_A = V_{AB} \cdot I_A \cdot \cos \theta_A \tag{2.3.2}$$

where θ_A is the angle between the phasors $\overline{V}_{AB}$ and $\overline{I}_A$, and

$$W_C = V_{CB} \cdot I_C \cdot \cos \theta_C \tag{2.3.3}$$

where θ_C is the angle between the phasors $\overline{V}_{CB}$ and $\overline{I}_C$.

The two-wattmeter method, when applied to the *balanced* loads, yields interesting results. Considering either balanced wye- or delta-connected loads, with the aid of the corresponding phasor diagrams drawn earlier for the phase sequence *A-B-C* (Figures 2.2.2 and 2.2.3), it can be seen that the angle between $\overline{V}_{AB}$ and $\overline{I}_A$ is $(30° + \phi)$ and that between $\overline{V}_{CB}$ and $\overline{I}_C$ is $(30° - \phi)$, where ϕ is the load power-factor angle or the angle associated with the load impedance. Thus we have

$$W_A = V_L I_L \cos(30° + \phi) \qquad (2.3.4)$$

and

$$W_C = V_L I_L \cos(30° - \phi) \qquad (2.3.5)$$

where V_L and I_L are the magnitudes of the line-to-line voltage and line current respectively. Simple manipulations yield the following:

$$W_A + W_C = \sqrt{3}\ V_L I_L \cos \phi \qquad (2.3.6)$$

and

$$W_C - W_A = V_L I_L \sin \phi \qquad (2.3.7)$$

from which

$$\tan \phi = \sqrt{3}\ \frac{W_C - W_A}{W_C + W_A} \qquad (2.3.8)$$

When the load power factor is unity, corresponding to a purely resistive load, both wattmeters will indicate the same wattage. In fact, both of them should read positive; if one of the wattmeters has a below-zero indication in the laboratory, an upscale deflection can be obtained by simply reversing the leads of either the current or the potential coil of the wattmeter. The sum of the wattmeter readings gives the total power absorbed by the load.

At zero power factor, corresponding to a purely reactive load, both wattmeters will again have the same wattage indication but with opposite sign, so that their algebraic sum will yield zero power absorbed, as it should. The transition from a negative to a positive value occurs when the load power factor is 0.5 (i.e., ϕ is equal to 60°); at this power factor, one wattmeter reads zero while the other one reads the total real power delivered to the load.

For power factors (leading or lagging) greater than 0.5, both wattmeters read positive, and the sum of the two readings gives the total power. For a power factor less than 0.5 (leading or lagging), the smaller reading wattmeter should be given a negative sign and the total real power absorbed by the load (which has to be positive) is given by the difference of the two wattmeter readings.

FIGURE 2.3.2   A plot of load power factor versus W_l/W_h.

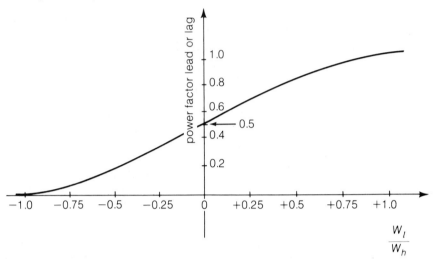

Figure 2.3.2 shows a plot of the load power factor versus the ratio (W_l/W_h), where W_l and W_h are the lower and higher readings of the wattmeters, respectively.

Another method that is sometimes useful in a laboratory environment for determining whether the total power is the sum or difference of the two wattmeter readings is described here. To start with, make sure that both wattmeters have an upscale deflection. To perform the test, remove the lead of the potential coil of the lower-reading wattmeter from the common line that has no current coil, and touch the lead to the line that has the current coil of the higher-reading wattmeter. If the pointer of the lower-reading wattmeter deflects upward, the two wattmeter readings should be added; if the pointer deflects in the below-zero direction, the wattage reading of the lower-reading wattmeter should be subtracted from that of the higher-reading wattmeter.

Given the two wattmeter readings from the two-wattmeter method used on a three-phase *balanced* load, it is possible to find the tangent of the phase-impedance angle as $\sqrt{3}$ times the ratio of the difference between the two wattmeter readings and their sum, based on Equation 2.3.8. If one knows the system sequence and the lines in which the current coils of the wattmeters are located, the sign for the angle can be determined with the aid of the following expressions.

For sequence *A-B-C*,

$$\tan \phi = \sqrt{3}\ \frac{W_C - W_A}{W_C + W_A} = \sqrt{3}\ \frac{W_A - W_B}{W_A + W_B}$$
$$= \sqrt{3}\ \frac{W_B - W_C}{W_B + W_C} \tag{2.3.9}$$

and for sequence *C-B-A*,

$$\tan \phi = \sqrt{3}\, \frac{W_A - W_C}{W_A + W_C} = \sqrt{3}\, \frac{W_B - W_A}{W_B + W_A}$$
$$= \sqrt{3}\, \frac{W_C - W_B}{W_C + W_B} \tag{2.3.10}$$

The two-wattmeter method discussed above for measuring three-phase power makes use of single-phase wattmeters. It may be noted, however, that three-phase wattmeters are also available that, when connected appropriately, indicate the total real power absorbed. The total reactive power associated with the three-phase *balanced* load is given by

$$Q = \sqrt{3}\, V_L I_L \sin \phi = \sqrt{3}\, (W_C - W_A) \tag{2.3.11}$$

based on the two wattmeter readings of the two-wattmeter method.

With the generator action of the source assumed, $+P$ for the real power indicates that the source is supplying real power to the load; $+Q$ for the reactive power shows that the source is delivering inductive vars while the current lags the voltage (i.e., the power factor is lagging), and $-Q$ for the reactive power indicates that the source is delivering capacitive vars or absorbing inductive vars, while the current leads the voltage (i.e., the power factor is leading).

BIBLIOGRAPHY

Cruz, J. B., Jr., and Van Valkenburg, M. E. *Signals in Linear Circuits*. Boston: Houghton Mifflin, 1974.
Gingrich, H. W. *Electrical Machinery, Transformers, and Control*. Englewood Cliffs, N.J.: Prentice-Hall, 1979.

Hayt, W. H., Jr., and Kemmerly, J. E. *Engineering Circuit Analysis*. 3d ed. New York: McGraw-Hill, 1978.
Johnson, D. E.; Hilburn, J. L.; and Johnson, J. R. *Basic Electric Circuit Analysis*. Englewood Cliffs, N.J.: Prentice-Hall, 1978.

PROBLEMS

2–1.

The line-to-line voltage of a balanced wye-connected three-phase source is given as 100 V. Choose $\overline{V}_{AB}$ as the reference.

 a. For the phase sequence *A-B-C*, sketch the phasor diagram of voltages and find the expressions for the phase voltages.
 b. Repeat Part *(a)* for the phase sequence *C-B-A*.

2–2.

A three-phase, three-wire, 208-volt system is connected to a balanced three-phase load. The line currents $\overline{I}_A$, $\overline{I}_B$, and $\overline{I}_C$ are given to be in phase with the line-to-line voltages $\overline{V}_{BC}$, $\overline{V}_{CA}$, and $\overline{V}_{AB}$, respectively. If the line current is measured to be 10 A, find the per-phase impedance of the load

 a. if the load is wye-connected.
 b. if the load is delta-connected.

2–3.

Consider a three-phase, 25-kVA, 440-volt, 60-Hz alternator operating at full load (i.e., delivering its rated kVA) under balanced steady-state conditions. Find the magnitudes of the alternator line current and the phase current

    a. if the generator windings are wye-connected.
    b. if the generator windings are delta-connected.

2–4.

Let the alternator of Problem 2–3 supply a line current of 20 A at a power factor of 0.8 lagging. Determine

    a. the kVA supplied by the machine.
    b. the real power kW delivered by the generator.
    c. the reactive power kvar delivered by the alternator.

2–5.

A balanced wye-connected load with a per-phase impedance of $(4 + j3)$ ohms is supplied by a 173-volt, 60-Hz, three-phase source.

    a. Find the line current, the power factor, and the total voltamperes, the real power, and the reactive power absorbed by the load.
    b. Sketch the phasor diagram showing all the voltages and currents, with $\overline{V}_{AB}$ as the reference.
    c. If the star point of the load is connected to the system neutral through an ammeter, what would the meter read?

2–6.

A balanced delta-connected load with a per-phase impedance of $(12 + j9)$ ohms is supplied by a 173-volt, 60-Hz, three-phase source.

    a. Determine the line current, the power factor, and the total volt-amperes, the real power, and the reactive power absorbed by the load.
    b. Compare the results of Problem 2.6(a) to those obtained for Problem 2.5(a) and explain why they are the same.
    c. Sketch the phasor diagram showing all voltages and currents, with $\overline{V}_{AB}$ as the reference.

2–7.

A 60-Hz, 440-volt, three-phase system feeds two balanced wye-connected loads in parallel. One load has a per-phase impedance of $(8 + j3)$ ohms, and the other, $(4 - j1)$ ohms. Compute the real power in kW delivered to

    a. the inductive load.
    b. the capacitive load.

2-8.

Balanced wye-connected loads drawing 10 kW at 0.8 pf lagging and 15 kW at 0.9 pf leading are connected in parallel and supplied by a 60-Hz, 300-volt, three-phase system. Find the line current delivered by the source.

2-9.

A balanced delta-connected load with a per-phase impedance of $(30 + j10)$ ohms is connected in parallel with a balanced wye-connected load with a per-phase impedance of $(40 - j10)$ ohms. This load combination is connected to a balanced three-phase supply through three conductors, each of which has a resistance of 0.4 ohm. The line-to-line voltage at the terminals of the load combination is measured to be 2,000 V. Determine the power in kW

    a. delivered to the delta-connected load.
    b. delivered to the wye-connected load.
    c. lost in the line conductors.

2-10.

Three identical impedances of $30\angle30°$ ohms are connected in delta to a three-phase, 173-volt system by conductors that have impedances of $(0.8 + j0.6)$ ohm each. Compute the magnitude of the line-to-line voltage at the terminals of the load.

2-11.

Repeat Problem 2–10 for a delta-connected set of capacitors with a per-phase reactance of $(-j60)$ ohms connected in parallel with the load, and compare the result with that obtained for Problem 2–10.

2-12.

Derive the relationships given in Figure 2.2.1 for the wye-delta and the delta-wye transformations.

2-13.

Two three-phase generators are supplying to a common balanced three-phase load of 30 kW at 0.8 power factor lagging. The per-phase impedance of the lines connecting the generator G_1 to the load is $(1.4 + j1.6)$ ohms, while that of the lines connecting the generator G_2 to the load is $(0.8 + j1)$ ohms. If the generator G_1, operating at a terminal voltage of 800 volts (line-to-line), supplies 15 kW at 0.8 pf lagging, find the voltage at the load terminals, the terminal voltage of the generator G_2, and the real power as well as the reactive power output of the generator G_2.

2-14.

Determine the wattmeter readings when the two-wattmeter method is applied to Problem 2–5, and check the total power obtained.

2–15.

When the two-wattmeter method for measuring three-phase power is used on a certain balanced load, readings of 1,200 W and 400 W are obtained (without any reversals). Determine the delta-connected load impedances if the system voltage is 440 V. With the information given above, is it possible to find whether the load impedance is capacitive or inductive in nature?

2–16.

The two-wattmeter method for measuring three-phase power is applied on a balanced wye-connected load as shown in Figure 2.3.1, and the readings are given by

$$W_C = 836 \text{ W and } W_A = 224 \text{ W}.$$

If the system voltage is 100 V, find the per-phase impedance of the load. In this problem, is it possible to specify the capacitive or inductive nature of the impedance?

2–17.

Two wattmeters are used as shown in Figure 2.3.1 to measure the power absorbed by a balanced delta-connected load. Determine the total power in kW, the power factor, and the per-phase impedance of the load if the supply-system voltage is 120 V and the wattmeter readings are given by

a. $W_A = -500$ W; $W_C = 1,300$ W.
b. $W_A = 1,300$ W; $W_C = -500$ W.

3

THE MAGNETIC ASPECT

In engineering practice it is generally required to find solutions to problems in forms that are convenient and easy to interpret and visualize. This accounts for the widely accepted use of the *circuit concept* as distinct from the *field concept*. Although all phenomena occur in space and time, and these are field phenomena, it is possible to approximate them by time relations only, summarizing the actual physical quantities by means of a few parameters, preferably of constant value. However, it becomes necessary sometimes to re-examine the basis of the circuit concept through the study of the basic electromagnetic fields involved. For producing the simplest, most economical, and most efficient device, an accurate knowledge of the fields present in the device, especially the magnetic field, is desirable. It is necessary to analyze the fields and modify the design for optimization. Thus the *field theory* as opposed to the *circuit theory* is becoming increasingly important in many engineering problems. A significant part of designing an electromechanical device centers on the design of the magnetic system. The purpose of this chapter is to review the basic concepts of the field theory and to present some aspects of magnetism of value in the design and analysis of electromagnetic devices.

3.1   REVIEW OF ELECTROMAGNETIC FIELD THEORY

The fundamental laws governing electromagnetic phenomena at any point in space can be expressed by the Maxwell equations. Assuming that the dielectric properties of any of the materials are of no significance in comparison with the conduction properties, as is true in low-frequency problems, the dielectric effects or the displacement currents can be neglected, and the material regions can be considered as void of the volume charge density. Then the relevant equations in differential and integral forms may be written as follows:

$$\nabla \cdot \overline{B} = 0; \quad \oiint_S \overline{B} \cdot d\overline{S} = 0 \qquad\qquad (3.1.1;\ 3.1.2)$$

$$\nabla \times \overline{H} = \overline{J}; \quad \oint_c \overline{H} \cdot d\overline{l} = I \qquad\qquad (3.1.3;\ 3.1.4)$$

$$\nabla \times \overline{E} = -\frac{\partial \overline{B}}{\partial t}; \quad \oint_c \overline{E} \cdot d\overline{l} = -\iint_S \frac{\partial \overline{B}}{\partial t} \cdot d\overline{S} \qquad (3.1.5;\ 3.1.6)$$

In the standard international (SI) system of units, $\overline{B}$ is the magnetic flux density in webers per square meter (Wb/m²) or tesla (T); $\overline{H}$ is the magnetic field intensity in amperes (or ampere-turns) per meter (A/m or At/m); $\overline{J}$ is the current density in amperes per square meter (A/m²); I is the current in amperes (A); $\overline{E}$ is the electric field intensity in volts per meter (V/m); and t is the time in seconds (s). Other systems of units such as the CGS system and the English system have been widely used by practitioners, and the relationships among these systems are given in Appendix A (Table A–4).

Equation 3.1.2 is known as Gauss's law for the magnetic field, which states that the total magnetic flux passing through a closed surface is zero. Equation 3.1.4 is Ampere's law, which states that the line integral of the magnetic field intensity around a closed line contour is equal to the current enclosed by the contour. The current-continuity relation, known as Kirchhoff's current law, is a consequence of Equation 3.1.3 given by

$$\nabla \cdot \overline{J} = 0; \quad \oint\!\!\!\oint_S \overline{J} \cdot d\overline{S} = 0 \qquad \text{(3.1.7; 3.1.8)}$$

Equation 3.1.6 is Faraday's law, according to which an electric field is produced in space when the magnetic field is allowed to vary with time; the line integral is taken along the closed contour around the surface through which the magnetic flux density passes. As for the direction, if the line-contour integral is taken in the counterclockwise (positive) direction along the fingers of the right hand oriented in the direction of the closed path, the thumb will then indicate the positive direction for the right-hand sides of Equations 3.1.4 and 3.1.6. A flux density $\overline{B}$ in the direction of $d\overline{S}$, *decreasing* with time, results in a direction for $\overline{E}$ along the positive direction about the closed path.

The scalar form of Faraday's law is given by

$$e = -\frac{d\phi}{dt} \text{ or } -\frac{d\lambda}{dt} \qquad \text{(3.1.9)}$$

where

$$\phi = \iint_S \overline{B} \cdot d\overline{S} \qquad \text{(3.1.10)}$$

and

$$\lambda = N\phi \qquad \text{(3.1.11)}$$

ϕ is the magnetic flux in webers (Wb) passing across an arbitrary surface; λ is the flux linkage in weber-turns given by the product of a number of turns *(N)* and the flux *(ϕ)* linking N.

One can see from Equation 3.1.9 that a time-varying flux linking a stationary coil yields a time-varying voltage. Also, a conductor or a coil moving through a magnetic field will have an induced voltage, which is known as the *motional emf* given by

$$\text{motional emf} = \oint_c (\overline{U} \times \overline{B}) \cdot d\overline{l} \qquad \text{(3.1.12)}$$

where $\overline{U}$ is the velocity vector of the conductor. The voltage induced in a conductor of length l moving perpendicular to the direction of a non-time-varying uniform magnetic field of flux density $\overline{B}$ with a constant velocity $\overline{U}$ can be expressed as

$$\text{motional emf} = B\,l\,U \qquad \text{(3.1.13)}$$

which is often known as the *cutting-of-flux* equation, where $\overline{B}$, $\overline{l}$, and $\overline{U}$ are mutually perpendicular. The direction for the motional emf can be worked out either from the right-hand side of

Equation 3.1.12 or from the simple *right-hand-rule* in cases where Equation 3.1.13 is applicable. By extending the thumb, first, and second fingers of the right hand so that they are mutually perpendicular to each other, if the thumb represents the direction of $\overline{U}$ and the first finger the direction of $\overline{B}$, the second finger will then represent the direction of the emf along $\overline{l}$. Equation 3.1.12 is quite useful in analyzing rotating machines.

Thus we see that the voltage effect can be caused by a time-varying flux with stationary coils, as seen later to be the case in a transformer; the same effect would result from the relative motion of a coil and a constant-amplitude spatial flux-density wave, as seen later to be the case in a rotating machine. Rotation, in effect, introduces the time element and transforms a space distribution of flux density into a time variation of voltage.

Besides Ampere's law and Faraday's law, another significant field relationship is the Lorentz force equation given by

$$dF = I \, d\overline{l} \times \overline{B} \qquad (3.1.14)$$

where I is the current flowing through a differential conductor of length $d\overline{l}$. Only the effect of the magnetic field is considered here, while that of the electric field, if any, is neglected. When Equation 3.1.14 is applied to a simple situation in which a current-carrying conductor is located in a uniform magnetic field and the direction of the conductor is perpendicular to that of the magnetic field, the force is then given by

$$F = B \, l \, I \qquad (3.1.15)$$

which is often used in machine analysis. The direction of the force is orthogonal to the direction of both the conductor and the magnetic field.

Another law that is of use for analytical purposes is known as the Biot-Savart law, given by

$$d\overline{H} = \frac{I \, d\overline{l} \times \overline{a}_R}{4\pi \, R^2} \, ; \overline{H} = \oint \frac{I \, d\overline{l} \times \overline{a}_R}{4\pi \, R^2} \qquad (3.1.16; \; 3.1.17)$$

from which the magnetic field intensity $d\overline{H}$ can be evaluated at a point in space due to a differential current-carrying element of length $d\overline{l}$ carrying a current I at a distance R in the direction from the current element given by the unit vector $\overline{a}_R$.

Through the manipulation of Maxwell's field equations one can come up with partial differential equations of Laplace's or Poisson's type in terms of either the scalar or the vector potential, and solve those subject to the boundary conditions either analytically or numerically on the digital computer. Since such methods of finding the flux distribution are quite involved, they are not considered here in view of the size and the scope of this book.

3.2 MAGNETIC MATERIALS

The magnetic flux density and the field intensity are related through the relationship

$$\overline{B} = \mu\overline{H} \qquad (3.2.1)$$

for material media. In free space, the permeability is given by μ_0, the free-space permeability, which is

$$\mu_0 = 4\pi \times 10^{-7} \text{ henry/meter} = 1.257 \times 10^{-6} \text{ H/m} \qquad (3.2.2)$$

For linear (i.e., with a straight-line relationship between B and H), homogeneous (i.e., having a uniform quality), and isotropic (i.e., with the same properties in any direction) materials, the permeability is a constant given by the slope of the linear *B-H* characteristic, and is related to the free-space permeability as

$$\mu = \mu_r\mu_0 \qquad (3.2.3)$$

where μ_r is known as the relative permeability, a dimensionless constant. If the *B-H* characteristic is nonlinear, as is true for a number of common materials, then the permeability varies as a function of the magnetic induction.

Some classifications of materials are given below along with their characteristic features:

paramagnetic: materials with a relative permeability slightly greater than unity.

diamagnetic: materials with a relative permeability slightly less than unity.

nonmagnetic: materials with a relative permeability essentially equal to 1.0.

superconducting: materials with essentially zero relative permeability, exhibiting "perfect diamagnetism" at temperatures near absolute zero. No magnetic field can be established in the superconducting material.

ferromagnetic: materials exhibiting a high degree of magnetizability, which are generally subdivided into *hard* and *soft* materials. Soft ferromagnetic materials include most of the soft steels and many four-component alloys of iron, nickel, cobalt, and a rare-earth element. Hard ferromagnetic materials include the permanent-magnet materials such as the alnicos, the alloys of cobalt with a rare-earth element such as samarium, the chromium steels, the copper-nickel alloys, and other metal alloys.

ferrimagnetic: materials that are ferrites (which are composed of iron oxides), which are subdivided into *hard* and *soft* ferrites. Hard ferrimagnetic materials include permanent magnetic ferrites, usually of barium or strontium ferrites. Soft ferrimagnetic materials include nickel-zinc and manganese-zinc ferrites.

super-paramagnetic: magnetic materials made from powdered iron particles or other magnetic material suspended in a nonferrous matrix such as the epoxy or other plastic. Of the powdered materials, molybdenum-nickel-iron powder, known as permalloy, is one of the best known.

ferro-fluidic: magnetic fluids consisting of three components: a carrier fluid, iron oxide (Fe_3O_4) particles suspended in the fluid, and a stabilizer.

TABLE 3.2.1   TYPICAL CHARACTERISTICS OF SOME SOFT MAGNETIC MATERIALS

Trade Name	Saturation Flux Density (T)	Maximum Relative Permeability ($\times 10^3$)	Electrical Resistivity (ohm-meter) $\times 10^{-6}$	Curie Temperature (°C)
Carpenter				
Silicon core iron	2–2.1	4–5	0.25–0.60	800
Electrical iron	2.15	2.2–5.5	0.10	760
430F solenoid quality	1.47	1.1–1.6	0.60	671
High permeability 49	1.6	30–120	0.48	450
Hy mu 80	0.78	70–75	0.58	460
Hiperco 27	2.36	2.8	0.19	925
Silectron	1.97	10–20	0.50	732
2 V permendur	2.30	8	0.40	932
Monimax	1.45	40–100	0.65	398
Deltamax	1.60	100–200	0.45	499
4-79 Molybdenum permalloy	0.80	100–400	0.55	454
Ferrite	0.22–0.45	0.16–10	0.1×10^6–10×10^6	135–500

The permeability ratio between good and poor magnetic materials over a typical working range of operation is of the order of 10^4, while the electrical conductivity ratio between a good electrical conductor (such as copper) and a good insulator (such as polystyrene) is of the order of 10^{24}. Thus, no material that can be described as a good magnetic insulator exists except for the superconductors.

For proper use of the magnetic materials in various electromagnetic systems, a number of magnetic properties (besides the mechanical properties) need to be considered: the magnetization characteristic giving the magnetic field intensity *(H)* at various levels of the magnetic flux density *(B);* the saturation flux density; the hysteresis characteristics; the permeability at various levels of the flux density; the variation of permeability with temperature; the electrical resistivity; the Curie temperature; and the loss coefficients. The *Curie temperature* or the *Curie point* is the critical temperature beyond which a ferromagnetic material loses its essential magnetic properties. Typical characteristics of some of the soft magnetic materials are given in Table 3.2.1.

Ferromagnetic materials are quite easily magnetized since they have a high value of relative permeability. The magnetization characteristic (or the *B-H* curve) is a result of the domain changes within the magnetic material. A typical *B-H* characteristic is shown in Figure 3.2.1. The application of weak magnetic fields causes the domains to undergo a boundary displacement. An increase in the magnetic field intensity produces a sudden orientation of the domains toward the direction of the applied field. Further increases result in a slower orientation of the domains; the material is said to be *saturated* when the *B-H* curve is practically flat. Beyond the *saturation flux density* the small increase that occurs in the flux density is exhibited by the linear relationship, $B = \mu_0 H$.

Figure 3.2.1 shows three regions of a typical nonlinear magnetization curve, which is single-valued. The ratio of *B* to *H* at any point on the curve is known as *amplitude permeability*. The

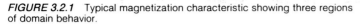

FIGURE 3.2.1 Typical magnetization characteristic showing three regions of domain behavior.

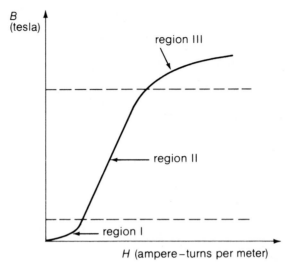

permeability in Region I is said to be the *initial permeability;* in Region II, the *B-H* curve for many materials is relatively straight, so that linear theory can be applied if a magnetic device is operated only in this region.

Nonlinearity of the *B-H* relationship of magnetic materials stands out as a principal impediment to the accurate analysis of electromechanical systems. Some analytical relations between *B* and *H* may be more convenient for digital computations. Typical expressions (out of the many proposed) are given below:

$$B = \frac{aH}{1 + bH} \tag{3.2.4}$$

$$B = \frac{a_0 + a_1H + a_2H^2 + \cdots}{1 + b_1H + b_2H^2 + \cdots} \tag{3.2.5}$$

$$H = [k_1 \exp(k_2B^2) + k_3]B \tag{3.2.6}$$

where $a, b, a_0, a_1, a_2, \ldots, b_1, b_2, \ldots, k_1, k_2,$ and k_3 are constants to be determined from the experimental points on a measured *B-H* curve for a given material.

Alternatively, by choosing a sufficiently large number of points on the *B-H* curve and storing the data in the computer, either linear interpolation or some other curve-fitting technique could be used for obtaining the intermediate values of the magnetization characteristic.

The magnetic permeability of ferromagnetic materials varies with the magnetic field intensity, starting at a relatively low value, increasing to a maximum, and then falling off with increasing

FIGURE 3.2.2   DC magnetization curves for three magnetic materials.

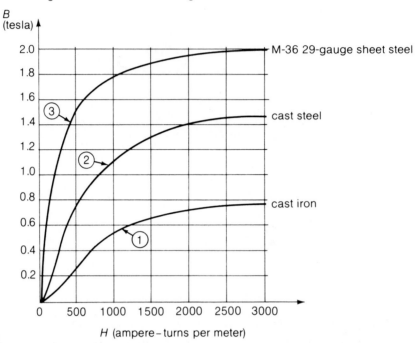

saturation. Typical dc magnetization curves for three magnetic materials are given in Figure 3.2.2. The characteristic can be obtained in two ways: the virgin *B-H* curve, obtained from a totally demagnetized sample; or the normal *B-H* curve, obtained by joining the tips of hysteresis loops of increasing magnitude. Slight differences between the two methods are not of sufficient importance to be discussed here.

When the magnetization is carried through a complete cycle, the permeability for increasing flux density generally differs from that for decreasing flux density at the same value of the magnetic field intensity, as exhibited by the typical *hysteresis loop* shown in Figure 3.2.3. As seen from the figure, when H is zero there is a residual value of the flux density, B_r, which depends on the material, its crystal structure, and the value of B_m. To bring the flux density down to zero from B_r requires a demagnetizing coercive force whose magnitude is given by H_c. As H is varied unidirectionally from $+H_m$ to $-H_m$, and back again unidirectionally to $+H_m$, a complete hysteresis loop is traced out. The normal single-valued *B-H* curve follows the locus of (B_m, H_m) in the first quadrant for a series of steady-state hysteresis loops, taken up to various values of the peak flux density, as shown in Figure 3.2.3.

The area of the loop is a function of B_m and is found experimentally to vary as B_m^n up to moderate values of the flux density; the index n ranges from 1.5 to 2.5. The loop area is related to the

FIGURE 3.2.3 Typical hysteresis loop.

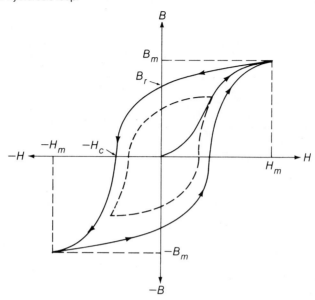

energy required to reverse the magnetic domain walls as the magnetizing force is reversed. This is an irreversible energy and results in an energy loss known as the *hysteresis loss*. The area of the loop varies with the temperature and the frequency of the reversal of *H* for a given material.

Square-loop ferrites exhibit nearly square hysteresis loops at low frequencies with relatively low values of the coercive force and high resistivities. Such ferrite materials are found to be suitable for high-speed memory storage and switching elements.

The second quadrant of the hysteresis loop, known as the *demagnetization curve,* is utilized for the analysis of devices containing *permanent magnets*. The characteristic for Alnico 5 is shown in Figure 3.2.4. The *coercive force, H_c,* is the magnitude of the demagnetizing force corresponding to zero flux density and gives a measure of the material's resistance to demagnetization. The magnitude of *B* on the vertical axis (for $H = 0$) is known as the *residual flux density* or the *residual induction B_r.* Also shown in Figure 3.2.4 is the *energy product,* which is the product of the flux density *(B)* and the magnitude of the demagnetizing force *(H)* for any point on the demagnetization curve. This energy product becomes a maximum at a point on the demagnetization curve. The maximum energy product is often used to compare the permanent-magnet materials on the basis of the amount of energy available to an external magnetic circuit.

The units, symbols, constants, and conversion factors for the SI system are given in Appendix A.

FIGURE 3.2.4   Normal demagnetization curve for Alnico 5.

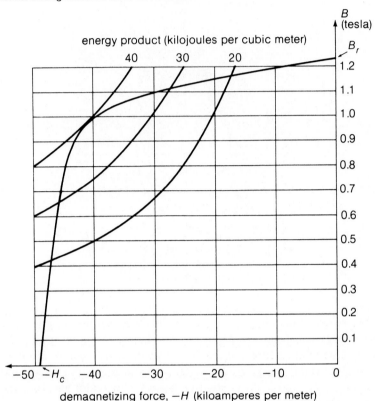

IRON LOSSES   Iron core losses are usually divided into two components: the *hysteresis loss* and the *eddy-current loss*. Time-varying fluxes produce core losses in ferromagnetic materials. No such losses occur in iron cores carrying flux that does not vary with time.

The area of the hysteresis loop represents the energy loss during one cycle in a unit cube of the core material. The hysteresis loss per cycle in a core of volume (Vol) that has a uniform flux density B throughout its volume is given by

$$p_h \text{ per cycle} = \text{Vol} \oint H \, dB \qquad\qquad (3.2.7)$$

where the closed line integral represents the area of the hysteresis loop. The hysteresis loss per second is commonly expressed empirically by

$$p_h = k_h \cdot \text{Vol} \cdot f \cdot B_m^{\,n} \qquad\qquad (3.2.8)$$

since a cyclic variation of the flux at f Hz results in f hysteresis loops per second; k_h is a proportionality constant dependent on the characteristics of iron and the units used; the exponent n, as

PHOTO 3.1   Typical stator- and rotor-core laminations for induction motors.   Photo courtesy of Westinghouse Electric Corporation.

indicated earlier, ranges from 1.5 to 2.5 (a value of 2.0 is often used for estimating purposes); and B_m is the maximum value of the flux density.

Since iron is a conductor, a changing flux induces voltages and currents, called the eddy currents, that circulate within the iron mass. These eddy currents produce losses, heating, and demagnetization. If the current paths are assumed to be wholly resistive and the redistribution of flux due to demagnetization is neglected so that the flux density can be taken as being uniform, a simple expression for the eddy-current loss can be obtained. Neglecting saturation, the eddy-current loss in a sinusoidally excited material is usually expressed by

$$P_e = k_e \cdot \text{Vol} \cdot f^2 \cdot \tau^2 \cdot B_m{}^2 \qquad (3.2.9)$$

where k_e is a proportionality constant dependent on the characteristics of iron and the units used; Vol is the volume of iron; f is the frequency; τ is the lamination thickness; and B_m is the maximum value of the flux density.

To reduce eddy-current loss, ferromagnetic materials that carry pulsating flux are *laminated* (built up from thin sheets electrically insulated from one another). To reduce the eddy currents, the thin sheets must be oriented in a direction parallel to the flow of magnetic flux. The eddy current loss is inversely proportional to the electrical resistivity of the ferromagnetic material and directly proportional to the square of the frequency, the square of the lamination thickness, and the square of the maximum flux density, as seen from Equation 3.2.9.

A common thickness of lamination for continuous operation at 60 Hz is 0.35 mm. Lamination thickness varies usually from 0.3 to 5 mm in electromagnetic devices meant for power applications, while it ranges from 0.01 to 0.5 mm for devices employed in electronics applications. See Photo 3–1 for typical stator- and rotor-core laminations.

Laminating a magnetic material generally results in an increase of the overall volume, the increase depending upon the method used to bond the laminations together. The *stacking factor* is defined as the ratio of the volume actually occupied only by the magnetic material to the overall (or gross) volume of the magnetic part. As the lamination thickness increases, the stacking factor approaches unity.

While Equations 3.2.8 and 3.2.9 are to be regarded as guides to the relative effects of the various factors, the core loss should be determined from experimental data. Most manufacturers of magnetic materials provide core-loss data for most of their products corresponding to the condition of sinusoidal excitation. However, such data are not generally adequate when the electromagnetic devices are applied in circuits in which the voltages and currents have nonstandard waveshapes, such as those caused frequently by the switching action of semiconductors. In general, it is a good idea to have the core-loss measurements made with the device excited by a source whose waveform is as close as possible to that under which the device will be actually operated.

The important characteristics in producing an alternating flux in minimum section with minimum loss are the permeability and the core loss. Magnetic sheet materials are alloy steels in which the chief alloying constituent is silicon, which raises the permeability at low flux densities and reduces the hysteresis and eddy-current losses. The silicon content, however, affects the mechanical properties; it increases the tensile strength, but impairs the ductility. A 5 percent silicon steel cannot be punched or sheared easily; but addition of nickel makes *cold-rolling* possible with some important resulting advantages. Where core loss is not a primary consideration, a low silicon content is used. A high-grade, low-loss steel is utilized if the losses are to be minimized and the cooling is critical. The directional properties of *cold-rolled grain-oriented* sheet and strip materials are highly advantageous in some cases, because the magnetic properties in the rolling direction are far superior to those on any other axis; such materials can sustain higher flux densities than the high-resistance steel of 4–5 percent silicon content.

EXAMPLE 3.2.1

a. Estimate the hysteresis loss at 60 Hz for a toroidal core of 300-mm mean diameter and a square cross section of 50 mm × 50 mm. The symmetrical hysteresis loop for M-36 electric sheet steel (of which the torus is made) is given in Figure 3.2.5.

b. Now suppose all the linear dimensions of the core are doubled. How does the hysteresis loss change?

c. Next, suppose that the torus (which was originally laminated for reducing the eddy-current losses) is redesigned with one-half of the original lamination thickness. Assume the stacking factor to be unity in either case. What would be the effect of such a change in design on the hysteresis loss?

d. If the toroidal core of part *(a)* is used on 50-Hz supply, estimate the change in the hysteresis loss if all the other conditions of operation remain unchanged.

FIGURE 3.2.5 Symmetrical hysteresis loop for M-36 electric sheet steel.

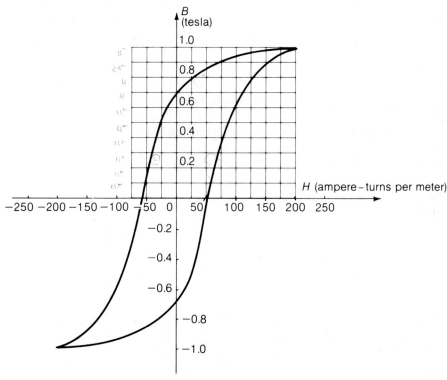

SOLUTION

a. Referring to Figure 3.2.5, the area of each square represents

$$(0.1 \text{ tesla}) \times (25 \text{ amperes/meter}) = 2.5 \frac{\text{weber}}{\text{meter}^2} \times \frac{\text{ampere}}{\text{meter}}$$

$$= 2.5 \frac{\text{volts} \times \text{seconds} \times \text{amperes}}{\text{meter}^3} = 2.5 \frac{\text{joule}}{\text{meter}^3}$$

By counting, the number of squares in the upper half of the loop is found to be 43; the area of the hysteresis loop is given by

$$2 \times 43 \times 2.5 = 215 \text{ J/m}^3$$

The toroidal volume is

$$\pi \times 0.3 \times 0.05^2 = 2.36 \times 10^{-3} \text{ m}^3$$

Hysteresis loss in the torus is then given by

$$2.36 \times 10^{-3} \times 60 \times 215 = 30.44 \text{ watts}$$

b. When all the linear dimensions of the core are doubled, its volume will be eight times the previous volume. Hence the new core hysteresis loss will be

$$8 \times 30.44 = 243.52 \text{ watts}$$

c. The lamination thickness has no bearing on the hysteresis loss. It changes only the eddy-current loss. Hence, the hysteresis loss remains unchanged.
d. Since the hysteresis loss is directly proportional to the frequency, the loss on the 50-Hz supply will be

$$30.44 \times \frac{50}{60} = 25.37 \text{ watts}$$

3.3 MAGNETIC CIRCUITS

The concept of a magnetic circuit is useful in estimating the flux produced by coils wound on ferromagnetic materials. Magnetic circuit analysis follows the lumped-parameter approach of simple dc electrical circuit analysis and is applicable to systems excited by dc sources or, by means of an incremental approach, to systems with a low-frequency ac excitation. Calculations of excitation are usually based on *Ampere's law,* given by

$$\oint \overline{H} \cdot d\overline{l} = \text{current enclosed (or ampere-turns} \atop \text{enclosed)} \qquad (3.3.1)$$

which is strictly valid for steady (or time-invariant) currents, and is approximately valid when the time variation of current is so slow that electromagnetic radiation is negligible. The direction of the current enclosed by the contour and the reference direction of the contour are related by the *right-hand rule.*

Equation 3.3.1 can be written in a scalar form as

$$\oint (\cos \theta) H \, dl = \text{current enclosed} \qquad (3.3.2)$$

in which θ is the angle between the vectors $\overline{H}$ and $\overline{dl}$. If the closed contour is considered to be made up of n small incremental lengths dl_k ($k = 1$ to n), the integral may be replaced by a summation:

$$\sum_{k=1}^{n} \cos \theta_k \, H_k \, dl_k = \text{current enclosed} \qquad (3.3.3)$$

where H_k can be assumed to be a constant over the corresponding dl_k. Further, if the corresponding H_k and dl_k are in the same direction, $\cos \theta_k$ becomes unity, simplifying the calculations.

The *magnetomotive force (mmf)* or the *magnetic potential difference* that tends to produce a magnetic field has the unit of amperes or ampere-turns in the SI system of units; note that the number of turns is a dimensionless constant. The two types of sources of mmf in magnetic circuits are the electric current and the permanent magnet. The current source commonly consists of a coil of N turns carrying a current known as the exciting current.

Analogous to Ohm's law for dc electric circuits, we have the following relation for magnetic circuits:

$$\phi = \frac{\mathbf{F}}{\mathbf{R}} \qquad (3.3.4)$$

ϕ is the magnetic flux; $\mathbf{F}$ is the mmf; and $\mathbf{R}$ is the *reluctance* given by

$$\mathbf{R} = \frac{l}{\mu A} \qquad (3.3.5)$$

for uniform permeability μ and the cross-sectional area A; l stands for the corresponding portion of the length of the magnetic circuit. The reciprocal of the reluctance is known as the *permeance* with the SI unit of henry. The permeance may also be expressed as the ratio of two quantities; the first is the flux through any cross section of a tubular portion of a magnetic circuit bounded by lines of force and by two equipotential surfaces; the second is the magnetic potential difference between the surfaces, taken within the portion under consideration.

The analogy between the electric and magnetic circuits holds in a number of aspects, and it is summarized in Table 3.3.1. One should not infer, however, that electric and magnetic circuits are analogous in all respects. For example, there are no magnetic insulators similar to the electrical insulators known to exist for electric circuits. Also, energy must be continuously supplied when a direct current is established and maintained in an electric circuit; a similar situation does not prevail in the case of a magnetic circuit in which a flux is established and maintained constant. The reluctance of a magnetic circuit is somewhat analogous to the resistance in an electric circuit with the exception that the reluctance is not an energy-loss component.

The magnetic circuit calculations can be carried out in any one of the systems of units that have been in practice from time to time. The student should refer to Table A–4 of Appendix A for the conversion factors associated with different systems of units.

TABLE 3.3.1   ANALOGY BETWEEN ELECTRIC AND MAGNETIC CIRCUITS

	Electric Circuit	Magnetic Circuit
Driving Force	emf	mmf
Response	current	flux
Impedance	resistance	reluctance
Equivalent Circuit	$V = IR$	$F = \phi R$
Field Intensity Relationship	$\oint \overline{E} \cdot d\overline{l} = V$	$\oint \overline{H} \cdot d\overline{l} = I$
Potential Difference	$V = IR$	$F = \phi R$
Other Relations	$J = \dfrac{I}{A} = \dfrac{V}{AR} = \dfrac{EI}{A(\rho l/A)} = \dfrac{E}{\rho}$ or $E = \rho J$ or $J = \sigma E$ where ρ is the resistivity and σ is the conductivity	$B = \dfrac{\phi}{A} = \dfrac{F}{AR} = \dfrac{HI}{A(l/\mu A)} = \mu H$ or $H = B/\mu = \nu B$ or $B = \mu H$ where μ is the permeability and ν is the reluctivity

In the case of ferromagnetic systems containing air gaps, a useful approximation for quick estimates is to consider the ferromagnetic material to have infinite permeability; this is somewhat justified since the permeability of ferromagnetic materials used in electromagnetic devices is of the order of a few thousand times that of air. The relative permeability of the iron is considered to be so high that all the ampere-turns of the winding are practically consumed in the air gaps alone.

Two other terms that are commonly associated with the analysis of magnetic circuits are *leakage* and *fringing*. That portion of the flux that does not pass through the air gap or the useful part of the magnetic circuit is generally known as leakage flux. When used in conjunction with coupled circuits with two or more windings, it refers to that part of the flux that links one coil but not the other. Leakage, being a characteristic of all practical magnetic circuits, can never be completely eliminated. The flux lines can be better directed along a chosen path by the use of magnetic shielding, which consists of thin sheets of high-permeability material effectively used at dc or very low ac excitation frequencies. At higher frequencies of excitation, electrical shielding by the use of aluminum foil or some other conducting material is employed to reduce the leakage flux by dissipating its energy in the form of induced currents in the shield.

FIGURE 3.3.1   Fringing of magnetic flux at an air gap.

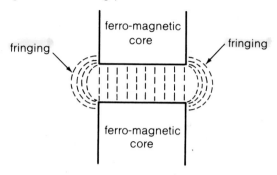

Whenever there is an air gap in a ferromagnetic system, there is a tendency for the flux to bulge outward along the edges of the air gap, instead of passing straight through the air gap parallel to the edges, as shown in Figure 3.3.1; this is referred to as *fringing*. Fringing occurs because the reluctances of different paths available near the air gap are quite comparable to each other, and the flux lines spread out just as the current divides into parallel branches in electric circuits. The effect of fringing is usually taken into account in magnetic circuit calculations by recognizing that the effective area of the air gap is greater than the actual area. Note that the relative effect of fringing increases with the length of the air gap.

Ampere's law, discussed earlier, when applied to the analysis of a magnetic circuit, states that the algebraic sum of the magnetic potentials around any closed path is equal to zero, which is analogous to the Kirchhoff's voltage relationship in electrical circuits. Series and parallel magnetic circuits can be analyzed by means of their corresponding electric-circuit analogs. A few examples with numerical values will illustrate these statements.

The calculation of the mmf or the excitation ampere-turns for simple electromagnetic structures is rather straightforward for a given value of the flux or the flux density. However, it will be seen to be not so simple to determine the flux or the flux density when the mmf is given, because of the nonlinear characteristic of the ferromagnetic material.

Graphical methods or successive approximation by iterative methods (best suited for the digital computer) can be used. The more complicated general field problems in two or three dimensions are solved by graphical or experimental techniques, or most recently by numerical methods with iterative relaxation procedures on the digital computer. These are not presented here. A few examples with simple structures are considered next.

FIGURE 3.3.2 Laminated electromagnet with an air gap for Example 3.3.1 (all dimensions in mm).

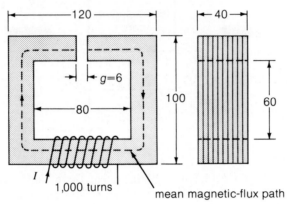

mean magnetic-flux path

EXAMPLE 3.3.1

The magnetic circuit shown in Figure 3.3.2 consists of a laminated core of USS Transformer 72, 29-gauge steel with an air gap g of 6 mm and an exciting winding of 1,000 turns. The stacking factor for the core is given to be 0.94. Neglect leakage but correct for fringing by adding the length of the air gap to each of the other two dimensions of the cross section. The dc magnetization characteristic for USS Transformer 72, 29-gauge steel is given in Figure 3.3.3.

Determine the current in the exciting winding that produces a core flux of 1×10^{-3} Wb.

SOLUTION

Since the leakage is neglected, the magnetic flux will be confined to the paths through the iron and the air gap in series; the mean path along the direction of the flux is shown in Figure 3.3.2. The equivalent magnetic circuit and its electric-circuit analog are given in Figure 3.3.4.

Net cross-sectional area of the core $= 20 \times 10^{-3} \times 40 \times 10^{-3} \times 0.94$
$$= 0.752 \times 10^{-3} \ m^2$$

Length of the mean path in iron $= 2(100 + 80) - 6 = 354$ mm
$$= 0.354 \ m$$

Uniform flux density in iron,

$$B_i = \frac{\phi}{A_i} = \frac{1 \times 10^{-3}}{0.752 \times 10^{-3}} = 1.33 \ T$$

FIGURE 3.3.3   DC magnetization characteristic for USS Transformer 72, 29-gauge steel.

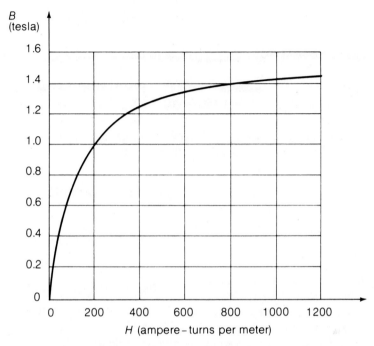

FIGURE 3.3.4   (a) Equivalent magnetic circuit for Example 3.3.1, (b) electric-circuit analog for Example 3.3.1.

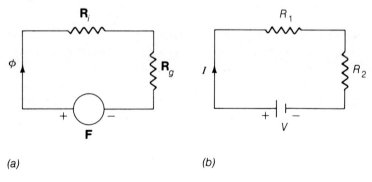

The corresponding H_i for iron from Figure 3.3.3 is 560 At/m.

The mmf for the iron portion of the magnetic circuit is given by

$$\mathbf{F}_i = H_i \, l_i = 560 \times 0.354 = 198 \text{ At}$$

The cross-sectional area of the air gap corrected for fringing is

$$A_g = (20 + 6) \, 10^{-3} (40 + 6) \, 10^{-3} = 26 \times 46 \times 10^{-6}$$
$$= 1.196 \times 10^{-3} \text{ m}^2$$

Note that the stacking factor does not apply to the air-gap region; it applies only to the laminated iron. The air-gap flux density is then

$$B_g = \frac{\phi}{A_g} = \frac{1 \times 10^{-3}}{1.196 \times 10^{-3}} = 0.836 \text{ T}$$

$$H_g = \frac{B_g}{\mu_0} = \frac{0.836}{1.257 \times 10^{-6}} = 0.665 \times 10^6 \text{ At/m}$$

The mmf for the air gap is given by

$$\mathbf{F}_g = H_g \cdot g = 0.665 \times 10^6 \times 6 \times 10^{-3} = 3{,}990 \text{ At}$$

The total mmf for the iron portion and the air gap in series is

$$\mathbf{F}_t = \mathbf{F}_i + \mathbf{F}_g = 198 + 3{,}990 = 4{,}188 \text{ At}$$

and the corresponding current in the exciting winding is

$$I = \frac{\mathbf{F}_t}{N} = \frac{4{,}188}{1{,}000} = 4.2 \text{ A}$$

Note that the mmf $\mathbf{F}_i$ for the iron portion is less than 5 percent of the total ampere-turns $\mathbf{F}_t$. A good first approximation for the ampere-turns could have been obtained by considering just the air gap.

EXAMPLE 3.3.2

Given the total mmf in Example 3.3.1, suppose you are asked to find the resultant flux in the core. Describe a graphical method of solution.

FIGURE 3.3.5   Graphical method for solution of Example 3.3.2.

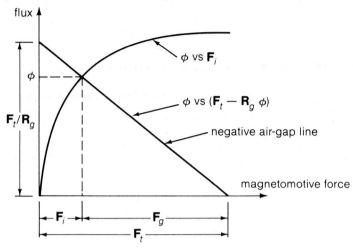

SOLUTION

The total mmf is given by

$$\mathbf{F}_t = \mathbf{F}_i + \mathbf{F}_g$$

or

$$\mathbf{F}_i = \mathbf{F}_t - \mathbf{F}_g = \mathbf{F}_t - \mathbf{R}_g\, \phi.$$

where $\mathbf{R}_g$ is the reluctance of the air gap.

Based on the numerical values of Example 3.3.1, we have

$$\mathbf{F}_i = 0.354\, H_i$$

$$\phi = B_i A_i = 0.752 \times 10^{-3}\, B_i$$

From the magnetization characteristic, by making use of the above relations, plot the flux ϕ as a function of $\mathbf{F}_i$, as shown in Figure 3.3.5.

$$\mathbf{R}_g \text{ (from Ex. 3.3.1)} = \frac{l_g}{\mu_0 A_g}$$

$$= \frac{6 \times 10^{-3}}{1.257 \times 10^{-6} \times 1.196 \times 10^{-3}}$$

$$= 3.991 \times 10^6$$

Next, plot the flux ϕ as a function of $(\mathbf{F}_t - \mathbf{R}_g \, \phi)$, knowing that

$$\mathbf{F}_t = 4,188$$

and

$$\mathbf{R}_g = 3.991 \times 10^6$$

It will be noted that the graph is a straight line as shown in Figure 3.3.5, with the intercept on the flux-axis given by

$$\frac{\mathbf{F}_t}{\mathbf{R}_g} = \frac{4,188}{3.991 \times 10^6} = 1.049 \times 10^{-3} \text{ Wb}$$

and the intercept on the mmf axis given by the total mmf

$$\mathbf{F}_t = 4,188 \text{ At}$$

The resultant core flux in the magnetic circuit is then determined by the intersection of the two plots, given by 1×10^{-3} Wb. The mmf for the iron portion $(\mathbf{F}_i)$ and that for the air gap $(\mathbf{F}_g)$ can be found as shown in Figure 3.3.5.

The student is encouraged to follow the graphical procedure with numerical figures and obtain the results to gain confidence.

The straight-line graph of ϕ versus $(\mathbf{F}_t - \mathbf{R}_g \, \phi)$ is sometimes referred to as the *negative air-gap line.* For the same magnetic structure, for other values of total mmf, the resultant flux is simply obtained by shifting the air-gap line parallel to itself so that it intersects the selected value of mmf on the horizontal mmf axis. On the other hand, to find the resultant flux for a given value of mmf and varying values of air-gap length, you need to adjust only the slope of the negative air-gap line corresponding to the reluctance of the air gap.

The procedure outlined above is applicable to fairly simple magnetic structures; more involved calculations may be needed for analyzing complex systems.

EXAMPLE 3.3.3

Figure 3.3.6 shows an electromagnet with two air gaps in parallel. Note that the magnetic circuit has two parallel paths linking the winding. Neglecting leakage, the reluctance of the iron, and fringing at the air gaps, calculate the flux and the flux density in each of the legs of the magnetic circuit when the current in the 1,000-turn exciting winding is 0.25 ampere. Assume the stacking factor to be unity.

FIGURE 3.3.6 Parallel-path magnetic circuit with two air gaps in parallel (all dimensions in inches).

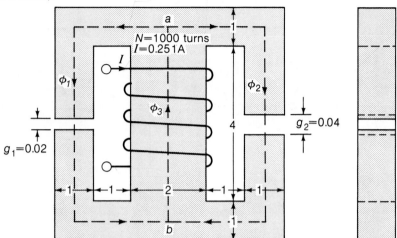

SOLUTION

The mean paths of the flux are shown in Figure 3.3.6. The equivalent magnetic circuit and its electric-circuit analog are given in Figure 3.3.7.

With two air gaps in parallel and with negligible reluctance of the iron, the entire mmf is expended across each of the air gaps. We shall convert to the SI system of units right from the beginning and solve the problem.

The cross-sectional area of the air gaps is

$$A_{g1} = A_{g2} = 1 \text{ in.}^2 \times \frac{1 \text{ m}^2}{(39.37 \text{ in.})^2} = 0.645 \times 10^{-3} \text{ m}^2$$

The air-gap lengths are

$$g_1 = 0.020 \text{ in.} \times \frac{1 \text{ m}}{39.37 \text{ in.}} = 0.508 \times 10^{-3} \text{ m}$$

$$g_2 = 0.040 \text{ in.} \times \frac{1 \text{ m}}{39.37 \text{ in.}} = 1.016 \times 10^{-3} \text{ m}$$

FIGURE 3.3.7 *(a)* Equivalent magnetic circuit for Example 3.3.3, *(b)* electric circuit analog for Example 3.3.3.

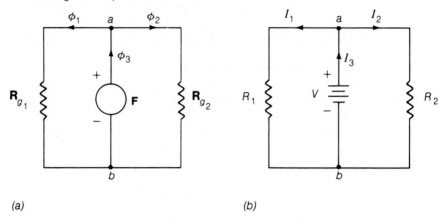

(a)   (b)

The reluctances of the air gaps are given by

$$R_{g1} = \frac{g_1}{\mu_0 A_{g1}} = \frac{0.508 \times 10^{-3}}{(1.257 \times 10^{-6})(0.645 \times 10^{-3})}$$
$$= 0.626 \times 10^6 \text{ At/Wb}$$

$$R_{g2} = \frac{g_2}{\mu_0 A_{g2}} = \frac{1.016 \times 10^{-3}}{(1.257 \times 10^{-6})(0.645 \times 10^{-3})}$$
$$= 1.252 \times 10^6 \text{ At/Wb}$$

The fluxes in the legs of the magnetic circuit are then calculated as

$$\phi_1 = \frac{F}{R_{g1}} = \frac{1{,}000 \times 0.25}{0.626 \times 10^6} = 0.4 \times 10^{-3} \text{ Wb}$$

$$\phi_2 = \frac{F}{R_{g2}} = \frac{1{,}000 \times 0.25}{1.252 \times 10^6} = 0.2 \times 10^{-3} \text{ Wb}$$

$$\phi_3 = \phi_1 + \phi_2 = 0.6 \times 10^{-3} \text{ Wb}$$

The cross-sectional areas of the different legs of the magnetic circuit are

$$A_{l1} = A_{l2} = 0.645 \times 10^{-3} \text{ m}^2$$

and

$$A_{l3} = 1.29 \times 10^{-3} \text{ m}^2$$

Then the corresponding flux densities are given by

$$B_{l1} = \frac{\phi_1}{A_{l1}} = \frac{0.4 \times 10^{-3}}{0.645 \times 10^{-3}} = 0.619 \text{ Wb/m}^2$$

$$B_{l2} = \frac{\phi_2}{A_{l2}} = \frac{0.2 \times 10^{-3}}{0.645 \times 10^{-3}} = 0.31 \text{ Wb/m}^2$$

$$B_{l3} = \frac{\phi_3}{A_{l3}} = \frac{0.6 \times 10^{-3}}{1.29 \times 10^{-3}} = 0.464 \text{ Wb/m}^2$$

Note that the total flux ϕ_3 in the center leg can also be found from the total permeance and the mmf as follows:

$$P_{g1} = \frac{1}{R_{g1}} = \frac{1}{0.626 \times 10^6} = 1.6 \times 10^{-6} \, H$$

$$P_{g2} = \frac{1}{R_{g2}} = \frac{1}{1.252 \times 10^6} = 0.8 \times 10^{-6} \, H$$

With two air gaps in parallel, their combined permeance is the sum of their individual permeances:

$$P_t = P_{g1} + P_{g2} = 2.4 \times 10^{-6} \, H$$

Then the total flux ϕ_3 is given by

$$\phi_3 = F \, P_t = 1{,}000 \times 0.25 \times 2.4 \times 10^{-6}$$
$$= 0.6 \times 10^{-3} \text{ Wb}$$

Note that the reluctances of n components in *series* have a resultant total reluctance of

$$\mathbf{R}_t = \mathbf{R}_1 + \mathbf{R}_2 + \ldots + \mathbf{R}_n$$

just as do resistances in series.

The permeances of n components in *parallel* have a total permeance of

$$\mathbf{P}_t = \mathbf{P}_1 + \mathbf{P}_2 + \ldots + \mathbf{P}_n$$

The different parts of a magnetic structure may be made out of different magnetic materials and may have different cross sections. Analogous to electric circuits, the magnetic circuits may be arranged either in series, parallel, or series-parallel configuration.

EXAMPLE 3.3.4

The magnetic circuit of Example 3.3.3 is solved under the assumption that the permeability of iron is infinite and hence the mmf of the iron parts of the magnetic path is zero. Now suppose that the core material is made of USS Transformer 72, 29-gauge steel, for which the dc magnetization characteristic is given in Figure 3.3.3. Assuming the stacking factor to be unity, compute the flux densities in the three legs of the magnetic structure. Neglect leakage and fringing.

SOLUTION

The problem can be solved graphically by following the procedure of Example 3.3.2. This is left as a desirable exercise for the student. The successive-approximation method is illustrated here for solving this problem, taking the solution of Example 3.3.3 as the first approximation.

The mean magnetic-path lengths between points a and b are

$$l_1 \simeq 10 \text{ in.} = 0.254 \text{ m}$$

$$l_2 \simeq 10 \text{ in.} = 0.254 \text{ m}$$

$$l_3 \simeq 5 \text{ in.} = 0.127 \text{ m}$$

The values of the magnetizing force, from Figure 3.3.3, for the *first* approximation are

$$\text{for } B_1 = 0.619 \text{ Wb/m}^2$$

$$H_1 = 80 \text{ At/m}$$

$$\text{for } B_2 = 0.31 \text{ Wb/m}^2$$

$$H_2 = 42 \text{ At/m}$$

$$\text{for } B_3 = 0.464 \text{ Wb/m}^2$$

$$H_3 = 60 \text{ At/m}$$

The mmfs for the paths are then given by

$$\mathbf{F_1} = H_1\, l_1 = 80 \times 0.254 = 20 \text{ At}$$

$$\mathbf{F_2} = H_2\, l_2 = 42 \times 0.254 = 11 \text{ At}$$

$$\mathbf{F_3} = H_3\, l_3 = 60 \times 0.127 = 8 \text{ At}$$

For the *second* approximation of the flux densities, we shall reduce the winding ampere-turns by the mmfs of the core calculated above and apply the resultant to the air gaps:

$$B_1 = \frac{\mu_0}{g_1}(\mathbf{F_t} - \mathbf{F_1} - \mathbf{F_3}) = \frac{1.257 \times 10^{-6}}{0.508 \times 10^{-3}}(250 - 20 - 8)$$

$$= 0.549 \text{ Wb/m}^2$$

$$B_2 = \frac{\mu_0}{g_2}(\mathbf{F_t} - \mathbf{F_2} - \mathbf{F_3}) = \frac{1.257 \times 10^{-6}}{1.016 \times 10^{-3}}(250 - 11 - 8)$$

$$= 0.286 \text{ Wb/m}^2$$

The flux ϕ_3 in the center leg of the magnetic structure is

$$\phi_3 = B_1\, A_1 + B_2\, A_2$$

and

$$B_3 = \frac{\phi_3}{A_3} = \frac{B_1\, A_1 + B_2\, A_2}{A_3} = \frac{B_1 + B_2}{2} = 0.418 \text{ Wb/m}^2$$

The corresponding new values of the magnetizing force are read from Figure 3.3.3, as follows:

$$H_1 = 70 \text{ At/m}$$

$$H_2 = 40 \text{ At/m}$$

$$H_3 = 50 \text{ At/m}$$

Using this second set of H-values to calculate the *third*-round approximation to the B-values gives

$$B_1 = 0.549 \frac{250 - 24}{250 - 28} = 0.559 \text{ Wb/m}^2$$

$$B_2 = 0.286 \frac{250 - 16}{250 - 19} = 0.29 \text{ Wb/m}^2$$

$$B_3 = \frac{B_1 + B_2}{2} = 0.425 \text{ Wb/m}^2$$

The values of B converge rapidly to the final values. Note that the values above are within 10 percent of the values calculated in Example 3.3.3 with negligible reluctance of the core material.

3.4 ENERGY STORAGE

We have mentioned earlier that the greater ease with which energy may be stored in magnetic fields accounts largely for the common use of electromagnetic devices for electromechanical energy conversion. The volume energy density stored in a magnetic field is given by

$$w_m = \frac{1}{2} \overline{B} \cdot \overline{H} \qquad (3.4.1)$$

The stored potential energy in a magnetic field is expressed by the volume integral

$$W_m = \frac{1}{2} \int_{\text{Vol}} \overline{B} \cdot \overline{H} \, dv \qquad (3.4.2)$$

The energy associated with the field is distributed throughout the space occupied by the field. For a magnetic medium with no losses and a constant permeability, the magnetic energy density is given by

$$w_m = \frac{1}{2} B H = \frac{1}{2} \frac{B^2}{\mu} = \frac{1}{2} \frac{B^2}{\mu_r \mu_0} \qquad (3.4.3)$$

expressed in joules per cubic meter in the SI system of units. Equation 3.4.3 shows that the greater the value of relative permeability μ_r, the less energy is stored in the magnetic field for a given value of the flux density B. It is easy to see why the energy stored in an air gap may be several times that stored in a much greater volume of iron.

In a linear region such as that of an air gap, the energy stored can be expressed as

$$W_{m\ air} = \frac{1}{2}\left(\frac{B_g^2}{\mu_0}\right) \text{Vol} = \frac{1}{2} \mathbf{F}\phi = \frac{1}{2} \mathbf{R}\phi^2 = \frac{1}{2} i\lambda \qquad (3.4.4)$$

The above equation expresses the amount of energy stored upon increasing the flux density from zero to B_g. With nonlinearity and hysteresis neglected, the reluctance and permeance are constant with ϕ and $\mathbf{F}$; the flux and mmf are directly proportional, as in air, for the entire magnetic circuit; in practical electromagnetic devices built with air gaps, the magnetic nonlinearity and the core losses may often be neglected and a linear analysis may be justified.

In Example 3.3.1 the energy stored in the air gap can be calculated as

$$W_{m\ air} = \frac{1}{2} \frac{(B_g)^2}{\mu_0} \text{Vol}$$

$$= \frac{1}{2} \frac{(0.836)^2}{1.257 \times 10^{-6}} \times 1.196 \times 10^{-3} \times 6 \times 10^{-3}$$

$$= 1.995 \text{ J}$$

or it can be alternatively calculated as

$$W_{m\ air} = \frac{1}{2} \mathbf{F}_g \phi = \frac{1}{2} (3{,}990) \ 1 \times 10^{-3} = 1.995 \text{ J}$$

In the magnetic material the energy stored is given by

$$W_{m\ iron} = \frac{1}{2} \mathbf{F}_i \phi = \frac{1}{2} (198) \ 1 \times 10^{-3} = 0.1 \text{ J}$$

The energy stored in the air gap in this case is about 20 times that stored in the iron, even though the volume of the air gap (even after accounting for fringing) is only about 1/37 that of the iron. It is obvious that most of the energy is required to establish the flux in the air gap.

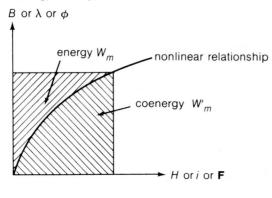

FIGURE 3.4.1   Graphical interpretation of energy and coenergy in a singly excited nonlinear system.

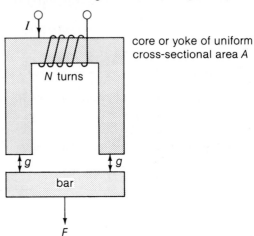

FIGURE 3.4.2   Magnetic circuit for weight lifting.

In a singly excited system, a change in flux density from a value of zero initial flux density to B requires an energy input to the field occupying a given volume:

$$W_m = \text{Vol} \int_0^B H \, dB \qquad (3.4.5)$$

which can also be expressed by

$$W_m = \int_0^\lambda i\,(\lambda) \, d\lambda = \int_0^\phi \mathbf{F}(\phi) \, d\phi \qquad (3.4.6)$$

Note that the current is a function of the flux linkages and the mmf is a function of the flux; their relations depend on the geometry of the coil, the magnetic circuit, and the magnetic properties of the core material. Equations 3.4.5 and 3.4.6 may graphically be interpreted as the area shown in Figure 3.4.1 labeled as *energy*. The other area, labeled as *coenergy* in Figure 3.4.1, can be expressed as

$$W_m' = \text{Vol} \int_0^H B \, dH = \int_0^i \lambda(i) \, di = \int_0^\mathbf{F} \phi(\mathbf{F}) \, d\mathbf{F} \qquad (3.4.7)$$

For a linear system in which B and H or λ and i or ϕ and $\mathbf{F}$ are proportional, it is easy to see that the energy and coenergy are numerically equal. However, for a nonlinear system the energy and coenergy will differ as shown in Figure 3.4.1; but the sum of the energy and coenergy for a singly excited system is given by

$$W_m' + W_m = \text{Vol} \cdot B \, H = \lambda i = \phi \mathbf{F} \qquad (3.4.8)$$

Magnetic circuit principles can be usefully applied to weight lifting by means of a magnetic crane. Let us consider a magnetic circuit configuration consisting of two distinct pieces of the same magnetic material with two air gaps, as shown in Figure 3.4.2. We shall calculate the weight that can be lifted by the crane for a given set of parameters, while making reasonable approximations.

The magnetic energy stored in an incremental volume of space is given by

$$dW_m = \frac{1}{2} (\overline{B} \cdot \overline{H}) \, dv \qquad (3.4.9)$$

In view of the symmetry, working with one pole of the magnetic circuit, the incremental change in volume is

$$dv = A \, dg \qquad (3.4.10)$$

Since the directions of $\overline{B}$ and $\overline{H}$ coincide, neglecting leakage and fringing, we have

$$dW_m = \frac{1}{2} B H \, dv = \frac{1}{2} B H A \, dg = \frac{1}{2} \left(\frac{B^2}{\mu_0}\right) A \, dg \quad (3.4.11)$$

The definition of work gives us

$$dW = \overline{F} \cdot d\overline{L} = F \, dg \qquad (3.4.12)$$

While a magnetic pull is exerted upon the bar, an energy dW equal to the magnetic energy dW_m stored in the magnetic field is expended. Thus

$$dW_m = dW \qquad (3.4.13)$$

or

$$\frac{1}{2} \left(\frac{B^2}{\mu_0}\right) A \, dg = F \, dg \qquad (3.4.14)$$

which yields the pulling force *per pole* on the bar to be

$$F = \frac{1}{2} \left(\frac{B^2}{\mu_0}\right) A \qquad (3.4.15)$$

The total pulling force on the bar is then given by

$$F_{total} = 2\left(\frac{1}{2}\right)\left(\frac{B^2}{\mu_0}\right) A = \left(\frac{B^2}{\mu_0}\right) A \text{ newtons} \qquad (3.4.16)$$

3.5   INDUCTANCE

An electric circuit in which the current links the magnetic flux is said to have *inductance*. A single loop or an inductor carrying a current i is linked by its own flux, as shown in Figure 3.5.1. If the medium in the flux path has a linear magnetic characteristic (i.e., constant permeability), then the relationship between the flux linkages λ and the current i is linear, and the slope of the linear λ-i characteristic gives the *self-inductance*, defined as flux linkage per ampere:

$$L = \frac{\lambda}{i} = \frac{N\phi}{i} = \frac{N^2}{\mathbf{R}} = N^2\,\mathbf{P} \qquad (3.5.1)$$

The above equation illustrates that the inductance is a function of the geometry and permeability, and that in a linear system, it is independent of voltage, current, and frequency. The energy stored in the magnetic field of an inductor in a linear medium is given by

$$W_m = \frac{1}{2}\,i\lambda = \frac{Li^2}{2} \qquad (3.5.2)$$

When more than one loop or circuit is present, the flux produced by the current in one loop may link another loop, thereby inducing a current in that loop; such loops are said to be mutually coupled, and there exists a *mutual inductance* between such loops. The mutual inductance between two circuits is defined as the flux linkage produced in one circuit by a current of one ampere in the other circuit. Let us now consider a pair of mutually coupled inductors, as shown in Figure 3.5.2. The self-inductances L_{11} and L_{22} of inductors 1 and 2 respectively are given by

$$L_{11} = \frac{\lambda_{11}}{i_1} \qquad (3.5.3)$$

and

$$L_{22} = \frac{\lambda_{22}}{i_2} \qquad (3.5.4)$$

where λ_{11} is the flux linkage of inductor 1 produced by its own current i_1, and λ_{22} is the flux linkage of inductor 2 produced by its own current i_2. The mutual inductances L_{12} and L_{21} are given by

$$L_{12} = \frac{\lambda_{12}}{i_2} \qquad (3.5.5)$$

$$L_{21} = L_{12} = M$$

and

$$L_{21} = \frac{\lambda_{21}}{i_1} \qquad (3.5.6)$$

where λ_{12} is the flux linkage of inductor 1 produced by the current i_2 in inductor 2, and λ_{21} is the flux linkage of inductor 2 produced by the current i_1 in inductor 1.

FIGURE 3.5.1 A single inductive loop.

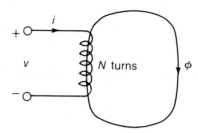

FIGURE 3.5.2 Mutually coupled inductors.

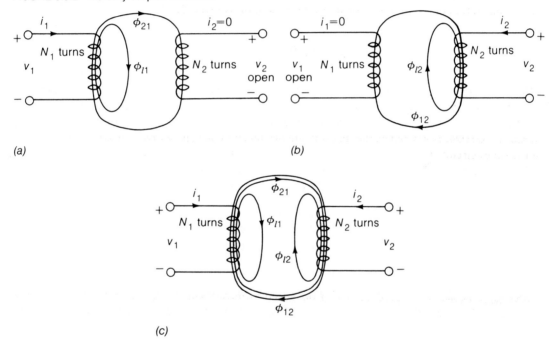

If a current of i_1 flows in inductor 1 while the current in inductor 2 is zero, the equivalent fluxes are given by

$$\phi_{11} = \frac{\lambda_{11}}{N_1} \tag{3.5.7}$$

and

$$\phi_{21} = \frac{\lambda_{21}}{N_2} \tag{3.5.8}$$

where N_1 and N_2 are the number of turns of inductors 1 and 2 respectively. That part of the flux of inductor 1 that does not link any turn of inductor 2 is known as the equivalent *leakage flux* of inductor 1:

$$\phi_{l1} = \phi_{11} - \phi_{21} \tag{3.5.9}$$

Similarly,

$$\phi_{l2} = \phi_{22} - \phi_{12} \tag{3.5.10}$$

The *coefficient of coupling* is given by

$$k = \sqrt{k_1 k_2} \tag{3.5.11}$$

where

$$k_1 = \phi_{21}/\phi_{11} \text{ and } k_2 = \phi_{12}/\phi_{22}$$

When k approaches unity, the two inductors are said to be tightly coupled; and when k is much less than unity, they are said to be loosely coupled. While the coefficient of coupling can never exceed unity, it may be as high as 0.998 for iron-core transformers; it may be smaller than 0.5 for air-core transformers.

When there are only two inductively coupled circuits, the symbol M is frequently used to represent the mutual inductance; it can be shown that the mutual inductance between two electric circuits coupled by a homogeneous medium of constant permeability is reciprocal:

$$M = L_{12} = L_{21} = k\sqrt{L_{11} L_{22}} \tag{3.5.12}$$

The energy considerations that lead to such a conclusion are taken up in a problem at the end of the chapter as an exercise for the student.

Let us next consider the energy stored in a pair of mutually coupled inductors:

$$W_m = \frac{i_1 \lambda_1}{2} + \frac{i_2 \lambda_2}{2} \tag{3.5.13}$$

where λ_1 and λ_2 are the total flux linkages of inductors 1 and 2 respectively. Equation 3.5.13 may be rewritten as follows:

$$W_m = \frac{i_1}{2} (\lambda_{11} + \lambda_{12}) + \frac{i_2}{2} (\lambda_{22} + \lambda_{21})$$

$$= \frac{1}{2} L_{11} i_1{}^2 + \frac{1}{2} L_{12} i_1 i_2 + \frac{1}{2} L_{22} i_2{}^2 + \frac{1}{2} L_{21} i_1 i_2$$

or

$$W_m = \frac{1}{2} L_{11} i_1{}^2 + M i_1 i_2 + \frac{1}{2} L_{22} i_2{}^2 \tag{3.5.14}$$

Equation 3.5.14 is valid whether the inductances are constant or variable, so long as the magnetic field is confined to a uniform medium of constant permeability.

When there are *n* coupled circuits, the energy stored in the magnetic field can be expressed as

$$W_m = \sum_{j=1}^{n} \sum_{k=1}^{n} \frac{1}{2} L_{jk} i_j i_k \tag{3.5.15}$$

Going back to the pair of mutually coupled inductors shown in Figure 3.5.2, the flux-linkage relations and the voltage equations for circuits 1 and 2 are given by the following, while neglecting the resistances associated with the coils:

$$\lambda_1 = \lambda_{11} + \lambda_{12} = L_{11} i_1 + L_{12} i_2 = L_{11} i_1 + M i_2 \tag{3.5.16}$$

$$\lambda_2 = \lambda_{21} + \lambda_{22} = L_{21} i_1 + L_{22} i_2 = M i_1 + L_{22} i_2 \tag{3.5.17}$$

$$v_1 = p\lambda_1 = \frac{d\lambda_1}{dt} = L_{11} \frac{di_1}{dt} + M \frac{di_2}{dt} \tag{3.5.18}$$

$$v_2 = p\lambda_2 = \frac{d\lambda_2}{dt} = M \frac{di_1}{dt} + L_{22} \frac{di_2}{dt} \tag{3.5.19}$$

where *p* is the derivative operator *d/dt*.

FIGURE 3.5.3   Dot notation for a pair of mutually coupled inductors:
(a) dots on upper terminals, (b) dots on lower terminals.

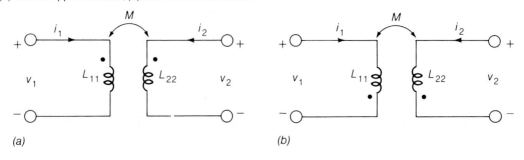

(a)   (b)

FIGURE 3.5.4   Polarity markings for complicated magnetic coupling
situations.

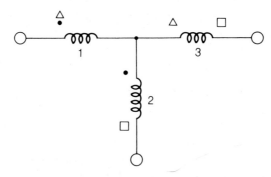

For the terminal voltage and current assignments as shown in Figure 3.5.2, the coil windings
are such that the fluxes produced by currents i_1 and i_2 are additive in nature, and in such a case
the algebraic sign of the mutual voltage term is positive as in Equations 3.5.18 and 3.5.19.

In order to avoid drawing detailed sketches of windings showing the sense in which each coil is
wound, a *dot convention* is developed, according to which the pair of mutually coupled inductors
of Figure 3.5.2 are represented by the system shown in Figure 3.5.3. The notation is such that a
current i entering a dotted (undotted) terminal in one coil induces a voltage $M[di/dt]$ with a positive
polarity at the dotted (undotted) terminal of the other coil. If the two currents i_1 and i_2 were to
be entering (or leaving) the dotted terminals, the adopted convention is such that the fluxes produced by i_1 and i_2 will be aiding each other, and the mutual and self-inductance terms for each
terminal pair will have the same sign; otherwise, they will have opposite signs.

Although just a pair of mutually coupled inductors are considered above for the sake of simplicity, complicated magnetic coupling situations do occur in practice; for example, Figure 3.5.4
shows the coupling between coils 1 and 2, 1 and 3, and 2 and 3.

For nonlinear models, the fluxes and currents are not related by linear equations. For example, for current-controlled nonlinear coupled coils, the flux-linkage relationships have the general form given by

$$\lambda_1 = f_1(i_1, i_2) \tag{3.5.20}$$

and

$$\lambda_2 = f_2(i_1, i_2) \tag{3.5.21}$$

For the time-varying case, the functions f_1 and f_2 will also depend explicitly on time t.

3.6 PERMANENT MAGNETS Skip

A permanent magnet is a body that is capable of maintaining a magnetic field at other than cryogenic temperatures with no expenditure of power. The basic purpose of any magnet is to store energy or to convert energy from one form to another. This is generally achieved by setting up a magnetic field at some region in space commonly called the air gap, or by storing the energy within the magnetic material. A magnetic field can be established either by an electrical excitation coil energized from a source or by a permanent magnet. Permanent magnets are magnetic circuit components widely used in a variety of industrial equipments and devices. They are utilized in such simple devices as latches for equipment doors to more complex applications such as field magnets for electrical rotating machines and magnetic focusing of electron beams.

Various classes of permanent-magnet materials are available these days: ductile metallic magnets, typified by Cunife; ceramic magnets, such as Indox; brittle metallic magnets, such as Alnico (cast and sintered); and rare-earth cobalt magnets, such as samarium-cobalt. The size, shape, physical properties, and other design limitations must be considered in selecting the most practical and economical magnetic material for a specific type of application. In a practical design, the choice will be based on cost factors, availability, structural requirements from the viewpoint of mechanical design, space limitations, and the environment in which the magnet will be situated, as well as the magnetic and electrical performance specifications. Since most of the permanent magnets are not machinable, they may have to be used in the devices in the form available from the manufacturer.

In magnet design, the shape of a permanent magnet influences the amount of flux produced. Equal volumes of magnetic material will produce different amounts of flux, depending on their shape. In a circuit excited by a permanent magnet, the operating conditions of the permanent magnet are determined to a great extent by the external magnetic circuit. Since there are no magnetic insulators, it is rather difficult to direct the flux to those parts of the magnetic circuit where it can be usefully applied. If one can think of the flux produced by a permanent magnet as consisting of useful flux and leakage flux, the fundamental problem in magnetic circuit design is to minimize the unavoidable flux leakage to improve the efficiency of the magnetic circuit.

The operating point and subsequent performance of a permanent magnet are a function of how the magnet is physically located in the magnetic circuit and whether it is magnetized before or after installation. In a number of applications, the magnet must be put through a stabilizing process prior to its use. In addition to the intrinsic magnetic properties of the material, other factors that influence the stability of a magnet include external magnetic field effects, temperature effects, structural stability, mechanical stress, and radiation.

Practical magnet specification involves a knowledge of available magnet materials, their magnetic characteristics, magnet test procedures, and various factors that affect the flux produced by a magnet. Too tight specifications may require extra manufacturing, testing, and quality control operations that will increase costs. On the other hand, a loose specification could allow shipment of magnets that would not work in the specified application. In addition to magnetic properties, specifications on surface finishes and physical dimensions should be evaluated in terms of each specific application.

Demagnetization curves are an essential tool in selecting the magnetic material that best meets the designer's requirements. Figure 3.2.4 shows the normal second-quadrant demagnetization curve for Alnico-5 permanent-magnet material.

The permanent magnet designer uses another important graph, in which the left half carries the second-quadrant normal demagnetization curve and the right half has a curve called the *BH* or energy product curve. These two curves are in mutual correspondence, as shown in Figure 3.6.1. A straight line drawn from the origin to the point with coordinates $(B_r, -H_c)$ will intersect the demagnetization curve at the point of maximum energy. Ideally, one would like to design the permanent magnet to operate at this optimum point (where the available external energy is at its maximum) by using the corresponding magnitudes of H_d and B_d. In practice, while this is usually not the case, the majority of the magnet material in a good design will operate in the vicinity of the maximum energy point on the knee of the *B-H* curve.

The solution to the magnet design problem is achieved by first designing a magnet that provides a specified flux across a specified air gap. Then the magnet is proportioned to have minimum volume. In practice, different magnetic materials will be considered in arriving at an efficient design in the space allocated for the permanent magnet. Thus a wide variety of possible solutions could be obtained prior to the final selection of the material to be used.

For a specific material, if the magnet is allowed to operate at an arbitrarily chosen point $(B_d, -H_d)$, the first step in a design is to determine the length of the magnet l_m. Figure 3.6.2 shows a typical configuration consisting of a permanent magnet, two L-shaped soft-iron pole pieces and an air gap. Neglecting the reluctance of the soft-iron pieces, from the application of Ampere's law, one obtains the following:

$$H_d l_m = H_g l_g \qquad (3.6.1)$$

where

H_d is the magnitude of the demagnetizing force of the magnet,
l_m is the length of the magnet,
H_g is the air-gap magnetic field intensity,
l_g is the length of the air gap parallel to the line of flux.

FIGURE 3.6.1 Second-quadrant normal demagnetization curve and energy-product curve.

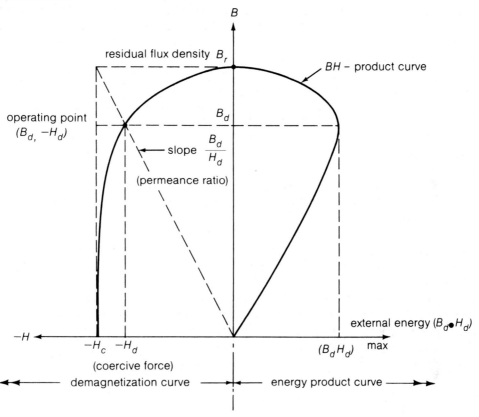

FIGURE 3.6.2 Typical configuration consisting of a permanent magnet, soft-iron pole pieces, and an air gap.

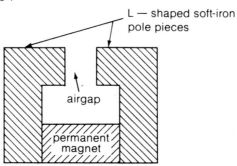

Equation 3.6.1 is true only if there is no leakage and fringing, and if the reluctance drop in soft-iron pole pieces can be neglected. In order to account for these in practical systems, one may introduce a *reluctance factor, f,* defined as the ratio of the total mmf produced by the permanent magnet to the mmf in the air gap. Then it follows that

$$H_d l_m = f H_g l_g = f B_g l_g / \mu_0$$

or

$$l_m = \frac{f B_g l_g}{\mu_0 H_d} \qquad (3.6.2)$$

where all the quantities are in the SI system of units. The reluctance factor ranges from 1.1 to 1.5 in different magnetic circuit configurations; normally a value of about 1.25 is quite adequate. A large reluctance factor may be caused either by magnetic saturation in some portion of the magnetic circuit or by disproportionate gap geometry.

The cross-sectional area of the magnet can be calculated from the flux required in the air gap upon introduction of a *leakage coefficient, F:*

$$B_d A_m = B_g A_g F$$

or

$$A_m = \frac{F B_g A_g}{B_d} \qquad (3.6.3)$$

where

      B_d is the flux density in the magnet,
      A_m is the cross-sectional area of the magnet (orthogonal to the direction of
        magnetization),
      B_g is the air-gap flux density,
      A_g is the cross-sectional area of the gap (orthogonal to the line of flux),
      F is the leakage factor, which is the ratio of the flux produced by the magnet to the flux
        in the air gap.

It is difficult to evaluate the leakage factor, F, which may range from 1.2 to more than 60. It may be kept small by placing the magnet material as close as possible to the air-gap region. Large steel surfaces between the magnet and the gap tend to increase the leakage factor considerably. The leakage factor for a particular type of magnetic system may be determined either by field computation or by estimation on the basis of design experience with similar systems, or by making measurements on a model. In a number of magnetic-circuit problems, the leakage factor can be estimated by calculating the permeance of various flux paths about the gap and magnet. The ratio of the sum of all the permeances to the gap permeance gives the leakage factor. In problems where it is critical, one may have to resort to detailed flux distribution maps, which can be obtained these days by numerical methods on digital computers or by other experimental methods such as the electrolytic tank.

FIGURE 3.6.3 Different operating conditions of a permanent magnet.

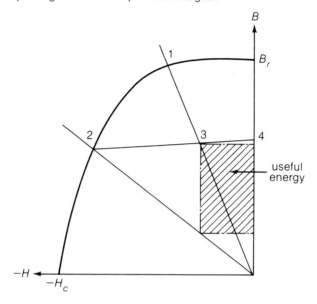

The magnet volume can then be obtained from Equations 3.6.2 and 3.6.3 as

$$V_m = A_m\, l_m = \frac{f\, F\, B_g^2\, V_g}{\mu_0\, B_d\, H_d} \tag{3.6.4}$$

where $V_g = A_g\, l_g$, the volume of the air gap, and all the quantities are expressed in the SI system of units.

The *permeance coefficient, P_c,* is defined by the slope (B_d/H_d) shown in Figure 3.6.1 and is obtained from Equations 3.6.2 and 3.6.3 as

$$P_c = \frac{B_d}{H_d} = \frac{F}{f}\cdot\frac{l_m}{A_m}\cdot\frac{A_g}{l_g}\cdot\mu_0 \tag{3.6.5}$$

The line joining the origin and the operating point $(B_d, -H_d)$ on the normal demagnetization curve is known as the *load line*. The slope of the load line depends only on the physical dimensions of the magnetic circuit and is independent of the properties of the magnetic material. Even though different sections of a magnet may practically operate on different permeance coefficients, generally a specified load line refers to the central or neutral section of the magnet.

Some magnets are used in open-circuited condition without steel pole pieces, in which case the permeance is strictly a function of the magnet geometry. When assembled in a magnetic circuit where soft pole pieces direct the flux path, one needs to consider the permeance of the complete magnetic circuit. Let the permeance coefficient line for the assembly shown in Figure 3.6.2 cause the magnet to operate at a point 1 as shown in Figure 3.6.3 when the magnet is magnetized in place. When the magnet is removed or open-circuited, its operating point will drop to point 2 (shown in Figure 3.6.3), which is determined by the open-circuited permeance coefficient of the magnet.

If the magnet should be replaced in the assembly, the operating point will shift from point 2 to some point 3 rather than to point 1, as shown in Figure 3.6.3. It moves up a *recoil line* until it intersects the assembly permeance coefficient line at point 3. This recoil line is approximately parallel to the slope of the demagnetization curve at the point B_r. From point 1 to point 3, there has been a loss of flux; it depends greatly on the shape of the demagnetization curve and the position of the operating point on the curve. The hatched rectangular area represents the total useful energy. It can be shown graphically that this area will be a maximum when point 3 is halfway between points 2 and 4.

The slope of the recoil line, called the *recoil permeability*, is an intrinsic property of the magnetic material and is one of the most important factors to be considered in designing variable-gap applications. Changes in magnetic circuit reluctance occur in many applications, such as holding magnets, electromechanical energy converters, and magnetic drives. Magnetic materials recommended for such applications should be those whose demagnetization curves most closely approximate straight lines. Alternatively, one may design so that the lowest operating point is above the knee of the demagnetization curve.

In many Indox and samarium-cobalt magnet assemblies, the loss may be essentially zero because these magnets have approximately straight demagnetization curves. On the other hand, in loudspeakers using Alnico-5 magnets, the loss in flux may be as much as 70 percent. To avoid the loss due to reluctance changes and for efficient magnet usage, in-circuit magnetization is used in many applications. To protect the magnets against demagnetizing influences either in shipping or in storage, *keepers* (i.e., magnetic conductors) are used to complete the magnetic circuit of the permanent magnet.

Table 3.6.1 gives some of the pertinent characteristics of some common permanent magnet materials. Figures 3.2.4 and 3.6.4 show the normal demagnetization curves for Alnico-5 and samarium-cobalt, respectively.

The most striking aspects of the rare-earth magnet materials, developed during the late sixties, are their high coercivity and possible maximum energy level, while the most gratifying aspect is their predictable performance. They constitute a major step forward in the art of making magnets, much as the Alnicos did in the 1930s and the ferrites in the 1950s. The cost of rare-earth magnets is relatively high, but is decreasing. In order to make the best use of a high-quality and relatively expensive magnetic material such as $SmCo_5$, one should strive for optimized designs. For accurate analysis of difficult applications and increased magnet efficiency, computer-aided design that can consider the effects of a large number of variables has become an important tool for the magnet designer.

Let us now take up a simple illustrative example of a permanent-magnet design.

TABLE 3.6.1   PERMANENT MAGNET MATERIAL CHARACTERISTICS (Typical Values)

Type		Residual Flux Density B_r (Tesla)	Coercive Force H_c (kA/m)	Maximum Energy Product (kJ/m³)	Average Recoil Permeability (H/m $\times$ 10⁻⁶)
Cast Alnico	5	1.28	51	44	2.1
	5–7	1.34	58	60	1.9
	6	1.05	62	31	5.0
Sintered Alnico	5	1.09	49	31	2.0
	6	0.94	64	23	5.0
Cunife		0.55	42	11	1.7
Indox	1	0.22	145	8	1.15
	2	0.29	193	14	1.15
	3	0.335	187	21	1.1
	4	0.255	183	12	1.1
Rare-earth cobalt	18	0.87	637	143	1.05
Incor	16	0.81	629	127	1.05
36% cobalt steel		1.04	18	8	10

FIGURE 3.6.4   Normal demagnetization curve for samarium-cobalt permanent-magnet material.

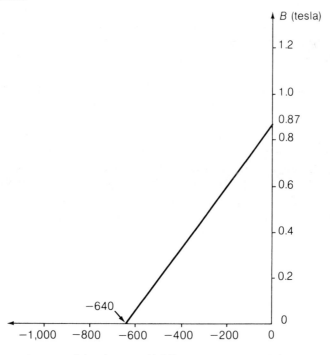

EXAMPLE 3.6.1

The problem is to find the length and the area of a permanent magnet that can establish a specified flux density through an air gap of known dimensions. For the sake of simplicity, let us neglect leakage and fringing, and let the operating point on the normal demagnetization curve of Alnico-5, as shown in Figure 3.2.4, correspond to a point ($B_d = 0.96$ tesla and $-H_d = -42$ kA/m) at which the product ($B_d H_d$) is at its maximum. It is further given that

$$B_g = 0.5 \text{ tesla; } A_g = 2 \text{ cm}^2; l_g = 0.5 \text{ cm}$$

SOLUTION

From Equation 3.6.2 one has

$$l_m = \frac{B_g l_g}{\mu_0 H_d} = \frac{0.5 \times 0.5 \times 10^{-2}}{4\pi \times 10^{-7} \times 42 \times 10^3} = 4.74 \times 10^{-2} \text{ m}$$
$$= 4.74 \text{ cm}$$

and from Equation 3.6.3 one gets

$$A_m = \frac{B_g A_g}{B_d} = \frac{0.5 \times 2 \times 10^{-4}}{0.96} = 1.04 \times 10^{-4} \text{ m}^2$$
$$= 1.04 \text{ cm}^2$$

BIBLIOGRAPHY

Durney, C. H., and Johnson, C. C. *Introduction to Modern Electromagnetics*. New York: McGraw-Hill, 1969.

Indiana General Corporation. *Manual No. 9: Permanent Magnet Design Manual*. Valparaiso, Ind., 1968.

Magnetic Materials Producers Association. *Standard No. 0100-78: Standard Specifications for Permanent Magnet Materials*. Chicago, 1978.

Parker, R. J., and Studders, R. J. *Permanent Magnets and Their Applications*. New York: John Wiley & Sons, 1962.

Roters, H. C. *Electromagnetic Devices*. New York: John Wiley & Sons, 1941.

Strnat, K. J., ed. "Proceedings of the Second International Workshop on Rare-Earth Cobalt Permanent Magnets and Their Applications." Dayton, Ohio, 1976.

Thomas & Skinner, Inc. *Bulletin No. M303: Permanent Magnet Design*. Indianapolis, 1967.

Woodson, H. H., and Melcher, J. R. *Electromechanical Dynamics, Part I: Discrete Systems*. New York: John Wiley & Sons, 1968.

PROBLEMS

3-1.

Write down the general Maxwell equations that govern the electromagnetic phenomena at any point in space and specify the assumptions that lead to Equations 3.1.1, 3.1.3, and 3.1.5. What is the unmodified current-continuity relation?

3-2.

Show that the magnetic field associated with a long, straight current-carrying wire is given by

$$B_\phi = \frac{\mu_0 I}{2\pi r}$$

where subscript ϕ denotes the ϕ-component in the circular cylindrical coordinate system, μ_0 is the free-space permeability, I is the current carried by the wire, and r is the radius from the current-carrying wire. Solve the problem by the application of *(a)* Ampere's circuital law, and *(b)* the Biot-Savart law.

3-3.

A magnetic force exists between two adjacent, parallel current-carrying wires. Let I_1 and I_2 be the currents carried by the wires and r the separation between them. Making use of the result of Problem 3-2, find the force between the wires.

Discuss the nature of the force when the wires carry currents in the same direction, and in opposite directions.

3-4.

Consider the arrangement shown in the following figure, in which a conductor-bar of length L is free to move along a pair of conducting rails. The bar is driven by an external force at a constant velocity of U meters per second. A constant uniform magnetic field $\bar{B}$ is present pointing into the paper. Neglect the resistance of the bar and rails, as well as the friction between the bar and the rails.

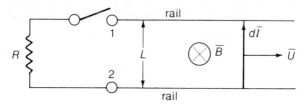

(a) Determine the expression for the motional voltage across terminals 1 and 2. Is terminal 1 positive with respect to terminal 2?

(b) If an electrical load resistance R is connected across the terminals, what is the current and power dissipated in the load resistance? Show the direction of current on the figure.

(c) Find the magnetic field force exerted on the moving bar, and the mechanical power required to move the bar. How is the principle of energy conservation satisfied?

(d) Since the moving bar is not accelerating, the net force on the bar must be equal to zero. How would you justify?

3–5.

A toroidal coil is wound on a plastic ring of rectangular cross section with inside diameter 30 cm, outside diameter 40 cm, and height 10 cm. The coil has 200 turns of round copper wire of 3-mm diameter.

(a) Calculate the magnetic flux density at the mean diameter of the coil when the coil current is 50 A.

(b) Assuming the answer of part *(a)* to be the uniform flux density within the coil, compute the flux linkages of the coil.

(c) Evaluate the percentage error involved in assuming uniform flux density in the coil.

(d) Find the resistance of the wire, given that the volume resistivity of copper is 17.2×10^{-9} ohm-meter.

3–6.

A toroid is made up of three ferromagnetic materials, whose segments are arranged in series. One material is a nickel-iron alloy having a mean arc length of 0.3 m. The second

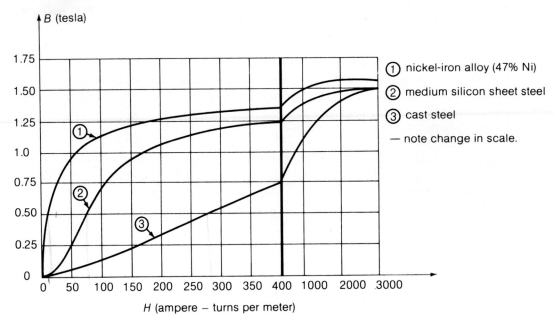

① nickel-iron alloy (47% Ni)

② medium silicon sheet steel

③ cast steel

— note change in scale.

H (ampere − turns per meter)

material is medium silicon steel with a mean arc length of 0.2 m. The third material is cast steel with a mean arc length of 0.1 m. Each of the materials has a cross-sectional area of 0.001 m². The excitation coil has 100 turns. The figure given here shows the magnetization curves for the materials. The leakage may be neglected.

(a) Determine the excitation current to establish a magnetic flux of 0.5×10^{-3} Wb.

(b) Calculate the relative permeability and the reluctance of each segment.

3–7.

For the toroid of Problem 3–6, compute the magnetic flux established by an applied mmf of 50 ampere-turns, while neglecting the leakage. Use a cut-and-try procedure consisting of successive approximations.

3–8.

Determine the mmf to establish an air-gap flux of 0.7×10^{-3} Wb in the magnetic circuit shown below, consisting of two different ferromagnetic materials in series with an air gap. Neglect leakage and fringing. The permeability of the materials at a working flux density of 1.4 tesla are for rolled steel, $\mu = 1,400\mu_0$; for cast steel $\mu = 620\mu_0$. Assume a uniform cross-sectional area of 5 cm² throughout.

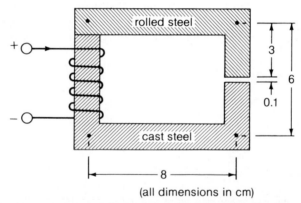

(all dimensions in cm)

3–9.

The configuration of a magnetic circuit is given in the figure that follows. Assume the permeability of the ferromagnetic material to be $\mu = 1,000\mu_0$. Neglect leakage, but correct for fringing by increasing each of the linear dimensions of the cross-sectional area by the length of the air gap. Find the air-gap flux, the air-gap flux density, and the magnetic field intensity in the air gap.

The magnetic material has a square cross-sectional area of 4 cm².

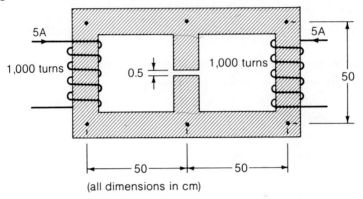

5A 1,000 turns 0.5 1,000 turns 5A 50

50 50

(all dimensions in cm)

3–10.

Compute the excitation current needed for a 1,000-turn coil in order to establish an air-gap flux of 0.4×10^{-3} Wb for the magnetic circuit of configuration shown in the figure below.

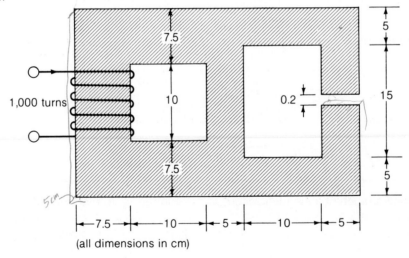

1,000 turns 7.5 10 0.2 15 5 5

7.5 7.5 10 5 10 5

(all dimensions in cm)

Neglect leakage and fringing; assume that the magnetic material (cast steel) has a thickness of 5 cm.

3–11.

The λ-i characteristic of a magnetic circuit is given in the figure below, consisting of two straight-line segments. Compute the energy W_m and coenergy W_m' for the magnetic circuit at point a and at point b.

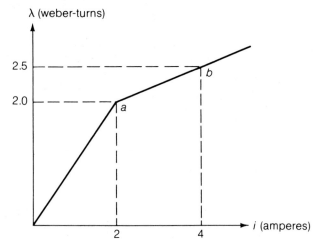

3–12.

The self-inductances of two coupled coils are L_{11} and L_{22}, and the mutual inductance between them is M. Show that the effective inductance of the two coils in series is given by

$$L_{series} = L_{11} + L_{22} \pm 2M$$

and the effective inductance of the two coils in parallel is given by

$$L_{parallel} = \frac{L_{11} L_{22} - M^2}{L_{11} \mp 2M + L_{22}}$$

Specify the conditions corresponding to different signs of the term $(2M)$.

3-13.

Referring to the circuit of Figure 3.5.3, let

$$L_{11} = L_{22} = 0.1 \ H$$

and

$$M = 10 \ mH$$

Determine v_1 and v_2 if

  a. $i_1 = 10$ mA and $i_2 = 0$,
  b. $i_1 = 0$ and $i_2 = (10 \sin 100t)$ mA,
  c. $i_1 = (0.1 \cos t)$ A and $i_2 = [0.3 \sin(t + 30°)]$ A

Also find the energy stored in each of the above cases at $t = 0$.

3-14.

Consider a pair of coupled coils as shown in Figure 3.5.3 with currents, voltages, and polarity dots as indicated. Show that the mutual inductance is $L_{12} = L_{21} = M$ by following the steps indicated below:

  a. Starting at time t_0 with $i_1(t_0) = i_2(t_0) = 0$, maintain $i_2 = 0$ and increase i_1 until, at time t_1, $i_1(t_1) = I_1$ and $i_2(t_1) = 0$. Determine the energy accumulated during this time. Now maintaining $i_1 = I_1$, increase i_2 until at time t_2, $i_2(t_2) = I_2$. Find the corresponding energy accumulated and the total energy stored at time t_2.
  b. Repeat the process in the reverse order, allowing the currents to reach their final values. Compare the expressions obtained for the total energy stored and obtain the desired result.

3-15.

For the coupled coils shown in the figure below, a dot has been arbitrarily assigned to a terminal as indicated. Following the dot convention presented in the text, place the other dot in the remaining coil and justify your answer with explanation.

Comment on whether the polarities are consistent with Lenz's law.

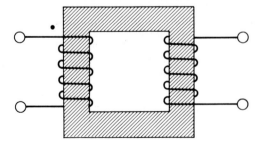

3–16.

For the configurations of the coupled coils shown in the figure below, obtain the voltage equations for v_1 and v_2.

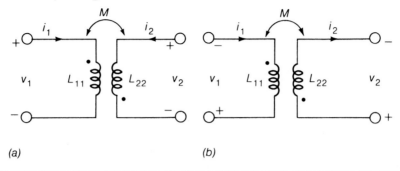

(a)   (b)

3–17.

A long, straight cylindrical conductor of radius R meters and length l meters has a relative permeability of unity, and it carries a current of I amperes distributed uniformly over its cross section.

Determine the following in terms of I, R, and l:

a. The total magnetic flux within the conductor.
b. The magnetic energy density within the conductor as a function of the distance ρ from the center.
c. The energy stored in the magnetic field within the conductor.
d. The corresponding self-inductance due to the magnetic field within the conductor.

3–18.

The λ-i characteristic of a nonlinear singly-excited magnetic circuit is given by

$$\lambda = \frac{2.8i}{0.12 + i}$$

Find the energy and coenergy for the magnetic circuit corresponding to a current of 0.10 amperes.

3–19.

Suppose in Example 3.6.1 given in the text, the dimensions of the Alnico-5 magnet and the air gap are known, and it is required to calculate the air-gap flux density. Neglect leakage and fringing as before, and let

$$l_m = 4.74 \text{ cm}; A_m = 1.04 \text{ cm}^2; A_g = 2 \text{ cm}^2; l_g = 0.5 \text{ cm}$$

3–20.

An Alnico-5 permanent magnet has the shape shown in the figure below, with a mean diameter of 15 cm and a cross-sectional area of 5 cm². The demagnetization curve is shown in Figure 3.2.4 for Alnico-5. Neglect fringing and determine the length of the air gap across which this magnet can produce a flux density of 1.05 tesla.

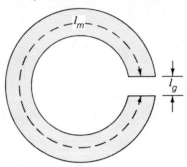

3–21.

A permanent magnet is required to produce a flux density of 0.6 tesla in an air gap of length 0.4 cm and cross-sectional area of 20 cm². A magnet of the shape shown in Figure 3.6.2 is to be used.

    a. Neglecting the reluctance in the L-shaped soft-iron portions of the magnetic circuit, and ignoring fringing and leakage flux, determine the minimum volume of Alnico-5 material that is required; specify the dimensions of the magnet.

    b. Repeat part *(a)* assuming a leakage factor F of 2.5 and a reluctance factor f of 1.2.

    c. Suppose that the air gap in the magnet system is closed by a keeper of ferromagnetic material whose permeability may be assumed to be infinite. If the recoil permeability of the magnetic material is 2×10^{-6} H/m, find the flux density in the magnet.

    d. Repeat part *(a)* using Index V material (for which $B_d = 0.2$ tesla and $H_d = 140$ kA/m, corresponding to the maximum energy product of 28 kJ/m³) and compare the shape and volume of this magnet with that made of Alnico-5.

4

TRANSFORMERS

The transformer is basically a static device in which two or more stationary electric circuits are coupled magnetically, the windings being linked by a common time-varying magnetic flux. Its analysis involves many of the principles essential to the study of rotating electrical machinery. Even though the static transformer is not an energy-conversion device and involves only the interchange of electrical energy between two or more electrical systems, it is an extremely important component in many energy-conversion systems. Thus the transformer deserves serious study.

The transformer performs many useful functions in several fields of electrical engineering. In the widespread ac power systems, the transformer provides the much-needed capability to change the voltage and current levels rather easily. Differing requirements of generation, transmission, and utilization of electric power dictate different optimum values for the voltage and current combinations. The transformer is utilized in stepping up the generator voltage to an appropriate transmission voltage for power transfer, and in stepping down the transmission voltage at different levels for proper distribution and power utilization. In communication systems ranging in frequency from audio to radio to video levels, the transformer is used for performing such functions as an input transformer, interstage transformer, output transformer, impedance-matching device for improved power transfer, insulation device between electric circuits, and isolation device for blocking the dc signals while maintaining the ac continuity between circuits.

Transformers are utilized in circuits of various voltage levels, from the microvolts of electronic circuits to the very high power-system transmission-voltage levels of 765 kV. The application of transformers can be seen throughout the entire frequency spectrum utilized in electrical circuits, from near dc to several hundreds of megahertz, with both continuous sinusoidal and pulse waveforms.

All that is really necessary for transformer action to take place is for two coils to be so positioned that some of the flux produced by a current in one coil links some of the turns of the other coil. Some air-core transformers employed in communications equipment are no more elaborate than this. However, the construction of transformers utilized in power-system networks is usually much more elaborate to minimize energy loss, to produce a large flux in the ferromagnetic core by a current in any one coil, and to see that as much of that flux as possible links as many of the turns as possible of the other coils on the core. Thus, transformers come in various sizes, from very small ones weighing only a few ounces to very large ones weighing several tons.

The ratings of transformers cover a very wide range. While electronics transformers usually have ratings of 300 volt-amperes or less, power-system transformers utilized in transmission and distribution of electric power have the highest voltampere ratings (a few kVA to several MVA) along with the highest continuous voltage ratings. Also, transformers (potential transformers and current transformers) with very small voltampere ratings are used in instrumentation for sensing voltages or currents.

Transformers may be classified by the frequency range: power transformers operating usually at a fixed frequency; audio and ultra-high frequency transformers; wide-band and narrow-band electronics transformers; and pulse transformers. Electronics transformers employed in supplying power to other electronic systems are known as "power transformers." However, in power-system applications, the term "power transformers" denotes those transformers used to transmit power in ratings larger than those associated with the distribution transformers, usually over 500 kVA at the voltage levels of 67 kV and above.

FIGURE 4.1.1 Elementary model of a transformer.

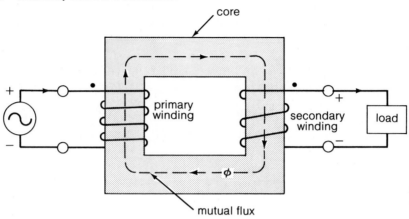

While a conventional transformer has two windings, there are those with only one winding (known as autotransformers) and also those with multiwindings (such as three-winding transformers). Transformers that are used in polyphase circuits are known as polyphase transformers. In the most popular three-phase systems, the most common connections are the wye (star or Y) and the delta (mesh or Δ) connections.

A two-winding transformer essentially consists of two windings interlinked by a mutual magnetic field. The winding that is excited or energized by connecting it to an input source is usually referred to as the *primary* winding, while the other, to which the electrical load is connected and from which the output energy is taken, is known as the *secondary* winding. Depending on the voltage level at which the winding is operated, the windings are classified as HV *(high voltage)* and LV *(low voltage)* windings. The terminology of *step-up* or *step-down* transformer is also common if the main purpose of the transformer is to raise or lower the voltage level. In a step-up transformer, the primary is a low-voltage winding while the secondary is a high-voltage winding. The opposite is true for a step-down transformer.

4.1 CONSTRUCTIONAL FEATURES OF TRANSFORMERS

An elementary model of a transformer is shown in Figure 4.1.1. While it is a convenient model for purposes of analysis and illustrates the principles involved, practically it will be a very unacceptable design, principally because of the large leakage flux. In order to minimize the leakage flux and improve the magnetic coupling between the two windings, a variety of core and winding configurations have been adopted. Basically these may be classified as the *core type* and the *shell type*.

FIGURE 4.1.2 Core-type transformers: *(a)* with a core of stacked laminated sheets *(b)* with a wound core.

FIGURE 4.1.3 Shell-type transformers: *(a)* with a core of stacked laminated sheets *(b)* with wound cores.

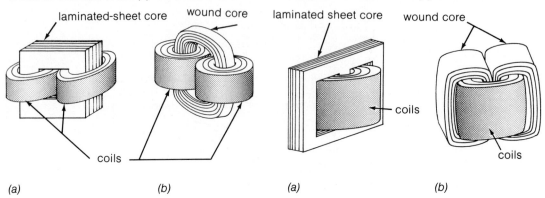

(a)   (b)   (a)   (b)

In order to confine most of the flux to a definite path linking both windings, the core is usually made of some ferromagnetic material. In order to reduce the losses caused by eddy currents in the core, the core is commonly comprised of a stack of thin laminations.

The winding and core arrangements for a core-type transformer are shown in Figure 4.1.2, while those for a shell-type transformer are shown in Figure 4.1.3. In a core-type transformer, the magnetic circuit takes the form of a single ring encircled by two or more groups of primary and secondary windings distributed around the periphery of the ring. In a shell-type transformer, the primary and secondary windings take the form of a common ring that is encircled by two or more rings of magnetic material distributed around its periphery.

The core may consist of stacks of steel laminations comprised of flat punchings of various shapes, or it may be wound of a long continuous strip of sheet steel in the direction in which the steel is rolled during manufacture. Magnetizing the wound cores in the direction of rolling results in a lower core loss and requires a lower exciting current. Grain-oriented steel materials are also used in large transformers in the form of flat strips stacked so that as much of the flux path as possible coincides with the direction of rolling.

Silicon-steel laminations 0.35 mm thick are generally used for transformers operating at frequencies below a few hundred hertz. Desirable properties of the silicon steel include its low cost, its low core loss, and its high permeability at flux-density levels of 1 to 1.4 tesla. While power-transformer cores are generally constructed from soft magnetic materials in the form of punched laminations or wound tapes of appropriate thickness (depending upon the operating frequency), cores of high-frequency electronics transformers are often constructed of soft ferrites. Some communication transformer cores are also made of compressed powdered ferromagnetic alloys such as permalloy.

In order to reduce the leakage, the windings are subdivided into sections placed as close together as possible. In the core-type construction, as shown in Figure 4.1.4*(a)*, each winding consists of two sections; one section is placed on each of the two legs of the core; the primary and secondary windings are arranged to be concentric coils. In the shell-type construction, as shown in Figure 4.1.4*(b)*, the windings usually consist of a number of thin "pancake" coils assembled in a stack with primary and secondary coils interleaved.

FIGURE 4.1.4   Transformer winding arrangements: *(a)* core-type, *(b)* shell-type.

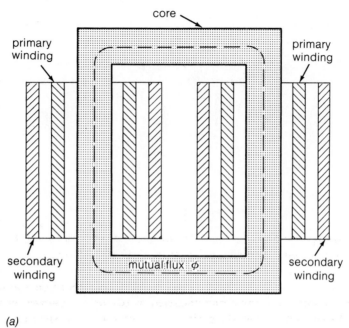

(a)

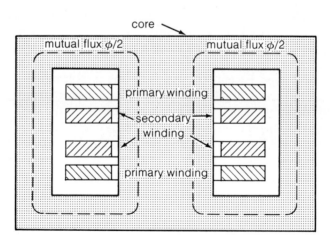

(b)

PHOTO 4.1 Rigidly braced core and coil assemblies of a typical medium-power core-type transformer. Photo courtesy of Westinghouse Electric Corporation.

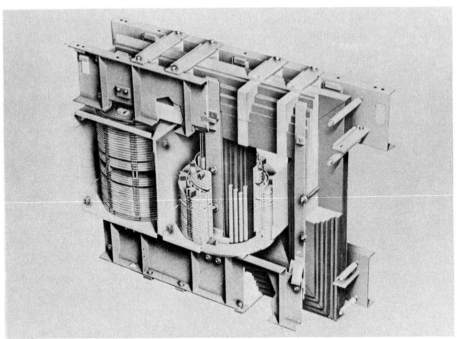

The characteristic features of the core-type transformer are a long mean length of the magnetic circuit and a short mean length of the windings, while those of the shell-type are a short mean length of the magnetic circuit and a long mean length of the windings. Thus, for a given output and performance, the core type will have a smaller area of core and a larger number of turns than the corresponding shell type. Except for certain extreme current ratings, the choice between core- and shell-type constructions is largely a matter of manufacturing facilities and of individual preference. Generally, the steel-to-copper weight ratio is greater in the shell-type transformer.

The constructional details of transformers vary greatly depending upon the specific applications, the winding voltage and current ratings, and the operating frequencies. While many electronics transformers are quite simple in their construction, transformers used in electric power systems are usually much more complicated. The main constructional elements of a power-system transformer include the following: cores, comprising limbs, yokes and clamping devices; primary, secondary (and sometimes also tertiary) phase windings, coil formers, spacers and conductor insulation; interwinding and winding/earth insulation and bracing; tanks, coolers, dryers, conservators and other auxiliaries; terminals and bushings, connections and tapping switches. See Photos 4–1, 4–2, and 4–3.

PHOTO 4.2 Cutaway view of a typical self-cooled/forced oil/forced air-cooled medium-power core-type transformer. Photo courtesy of Westinghouse Electric Corporation.

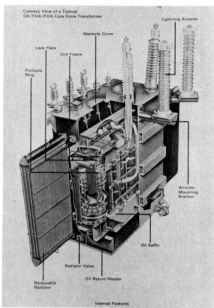

PHOTO 4.3 Typical single-phase, oil-filled, pole-mounted, overhead distribution transformer with surge arrester. Photo courtesy of Westinghouse Electric Corporation.

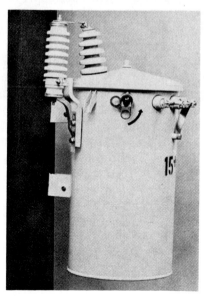

FIGURE 4.1.5   Stepped transformer-core cross sections.

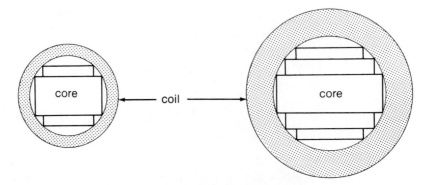

In small transformers, cores with limbs of square cross section may be used; either square or circular coils are fitted on them. In large transformers, however, either of these arrangements is wasteful of copper, thereby increasing the copper loss. Stepped cores of cross sections such as those shown in Figure 4.1.5 are then employed, whenever economically justified, to embrace the maximum cross-sectional area of steel with the minimum length of copper conductor.

In order to avoid overheating and consequent breakdown of insulation, the heat produced in a transformer must be removed. Some transformers are air-cooled; the heat is removed by radiation and convection either immediately from the core and coil surfaces or from the protective enclosure surrounding them. There are what are known as "dry transformers," which may have simple natural air cooling or be sealed into cabinets filled with nitrogen. The great majority of industrial power-system transformers are immersed in oil for cooling, insulation, and mechanical protection. The transformer oil removes the heat by convection and conveys it to the walls of the tank (and the radiators) from which it is transferred to the atmosphere. Conservator tanks may be required to take up the cyclic expansion and contraction of the transformer oil without allowing it to come in contact with the ambient air, from which it may absorb moisture. In very large power-system transformers, radiators and forced circulation of oil in the tank as well as forced air cooling are employed. Recent studies have shown that it is technically feasible to build large power transformers with sheet-wound coils that are insulated by compressed gas (such as sulfur hexafluoride, SF_6) and cooled by forced circulation of a liquid.[1]

The windings for electronics transformers are usually made out of "magnet wire" designated by the American Wire Gauge, AWG (in which an increasing gauge number corresponds to a decreasing conductor cross section) and by the insulation class A, B, C, F, and H, corresponding to the safe operating temperatures of 105° C to 180° C. The windings of power-system transformers, however, generally are conductors with heavier insulation and are assembled with much

[1]"Evaluation of Advanced Technologies for Power Transformers," the General Electric Company Report prepared for the U.S. Dept. of Energy, DOE/RA/ 2134-01, June 1980.

FIGURE 4.2.1   Ideal transformer.

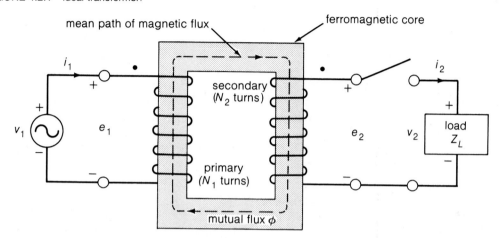

greater mechanical support. Often, the windings are preformed and the transformer core is built around them by stacking the laminations. The main constructional features include the following:

a. Provision of insulation strength adequate to withstand the voltages occurring in service or in tests, both to ground and between coils, turns, and phases, as well as full-wave and chopped-impulse voltages imposed to prove capability against switching and lightning surges.
b. Design to limit the losses in order to obtain better efficiency.
c. Provision of cooling means to meet the set limits on the temperature rise of the windings.
d. Provision of adequate leakage reactance.
e. Provision of adequate mechanical strength to withstand short-circuit forces.

4.2 TRANSFORMER THEORY

The most important aspects of transformer action can be brought out by idealizing the transformer. Figure 4.2.1 shows schematically a transformer having two windings with N_1 and N_2 turns, respectively, on a common magnetic circuit. This is very similar to Figure 4.1.1, repeated here for convenience. The *ideal transformer* is one that has no losses (associated with iron or copper) and no leakage fluxes; it has a core of infinite magnetic permeability and of infinite electrical resistivity. That is to say, all the flux is confined to the core linking both windings; the winding resistances are negligible; the core losses are negligible; only a negligible mmf is required to establish the flux; and the electric fields produced by the windings are negligible. While these properties are never actually achieved in practical transformers, they are, however, closely approached.

When a time-varying voltage v_1 is applied to the primary winding (assumed to have zero resistance), a core flux ϕ is established and a counter emf e_1 is developed such that

$$v_1 = e_1 = \frac{d\lambda_1}{dt} = N_1 \frac{d\phi}{dt} \qquad (4.2.1)$$

where λ_1 $(= N_1 \phi)$ is the flux linkage with the primary N_1-turn winding. The polarity of the primary induced voltage e_1 with respect to that of the applied voltage v_1 is as shown in Figure 4.2.1, satisfying Lenz's law, which says that the current due to induced voltage produces flux in such a direction as to oppose the change of flux linkage.

Since there is no leakage flux for the ideal transformer, the core flux ϕ also links all N_2 turns of the secondary winding and produces an induced emf e_2.

$$e_2 = \frac{d\lambda_2}{dt} = N_2 \frac{d\phi}{dt} \qquad (4.2.2)$$

When the flux is increasing, the secondary induced voltage e_2 has a polarity as shown in Figure 4.2.1; the polarity marks are appropriately shown by dots. If a passive external load circuit is connected to the secondary-winding terminals, then the terminal voltage v_2 will cause a current i_2 to flow as shown in Figure 4.2.1. Since the resistance of the secondary winding is assumed to be zero, the secondary terminal voltage will be equal to the secondary induced voltage e_2.

$$v_2 = e_2 = \frac{d\lambda_2}{dt} = N_2 \frac{d\phi}{dt} \qquad (4.2.3)$$

From Equations 4.2.1 and 4.2.3 it follows that

$$\frac{v_1}{v_2} = \frac{e_1}{e_2} = \frac{N_1}{N_2} = a \qquad (4.2.4)$$

where a is the turns ratio. Thus, in an ideal transformer, voltages are transformed in the direct ratio of the turns.

The application of Ampere's law around the closed contour along the mean path of magnetic flux (shown in Figure 4.2.1) yields

$$N_1 i_1 - N_2 i_2 = 0 \qquad (4.2.5)$$

since the core is assumed to have infinite magnetic permeability, and as a consequence the magnetic field intensity is zero everywhere in the core, even though the magnetic flux and the magnetic flux density are finite. That is to say, the net mmf acting on the core of an ideal transformer at any instant is zero. Equation 4.2.5 implies that

$$\frac{i_1}{i_2} = \frac{N_2}{N_1} = \frac{1}{a} \tag{4.2.6}$$

Thus, for an ideal transformer, currents are transformed in the inverse ratio of the turns.

The instantaneous power input equals the instantaneous power output for the case of an ideal transformer:

$$v_1 i_1 = v_2 i_2 \tag{4.2.7}$$

Let us next consider the case where the waveforms of the applied voltage and flux are sinusoidal. If the flux as a function of time is given by

$$\phi = \phi_{max} \sin \omega t \tag{4.2.8}$$

where ϕ_{max} is the maximum value of the flux, and ω is $2\pi f$, f being the frequency, then the induced voltage e_1 is given by

$$e_1 = N_1 \frac{d\phi}{dt} = \omega N_1 \phi_{max} \cos \omega t \tag{4.2.9}$$

The induced emf leads the flux by 90° for the positive directions shown in Figure 4.2.1. The rms value of the induced emf is given by

$$E_1 = \frac{2\pi}{\sqrt{2}} f N_1 \phi_{max} = 4.44 f N_1 \phi_{max} \tag{4.2.10}$$

If the resistance drop in the winding is neglected, the counter emf equals the applied voltage. Thus

$$V_1 = E_1 = 4.44 f N_1 \phi_{max}$$

or

$$\phi_{max} = \frac{V_1}{4.44 f N_1} \tag{4.2.11}$$

where V_1 is the rms value of the applied voltage. The flux is thus determined solely by the applied voltage, the frequency of the applied voltage, and the number of turns in the winding.

FIGURE 4.2.2 Equivalent circuits viewed from source terminals when the transformer is ideal.

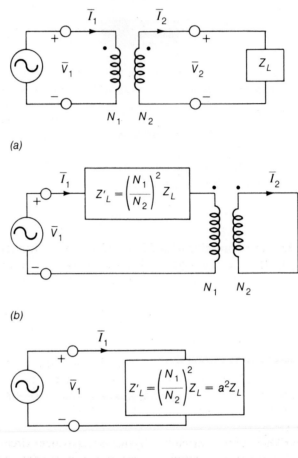

(a)

(b)

(c)

For the case of a sinusoidal voltage applied to the primary winding and a load impedance connected to the secondary winding, the phasor notation can conveniently be used. Equivalent circuits viewed from the source terminals, when the transformer is ideal, are shown in Figure 4.2.2 with the rms or effective magnitudes of the variables. The ends of the two windings at which the dots are placed become positive in potential simultaneously with respect to the other ends of the windings. The voltages $\overline{V}_1$ and $\overline{V}_2$ are in phase, and so also the currents $\overline{I}_1$ and $\overline{I}_2$. The directions of the currents are such that their net mmf is zero. In phasor form, the voltage and current relations are given by

$$\overline{V}_1 = \frac{N_1}{N_2} \overline{V}_2 = a \, \overline{V}_2 \qquad (4.2.12)$$

$$\overline{I}_1 = \frac{N_2}{N_1} \overline{I}_2 = \frac{1}{a} \, \overline{I}_2 \qquad (4.2.13)$$

and

$$\overline{V}_2 = \overline{I}_2 \, Z_L \qquad (4.2.14)$$

from which it follows

$$\frac{\overline{V}_1}{\overline{I}_1} = \left(\frac{N_1}{N_2}\right)^2 \frac{\overline{V}_2}{\overline{I}_2} = \left(\frac{N_1}{N_2}\right)^2 Z_L = a^2 \, Z_L \qquad (4.2.15)$$

where Z_L is the complex impedance of the load, as shown in Figure 4.2.2. The consequence of Equation 4.2.15 is that an impedance Z_L in the secondary circuit can be replaced by an equivalent impedance Z_L' in the primary circuit insofar as the effect at the source terminals is concerned.

$$Z_L' = \left(\frac{N_1}{N_2}\right)^2 Z_L = a^2 \, Z_L \qquad (4.2.16)$$

Thus the circuits shown in Figure 4.2.2 are indistinguishable viewed from the source terminals. Z_L' is known as the load impedance *referred to the primary side*. Transferring an impedance from one side of the transformer to the other in this manner is known as *referring the impedance* to the other side. With the use of Equations 4.2.12 and 4.2.13, voltages and currents may also be *referred* to one side or the other.

Based on our knowledge of circuit analysis, the equivalent circuit of Figure 4.2.2(c) suggests a way, known as *impedance matching,* to obtain the maximum power transfer from a source of internal impedance Z_S to a load impedance Z_L; and that is by choosing the turns ratio such that

$$Z_L' = \left(\frac{N_1}{N_2}\right)^2 Z_L = a^2 \, Z_L = Z_S{}^* \qquad (4.2.17)$$

where $Z_S{}^*$ is the complex conjugate of Z_S.

EXAMPLE 4.2.1

A 60-Hz, 100-kVA, 2400/240-V transformer is used as a step-down transformer from a transmission line to a distribution system. Consider the transformer to be ideal.

   a. Determine the turns ratio.
   b. What secondary load impedance will cause the transformer to be fully loaded, and what is the corresponding primary current?
   c. Find the value of the load impedance referred to the primary side of the transformer.

SOLUTION

   a. The turns ratio is the same as the voltage ratio specified in the rating of the transformer. The primary is the high-voltage winding and the secondary is the low-voltage winding for a step-down transformer.

$$a = \frac{N_1}{N_2} = \frac{V_1}{V_2} = \frac{2{,}400}{240} = 10$$

   b. When the transformer is fully loaded, it should be delivering 100 kVA based on the given rating. So

$$V_2 I_2 = 100 \times 10^3 \text{ VA}$$

or

$$I_2 = \frac{100 \times 10^3}{240} = 416.67 \text{ A}$$

The corresponding primary current may be obtained as

$$I_1 = \frac{I_2}{a} = \frac{416.67}{10} = 41.67 \text{ A}$$

or

$$V_1 I_1 = 100 \times 10^3$$

from which

$$I_1 = \frac{100 \times 10^3}{2,400} = 41.67 \text{ A}$$

Note that the values given above for the currents and voltages are the rms values.

The secondary load impedance, when the transformer is fully loaded, is given by

$$|Z_L| = \frac{V_2}{I_2} = \frac{240}{416.67} = 0.576 \ \Omega$$

In most of the practical situations, the load impedance is a combination of the resistance and the inductive reactance with a lagging power factor. However, other situations corresponding to unity power factor (i.e., purely resistive load) as well as the leading power factor (i.e., capacitive load) could occur.

c.
$$|Z_L'| = \left(\frac{N_1}{N_2}\right)^2 |Z_L| = a^2 |Z_L| = 10^2 \times 0.576 = 57.6 \ \Omega$$

TRANSFORMER ON NO-LOAD   When there is no current in the secondary, the transformer is said to be on *no-load*. Even when the transformer is on no-load, a current, known as the *exciting current*, flows in the primary because of the core losses and the finite permeability of any practical (realistic) transformer core. The exciting current can be considered as having two components, the *core-loss current* and the *magnetizing current*.

The core-loss current is in phase with the induced primary voltage and is given by

$$I_c = \frac{P_c}{E_1} \tag{4.2.18}$$

where P_c is the core loss given by the sum of the hysteresis and eddy-current losses; it manifests itself in the form of heat generated in the core.

In linear magnetic circuits, the *B–H* characteristic is a straight line and the permeability has a finite constant value. In such cases the magnetizing current is proportional to the flux and is in phase with the flux, thereby lagging the primary induced voltage by 90 degrees.

FIGURE 4.2.3 Transformer on no-load: *(a)* transformer with no-load
current, *(b)* no-load phasor diagram, *(c)* equivalent circuit taking the no-load
exciting current into account.

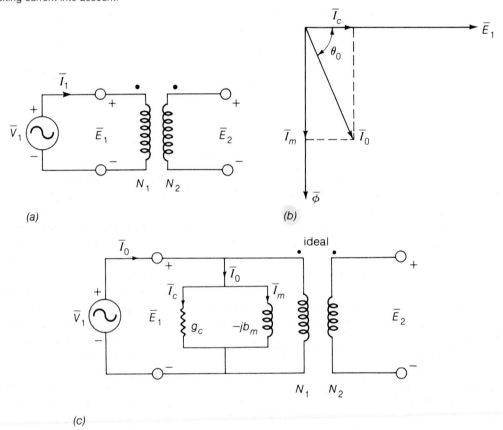

(a)   (b)

(c)

The case of a transformer on no-load is represented by Figure 4.2.3*(a)* and the corresponding
no-load phasor diagram is shown in Figure 4.2.3*(b)*. The no-load current is given by

$$I_0 = \sqrt{I_c^2 + I_m^2} \qquad\qquad (4.2.19)$$

and the no-load power factor is given by

$$\cos \theta_0 = \frac{I_c}{I_0} \qquad\qquad (4.2.20)$$

The equivalent circuit of a transformer on no-load, while taking the exciting current into account,
can then be constructed as in Figure 4.2.3*(c)*.

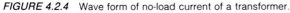

FIGURE 4.2.4   Wave form of no-load current of a transformer.

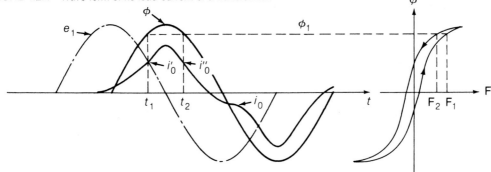

In the case of ferromagnetic circuits mostly used in transformers, the magnetic properties are nonlinear; hence the wave form of the exciting current differs from the wave form of the flux. Based on the hysteresis loop or the flux-mmf loop of the core material, it is possible to obtain graphically the wave form of the exciting current as a function of time. This graphical procedure is shown in Figure 4.2.4. The wave forms of the flux (Equation 4.2.8) and the voltage (Equation 4.2.9) as well as the flux-mmf loop of the ferromagnetic core material are drawn as shown. Values of mmf corresponding to various values of flux are then found. For example, at time t_1 the instantaneous value of the flux is ϕ_1 while the flux is increasing; the corresponding value of the mmf as read from the increasing-flux portion of the flux-mmf loop is $\mathbf{F_1}$, based on which a value of the exciting current i_0' is plotted at time t_1. At time t_2, while the flux has the same instantaneous value ϕ_1, the flux is decreasing; the corresponding values of the mmf and the exciting current are $\mathbf{F_2}$ and i_0''. In this manner the complete curve of the exciting current i_0 as a function of time can be plotted, as shown in Figure 4.2.4.

As seen from Figure 4.2.4 the wave form of the exciting current is not sinusoidal. However, it is symmetrical because $i_0(t + \pi) = -i_0(t)$. The wave form can therefore be represented by a series of odd harmonics including the fundamental. Analyzing by the Fourier-series method, it can be seen that the predominant harmonic is the third. It is usually of the order of 40 percent of the exciting current for typical power transformers, while the exciting current itself is about 5 percent of the full-load rated current. The nonsinusoidal nature of the exciting current wave form is not usually accounted for except in problems dealing directly with the effect of harmonics. The exciting current is then represented by its equivalent sine wave, which has the same effective value and frequency and causes the same average power as the actual wave. With such a representation it becomes possible to draw a phasor diagram as shown in Figure 4.2.3(b) even for the nonlinear ferromagnetic circuits. The phasor $\overline{I_0}$ now represents the equivalent sinusoidal exciting current.

The equivalent circuit of Figure 4.2.3(c) shows the exciting admittance added to the ideal transformer circuit as

$$y_0 = g_c - j\,b_m = \frac{\overline{I_0}}{\overline{E_1}} \tag{4.2.21}$$

where the conductance $g_c = I_c/E_1$ and the susceptance $b_m = I_m/E_1$.

The circuit element g_c absorbs a power corresponding to that dissipated in core losses such that

$$P_c = I_c E_1 = g_c E_1{}^2 \qquad\qquad (4.2.22)$$

EXAMPLE 4.2.2

A 200-turn winding on a laminated magnetic steel core is excited by a 60-Hz sinusoidal source of 200 V (rms).

    a. Neglecting the resistance of the winding and any leakage flux, find the maximum flux density in the core if the uniform cross-sectional area of the core is 5 cm × 5 cm.

    b. Corresponding to the maximum flux density of 1.5 Wb/m², the core loss and the exciting rms volt-amperes for the core are given to be

$$P_c = 50 \text{ W} ; \qquad (VI)_{rms} = 500 \text{ VA}$$

Determine the exciting current, the core-loss component, the magnetizing component, and the corresponding power-factor angle.

    c. Find the exciting admittance for part *(b)* and the corresponding equivalent circuit elements g_c and b_m.

SOLUTION

    a. Since the resistance drop in the winding is neglected, the induced emf is the same as the applied voltage.

$$E_1 = V_1 = 200 \text{ V}$$

From Equation 4.2.11 one obtains

$$\phi_{max} = \frac{200}{4.44 \times 60 \times 200} = 0.00375 \text{ Wb}$$

The corresponding flux density is given by

$$B_{max} = \frac{0.00375}{0.05 \times 0.05} = 1.5 \text{ Wb/m}^2$$

    b. Exciting current: I_2

$$I_0 = \frac{500}{200} = 2.5 \text{ A}$$

Core-loss component: $I_2 \cos \theta$

$$I_c = \frac{50}{200} = 0.25 \text{ A}$$

Magnetizing component:

$$I_m = \sqrt{I_0^2 - I_c^2} = \sqrt{(2.5)^2 - (0.25)^2} = 2.49 \text{ A}$$

Power factor:

$$\cos \theta_0 = \frac{P_c}{VI} = \frac{I_c}{I_0} = \frac{0.25}{2.5} = 0.1$$

and the corresponding power-factor angle θ_0 is 84.3°.

Note that θ_0 is nearly 90°; I_0 and I_m are nearly the same in magnitude.

c. The exciting admittance is given by

$$y_0 = \frac{\overline{I_0}}{\overline{E_1}} = \frac{2.5 \angle -84.3°}{200 \angle 0°} = 0.0125 \angle -84.3°$$
$$= (0.00125 - j0.01245) \text{ siemen}$$

g_c and b_m may also be found as follows:

$$g_c = \frac{I_c}{E_1} = \frac{0.25}{200} = 0.00125 \text{ S}$$

$$b_m = \frac{I_m}{E_1} = \frac{2.49}{200} = 0.01245 \text{ S}$$

The equivalent circuit is shown below:

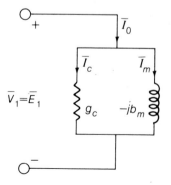

FIGURE 4.2.5   Mutual flux and leakage fluxes for the case of an
elementary core-type transformer.

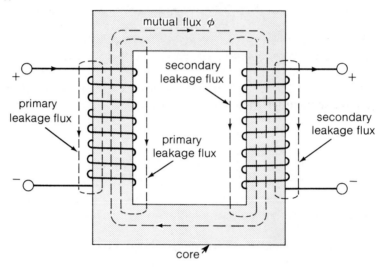

TRANSFORMER LEAKAGE IMPEDANCE AND EQUIVALENT CIRCUIT   As a departure
from the properties assumed for an ideal transformer, we have considered the exciting no-load
current; next we shall consider including the effects of the winding resistances and the magnetic
leakage. The winding capacitances may also have to be included to analyze high-frequency transformer problems, the transient conditions encountered in pulse transformers, and the effects on
power-system transformers of voltage surges caused by lightning or switching. However, we shall
neglect the capacitances of the windings for the time being.

In order to account for the magnetic leakage, the total flux linking the primary winding can be
considered to be made up of two components: the resultant mutual flux ϕ linking both windings
and the primary leakage flux ϕ_{l1} linking only the primary. For the case of an elementary core-type
transformer these components are shown in Figure 4.2.5. ϕ_{l2} is the secondary leakage flux linking
only the secondary winding. Under load conditions, the windings develop sizable magnetomotive
forces that act on spaces external to the core, thereby giving rise to leakage fluxes. The actual
paths taken by the leakage fluxes in practical transformers are more complex than those indicated
in Figure 4.2.5; however, the essential features remain the same. The resultant mutual flux linking
both windings is produced by the combined effect of the primary and secondary currents.

The primary mmf must not only counteract the demagnetizing effect of the secondary mmf but
also have sufficient additional mmf to create the resultant mutual flux. That is to say, the primary
current can be considered to be made up of two components: a load component that is the same
as the secondary current referred to the primary as in an ideal transformer, and the exciting no-
load component $\overline{I}_0$ discussed earlier.

In order to account for leakage fluxes, a leakage inductance is assigned to each winding on the basis of its own leakage-flux linkages per unit current, and the corresponding *leakage reactances* (X_{l1} and X_{l2}) are obtained by multiplying the leakage inductances by ($2\pi f$), to be introduced as additional external inductive reactances in the equivalent circuit. External resistances (R_1 and R_2) are added to represent the resistance of each winding. Assuming the ideal transformer to carry only the resultant mutual flux, the equivalent circuit accounting for the exciting current, the winding resistances, and the magnetic leakage fluxes is shown in Figure 4.2.6(a).

$\overline{I_2}'$ is the secondary current referred to the primary as shown in Figure 4.2.6(a). The primary current $\overline{I_1}$ is the phasor summation of the load component $\overline{I_2}'$ and the exciting no-load component $\overline{I_0}$.

$$\overline{I_1} = \overline{I_2}' + \overline{I_0} = \left(\frac{N_2}{N_1}\right)\overline{I_2} + \overline{I_0} = \frac{\overline{I_2}}{a} + \overline{I_0} \qquad (4.2.23)$$

The primary induced voltage $\overline{E_1}$ and the primary impressed voltage $\overline{V_1}$ will differ by an impedance drop as in the following phasor relationship:

$$\overline{V_1} = \overline{E_1} + \overline{I_1}(R_1 + jX_{l1}) = \overline{E_1} + \overline{I_1}\,Z_{l1} \qquad (4.2.24)$$

Similarly, on the secondary side one has

$$\overline{E_2} = \overline{V_2} + \overline{I_2}(R_2 + jX_{l2}) = \overline{V_2} + \overline{I_2}\,Z_{l2} \qquad (4.2.25)$$

The induced voltages $\overline{E_1}$ and $\overline{E_2}$ produced by the resultant mutual flux are related by the turns ratio:

$$\frac{\overline{E_1}}{\overline{E_2}} = \frac{N_1}{N_2} = a \qquad (4.2.26)$$

The actual transformer is thus equivalent to an ideal transformer plus external impedances, as shown in Figure 4.2.6(a). It is convenient to refer all quantities either to the primary or secondary, in which case the ideal transformer would be moved to the right or left, respectively, of the equivalent circuit as shown in Figures 4.2.6(b) and (c). R_2' and X_{l2}' are the secondary resistance and leakage reactance referred to the primary, while R_1' and X_{l1}' are the primary resistance and leakage reactance referred to the secondary. In Figure 4.2.6(b) the load can also be referred to the primary and the ideal transformer can be eliminated from the equivalent circuit as was done in Figure 4.2.2(c). When all quantities are referred to the secondary, the ideal transformer may be omitted from the diagram and remembered mentally; the exciting shunt branch conductance and susceptance should also be referred appropriately as shown in Figure 4.2.6(c). The phasor diagram for the equivalent circuit can be drawn as discussed in Chapter 1.

FIGURE 4.2.6 Transformer equivalent circuits: *(b)* referred to primary,
(c) referred to secondary.

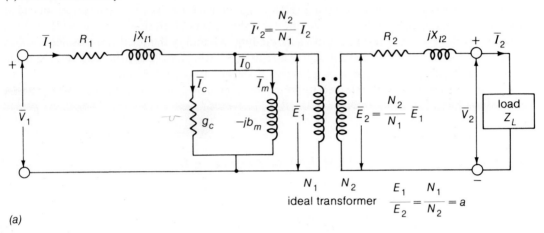

(a)

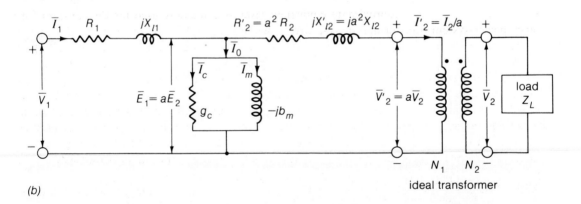

(b)

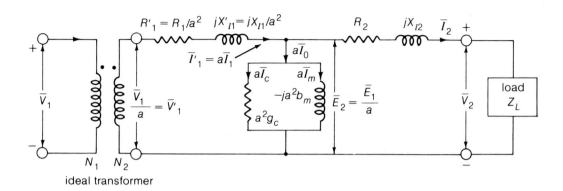

(c)

The commonly used equivalent circuit developed in Figure 4.2.6 and shown in Figure 4.2.7(a) is often known as the *T-circuit* of a transformer. Several modifications and simplifications of this basic complete equivalent circuit (with winding capacitances neglected here, of course) are used in practice, depending upon the requirements of the particular problem being solved and the degree of accuracy desired. Approximate equivalent circuits commonly used for the constant-frequency power-system transformer analysis are shown in Figures 4.2.7(b), (c), and (d). By moving the exciting admittance from the middle of the T-circuit to either the left (as shown in Figure 4.2.7(b) or the right, computational labor can be appreciably reduced with little error in the final analysis. The resistances R_1 and R_2' may then be combined in series to form an *equivalent resistance* R_{eq}; similarly, the leakage reactances X_{l1} and X_{l2}' may be combined in series to form an *equivalent reactance* X_{eq}, as shown in Figure 4.2.7(b). Further simplification is made by neglecting the exciting current altogether, as in Figure 4.2.7(c), in which the transformer is represented as an equivalent series impedance. When R_{eq} is small compared to X_{eq}, as in the case of large power-system transformers, R_{eq} may frequently be neglected; the transformer then is modelled by its equivalent reactance X_{eq} only, as shown in Figure 4.2.7(d), for power-system analysis.

While the equivalent circuits shown in Figure 4.2.7 are referred to the primary, the student should have no difficulty in drawing those referred to the secondary. The phasor diagrams corresponding to all parts of Figure 4.2.7 are drawn in Figure 4.2.8, following the step-by-step procedure indicated in Article 1.2 of Chapter 1, corresponding to a lagging load power factor (cos θ_L). For the cases of unity power-factor loads and leading power-factor loads, the phasor diagrams need to be appropriately modified.

FIGURE 4.2.7   Other equivalent circuits of a transformer.

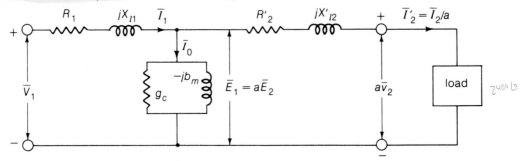

(a)

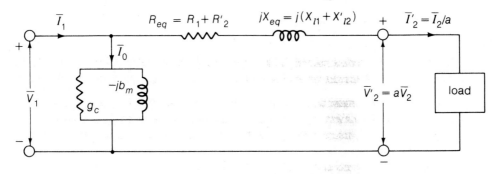

(b)

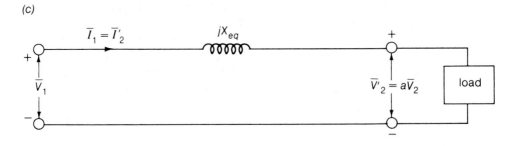

(c)

(d)

FIGURE 4.2.8 Phasor diagrams of a transformer on load.

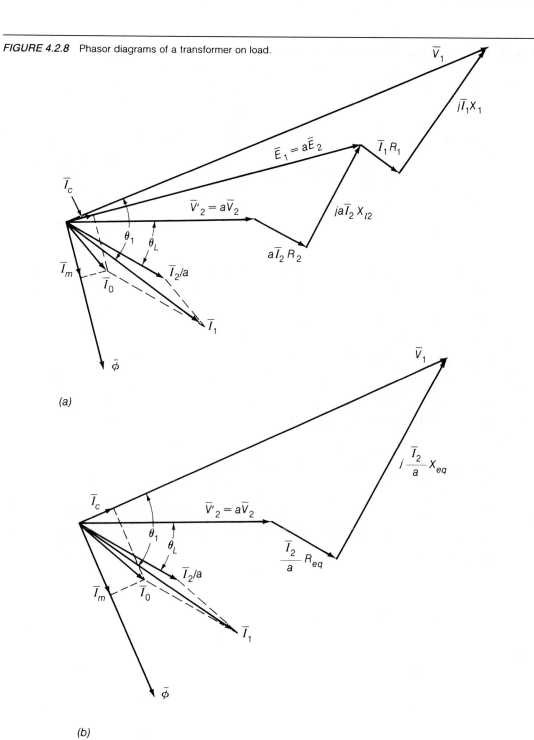

(a)

(b)

FIGURE 4.2.8   (Continued)

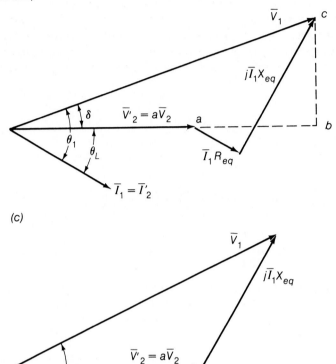

(c)

(d)

EXAMPLE 4.2.3

A single-phase, 50-kVA, 2,400:240-Volt, 60-Hz distribution transformer has the following parameters:

Resistance of the 2,400-volt winding $R_1 = 0.75 \ \Omega$
Resistance of the 240-volt winding $R_2 = 0.0075 \ \Omega$
Leakage reactance of the 2,400-volt winding $X_{l1} = 1 \ \Omega$
Leakage reactance of the 240-volt winding $X_{l2} = 0.01 \ \Omega$
Exciting admittance on the 240-volt side $y_0 = (0.003 - j0.02)$ S

a. Draw the equivalent circuit referred to the high-voltage side and referred to the low-voltage side; label the impedances numerically.

b. The transformer is used as a step-down transformer at the load end of a feeder whose impedance is $(0.5 + j2.0)$ ohms. Determine the voltage V_s at the sending end of the feeder when the transformer delivers rated load at rated secondary voltage and 0.8 lagging power factor. Neglect the exciting current of the transformer.

SOLUTION

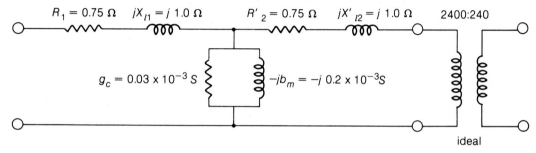

$R_1 = 0.75\ \Omega$ $jX_{l1} = j\ 1.0\ \Omega$ $R'_2 = 0.75\ \Omega$ $jX'_{l2} = j\ 1.0\ \Omega$ 2400:240

$g_c = 0.03 \times 10^{-3}\ S$ $-jb_m = -j\ 0.2 \times 10^{-3} S$

ideal

a. (i) The equivalent circuit referred to the high-voltage side is shown in the figure above. The quantities on the low-voltage side, referred to the high-voltage side, are calculated as follows:

$$R_2' = a^2 R_2 = \left(\frac{2{,}400}{240}\right)^2 (0.0075) = 0.75\ \Omega$$

$$X_{l2}' = a^2 X_{l2} = \left(\frac{2{,}400}{240}\right)^2 (0.01) = 1.0\ \Omega$$

The exciting branch conductance and susceptance referred to the high-voltage side are given by

$$\frac{1}{a^2}(0.003)\ \text{or}\ \frac{1}{100} \times 0.003 = 0.03 \times 10^{-3}\ S$$

and

$$\frac{1}{a^2}(0.02)\ \text{or}\ \frac{1}{100} \times 0.02 = 0.2 \times 10^{-3}\ S$$

a. *(ii)* The equivalent circuit referred to the low-voltage side is given below:

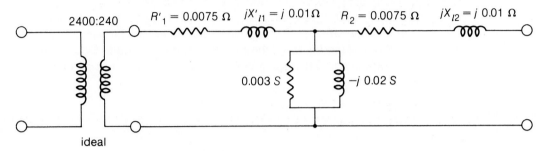

ideal

Note the following points:

The voltages specified on the nameplate of a transformer yield directly the turns ratio while neglecting the small leakage-impedance voltage drops under load. The actual voltages at the terminals of either the high-voltage or the low-voltage side depend on the particular conditions of the problem; they are generally near the specified voltages. The turns ratio in this problem is 10 to 1.

Impedances are referred by either multiplying or dividing by 100, which is the square of the turns ratio in this problem. The value of an impedance referred to the high-voltage side is larger than its value referred to the low-voltage side, while the value of an admittance referred to the high-voltage side is smaller than its value referred to the low-voltage side. Since admittance is the reciprocal of impedance, when referring admittance from one side to the other, the reciprocal of the referring factor for impedance must be used.

The ideal transformer may usually be omitted in the diagram while remembering it mentally.

b. The equivalent circuit of the transformer (along with the feeder impedance) referred to the high-voltage (primary) side is given below, neglecting the exciting current of the transformer:

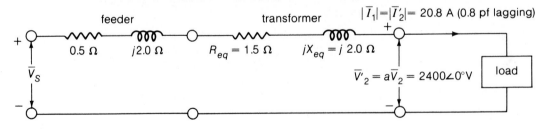

From the given conditions of the problem, the voltage at the load terminals referred to the primary or the high-voltage side is 2,400 V. Further, the load current corresponding to the rated load condition is given by

$$I = \frac{50 \times 10^3}{2,400} = 20.8 \text{ A}$$

which is also referred to the high-voltage side.

   The feeder impedance may be combined with the transformer impedance in series:

$$Z = (2.0 + j4.0) \, \Omega = R + jX$$

The phasor diagram corresponding to the lagging 0.8 power factor is drawn below.

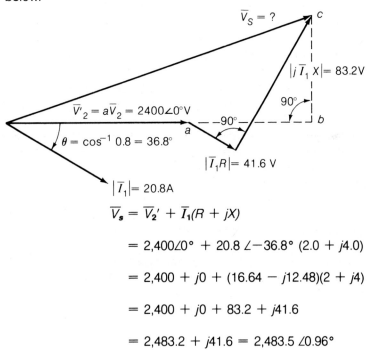

$$\overline{V}_s = \overline{V}_2' + \overline{I}_1(R + jX)$$

$$= 2,400\angle 0° + 20.8 \angle{-36.8°} \, (2.0 + j4.0)$$

$$= 2,400 + j0 + (16.64 - j12.48)(2 + j4)$$

$$= 2,400 + j0 + 83.2 + j41.6$$

$$= 2,483.2 + j41.6 = 2,483.5 \angle 0.96°$$

The voltage at the sending end of the feeder is 2,483.5 volts; and the power factor at the sending end is cos(36.8 + 0.96)° or 0.79. Alternatively, noting that

$$ab = I_1 R \cos\theta + I_1 X \sin\theta; \quad bc = I_1 X \cos\theta - I_1 R \sin\theta$$

V_s could be solved for.

PER-UNIT VALUES   A per-unit system, which is essentially a system of dimensionless parameters, is adopted extensively for computational convenience and for ready comparison of the performance of a set of transformers or of a set of rotating machines. Per-unit quantities simplify the analysis of complex power systems involving transformers of different ratios. When expressed in a per-unit system related to their rating, the parameters of transformers and machines lie in a reasonably narrow numerical range. The per-unit system is also useful in simulating complex power-system problems on analog and digital computers for analyzing the transient and dynamic behavior.

The process of normalization is carried out by dividing one physical parameter by another of the same dimension. The denominator is referred to as the *base* quantity. Base quantities are chosen because of the ways in which they characterize particular features of the physical system. The magnitudes of some base quantities may be chosen freely and quite arbitrarily. The magnitudes of others follow by dependence through the laws governing the physical nature of the system. The choice of bases should be so made that the computational effort is minimized, and the evaluation as well as the understanding of the main characteristics are made as simple and direct as possible. Normally, for any device, the principal per-unit variables assume unit magnitude under rated conditions.

The per-unit (pu) quantity is defined as follows:

$$\text{Quantity in pu} = \frac{\text{Actual quantity}}{\text{Base-value quantity}} \qquad (4.2.27)$$

Base values of apparent power $(VA)_{base}$ and voltage V_{base} are usually chosen first. The base current I_{base} for a single-phase system is then calculated as

$$I_{base} = \frac{(VA)_{base}}{V_{base}} \qquad (4.2.28)$$

Other base quantities are then established as follows for a single-phase system:

$$P_{base} = Q_{base} = |S|_{base} = (VA)_{base} \qquad (4.2.29)$$

$$R_{base} = X_{base} = |Z|_{base} = \frac{V_{base}}{I_{base}} \qquad (4.2.30)$$

$$G_{base} = B_{base} = |Y|_{base} = \frac{I_{base}}{V_{base}} \qquad (4.2.31)$$

The per-unit impedance is generally expressed by

$$|Z|_{pu} = \frac{\text{Actual impedance in ohms}}{\text{Base impedance in ohms}}$$
$$= |Z|_{ohms} \frac{(VA)_{base}}{(V_{base})^2} \qquad (4.2.32)$$

When only one electrical device, such as a transformer, is involved, its own rating is generally taken for the voltampere base. In order that the per-unit impedance of a transformer may have the *same value* whether referred to the high-voltage or the low-voltage side, the rated (or nominal) voltages of the respective sides of the transformer are chosen as the base voltages. Thus, for a transformer, the values of V_{base} are different on the two sides and are in the same ratio as are the turns on the transformer. The advantage of such a choice of base quantities is that the equivalent circuit in per-unit quantities will be the same whether referred to the high-voltage or the low-voltage side. The next example illustrates this with numerical values.

In studies of power systems involving several devices, an arbitrary choice of the voltampere base is usually made and the same base is used for the overall system. In such a case, per-unit values may have to be changed from one voltampere base to another, while the voltage base may or may not be the same. The following relations may then be applied:

$$(P,Q,VA)_{pu\ (base\ 2)} = (P,Q,VA)_{pu\ (base\ 1)} \frac{(VA)_{base\ 1}}{(VA)_{base\ 2}} \qquad (4.2.33)$$

$$(R,X,|Z|)_{pu\ (base\ 2)} = (R,X,|Z|)_{pu\ (base\ 1)}$$
$$\frac{(VA)_{base\ 2}}{(VA)_{base\ 1}} \left(\frac{V_{base\ 1}}{V_{base\ 2}}\right)^2 \qquad (4.2.34)$$

$$(G,B,|Y|)_{pu\ (base\ 2)} = (G,B,|Y|)_{pu\ (base\ 1)}$$
$$\frac{(VA)_{base\ 1}}{(VA)_{base\ 2}} \left(\frac{V_{base\ 2}}{V_{base\ 1}}\right)^2 \qquad (4.2.35)$$

EXAMPLE 4.2.4

Consider the transformer of Example 4.2.3 and draw its equivalent circuit in per-unit quantities referred to the high-voltage side and also to the low-voltage side.

SOLUTION

The base quantities for high-voltage and low-voltage sides are given below:

	High-voltage side	Low-voltage side
$(VA)_{base}$	50,000 VA	50,000 VA
V_{base}	2,400 V	240 V
I_{base}	$\dfrac{50,000}{2,400} = 20.8$ A	$\dfrac{50,000}{240} = 208$ A
$|Z|_{base}$	$\dfrac{2,400}{20.8} = 115.4\ \Omega$	$\dfrac{240}{208} = 1.154\ \Omega$
$|Y|_{base}$	$\dfrac{20.8}{2,400} = 0.0087$ S	$\dfrac{208}{240} = 0.87$ S

The equivalent circuit in per-unit quantities is given below:

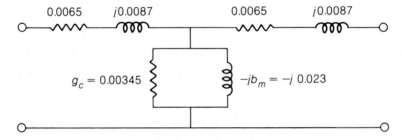

It comes out to be exactly the same whether referred to the high-voltage or the low-voltage side, since

$$R_{pu(1)} = R_{pu(2)} = \frac{0.75}{115.4} = \frac{0.0075}{1.154} = 0.0065 \text{ pu}$$

$$X_{pu(1)} = X_{pu(2)} = \frac{1.0}{115.4} = \frac{0.01}{1.154} = 0.0087 \text{ pu}$$

$$g_c = \frac{0.3 \times 10^{-4}}{0.0087} = \frac{0.003}{0.87} = 0.00345 \text{ pu}$$

$$b_m = \frac{2 \times 10^{-4}}{0.0087} = \frac{0.02}{0.87} = 0.023 \text{ pu}$$

in which the actual impedance or admittance values are taken from the solution of Example 4.2.3. The base system is so chosen that the per-unit values come out to be the same referred to either side. Note that the per-unit value is a dimensionless quantity.

EXAMPLE 4.2.5

In the single-phase electric system shown below, two transformers, whose ratings are given below, are used to interconnect three circuits A, B, and C.

Transformer A–B: 15 MVA, 13.2:132 kV, equivalent leakage reactance 0.1 pu.
Transformer B–C: 15 MVA, 66:132 kV, equivalent leakage reactance 0.08 pu.

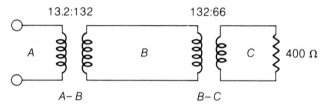

Choosing the bases in circuit B as 15 MVA and 132 kV, determine the per-unit value of the 400-Ω resistive load in Circuit C referred to the Circuits C, B, and A.

Draw the equivalent circuit diagram with per-unit quantities, neglecting the exciting currents of the transformers, the transformer resistances, and the line impedances.

SOLUTION

Since the base voltage for Circuit B is to be chosen as 132 kV, the base voltages for Circuits A and C will be 13.2 kV and 66 kV, respectively, based on the turns ratios of the transformers.

Base impedance of Circuit C:

$$\frac{V_{base}}{I_{base}} = \frac{(V_{base})^2}{(VA)_{base}} = \frac{(66 \times 10^3)^2}{15 \times 10^6} = 290.4 \ \Omega$$

Per-unit impedance of the load in Circuit C:

$$\frac{400}{290.4} = 1.38 \ \text{pu}$$

Base impedance of Circuit B:

$$\frac{(132 \times 10^3)^2}{15 \times 10^6} = 1{,}161.6 \ \Omega$$

Impedance of the load referred to Circuit B:

$$400\left(\frac{132}{66}\right)^2 = 400 \times 4 = 1{,}600 \ \Omega$$

Per-unit impedance of the load referred to B:

$$\frac{1{,}600}{1{,}161.6} = 1.38 \ \text{pu}$$

Base impedance of Circuit A:

$$\frac{(13.2 \times 10^3)^2}{15 \times 10^6} = 11.616 \ \Omega$$

Impedance of the load referred to Circuit A:

$$400\left(\frac{132}{66}\right)^2\left(\frac{13.2}{132}\right)^2 = 400 \times 4 \times 0.01 = 16 \ \Omega$$

Impedance of the load referred to Circuit A:

$$\frac{16}{11.616} = 1.38 \ \text{pu}$$

Note that the per-unit impedance of the load referred to any part of the system is the same because the voltage bases in different parts of the system are based on the turns ratios of the transformers.

The desired equivalent circuit diagram with per-unit quantities is shown below:

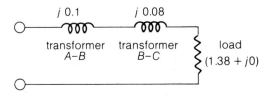

So far we have considered per-unit values related to a single-phase system. In dealing with balanced three-phase systems, the three-phase volt-ampere base and the line-to-line voltage base are usually chosen first. The base line current is then calculated as

$$I_{L \ base} = \frac{(VA)_{3\phi \ base}}{\sqrt{3} \ V_{L \ base}} = \frac{3(VA)_{base \ per \ phase}}{\sqrt{3} \ V_{L \ base}} \qquad (4.2.36)$$

Other base quantities are then established from the relations that hold for a balanced three-phase system:

$$P_{3\phi \ base} = Q_{3\phi \ base} = (VA)_{3\phi \ base} = 3(VA)_{base \ per \ phase} \qquad (4.2.37)$$

$$V_{L \ base} = \sqrt{3} \ V_{base \ line \ to \ neutral} \qquad (4.2.38)$$

$$I_{base \ per \ phase \ of \ \Delta} = \frac{1}{\sqrt{3}} I_{base \ per \ phase \ of \ Y} \qquad (4.2.39)$$

From Equations 4.2.38 and 4.2.39 it follows that

$$|Z|_{base \ per \ phase \ of \ \Delta} = 3 |Z|_{base \ per \ phase \ of \ Y} \qquad (4.2.40)$$

Equations 4.2.29, 4.2.30 and 4.2.31 still apply to the base values per phase. The base impedance per phase of Y is given by

$$|Z|_{base \ per \ phase \ of \ Y} = \frac{V_{base \ line \ to \ neutral}}{I_{L \ base}} = \frac{V_{L \ base}/\sqrt{3}}{I_{L \ base}} \qquad (4.2.41)$$

Multiplying the numerator and denominator by ($\sqrt{3} \ V_{L \ base}$), one obtains

$$|Z|_{base \ per \ phase \ of \ Y} = \frac{(V_{L \ base})^2}{(VA)_{3\phi \ base}} \qquad (4.2.42)$$

$$\frac{\left(\frac{V_L}{\sqrt{3}}\right)^2}{\frac{VA}{3} \ \phi ph}$$

The per-unit impedance per phase of Y is then given by

$$|Z|_{pu \ per \ phase \ of \ Y} = |Z|_{ohms \ line \ to \ neutral} \frac{(VA)_{3\phi \ base}}{(V_{L \ base})^2} \qquad (4.2.43)$$

Balanced three-phase problems can be solved in per unit as if they were single-phase problems with a single line and a neutral return.

If, for example, for a balanced three-phase system

$$(VA)_{3\phi \, base} = 10,000 \text{ VA}$$

and

$$V_{L \, base} = 1,000 \text{ V}$$

then it follows that

$$(VA)_{1\phi \, base} = \frac{10,000}{3} \text{ VA}$$

and

$$V_{base \, line \, to \, neutral} = \frac{1,000}{\sqrt{3}} \text{ V}$$

Now to express an actual line-to-line voltage of 900 V or the corresponding line-to-neutral voltage of $[900/\sqrt{3}]$ V, we have

$$\text{per-unit voltage} = \frac{900}{1,000} = \frac{900/\sqrt{3}}{1,000/\sqrt{3}} = 0.9 \text{ pu}$$

If the power per phase is 2,000 W, the total three-phase power is 6,000 W. The power in per unit is expressed as

$$\text{per-unit power} = \frac{6,000}{10,000} = \frac{2,000}{10,000/3} = 0.6 \text{ pu}$$

4.3 TRANSFORMER TESTS

Most of the transformers are supplied by the manufacturer with polarity markings such as the dot-marks. Following the dot-notation that has been already explained, if the primary current is flowing into the dotted terminal of one winding, the secondary load current will be flowing out of the dotted terminal of the other winding. Another type of polarity convention is used for power-system transformers. For single-phase transformer windings, the terminals on the high-voltage side are identified as H_1 and H_2, while those on the low-voltage side are labelled as X_1 and X_2. The terminals with subscript 1 (or 2) in this convention are equivalent to the dotted terminals in

FIGURE 4.3.1   *(a)* Polarity markings of a single-phase two-winding transformer, *(b)* polarity test.

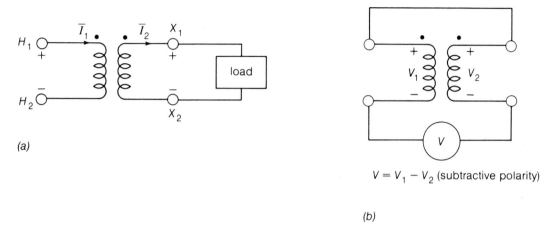

(a)

$V = V_1 - V_2$ (subtractive polarity)

(b)

the dot-polarity notation. Figure 4.3.1*(a)* shows the notation for a single-phase two-winding transformer. If there should be a doubt about the transformer polarity, it can easily be checked by the polarity test as shown in Figure 4.3.1*(b)* and the dot marks be made according to the convention. In the polarity test, terminals in the closest physical proximity, one from each winding, are connected as shown. By applying the rated voltage (say V_1) to one winding, the voltage V across the two remaining terminals is measured. If the measured voltage is larger than the input voltage, the polarity is said to be additive; if smaller, the polarity is subtractive, as shown in Figure 4.3.1*(b)*.

The parameters of the equivalent circuit of a transformer are usually determined by conducting two tests: (1) the open-circuit (or no-load) test, and (2) the short-circuit test.

OPEN-CIRCUIT TEST   The open-circuit test is conducted in order to determine the exciting admittance of the transformer equivalent circuit, the no-load loss, the no-load exciting current, and the no-load power factor. While one winding is open-circuited, a voltage (usually rated voltage at rated frequency) is applied to the other winding, and measurements of voltage, current, and real power are made with voltmeter, ammeter, and wattmeter, respectively, as shown in Figure 4.3.2*(a)*. Usually the high-voltage winding is open-circuited and the test is conducted by placing the instruments on the low-voltage side.

Let V be the applied voltage in volts as read by the voltmeter V, I_0 the exciting no-load current in amperes as read by the ammeter A, and P_{oc} be the real power in watts as read by the wattmeter W. The voltage drop in the leakage impedance of the winding (which is excited in the open-circuit

FIGURE 4.3.2   *(a)* Open-circuit test, *(b)* approximate equivalent circuit of a
transformer on open-circuit, *(c)* no-load phasor diagram.

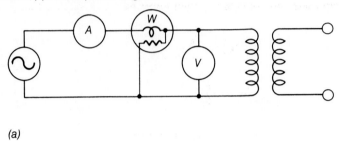

(a)

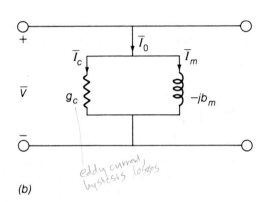

(b)

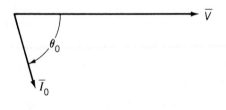

(c)

test) caused by the usually small exciting current is normally neglected; the approximate equivalent circuit is then as shown in Figure 4.3.2*(b)*. The exciting admittance in siemens and its conductance and susceptance components are then given by

$$|y_0| = \frac{I_0}{V} \qquad (4.3.1)$$

$$g_c = \frac{P_{oc}}{V^2} \qquad (4.3.2)$$

$$b_m = \sqrt{|y_0|^2 - g_c^2} \qquad (4.3.3)$$

The values so obtained are referred to the side on which the instruments are connected and the winding is excited in the open-circuit test. Neglecting the copper loss caused by the exciting no-load current I_0, the power input P_{oc} will be equal to the core loss P_c of the transformer. The no-load power factor is obtained as

$$\cos \theta_0 = \frac{P_{oc}}{V I_0} \qquad (4.3.4)$$

The no-load phasor diagram can then be easily drawn by showing the exciting current $\overline{I}_0$ lagging the applied voltage $\overline{V}$ by an angle of θ_0.

As a part of the open-circuit test, the voltage at the terminals of the open-circuited winding is sometimes measured to check the turns ratio of the transformer. Since the no-load power factor is generally very low (in the range of 0.05 to 0.2), it becomes very difficult to obtain accurate power measurements; electronic multiplier wattmeters are generally used for no-load transformer power measurements. Because of the small quantities involved in most communication transformers, ac bridges or other suitable devices may be used instead of voltmeters, ammeters, and wattmeters. Care should be exercised during the test procedure for possible high-voltage conditions.

SHORT-CIRCUIT TEST   The short-circuit test is conducted by short-circuiting one winding (usually the low-voltage winding) and applying a *reduced* voltage at rated frequency such that the rated current results. Figure 4.3.3*(a)* shows a connection diagram for performing the short-circuit test on power-system transformers. Let V_{sc}, I_{sc}, and P_{sc} be, respectively, the impressed voltage, the input short-circuit current, and the power input as measured by the indicating instruments.

FIGURE 4.3.3   (a) Short-circuit test, (b) approximate equivalent circuit of a
transformer on short-circuit, (c) phasor diagram under short-circuit
conditions.

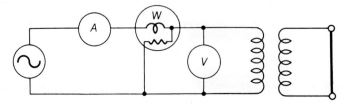

(a)

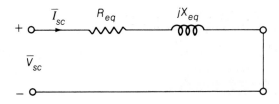

(b)

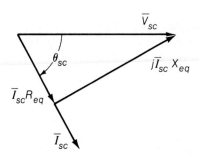

(c)

Neglecting the exciting current, which is usually small compared to the rated (i.e., full-load) current, the approximate equivalent circuit of a transformer under short-circuit condition is shown in Figure 4.3.3(b). The short-circuit impedance Z_{sc}, which is then equal to the equivalent series impedance of the transformer, is given by

$$|Z_{eq}| = |Z_{sc}| = \frac{V_{sc}}{I_{sc}}$$
(4.3.5)

Neglecting the core loss at the low value of V_{sc}, the equivalent series resistance and leakage reactance are given by

$$R_{eq} = R_{sc} = \frac{P_{sc}}{I_{sc}^2}$$
(4.3.6)

$$X_{eq} = X_{sc} = \sqrt{|Z_{eq}|^2 - R_{eq}^2}$$
(4.3.7)

The values so obtained are referred to the side on which the voltage is applied and the measuring instruments are connected. The short-circuit power factor is obtained as

$$\cos \theta_{sc} = \frac{P_{sc}}{V_{sc} I_{sc}}$$
(4.3.8)

The phasor diagram under short-circuit conditions is shown in Figure 4.3.3(c).

EXAMPLE 4.3.1

The following data were obtained when open-circuit and short-circuit tests were performed on a single-phase, 50-kVA, 2,400:240-volt, 60-Hz distribution transformer:

	Voltage (volts)	Current (amps)	Power (watts)
With high-voltage winding open-circuited	240	4.85	173
With low-voltage winding short-circuited	52	20.8	650

Determine the equivalent circuit of the transformer referred to its high-voltage side and also referred to its low-voltage side.

SOLUTION

The exciting admittance *referred to the low-voltage side* is obtained from the open-circuit test data as follows:

$$|y|_{LV} = \frac{4.85}{240} = 0.0202 \text{ S}$$

$$g_{LV} = \frac{173}{(240)^2} = 0.003 \text{ S}$$

and

$$b_{LV} = \sqrt{(0.0202)^2 - (0.003)^2} = 0.02 \text{ S}$$

These values can be referred to the high-voltage side by dividing by the square of the turns ratio.

The equivalent series impedance of the transformer *referred to its high-voltage side* is obtained from the short-circuit test data:

$$|Z_{eq}|_{HV} = \frac{52}{20.8} = 2.5 \text{ ohms}$$

$$R_{eq\,HV} = \frac{650}{(20.8)^2} = 1.5 \text{ ohms}$$

$$X_{eq\,HV} = \sqrt{(2.5)^2 - (1.5)^2} = 2.0 \text{ ohms}$$

The above values can be referred to the low-voltage side by dividing by the square of the turns ratio.

Approximate equivalent circuits with the exciting shunt branch moved to the high-voltage or low-voltage terminals may now be drawn:

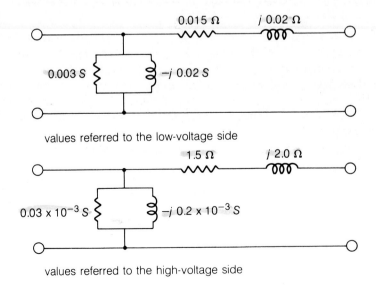

values referred to the low-voltage side

values referred to the high-voltage side

If an equivalent T-circuit is desired, the values of individual primary and secondary resistances and leakage reactances, *referred to the same side,* are usually assumed to be equal:

$$R_1 = R_2' = \frac{R_{eq\ HV}}{2} = 0.75\ \Omega$$

$$X_{l1} = X_{l2}' = \frac{X_{eq\ HV}}{2} = 1.0\ \Omega$$

$$R_1' = R_2 = \frac{R_{eq\ LV}}{2} = \frac{0.015}{2} = 0.0075\ \Omega$$

$$X_{l1}' = X_{l2} = \frac{X_{eq\ LV}}{2} = \frac{0.02}{2} = 0.01\ \Omega$$

The subscripts 1 and 2 refer to the high-voltage and low-voltage sides, respectively. The T-equivalent circuits referred to the high-voltage and low-voltage sides are then the same as those shown for Example 4.2.3.

4.4 TRANSFORMER PERFORMANCE

The losses in a transformer consist of *(a)* the core loss, *(b)* the copper loss, and *(c)* the stray loss. The core loss is practically independent of the load for a given voltage and frequency of operation. It is taken to be approximately equal to the no-load loss P_{oc} as read by the wattmeter in the no-load (open-circuit) test. The copper loss or the I^2R loss is due to the resistance of the windings; and the stray loss is largely due to eddy currents induced by the leakage flux in the tank and other structural parts. The sum of the copper loss and the stray loss is known as the load loss, being equal to I^2R_{eq} as determined from the short-circuit test by the wattmeter reading P_{sc}. The term "copper loss" itself is often used in place of the load loss, including the stray loss. The copper loss is a function of the load or the load current for constant voltage operation. In fact, it varies as the square of the load current.

The efficiency of any power-transmitting device including a transformer is given by

$$\text{Per-unit efficiency } \eta = \frac{\text{Power output}}{\text{Power input}} = \frac{\text{Input} - \text{losses}}{\text{Input}}$$

$$= 1 - \frac{\text{losses}}{\text{input}} \tag{4.4.1}$$

The efficiency of power-system transformers at rated load is quite high; it may be as high as 90 to 99 percent or 0.9 to 0.99 per unit; the higher the rating of a transformer, the greater is the value of its efficiency.

Most of the power-system transformers used to transmit power operate normally near their rated capacities and are switched out of circuit when not required. In such cases it makes sense to design the transformer for its maximum efficiency at or near rated output. On the other hand, for a distribution transformer that is in the system for 24 hours a day, but operates well below its rated output for a good number of those hours, it is desirable to design for maximum efficiency at or near the average output.

The ratio of core loss to copper loss at rated output is made to be different in transformers used for different types of service. It can be shown that the maximum efficiency of a transformer operating at a constant output voltage and power factor occurs at that load for which the copper loss equals the core loss. On the basis of the approximate equivalent circuit of Figure 4.2.7(b), it follows that

$$\eta = \frac{V_2' \, I_2' \, \cos \theta}{V_2' \, I_2' \, \cos \theta + P_c + (I_2')^2 \, R_{eq}} \qquad (4.4.2)$$

where the core loss $P_c \, (= V_1^2 \, g_c)$ is a constant, the copper loss $(I_2')^2 \, R_{eq}$ varies with the square of the load current, and $\cos \theta$ is the load power factor. Maximum efficiency occurs when $1/\eta$ is a minimum. For constant output voltage and power factor, differentiating with respect to I_2' and setting the derivative equal to zero, one obtains the condition for maximum efficiency to be

$$(I_2')^2 \, R_{eq} = P_c \qquad (4.4.3)$$

which states that the core loss should be equal to the copper loss for maximum operating efficiency at a given load.

For distribution transformers an all-day (or energy) efficiency becomes an important figure of merit given by

$$\eta_{AD} = \frac{\text{Energy output over 24 hours}}{\text{Energy input over 24 hours}} \qquad (4.4.4)$$

This can be determined once the load cycle of the transformer is known.

While the equivalent circuits may be used to calculate the losses for a given output of the transformer, it is usually more convenient to use the open-circuit and short-circuit test data directly.

EXAMPLE 4.4.1

Compute the efficiency of the transformer of Example 4.3.1 corresponding to *(a)* full load, 0.8 power factor lagging, and *(b)* one-half rated load, 0.6 power factor lagging.

SOLUTION

a. Corresponding to full load, 0.8 power factor lagging,

$$\text{Output} = 50{,}000 \times 0.8 = 40{,}000 \text{ watts}$$

The load loss (or the copper loss) at rated (full) load equals the real power measured in the short-circuit test *at rated current:*

$$\text{Copper loss} = I_{HV}^2 R_{eq\,HV} = P_{sc} = 650 \text{ W}$$

where subscript *HV* refers to the high-voltage side.

$$\text{Core loss} = P_{oc} = 173 \text{ W}$$

$$\text{Total losses at full load} = 650 + 173 = 823 \text{ W}$$

$$\text{Input} = 40{,}000 + 823 = 40{,}823 \text{ W}$$

The full-load efficiency at 0.8 power factor is given by

$$\eta = 1 - \frac{\text{Losses}}{\text{Input}} = 1 - \frac{823}{40{,}823} = 1 - 0.02 = 0.98$$

b. Corresponding to one-half rated load, 0.6 power factor lagging,

$$\text{Output} = \frac{1}{2} \times 50{,}000 \times 0.6 = 15{,}000 \text{ W}$$

$$\text{Copper loss} = \frac{1}{4} \times 650 = 162.5 \text{ W}$$

Note that the current at one-half rated load is one-half the full-load current, and the copper loss is one-quarter of that at rated current value.

$$\text{Core loss} = P_{oc} = 173 \text{ W}$$

which is considered to be unaffected by the load, as long as the secondary terminal voltage is at its rated value.

Total losses at one-half rated load:

$$162.5 + 173 = 335.5 \text{ W}$$

$$\text{Input} = 15,000 + 335.5 = 15,335.5 \text{ W}$$

Efficiency at one-half rated load and 0.6 power factor is given by

$$\eta = 1 - \frac{335.5}{15,335.5} = 1 - 0.022 = 0.978$$

EXAMPLE 4.4.2

The distribution transformer of Example 4.3.1 is supplying a load at 240 volts and 0.8 power factor lagging.

a. Determine the fraction of full-load rating at which the maximum efficiency of the transformer occurs, and compute the efficiency at that load.
b. The load cycle of the transformer operating at a constant power factor is 0.9 full load for 8 hours, 0.5 full load for 12 hours, and no load for 4 hours. Compute the all-day (or energy) efficiency of the transformer under specified conditions.

SOLUTION

a. Referred to the low-voltage side, the equivalent series impedance of the transformer (based on the solution of Example 4.3.1) is given by

$$|Z_{eq}|_{LV} = \frac{|Z_{eq}|_{HV}}{a^2} = \frac{2.5}{100} = 0.025 \text{ ohm}$$

$$R_{eq\,LV} = \frac{R_{eq\,HV}}{a^2} = \frac{1.5}{100} = 0.015 \text{ ohm}$$

$$X_{eq\,LV} = \frac{X_{eq\,HV}}{a^2} = \frac{2.0}{100} = 0.02 \text{ ohm}$$

For maximum efficiency, the core loss should be equal to the copper loss. Thus

$$P_{oc} = I_{LV}{}^2 R_{eq\,LV}$$

where subscript *LV* refers to the low-voltage side. That is,

$$173 = I_{LV}^2 (0.015)$$

or

$$I_{LV} = \sqrt{\frac{173}{0.015}} = 107.4 \text{ A}$$

Full-load current on the low-voltage side, as given by the rating, is

$$I_{FL\ LV} = \frac{50,000}{240} = 208.3 \text{ A}$$

from which the fraction of full-load rating at which the maximum efficiency occurs is given by

$$\frac{107.4}{208.3} = 0.516$$

Alternatively, it may be simply obtained as follows:

$$k^2 P_{sc} = P_{oc}$$

where *k* is the fraction of the full-load rating. Therefore,

$$k = \sqrt{\frac{P_{oc}}{P_{sc}}} = \sqrt{\frac{173}{650}} = 0.516$$

Output power corresponding to the above condition is

$$50,000 \times 0.516 \times 0.8 = 20,640 \text{ W}$$

where 0.8 is the power factor given in the problem statement.

$$\text{Core loss} = \text{Copper loss} = 173 \text{ W}$$

The maximum efficiency is then given by

$$\eta_{max} = \frac{\text{Output}}{\text{Output} + \text{losses}} = \frac{20,640}{20,640 + 173 + 173}$$
$$= \frac{20,640}{20,986} = 0.9835$$

b. Energy output in kilowatt-hours during 24 hours is

$$(8 \times 0.9 \times 50 \times 0.8) + (12 \times 0.5 \times 50 \times 0.8)$$
$$= 528 \text{ kwh}$$

Core loss during 24 hours is

$$24 \times 0.173 = 4.15 \text{ kwh}$$

Copper loss during 24 hours is given by

$$(8 \times 0.9^2 \times 0.65) + (12 \times 0.5^2 \times 0.65) = 6.16 \text{ kwh}$$

The all-day (or energy) efficiency of the transformer is given by

$$\eta_{AD} = \frac{528}{528 + 4.15 + 6.16} = 0.9808$$

REGULATION   The voltage regulation of a transformer is the change in the magnitude of secondary terminal voltage from no load to full load and is usually expressed as a percentage of the full-load value.

$$\text{Percentage voltage regulation}$$
$$= \frac{V_{2\,(no\ load)} - V_{2\,(full\ load)}}{V_{2\,(full\ load)}} \times 100 \tag{4.4.5}$$

The voltages in Equation 4.4.5 may be referred to either the low-voltage or the high-voltage side.

Referring to the approximate equivalent circuit of Figure 4.2.7(c) with an equivalent series impedance, and the corresponding phasor diagram for a lagging power-factor case shown in Figure 4.2.8(c),

$$\text{Percentage voltage regulation} = \frac{V_1 - V_2'}{V_2'} \times 100 \tag{4.4.6}$$

The magnitude of the secondary voltage referred to the primary side (V_2') would rise to a value of V_1, if V_1 were held constant and the load removed. For the cases of loads having lagging or unity power factor, the regulation is positive since V_1 is greater (in magnitude) that V_2'. However, for loads having leading power factor, the regulation may be negative.

From Figure 4.2.8(c), for small values of δ, it can be seen that the difference between the magnitudes of $\overline{V}_1$ and $\overline{V}_2'$, i.e., $(V_1 - V_2')$ is approximately equal to ab given by $(I_1 R_{eq} \cos \theta_L + I_1 X_{eq} \sin \theta_L)$. It is sometimes convenient to express the regulation in per-unit quantities corresponding to rated (i.e., full-load) conditions as

$$\epsilon = \epsilon_r \cos \theta_L + \epsilon_x \sin \theta_L \tag{4.4.7}$$

where ϵ_r is the per-unit equivalent resistance, or the per-unit resistance drop, or the per-unit full-load I^2R, given by

$$\epsilon_r = I_1^2 R_{eq}/(V_1 I_1) = I_1 R_{eq}/V_1 = R_{eq}/(V_1/I_1) \tag{4.4.8}$$

and ϵ_x is the per-unit equivalent reactance, or the per-unit reactance drop, given by

$$\epsilon_x = I_1^2 X_{eq}/(V_1 I_1) = I_1 X_{eq}/V_1 = X_{eq}/(V_1/I_1) \tag{4.4.9}$$

Most of the loads, such as motors, that are operated near rated voltage and frequency are to be supplied by transformers that have small values of regulation, of the order of a few percent. On the other hand, loads, such as welding arcs, that operate at nearly constant current are supplied by their own individual transformer, which has a high value of regulation.

Transformers are often designed with taps on one winding so that the turns ratio may be varied over a small range. The *tap-changing* is often done automatically in large power-system transformers so that a sensibly constant secondary terminal voltage is maintained as the magnitude and power factor of the load vary. Tap-changing is usually manually done for distribution transformers, adjusted to a setting that gives optimum voltage over the projected load cycle. Tap-changing may also be used for compensating the variations in primary terminal voltage due to the feeder impedance.

EXAMPLE 4.4.3

The transformer of Example 4.3.1 is supplying full load at 240 volts and 0.8 power factor lagging.

  a. Determine the voltage regulation of the transformer.
  b. Find the approximate change in turns ratio required if the primary terminal voltage is fixed at 2,400 V.

SOLUTION

a. The equivalent circuit of the transformer, referred to the high-voltage (primary) side, is given below, neglecting the exciting current of the transformer.

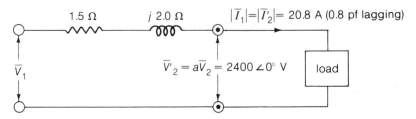

The corresponding phasor diagram is shown below.

$$\bar{V}_1 = \bar{V}_2' + \bar{I}_1(R + jX)$$
$$= 2{,}400\ \angle 0° + 20.8\ \angle{-}36.8°\ (1.5 + j2.0)$$

$$= 2{,}400 + j0 + (16.64 - j12.48)(1.5 + j2.0)$$

$$= 2{,}400 + j0 + 49.92 + j14.56 = 2{,}449.92 + j14.56$$
$$= 2{,}450\ \angle 0.34°$$

$$\text{Percentage voltage regulation} = \frac{2{,}450 - 2{,}400}{2{,}400} \times 100$$
$$= +2.08\%$$

The per-unit voltage regulation is given by 0.0208. The regulation may also be calculated by the approximate method indicated in the text.

b. Based on the result of part *(a),* the turns ratio of the transformer must be reduced on load by 2.08 percent if the secondary terminal voltage is to remain unaltered. That is,

$$\text{Approximate new turns ratio} = \frac{100 - 2.08}{100} \times 10 = 9.792$$

4.5 THREE-PHASE TRANSFORMERS

Most electrical energy is generated and transmitted using a three-phase system involving several three-phase voltage transformations. For transformation of three-phase power, either a bank of three identical single-phase transformers suitably connected or one three-phase transformer may be employed. Under conditions of balanced load and balanced voltages, with identical transformers in a given arrangement, each single-phase transformer will carry one-third of the total three-phase load.

A three-phase transformer for a given rating—compared to a bank of three single-phase transformers—weighs less, costs less, requires less floor space, and has somewhat higher efficiency. In addition, the number of external connections is reduced from twelve to six, and so also the consequent number of elaborate bushing arrangements. However, the provision of a spare stand-by three-phase transformer is more expensive than that of a single-phase spare transformer; if one of the three phases breaks down, the whole three-phase transformer has to be removed for repair.

A three-phase transformer is one in which cores and windings for all three phases are combined in a single structure. To visualize the development of a three-phase, core-type transformer, let us start with three identical single-phase core-type transformers arranged as in Figure 4.5.1*(a),* in which both the primary and secondary windings of each phase have been placed on only one leg of each transformer, and for simplicity, only the primary windings are shown in the figure. For balanced three-phase sinusoidal voltages, the fluxes ϕ_a, ϕ_b, and ϕ_c are 120° apart in time phase, and the sum of the three fluxes at any instant is zero; the leg carrying the sum by merging the three legs shown in Figure 4.5.1*(a)* can be omitted as in Figure 4.5.1*(b).* However, in practice, the construction is as shown in Figure 4.5.1*(c)* to make it more economical; the magnetic paths of the outer legs are somewhat longer than that of the center leg; but the resultant imbalance in the magnetizing currents is seldom significant. Figure 4.5.2*(a)* shows a conventional three-phase core with cruciform section to fit cylindrical coils. In order to reduce the overall height, a five-legged core is sometimes used, as shown in Figure 4.5.2*(b),* in which the phase coils are wound around the three center legs. See Photo 4–4 for a completed core- and coil-assembly with leads of a three-phase core-type transformer.

FIGURE 4.5.1 Evolution of a three-phase, two-winding, core-type transformer.

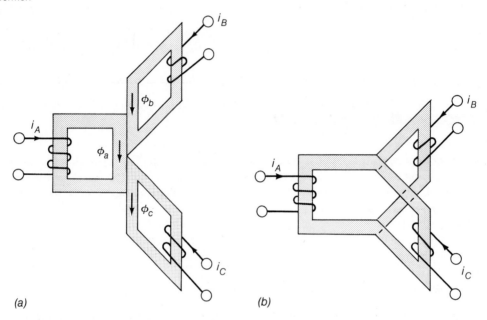

(a)

(b)

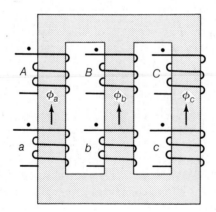

(c)

FIGURE 4.5.2 *(a)* Three-legged core of a three-phase core-type transformer; *(b)* five-legged core of a three-phase core-type transformer, the phase coils are wound around the three center legs.

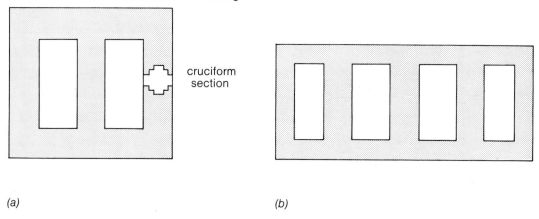

cruciform section

(a)   *(b)*

PHOTO 4.4 Completed core-and-coil assembly with leads of a 30-MVA, 69Y:13.2Δ-kV, three-phase core-type transformer. Photo courtesy of McGraw Edison Company.

FIGURE 4.5.3 Evolution of a three-phase, two-winding shell-type transformer.

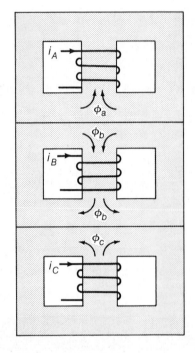

(a)

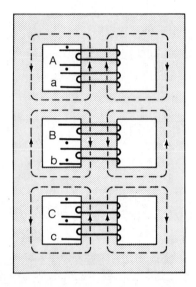

(b)

Next, to visualize the development of a three-phase, shell-type transformer, one can stack three single-phase shell-type units as in Figure 4.5.3(a), in which only the primary windings of the three phases are shown for simplicity. The direction of the winding on the center unit is made opposite to that of the outer units. That is to say, the polarities of both middle windings (primary and secondary) are reversed with respect to those of the outer windings. With this arrangement, for balanced three-phase sinusoidal voltages, the adjacent yoke sections of units a and b carry a combined flux whose magnitude is equal to the magnitude of each of its components. This follows from

$$\frac{\overline{\Phi}_a}{2} + \frac{\overline{\Phi}_b}{2} = \frac{\Phi_a}{2}(1 \angle 0° + 1 \angle 240°) = \frac{\Phi_a}{2} \angle -60° \qquad (4.5.1)$$

where $\overline{\Phi}_b$ is leading $\overline{\Phi}_a$ by 240° for the positive phase sequence a-b-c. Because of this it is possible to reduce the cross-sectional area of the combined yoke sections to the same value as that used for the outer legs, the top and bottom yokes, as shown in Figure 4.5.3(b), in which a conventional three-phase, shell-form core with the coils in section is given. The slight imbalance in the magnetic paths among the three phases of the shell-type construction is quite insignificant. See Photo 4–5 for an assembly of a typical shell-type transformer.

PHOTO 4.5   Assembly of a 400-MVA, 345Y:138Y:13.8Δ-kV shell-type transformer.   Photo courtesy of McGraw Edison Company.

THREE-PHASE TRANSFORMER CONNECTIONS   The windings of either core-type or shell-type three-phase transformers may be connected in either wye (star or Y) or delta (mesh or Δ). Four possible combinations of connections for the three-phase, two-winding transformers are

$$Y\text{-}\Delta,\ \Delta\text{-}Y,\ \Delta\text{-}\Delta,\ \text{and}\ Y\text{-}Y$$

These are shown in Figure 4.5.4 with the transformers assumed to be ideal; the windings on the left are the primaries while those on the right are the secondaries, and a primary winding of the transformer is linked magnetically with the secondary winding drawn parallel to it. Taking the primary-to-secondary turns ratio N_1/N_2 per phase to be a, also shown in the figure are the voltages and currents that result from balanced applied line-to-line voltages V and line currents I.

FIGURE 4.5.4   Wye-delta, delta-wye, delta-delta, and wye-wye transformer connections in three-phase systems.

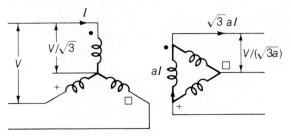

(a)

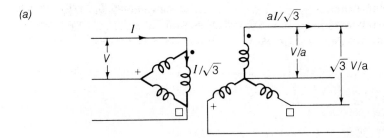

(b)

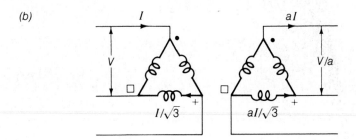

(c)

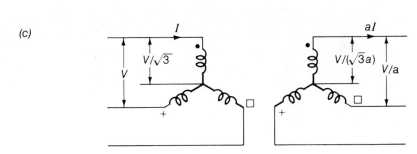

(d)

As in the case of three-phase circuits under balanced conditions, only one phase needs to be considered for circuit computations, because the conditions in the other two phases are the same except for the phase displacements associated with a three-phase system. It is usually convenient to carry out the analysis on a per-phase of Y (i.e., line-to-neutral) basis, and in such a case the transformer series impedance can then be added in series with the transmission-line series impedance. In dealing with Y-Δ or Δ-Y connections, all quantities may be referred to the Y-connected side; for Δ-Δ connections, it is convenient to replace the Δ-connected series impedances of the transformers by equivalent Y-connected impedances by making use of the relation

$$Z_{\text{per phase of Y}} = \frac{1}{3} Z_{\text{per phase of } \Delta} \qquad (4.5.2)$$

It can be shown that to transfer the ohmic value of impedance from the voltage level on one side of a three-phase transformer to the voltage level on the other side, the multiplying factor is the square of the ratio of line-to-line voltages, regardless of whether the transformer connection is Y-Y, Δ-Y, or Δ-Δ.

The Y-Δ connection is commonly used in high-voltage transmission systems to step down the voltage, with the high-voltage side connected in Y and the low-voltage side connected in Δ. A common arrangement in distribution circuits is the 208 (line-to-line)/120 (line-to-neutral)-volt system supplied by the Y-connection on the low-voltage side, with the high-voltage side of the transformer connected in Δ. The neutral point of the Y is usually grounded. The Δ-Y connection is popularly used for stepping up to a high voltage in transmission systems. The Y-connection on the high-voltage side, besides providing a neutral for grounding, reduces the insulation requirements.

The Δ-Δ connection is generally used in moderate voltage systems since the windings operate at full line-to-line voltage. In the case of a bank consisting of three single-phase transformer units, one of the advantages of the Δ-Δ connection is that if one transformer breaks down it can be removed from the circuit and reduced three-phase output supplied by the other two units connected in *open-delta*. The open-delta connection is also known as V-connection. For reasons of economy this connection is sometimes used for instrument transformers, and also used initially in growing load centers, the full growth of which may take several years, at which time a third unit is added to the existing open-delta connection to make it into delta-delta connection. The rating of the V-V connection will only be about 58 percent of that of Δ-Δ connection without overheating.

The Y-Y connection may be used in high-voltage applications because the voltage across the transformer winding is only $1/\sqrt{3}$ of the line-to-line voltage. However, it is seldom used because of the difficulties with the exciting current phenomena. In cases where Y-Y transformation is utilized, it is quite common to incorporate a third winding, known as a *tertiary winding,* connected in delta. See Photos 4–6 and 4–7 for views of a large step-up power transformer and a mobile-substation transformer.

PHOTO 4.6   Completely assembled 910-MVA, 20.5:500-kV step-up
transformer, 39 feet high, 35 feet long, 20 feet wide, weighing 562 tons.
Photo courtesy of Westinghouse Electric Corporation.

PHOTO 4.7   A 10-MVA, 66:6.6-kv mobile-substation transformer with the
high-voltage section, transformer section, and low-voltage section.   Photo
courtesy of Westinghouse Electric Corporation.

EXAMPLE 4.5.1

A one-line diagram of a part of a three-phase transmission system is shown in the figure below, along with necessary details. Draw the equivalent reactance diagram per phase (line-to-neutral) with all reactances marked in per unit, while choosing the bases of 30,000 kVA and 120 kV in the transmission line.

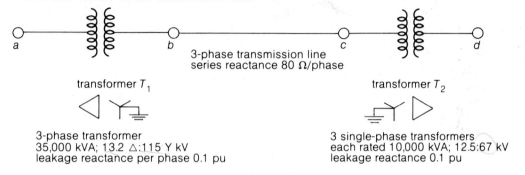

a   b

3-phase transmission line
series reactance 80 Ω/phase

c   d

transformer T_1

transformer T_2

3-phase transformer
35,000 kVA; 13.2 Δ:115 Y kV
leakage reactance per phase 0.1 pu

3 single-phase transformers
each rated 10,000 kVA; 12.5:67 kV
leakage reactance 0.1 pu

SOLUTION

The per-unit leakage reactance of a transformer is based on its own rating. The per-unit reactances of the transformers need to be converted to the chosen bases of 30,000 kVA and 120 kV in the transmission line. Making use of Equation 4.2.34, one obtains for transformer T_1:

$$X_I = 0.01 \frac{30,000}{35,000} \left(\frac{115}{120} \right)^2 = 0.0787 \text{ pu}$$

for transformer T_2:

$$X_I = 0.1 \frac{30,000}{30,000} \left(\frac{67\sqrt{3}}{120} \right)^2 = 0.0935 \text{ pu}$$

The base impedance per phase of Y of the transmission line is given by Equation 4.2.42:

$$\frac{(120,000)^2}{30,000 \times 10^3} = 480 \ \Omega$$

The per-unit series reactance of the transmission line per phase is

$$\frac{80}{480} = 0.167 \text{ pu}$$

The following figure gives the desired line-to-neutral reactance diagram with per-unit values:

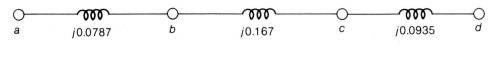

a   j0.0787   b   j0.167   c   j0.0935   d

neutral

Note that each of the transformers, as well as the transmission line, is represented by a corresponding equivalent series reactance (adjusted properly to the chosen base, of course); only a single-phase circuit needs to be considered for the analysis of balanced three-phase systems; in the specifications given for a three-phase transformer, the kVA is the three-phase kVA and the voltages are the line-to-line voltages on either side.

EXAMPLE 4.5.2

A one-line diagram of a three-phase distribution system is given below:

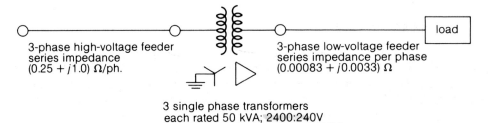

3-phase high-voltage feeder
series impedance
$(0.25 + j1.0)$ Ω/ph.

3-phase low-voltage feeder
series impedance per phase
$(0.00083 + j0.0033)$ Ω

load

3 single phase transformers
each rated 50 kVA; 2400:240V
identical with that of example 4.3.1

Determine the line-to-line voltage at the sending end of the high-voltage feeder when the transformer delivers rated load at 240 V (line-to-line) and 0.8 lagging power factor. Neglect the exciting current of the transformer.

SOLUTION

The analysis is carried out on a per-phase-of-Y basis by referring all quantities to the high-voltage Y-side of the transformer bank. The impedance of the low-voltage feeder—referred to the high-voltage side by means of the square of the ideal line-to-line voltage ratio of the transformer bank—is

$$(0.00083 + j0.0033)\left(\frac{2,400\sqrt{3}}{240}\right)^2 = (0.25 + j0.1)\ \Omega$$

Thus we have, referred to the high-voltage side, the combined series impedance of the high-voltage and low-voltage feeders as

$$Z_{f\,HV} = (0.25 + j1.0) + (0.25 + j1.0)$$
$$= (0.5 + j2.0)\ \Omega \text{ per phase of Y}$$

The equivalent series impedance of the transformer bank referred to its high-voltage Y-side is given from Example 4.3.1 as

$$Z_{eq\,HV} = (1.5 + j2.0)\ \Omega \text{ per phase of Y}$$

The equivalent circuit per phase of Y referred to the Y-connected primary side is then given by exactly the same circuit shown in Example 4.2.3(b). For the conditions given in the problem, the solution on a per-phase basis is exactly the same as the solution of Example 4.2.3(b), from which it follows that the voltage at the sending end of the high-voltage feeder is 2,483.5 volts. The actual line-to-line voltage, because of the primary Y-connection, is given by

$$2,483.5\sqrt{3} = 4,301.4 \text{ V}$$

4.6 AUTOTRANSFORMERS (VARIAC)

In a two-winding transformer, whether single-phase or three-phase, the primary and secondary windings, though magnetically linked, are isolated electrically. For the case of power-system transformers, in which large transformation ratios are involved and safety plays a dominant role, the electrical isolation of the primary and secondary windings is necessary. However, when the transformation ratio is close to unity and electrical isolation between the two windings is less critical, a device known as an *autotransformer,* consisting of a single tapped winding on a transformer core, can be used in place of a two-winding transformer with distinct advantages. The same fundamental considerations already discussed for transformers having two separate windings hold true for the analysis of an autotransformer, which is really nothing but a normal transformer connected in a special way.

Let us consider a two-winding transformer shown in Figure 4.6.1(a) and convert it to an autotransformer arrangement, as shown in Figure 4.6.1(b), by connecting the two windings electrically in series so that the polarities are additive. The N_1-turn winding is placed between points *a* and *b*, while the N_2-turn winding is placed between points *b* and *c*, as illustrated in Figure 4.6.1(b). Such an arrangement may be used for either stepping up or stepping down the voltage. Also, several other connections and selections of input and output terminals are possible. However, basic considerations for analysis will essentially remain the same.

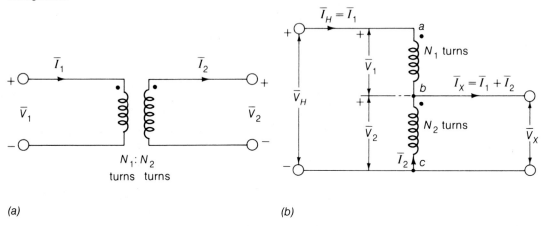

FIGURE 4.6.1 *(a)* Two-winding transformer, *(b)* an autotransformer arrangement.

(a)   (b)

Referring to the autotransformer arrangement of Figure 4.6.1*(b)*, the following relations apply:

$$\overline{V}_H = \overline{V}_1 + \overline{V}_2 = \overline{V}_1 + \overline{V}_x \tag{4.6.1}$$

$$\overline{I}_x = \overline{I}_1 + \overline{I}_2 \tag{4.6.2}$$

The phase angle between the primary and secondary voltages being very small (usually less than 10°), the phasor sum of Equation 4.6.1 can be taken to be practically equal to the arithmetic sum in magnitude.

$$V_H \simeq V_1 + V_2 \simeq V_1 + V_x \tag{4.6.3}$$

If the exciting current is neglected, the currents I_H and I_x are in phase, and the magnetomotive forces of the two windings are equal, so that

$$N_1 \overline{I}_1 = N_1 \overline{I}_H = N_2 \overline{I}_2 \tag{4.6.4}$$

or

$$\overline{I}_H = \overline{I}_1 = \frac{N_2}{N_1} \overline{I}_2 \tag{4.6.5}$$

The ratio of input and output currents is given by

$$\frac{\overline{I_H}}{\overline{I_X}} = \frac{\overline{I_1}}{\overline{I_1} + \overline{I_2}} = \frac{N_2}{N_1 + N_2} \qquad (4.6.6)$$

or

$$\frac{\overline{I_X}}{\overline{I_H}} = \frac{\overline{I_1} + \overline{I_2}}{\overline{I_1}} = \frac{N_1 + N_2}{N_2} = 1 + \frac{N_1}{N_2} = 1 + a \qquad (4.6.7)$$

where a is the turns ratio (N_1/N_2) of the two-winding transformer of Figure 4.6.1(a). The ratio of the terminal voltages is given by

$$\frac{\overline{V_H}}{\overline{V_X}} = \frac{\overline{V_1} + \overline{V_2}}{\overline{V_2}} = \frac{N_1 + N_2}{N_2} = 1 + \frac{N_1}{N_2} = 1 + a \qquad (4.6.8)$$

Thus, the autotransformer with a transformation ratio of $(1 + a)$ can be treated as an equivalent two-winding transformer with a turns ratio of a. The rating as an autotransformer is much higher than that of the two-winding transformer:

$$\frac{\text{Autotransformer rating}}{\text{Two-winding transformer rating}} = \frac{V_H I_H}{V_1 I_1} = \frac{V_H}{V_1}$$

$$= 1 + \frac{V_X}{V_1} \qquad (4.6.9)$$

$$= 1 + \frac{N_2}{N_1} = 1 + \frac{1}{a}$$

The above ratio becomes much larger than unity, since for many autotransformer applications a is rather small and the ratio of the terminal voltages $(1 + a)$ is very nearly unity. This is because most of the power in autotransformers is delivered to the secondary through direct *conduction*, and only a small portion is delivered through transformation or induction as in a two-winding transformer.

When the ratio of the terminal voltages does not differ too greatly from unity, autotransformers as compared with two-winding transformers of the same output will have lower leakage reactances, lower losses, smaller exciting current, and less cost. Since the primary and secondary of the autotransformer share one winding, and the current in the common winding is always the difference between the input and output currents, the cross-sectional area of the common winding is much smaller than the primary or secondary winding of a two-winding transformer of the same current density. Hence, significant savings in copper would result.

PHOTO 4.8   A 367-MVA, 525-kV extra-high-voltage autotransformer, 40 feet high, 28 feet wide, weighing 270 tons.   Photo courtesy of Westinghouse Electric Corporation.

For the same output, compared to a two-winding transformer, the autotransformer is smaller in size and has higher efficiency as well as superior voltage regulation because of the reduced resistance drop and lower leakage reactance drop. An important disadvantage of the autotransformer is the direct copper connection (i.e., no electrical isolation) between the high- and low-voltage sides. A common type of autotransformer that is found in laboratories is the variable-ratio autotransformer in which the tapped point *b,* shown in Figure 4.6.1*(b),* is movable. It is known as the *variac* (variable ac).

While we have considered here only the single-phase autotransformer for the sake of simplicity, three-phase autotransformers can be analyzed on a per-phase basis just as the single-phase case. See Photo 4–8 in which the extra-high-voltage autotransformer is shown on test at the plant.

EXAMPLE 4.6.1

The single-phase 50-kVA, 2,400:240-volt, 60-Hz, two-winding distribution transformer of Example 4.3.1 is connected as a step-up autotransformer as shown in the figure below.

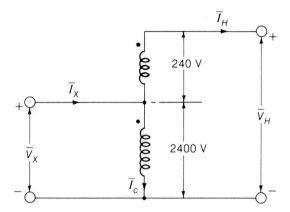

Assume that the 240-volt winding is provided with sufficient insulation to withstand a voltage of 2,640 volts to ground.

a. Find V_H, V_x, I_H, I_C, and I_x corresponding to rated (full-load) conditions.

b. Determine the kVA rating as an autotransformer, and find how much of that is the conducted kVA.

c. Based on the data given for the two-winding transformer in Example 4.3.1, compute the efficiency of the autotransformer corresponding to full load, 0.8 power factor lagging, and compare it with that of the two-winding transformer calculated in Example 4.4.1(a).

SOLUTION

a. The two windings are connected in series so that the polarities are additive. Neglecting the leakage-impedance voltage drops,

$$V_H = 2,400 + 240 = 2,640 \text{ V}$$

$$V_x = 2,400 \text{ V}$$

The full-load rated current of the 240-V winding, based on the rating of 50 kVA as a two-winding transformer, is

$$\frac{50,000}{240} = 208.33 \text{ A}$$

Since the 240-V winding is in series with the high-voltage circuit, the full-load current of this winding is the rated current

$$I_H = 208.33 \text{ A}$$

on the high-voltage side of the autotransformer.

Neglecting the exciting current, the mmf produced by the 2,400-V winding must be equal and opposite to that of the 240-V winding.

$$I_C = 208.33 \left(\frac{240}{2,400}\right) = 20.83 \text{ A}$$

in the direction shown in the figure. Then the current on the low-voltage side of the autotransformer is given by

$$I_x = I_H + I_C = 208.33 + 20.83 = 229.16 \text{ A}$$

b. The kVA rating as an autotransformer is

$$\frac{V_H I_H}{1,000} = \frac{(2,640)(208.33)}{1,000} = 550 \text{ kVA}$$

or

$$\frac{V_x I_x}{1,000} = \frac{(2,400)(229.16)}{1,000} = 550 \text{ kVA}$$

which is 11 times that of the two-winding transformer. The transformer has to boost the current I_H of 208.33 A through a potential rise of 240 volts; thus the kVA transformed by electromagnetic induction is given by

$$\frac{240 \times 208.33}{1,000} = 50 \text{ kVA}$$

The remaining 500 kVA is the conducted kVA.

c. With the currents and voltages shown for the autotransformer connections, the losses at full load will be the same as 823 W as in Example 4.4.1(a). However, the output as an autotransformer at 0.8 power factor is given by

$$550 \times 1,000 \times 0.8 = 440,000 \text{ W}$$

The efficiency of the autotransformer is then calculated as

$$\eta = 1 - \frac{\text{Losses}}{\text{Input}} = 1 - \frac{823}{440{,}823} = 0.9981$$

while that of the two-winding transformer is 0.98 as calculated in Example 4.4.1*(a)*. Because the losses are only those due to transforming 50 kVA, the higher efficiency results.

FIGURE 4.7.1   A distribution three-winding transformer.

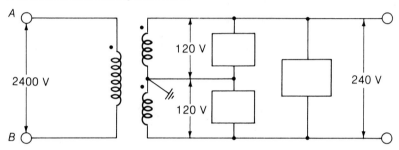

4.7 MULTIWINDING TRANSFORMERS

For the interconnection of three or more circuits that may have different voltages, *multicircuit* or *multiwinding* transformers consisting of three or more windings are often used for the sake of efficiency and economy in place of two or more two-winding transformers. In supplying power to electronic circuits, a transformer with a primary and two secondaries is commonly employed. A distribution transformer shown in Figure 4.7.1 with a center tap on the secondary side is also an example of a three-winding transformer. Loads consisting of 120-volt appliances and lighting load, divided roughly equally between the two 120-volt secondaries, are connected from line to neutral; appliances such as electric ranges and domestic hot-water heaters are supplied with 240-volt power from the series-connected secondaries.

Large power-system transformers in three-phase systems used for transmission and distribution of power are usually three-winding transformers. While the primary and secondary windings of the three-phase transformer bank are utilized to interconnect two transmission systems of different voltages, a suitable voltage for auxiliary power purposes in the substation or for supplying a local distribution system is provided by the third winding, known as the *tertiary* winding. While the primaries and secondaries are connected in Y-Y, the tertiary windings are connected in Δ to provide a path for the third harmonics and their multiples in the exciting current. The problems associated with multiwinding transformers, such as the effects of leakage impedances on voltage regulation, short-circuit currents, and division of load among different circuits, can be solved by an equivalent-circuit approach similar to that employed in solving the two-winding transformer problems. Since the leakage impedances associated with each pair of windings must be considered, the multiwinding-transformer equivalent circuits tend to be complicated. It is most convenient to express all quantities in per unit, referred to a common base, in these equivalent circuits.

Let us consider a three-winding transformer whose schematic diagram is shown in Figure 4.7.2. For the case of an ideal transformer,

$$\frac{\overline{V}_2}{\overline{V}_1} = \frac{N_2}{N_1} \tag{4.7.1}$$

$$\frac{\overline{V}_3}{\overline{V}_1} = \frac{N_3}{N_1} \tag{4.7.2}$$

$$N_1 \overline{I}_1 = N_2 \overline{I}_2 + N_3 \overline{I}_3 \tag{4.7.3}$$

The equivalent circuit that takes into account the leakage impedance associated with three windings on a common magnetic core and exciting admittance is shown in Figure 4.7.2(b), in which all quantities are expressed in per unit and referred to a common base.

The parameters of the equivalent circuit can be determined from open-circuit and short-circuit tests. The data for calculating the exciting admittance are obtained by the open-circuit test as in the case of a two-winding transformer. Since there are three windings, three short-circuit tests are needed to obtain the data and determine Z_1, Z_2, and Z_3 of the equivalent circuit. In analyzing the short-circuit test data, the exciting current is neglected. Thus, if Z_{12} is the short-circuit impedance of Circuits 1 and 2 with Circuit 3 open (i.e., by applying voltage to the primary Circuit 1 with the secondary circuit 2 short-circuited and the tertiary Circuit 3 open, the measured impedance is Z_{12}).

$$Z_{12} = Z_1 + Z_2 \tag{4.7.4}$$

Similarly,

$$Z_{13} = Z_1 + Z_3 \tag{4.7.5}$$

$$Z_{23} = Z_2 + Z_3 \tag{4.7.6}$$

where Z_{13} is the short-circuit impedance of Circuits 1 and 3 with secondary Circuit 2 open, and Z_{23} is the short-circuit impedance of Circuits 2 and 3 with primary Circuit 1 open. These imped-

FIGURE 4.7.2   Three-winding transformer: *(a)* schematic diagram,
(b) equivalent circuit with all quantities expressed in per unit, referred to a
common base.

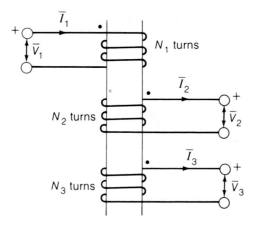

(a)

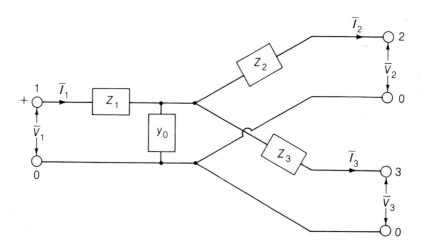

(b)

ances should be referred to a common base. From Equations 4.7.4, 4.7.5, and 4.7.6, the leakage impedances Z_1, Z_2, and Z_3 are then given by

$$Z_1 = \frac{Z_{12} + Z_{13} - Z_{23}}{2} \qquad\qquad (4.7.7)$$

$$Z_2 = \frac{Z_{23} + Z_{12} - Z_{13}}{2} \qquad\qquad (4.7.8)$$

$$Z_3 = \frac{Z_{13} + Z_{23} - Z_{12}}{2} \qquad\qquad (4.7.9)$$

The following example illustrates the computation of the equivalent circuit parameters in per unit referred to a common base.

EXAMPLE 4.7.1

A three-phase bank consisting of three single-phase three-winding transformers is used to step down the voltage of a three-phase 66-kV (line-to-line) transmission line. The transformers are connected Y-Y-Δ; the wye-connected secondary and the delta-connected tertiary have line-to-line voltages of 11 kV and 2.2 kV, respectively. The following is the test data obtained from open-circuit and short-circuit tests conducted on one of the single-phase three-winding transformers:

SHORT-CIRCUIT TEST DATA

Winding		Measured on the Excited-Winding Side	
Excited	Short-Circuited	Volts	Amperes
Primary(1)	Secondary(2)	3,000	800
Primary(1)	Tertiary(3)	5,800	650
Secondary(2)	Tertiary(3)	550	1,600

OPEN-CIRCUIT TEST DATA

Winding	Volts	Amperes
Tertiary (3)	2,200	150

Neglecting the resistance of the windings and the core losses, determine the values of reactance and admittance to be used in the equivalent circuit of Figure 4.7.2(b), while using a three-phase base of 90,000 kVA (or base of 30,000 kVA per single-phase transformer) for determining the per-unit values.

SOLUTION

The short-circuit test data yield the following, referred to the primary:

$$Z_{12} = j3,000/800 = j3.75 \text{ ohms}$$

$$Z_{13} = j5,800/660 = j8.79 \text{ ohms}$$

$$Z_{23} = (j550/1,600)(N_1/N_2)^2 = (j550/1,600)(66/11)^2 = j12.38 \text{ ohms}$$

From Equations 4.7.7, 4.7.8, and 4.7.9 one obtains

$$Z_1 = \frac{j3.75 + j8.79 - j12.38}{2} = j0.08 \text{ ohm}$$

$$Z_2 = \frac{j12.38 + j3.75 - j8.79}{2} = j3.67 \text{ ohms}$$

$$Z_3 = \frac{j8.79 + j12.38 - j3.75}{2} = j8.71 \text{ ohms}$$

Referred to the tertiary, the exciting admittance is given by

$$y_0 = -j\frac{150}{2,200} = -j0.068 \text{ siemen}$$

or, referred to the primary,

$$(-j0.068)(N_3/N_1)^2 = (-j0.068)\left(\frac{2.2}{66/\sqrt{3}}\right)^2 = -j0.00023 \text{ S}$$

Base impedance referred to the primary is given by

$$Z_{base\ pri} = \frac{(volts)^2}{volt\text{-}amperes} = \frac{(66,000/\sqrt{3})^2}{30,000 \times 1,000} = 48.4 \text{ ohms}$$

Hence, the per-unit values of the equivalent-circuit parameters of Figure 4.7.2(b) are given by the following:

$$Z_1 = j0.08/48.4 = j0.0017 \text{ pu}$$

$$Z_2 = j3.67/48.4 = j0.076 \text{ pu}$$

$$Z_3 = j8.71/48.4 = j0.18 \text{ pu}$$

$$y_0 = (-j0.00023)(48.4) = -j0.011 \text{ pu}$$

4.8 SOME OTHER TRANSFORMER TOPICS

There are several other important transformer topics, some of which are discussed below:

INSTRUMENT TRANSFORMERS   Instrument transformers are generally of two types, potential transformers and current transformers. These are designed in such a way that the former may be regarded as having an ideal potential ratio, while the latter has an ideal current ratio. For instrument transformers it is the accuracy of measurement that is important, but not the load-carrying capability. Instrument transformers are commonly used in ac power circuits to supply instruments, protective relays, and control circuits.

Potential transformers are employed to step down the voltage to about 110 volts for connecting to voltmeters, wattmeters, relays, and control devices. In three-phase installations, these may be connected line-to-neutral or line-to-line, depending upon the voltage.

Current transformers (connected in series with the line) are used to step down the current to about 5 amperes for metering purposes. Often the primary is not an integral part of the transformer itself, but is part of the line whose current is being measured. In addition to providing a desirable low current in the metering circuit, the current transformer isolates the meter from the line, which may be at a high potential. The secondary terminals of a current transformer should never be open-circuited under load, because there would then be no secondary mmf to oppose the primary mmf, the net mmf acting on the core would rise to a value many times greater than that for which the core is designed, and as a consequence very high voltage might be induced in the secondary. Besides endangering the user, this may damage the transformer insulation and also cause transformer overheating due to excessive core losses.

The connections of instrument transformers to measure current, voltage, and power in a single-phase power circuit are shown in Figure 4.8.1. Note the polarity markings indicated by ±, located on primary and secondary terminals of like polarity, as in laboratory transformers. Before connecting to a wattmeter, the terminal polarity of an instrument transformer must be known.

FIGURE 4.8.1 Instrument transformer connections.

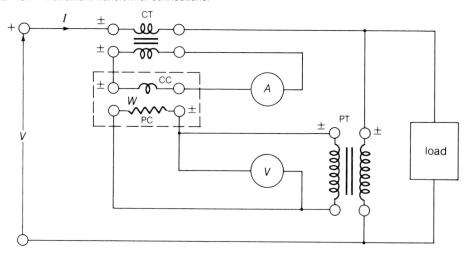

CT: current transformer   A: Ammeter
PT: potential transformer   V: Voltmeter
CC: current coil of wattmeter   W: Wattmeter
PC: potential coil of wattmeter

VARIABLE-FREQUENCY TRANSFORMERS In electronic and communication circuits, audio-frequency transformers are required to operate over the audible frequency range of about 20 to 20,000 Hz. Other examples of variable-frequency operations include variable-frequency sources containing transformers and supplying power to variable-speed motors. The *frequency characteristic* and the *phase characteristic* are the curves usually referred to the variable-frequency operation. The frequency characteristic (also known as the amplitude-frequency characteristic) is a curve of the ratio of the secondary-load voltage to the primary internal-source voltage, plotted as a function of the frequency. A flat characteristic is most desirable. The phase characteristic is a curve of the phase angle of the load voltage with respect to the source voltage, plotted as a function of the frequency. A small phase angle is desirable.

The analysis of a properly designed circuit including a transformer may usually be broken down into three frequency ranges:

1. In the mid-frequency range, the equivalent circuit reduces to a network of resistances, without any inductance, as shown in Figure 4.8.2(a). The variable frequency source (such as an electronic amplifier) is represented by E_S in series with an internal resistance R_S. The load, such as a speaker, is represented by the load resistance R_L. The leakage reactances and the magnetizing current are generally neglected in the mid-frequency range.

FIGURE 4.8.2 Equivalent circuits of a variable-frequency transformer:
(a) mid-frequency range equivalent circuit, *(b)* low-frequency range
equivalent circuit, *(c)* high-frequency range equivalent circuit (without
capacitance effects), *(d)* high-frequency range equivalent circuit with
capacitances.

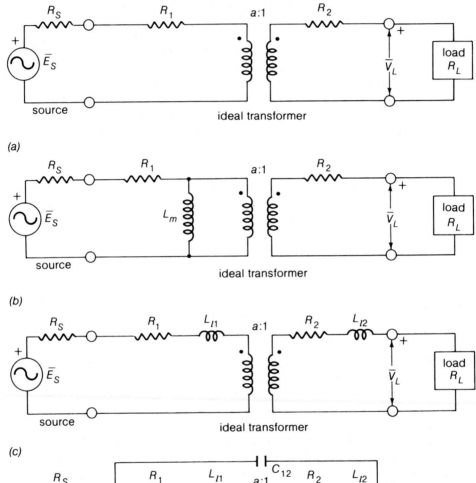

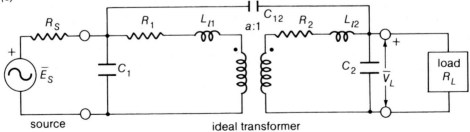

2. At low frequencies, the leakage reactances are negligible, but the magnetizing current of the transformer becomes significant, and the circuit of Figure 4.8.2*(b)* applies.
3. In the high-frequency range, the leakage reactances of the transformer become significant, and the equivalent circuit of Figure 4.8.2*(c)* applies. The effects of distributed capacitances across and between the windings may become significant, in which case the equivalent circuit of Figure 4.8.2*(d)* is applicable. A lumped circuit containing three capacitances is commonly used to describe the effect of the distributed capacitances. C_1 and C_2 are the primary (Winding 1) and secondary (Winding 2) capacitances connected across the transformer terminals, while C_{12} is the capacitance between the primary and secondary terminals. The values of these capacitances depend on the winding arrangements. It is preferable to refer all the parameters to one side of the ideal transformer for facilitating the analysis of the equivalent circuit.

An accurate analysis of the transient surge potential distribution in high-voltage power-system transformers requires a more elaborate circuit model that takes account of the distributed nature of the leakage inductance and interwinding capacitances. The analysis of the effects of lightning and switching surges on the power-system transformers is too involved to be taken up here. Electrostatic shielding is usually provided to increase the series capacitance of the winding and thereby prevent excessive concentration of transient voltages on the line-end turns.

PULSE TRANSFORMERS   Pulse transformers are used in radar, television, digital-computer circuits, and linear accelerators. Recently, they also have been used in the gate firing circuits of power thyristors and several other electronic and control applications. Most pulse transformers are physically small and have relatively few turns so as to minimize the leakage inductance. The cores are made out of ferrites or wound strips of high-permeability alloys such as permalloy. Because the time interval between pulses is longer than the pulse duration, the load-carrying duty on the pulse transformer is rather light; and hence, relatively high pulse-power levels can be handled by a small-size transformer.

Pulse excitation typically consists of a rectangular voltage pulse, though many other pulse shapes are also commonly used. The average transformer power is determined by the magnitude and the repetition rate of the pulse. It is important that the transformer is able to reproduce the input pulse as truly as possible at its secondary terminals. Figure 4.8.3*(a)* shows a rectangular pulse excitation, whose pulse width may be a fraction of a microsecond to about 20 microseconds with a relatively long time elapsing before the pulse repeats. Figure 4.8.3*(b)* is the waveform of the output (either a voltage or current) resulting from a rectangular pulse excitation. A transient rather than a steady-state analysis is required in order to determine the output waveform.

FIGURE 4.8.3 Input and output voltage waveforms of a pulse transformer: *(a)* rectangular input-voltage pulse excitation, *(b)* output-voltage waveform of a pulse transformer.

input voltage

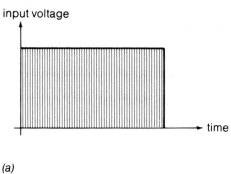

(a)

output voltage

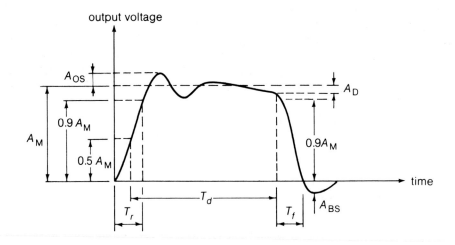

A_M: pulse amplitude T_d: pulse duration

A_{OS}: overshoot T_r: pulse rise time

A_D: droop or tilt T_f: pulse fall time

A_{BS}: back swing

(b)

Equivalent circuits such as those shown in Figure 4.8.2 are used for analysis of pulse transformers. Detailed analysis including the transformer nonlinearities requires computer simulation techniques. However, by employing simplified equivalent circuits during different portions of the pulse, it is possible to observe the response. The rise time associated with the leading edge of the output waveform can be found by the capacitive high-frequency equivalent circuit of Figure 4.8.2(d). The response to the flat-top portion of the input pulse can be determined by the low-frequency equivalent circuit of Figure 4.8.2(b). The fall time and the backswing associated with the trailing edge of the output response can be analyzed by a parallel RLC circuit in which the magnetizing inductance is discharging the energy of the decaying magnetic field through the capacitance and circuit resistance.

SATURABLE REACTORS   Saturable reactors are a means of controlling the power delivered to a load from a constant-potential ac source. They consist of a group of saturable magnetic elements whose core materials may be assumed to have an idealized *B-H* characteristic with infinite unsaturated permeability and zero saturated permeability, as shown in Figure 4.8.4(a). Consider an arrangement, shown in Figure 4.8.4(b), of a series saturable reactor, in which the primary windings are connected in series with each other and with the load. The control windings are connected in series opposition and supplied by a direct current *I*. To ensure that the current is essentially constant and free of ripples, a large inductance is introduced in the control winding, as shown in the figure.

With the control current *I*, Core 1 is unsaturated only if

$$i_L = \frac{N_2}{N_1} I \qquad (4.8.1)$$

and Core 2 is unsaturated only if

$$i_L = -\frac{N_2}{N_1} I \qquad (4.8.2)$$

The idealized relations of the flux linkages λ_1 of Core 1, the flux linkages λ_2 of Core 2, and the combined flux linkages $(\lambda_1 + \lambda_2)$ as a function of the current i_L are shown in Figures 4.8.4(c), (d), and (e), respectively. Since the combined flux linkages $(\lambda_1 + \lambda_2)$ can change only if the current i_L has one of the values given by Equations 4.8.1 and 4.8.2, it follows that the current i_L in the load R_L will be a square wave having the amplitudes given by Equations 4.8.1 and 4.8.2, when an ac voltage of $v = v_{max} \sin \omega t$ is impressed. Thus the saturable reactor acts as a source of constant rectangular-wave current, the magnitude of which is directly proportional to the control current. The response in practical situations is somewhat different because of departure from the idealized conditions assumed in this analysis.

FIGURE 4.8.4 A saturable reactor: *(a)* idealized *B-H* characteristic, *(b)* series saturable reactor with high-impedance control, *(c)* flux linkages λ_1 of core 1 as a function of the current i_L, *(d)* flux linkages λ_2 of core 2 as a function of the current i_L, *(e)* combined flux linkages $(\lambda_1 + \lambda_2)$ as a function of the current i_L, *(f)* waveforms of supply voltage v and the load voltage $(R_L i_L)$ as a function of ωt.

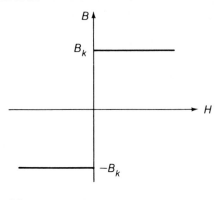

(a)

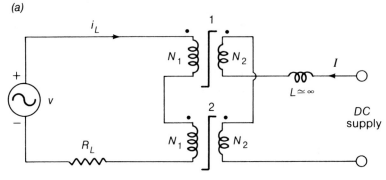

(b)

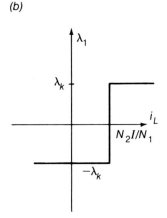

(c)

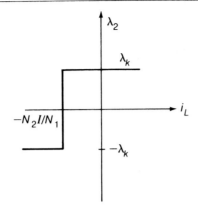

(d)

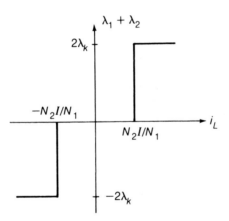

(e)

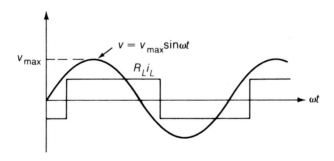

(f)

A series saturable reactor with an arrangement as shown in Figure 4.8.4*(b)*, but with low-impedance control, can also be considered. With a sinusoidal applied voltage, the waveforms of λ_1, λ_2, the control current i, and the load current i_L can be made to assume various desirable shapes, and thereby control the power delivered to the load R_L. The control current as well as the load current can be made to consist of only sections of waves. Saturable reactors can also be connected in an alternative arrangement in which the primary windings are in parallel instead of in series, as shown in Figure 4.8.4*(b)*.

INRUSH CURRENT   When the primary winding of an unloaded transformer is switched on to the normal voltage supply, it acts as a nonlinear inductor. Neglecting the core loss and the primary resistance, the applied voltage must be balanced at every instant by the emf induced by the magnetizing current. The response depends on the magnitude, the polarity, and the rate of change of applied voltage at the instant of switching. It may result in a current peak many times (about eight to ten times) the rated transformer current, or it may be practically unobservable. Such a transient effect is known as the *inrush current* phenomenon, even though the exciting current eventually will come down to its normal value (of about 5 percent or less of the rated transformer current) because of the winding resistance and core losses. The inrush current is a matter of great concern while energizing large power transformers, since it may unduly stress the transformer windings and may cause improper operation of protective devices. Thus, it is one aspect of transformer transient characteristics that should be looked into.

Consider the transformer core to be initially unmagnetized and the primary winding of the unloaded transformer to be switched on at a voltage peak, as shown in Figure 4.8.5*(a)*. Neglecting the core loss and the primary-winding resistance, the voltage and flux need to satisfy

$$v_1 = N_1 \frac{d\phi}{dt} \tag{4.8.3}$$

Hence, the flux is as shown in Figure 4.8.5*(a)* and the corresponding magnetizing current can be found from the *B-H* characteristic of the transformer core. Such conditions correspond to those of normal no-load operation, when there is no transient.

Next, consider the switch to be closed at a voltage zero at $t = 0$ as shown in Figure 4.8.5*(b)*, corresponding to which

$$v_1 = \sqrt{2}\ V \sin \omega t = N_1 \frac{d\phi}{dt} \tag{4.8.4}$$

The flux is then given by

$$\phi = \frac{\sqrt{2}\ V}{N_1} \int_0^t \sin \omega t\ dt = \frac{\sqrt{2}\ V}{\omega N_1} (1 - \cos \omega t)$$

or

$$\phi = -\phi_m \cos \omega t + \phi_m \tag{4.8.5}$$

FIGURE 4.8.5   Inrush current phenomenon.

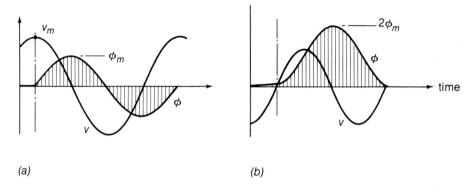

(a)   (b)

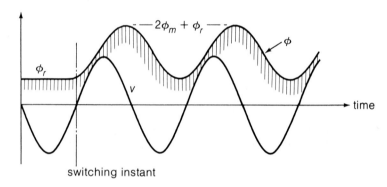

(c)

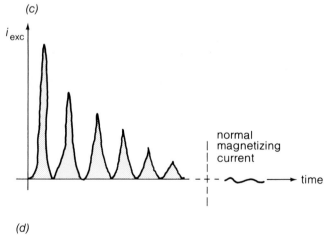

(d)

The instantaneous flux at $\omega t = \pi$ is then $2\phi_m$, as shown in Figure 4.8.5(b). The residual flux ϕ_r in the core can aggravate the condition, as shown in Figure 4.8.5(c), if the flux increases in the direction of the residual flux. The peak flux in the worst case could then be $(2\phi_m + \phi_r)$. As the time progresses, the effect of losses will rapidly reduce the flux waveform to a symmetrical one about the time axis.

Since the maximum steady-state flux occurs around the knee of the saturation curve for most of the transformers, a flux of two or more times the maximum steady-state flux may demand an exciting current many hundreds of times the normal excitation current because of supersaturation of the core. Depending upon the change in flux path and the transformer time constant (which may be a few milliseconds for small transformers and a few seconds for large transformers), the inrush current phenomenon may persist for a few seconds, as shown in Figure 4.8.5(d). Analytical methods of predicting the inrush current phenomenon are too involved to be addressed here.

PARALLEL OPERATION OF TRANSFORMERS   In paralleling single-phase transformers, besides observing the polarities in making physical connections, the following conditions are given in the order of their relative importance:

1. Equal primary and secondary voltage ratings.
2. Equal ratios of transformation.
3. Equivalent impedances (in ohms) inversely proportional to their current or kVA ratings.
4. Equal ratios of equivalent resistance to equivalent reactance.

If the voltage ratings are not the same, some windings will be operating on a higher and some on a lower voltage than that for which they are designed.

If the ratios of transformation are not the same, there will be circulating currents in the transformer windings in addition to the exciting currents, when there is no load on the system. They increase the copper loss, overload the transformer, and reduce the overall permissible load kVA.

If the impedances (in ohms) are not inversely proportional to the current outputs, the load does not get divided evenly between the transformers in proportion to their ratings. Such uneven load sharing may lead to overheating or reduced total rated capacity.

If the ratios of equivalent resistance to equivalent reactance are not equal, the currents delivered by the transformers will not be in phase with one another or the load current. That is to say, the transformers will be carrying kilowatt loads that are not proportional to their current loads. This will lead to increased copper loss and consequent overheating or reduced overall kilowatt output.

In paralleling three-phase transformers, the conditions given above for single-phase transformers apply equally, but with an added condition that the secondary voltages shall be at compatible phase angles. Not only must the transformer polarities be observed but also the phase sequence must be identical for paralleled transformers. Transformers with similar connections on either side can be connected in parallel. Also, three-phase transformer banks connected in Y-Y and Δ-Δ may be paralleled if the line-voltage ratios are identical. The phase shifts resulting from Δ-Y or Y-Δ connections generally preclude the use of either in parallel with a configuration that is different from its own. However, the secondary voltages of a Y-Δ system and a Δ-Y system are either in phase or are 60° out of phase, depending upon the way the connections are made; two such systems can therefore be paralleled with proper connections.

SUPERCONDUCTING TRANSFORMER WINDINGS   Superconducting metal films capable of carrying very high current densities when immersed in a magnetic field are being developed. The possibility of operating large power-system transformers under cryogenic conditions has been researched for the past two decades.[2] The superconducting windings would have to conduct in the presence of high leakage fields and carry very high current densities. With the superconducting windings, the conventional I^2R loss could be entirely eliminated, and also the core loss would be reduced because of the reduction in the core mass. However, refrigeration power would have to be supplied and capital costs would have to be taken into account. While there are small experimental superconducting transformers being built at this book's publication time, the research so far has not yet led to a commercially feasible and economical proposition.

BIBLIOGRAPHY

Blume, L. F., et al. *Transformer Engineering*. 2d ed. New York: John Wiley & Sons, 1951.

Fink, D. G., and Beaty, H. W., eds. *Standard Handbook for Electrical Engineers*. 11th ed. New York: McGraw-Hill, 1978.

Grossner, N. R. *Transformers for Electronic Circuits*. New York: McGraw-Hill, 1967.

Institute of Electrical and Electronics Engineers. *Standard No. 389–1978: Tests for Electronics Transformers and Inductors*. New York, 1978.

———. *Standard No. 390–1975: Low Power Pulse Transformers*. New York, 1975.

———. *Standard No. 393–1977: Test Procedures for Magnetic Cores*. New York, 1977.

Langsdorf, A. S. *Theory of Alternating-Current Machinery*. 2d ed. New York: McGraw-Hill, 1955.

Lawrence, R. R., and Richards, H. E. *Principles of Alternating-Current Machinery*. 4th ed. New York: McGraw-Hill, 1953.

Say, M. G. *Alternating Current Machines*. New York: John Wiley & Sons, Halsted Press, 1976.

Westinghouse Electric Corporation Central Station Engineers. *Electrical Transmission and Distribution Reference Book*. 4th ed. East Pittsburgh, Pa.: Westinghouse Electric Corporation, 1964.

[2]Feldman, J. M., B. Cogbill, and M. S. Sarma, "Superconducting Windings in a Power Transformer—An old question with a new answer—Part II: Some of the practical problems," IEEE Paper No. A77-020-1, February 1977; Wilkinson, K. J. R., "Superconductive Windings in Power Transformers," Proc. IEEE, vol. 110, p. 2271, 1963.

PROBLEMS

4-1.

The sinusoidal flux in the core of a transformer is given by

$$\phi = 0.002 \sin 377t \text{ Wb}$$

This flux links a primary winding of 100 turns.

      a. Determine the induced voltage in the coil as a function of time.
      b. Find the rms value of the induced voltage.

4-2.

p.99

A sinusoidal voltage $v = 100 \sin 377t$ is applied to a 200-turn transformer winding. Determine the rms values of the impressed voltage and the flux produced in the core, neglecting the leakage flux and the winding resistance.

4-3.

A transformer is rated 10 kVA, 220:110 V. Consider it as an ideal transformer.
      a. Find the turns ratio and the winding current ratings.
      b. If a 2-ohm load resistance is connected across the 110-volt winding, determine the current in the high-voltage and low-voltage windings.
      c. What is the equivalent load resistance referred to the 220-volt side?

4-4.

A transmission line is to be terminated with a resistance of 200 ohms to minimize standing waves. If the actual load resistance is 500 ohms, find the required transformer ratio to match the actual load to the transmission line.

4-5.

The coupled-circuit model shown in the figure below is often used to represent a two-winding air-core transformer with negligible core loss:

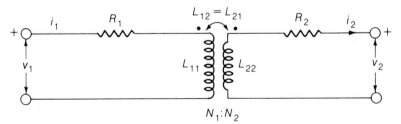

L_{11} and L_{22} are the self-inductances of the primary and secondary windings, respectively; $M = L_{12} = L_{21}$ is the mutual inductance between the two windings.

      a. Write the flux-linkage equations (for λ_1 and λ_2) and the voltage equations (for v_1 and v_2) as a function of the currents (i_1 and i_2) and circuit parameters.

b. Show that the above circuit is equivalent to the linear circuit given below if

$$L_{11} = L_{l1} + L_{m1}; L_{22} = L_{l2} + L_{m2}$$

and

$$L_{m1} = \frac{N_1}{N_2} L_{12}; L_{m2} = \frac{N_2}{N_1} L_{12}$$

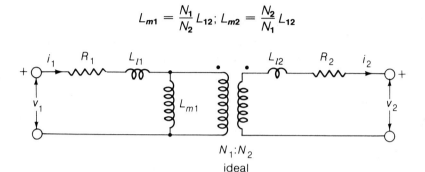

$N_1:N_2$
ideal

c. The coupled-circuit model is rarely used with iron-core transformers because the high relative permeability of the core will lead to large values of L_{11}, L_{22}, and L_{12}. What will happen to all inductances in the ideal case of an infinitely permeable core?

4-6.

use 4.2.7

For a single-phase, 60-Hz transformer rated 500 kVA, 2,400:480 V, following the notation of Figure 4.2.6 in the text, the equivalent circuit impedances in ohms are given below:

$R_1 = 0.06$	$R_2 = 0.003$	$R_c = 2,000$
$X_{l1} = 0.3$	$X_{l2} = 0.012$	$X_m = 500$

The load connected across the low-voltage terminals draws rated current at 0.8 lagging power factor with rated voltage at the terminals.

a. Calculate the high-voltage winding terminal voltage, current, and power factor.
b. Compute the transformer efficiency for the given condition of load.
c. Determine the transformer series equivalent impedance in terms of the high-voltage and low-voltage sides.
d. Considering the T-equivalent circuit, find the Thévenin equivalent impedance of the transformer under load as seen from the primary high-voltage terminals.

4-7.

A 20-kVA, 2,200:220-V, 60-Hz single-phase transformer has the following parameters:

Resistance of the 2,200-volt winding $R_1 = 2.50\ \Omega$
Resistance of the 220-volt winding $R_2 = 0.03\ \Omega$
Leakage reactance of the 2,200-volt winding $X_{l1} = 10\ \Omega$
Leakage reactance of the 220-volt winding $X_{l2} = 0.1\ \Omega$
Magnetizing reactance on the 2,200-volt side $x_m = 25,000\ \Omega$

a. Draw the equivalent circuit of the transformer referred to the high-voltage side and referred to the low-voltage side. Label impedances numerically in ohms.
b. Draw the equivalent circuit in per-unit quantities referred to the high-voltage and low-voltage sides.
c. The transformer is supplying 15 kVA at 220 volts and a lagging power factor of 0.85. Determine the required voltage at the high-voltage terminals of the transformer.

4-8.

A single-phase, 60-Hz transformer is rated 110/440 V, 2.5 kVA. The leakage reactance of the transformer measured from its low-voltage side is given as 0.08 ohm.

a. Express the leakage reactance in per unit.
b. If the leakage reactance had been measured on the high-tension side, what would be its ohmic value and per-unit value?

4-9.

The no-load test on a certain transformer with its secondary terminals open yields the following data: 400 volts, 2.5 amps, 150 watts.
Neglecting the winding resistance and leakage flux, calculate

a. the core-loss current.
b. the magnetizing current.
c. the no-load power factor.

4-10.

The short-circuit test on a certain transformer with its low-voltage winding short-circuited yields the following data: 220 volts, 1.4 amps, 200 watts. Neglecting the exciting current of the transformer, determine the series equivalent impedance referred to its high-voltage and low-voltage sides, if the voltage rating is 7,200:120 V.

4-11.

A 3-kVA, 220:110-V, 60-Hz single-phase transformer yields the following test data:

Open-circuit test: 200 V, 1.4 A, 50 W
Short-circuit test: 4.5 V, 13.64 A, 30 W

Determine the efficiency when the transformer delivers a load of 2 kVA at 0.85 power factor lagging.

4-12.

The following data were obtained from tests carried out on a 10-kVA, 2,300:230-V, 60-Hz distribution transformer:

Open-circuit test, with the low-voltage winding excited:
Applied voltage: 230 V
Current: 0.45 A
Input power: 70 W

Short-circuit test, with the high-voltage winding excited:
Applied voltage: 120 V
Current: 4.5 A
Input power: 240 W

a. Determine the equivalent circuit of the transformer referred to its high-voltage side and also referred to its low-voltage side, in actual ohmic values for impedances as well as in per unit.

b. Express the exciting current of the transformer in per unit on the basis of the rated full-load current.

c. Compute the efficiency of the transformer when it is delivering full load at 230 volts and 0.8 power factor lagging.

d. Find the efficiency of the transformer when it is delivering 7.5 kVA at 230 volts and 0.85 power factor lagging.

e. Determine the per-unit rating at which the transformer efficiency is a maximum, and calculate the efficiency corresponding to that load, if the transformer is delivering the load at 230 volts and a power factor of 0.85.

f. The transformer is operating at a constant load power factor of 0.85 on the following load cycle: 0.85 full load for 8 hours, 0.60 full load for 12 hours, no load for 4 hours. Compute the all-day (or energy) efficiency of the transformer.

g. If the transformer is supplying full load at 230 volts and 0.8 lagging power factor, determine the voltage regulation of the transformer. Also, find the power factor at the high-voltage terminals.

4-13.

A 300-kVA transformer has a core loss of 1.5 kW and a full-load copper loss of 4.5 kW.

a. Calculate its efficiency corresponding to 0.25, 0.5, 0.75, 1.00, and 1.25 per-unit loads at unity power factor.

b. Repeat the efficiency calculations for the loads at power factors of 0.8 and 0.6.

c. Determine the per-unit load for which the efficiency is a maximum, and calculate the corresponding efficiencies for the power factors of unity, 0.8, and 0.6.

4-14.

Consider Example 4.4.3, which was worked out in the text. By means of a phasor diagram, determine the load power factor for which the regulation is maximum (i.e., the poorest), and find the corresponding regulation.

4–15.

A single-phase, 3-kVA, 220:110-V, 60-Hz transformer has a high-voltage winding resistance of 0.3 ohm, a low-voltage winding resistance of 0.06 ohm, a leakage reactance of 0.8 ohm on its high-voltage side, and a leakage reactance of 0.2 ohm on its low-voltage side. The core loss at rated voltage is 45 W, and the copper loss at rated load is 100 W. Neglect the exciting current of the transformer.

   a. Find the per-unit voltage regulation when the transformer is supplying full load at 110 volts and 0.9 lagging power factor.

   b. Calculate the efficiency when the transformer delivers 3 kW at 0.95 power factor lagging.

   c. Find the all-day efficiency for the following loading: 3.5 kVA, 0.95 pf, for 3 hours; 3.0 kVA, 0.90 pf, for 8 hours; 2.4 kVA, 0.80 pf, for 4 hours; 1.0 kVA, 0.95 pf, for 9 hours.

✗ 4–16.

A 5-MVA, 66:13.2-kV, three-phase transformer supplies a three-phase resistive load of 4,500 kW at 13.2 kV. What is the load resistance in ohms as measured from the line-to-neutral on the high-voltage side of the transformer, if it is

   a. connected in Y-Y.

   b. reconnected in Y-Δ, with the same high-tension voltage supplied and the same load resistors connected.

4–17.

A three-phase transformer bank, consisting of three 10-kVA, 2,300:230-V, 60-Hz, single-phase transformers connected in Y-Δ, is used to step down the voltage. The loads are connected to the transformers by means of a common three-phase low-voltage feeder whose series impedance is $(0.005 + j0.01)$ ohms per phase. The transformers themselves are supplied by means of a three-phase high-voltage feeder whose series impedance is $(0.5 + j5.0)$ ohms per phase. The equivalent series impedance of the single-phase transformer referred to the low-voltage side is $(0.12 + j0.24)$ ohms. The star point on the primary side of the transformer bank is grounded. The load consists of a heating load of 2 kW per phase and a three-phase induction-motor load of 20 kVA with a lagging power factor of 0.8, supplied at 230 volts line-to-line.

   a. Draw a one-line diagram of the three-phase distribution system described above.

   b. Neglecting the exciting current of the transformer bank, draw the per-phase equivalent circuit of the distribution system.

   c. Determine the line current and the line-to-line voltage at the sending end of the high-voltage feeder.

× 4-18.

Three single-phase transformers are used to form a three-phase transformer bank rated 18 MVA, 13.8Δ-120Y kV. One side of the transformer bank is connected to a 120-kV transmission line, and the other side is connected to a three-phase load of 10 MVA at 13.2 kV and 0.8 lagging power factor.

   a. Determine the MVA and voltage ratings of each of the three single-phase transformers.
   b. Using the base in the transmission line of 120 kV and 20 MVA, find the load impedance in per unit to be used in the line-to-neutral impedance diagram.

$$\frac{V^2}{P} = 720\,\Omega$$

$$\frac{120^2}{20M}$$

4-19.

A single-line diagram of a three-phase transformer bank connected to a load is given below.

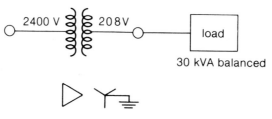

2400 V   208 V   load

30 kVA balanced

3 single-phase transformers
each rated 10 kVA, 2400:120V, 60 Hz.

Find the magnitudes of the line-to-line voltages, line currents, phase voltages, and phase currents on either side of the transformer bank. Determine the primary to secondary ratio of the line-to-line voltages and the line currents.

× 4-20.

A single-line diagram of a part of a three-phase transmission system is given below. Choose a base of 20 MVA, 12.5 kV in the load circuit, and draw the equivalent line-to-neutral impedance diagram in per-unit quantities. Show the base kV in the three parts of the system.

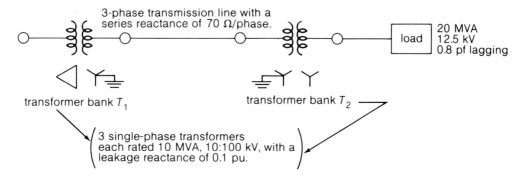

3-phase transmission line with a
series reactance of 70 Ω/phase.

load   20 MVA
12.5 kV
0.8 pf lagging

transformer bank T_1

transformer bank T_2

3 single-phase transformers
each rated 10 MVA, 10:100 kV, with a
leakage reactance of 0.1 pu.

X **4–21.**

A 400-kVA, 11 kV/415-V, three-phase, Δ-Y connected transformer has on rated load an I^2R loss of 2.5 kW for its high-voltage windings and 2.0 kW for its low-voltage windings. The total leakage reactance is 0.06 per unit. Neglect the exciting current.

Determine the per-phase equivalent circuit of the transformer referred to the high-voltage and the low-voltage sides, both in ohmic values of impedances and in per unit.

X **4–22.**

Three single-phase transformers, each rated 15 MVA, 13.2:66 kV, with a leakage reactance of 0.1 pu, are connected to form a three-phase Y-Δ bank, one side of which is connected to a 120-kV transmission line and the other side is connected to three 12-ohm resistors in delta. The star point of the transformer bank is grounded. The series reactance per phase of the transmission line is 80 ohms.

Draw a single-line diagram of the system and obtain the line-to-neutral impedance diagram with per-unit values, for a base voltage of 120 kV in the transmission line and a three-phase voltampere base of 50 MVA.

4–23.

A three-phase, Δ-Y, 11 kV/440-V, 60-Hz, 300-kVA transformer has the following test data:

No-load test with high-voltage side open: 440 V, 20 A, 1.3 kW
Short-circuit test with low-voltage side short-circuited: 630 V, 15.75 A, 3.1 kW

Determine the efficiency and regulation of the transformer when it is delivering full load at 440 volts and 0.8 lagging power factor.

4–24.

A single-phase 10-kVA, 2,300:230-V, 60-Hz, two-winding distribution transformer is connected as an autotransformer to step up the voltage from 2,300 volts to 2,530 volts.

a. Draw a schematic diagram of the arrangement showing all the voltages and currents, while delivering full load.
b. Find the permissible kVA rating of the autotransformer, if the winding currents are not to exceed those for full-load operation as a two-winding transformer. How much of that is transformed by electromagnetic induction?
c. Based on the data given for the two-winding transformer in Problem 4–12, compute the efficiency of the autotransformer corresponding to full load and 0.8 lagging power factor. Comment on why the efficiency of the autotransformer is higher than that of the two-winding transformer.

4–25.

A two-winding, single-phase transformer, rated 3 kVA, 220:110 V, 60-Hz, is connected as an autotransformer to transform a line input voltage of 330 volts to a line output voltage of 110 V, and deliver a load of 2 kW at 0.8 lagging power factor.

Draw the schematic diagram of the arrangement, label all the currents and voltages, and calculate all the quantities involved.

4-26.

The results of three short-circuit tests on a 9,600:2,400:600-V, 60-Hz, single-phase, three-winding transformer are given below:

Winding		Measured on the Excited-Winding Side	
Excited	Short-Circuited	Volts	Amperes
Primary(1)	Secondary(2)	250	52
Primary(1)	Tertiary(3)	800	52
Secondary(2)	Tertiary(3)	200	208

The kVA-ratings of the primary, secondary, and tertiary windings are 1,000, 500, and 500 respectively. Neglect the winding resistances and the exciting current of the transformer.

   a. Determine the per-unit values of the equivalent circuit impedances Z_1, Z_2, and Z_3 of Figure 4.7.2(b) on a 1-MVA rated-voltage base.
   b. Three of these transformers are used to form a three-phase Y-Δ-Δ connected bank. The wye-connected primaries are connected to a 16.6-kV, three-phase supply. If a three-phase short circuit should occur at the terminals of the tertiary windings, find the per-unit values of the steady-state short-circuit currents and of the voltage at the secondary terminals, on a three-phase, 3-MVA rated-voltage base.

4-27.

The rating of a three-winding, three-phase transformer is given below:

   Primary (1): Y-connected, 6.6 kV, 15 MVA
   Secondary (2): Y-connected, 33 kV, 10 MVA
   Tertiary (3): Δ-connected, 2.2 kV, 5 MVA

Neglecting the resistances, the leakage impedances calculated from the short-circuit tests are:

   $Z_{12} = j0.25$ ohm; $Z_{13} = j0.3$ ohm (measured from the primary side)
   $Z_{23} = j9.0$ ohms (measured from the secondary side)

Determine the per-unit impedances Z_1, Z_2, and Z_3 of the equivalent circuit of Figure 4.7.2(b), on a 15-MVA, 6.6-kV base in the primary circuit.

4-28.

The three-phase ratings of a three-winding transformer are given by the following:

   Primary (1): Y-connected, 66 kV, 15 MVA
   Secondary (2): Y-connected, 13.2 kV, 10 MVA
   Tertiary (3): Δ-connected, 2.3 kV, 5 MVA

Neglecting the resistance, the per-unit leakage impedances are:

$$Z_{12} = j0.08 \text{ on 15-MVA, 66-kV base}$$

$$Z_{13} = j0.10 \text{ on 15-MVA, 66-kV base}$$

$$Z_{23} = j0.09 \text{ on 10-MVA, 13.2-kV base}$$

a. Determine the per-unit impedances Z_1, Z_2, and Z_3 of the equivalent circuit of Figure 4.7.2(b), on a 15-MVA, 66-kV base in the primary circuit.
b. Purely resistive loads of 7.5 MW at 13.2 kV and 5 MW at 2.3 kV are connected to the secondary and tertiary sides of the transformer, respectively. Draw the impedance diagram of the system, showing the per-unit impedances for a 66-kV, 15-MVA base in the primary circuit.

4-29.

A three-phase, three-winding, 6,600/400/110-V, Y-Y-Δ connected transformer with a magnetizing current of 5 amps is delivering three-phase loads of 1 MVA at 0.8 lagging power factor connected to the secondary, and 200 kVA at 0.5 leading power factor connected to the tertiary winding. The per-unit impedances of the equivalent circuit of Figure 4.7.2(b) on a 6.6-kV, 1-MVA base are given by

$$Z_1 = 0.005 + j0.03$$

$$Z_2 = 0.006 + j0.025$$

$$Z_3 = 0.008 + j0.033$$

a. Determine the approximate per-unit regulation component associated with the primary alone (ϵ_1), the secondary alone (ϵ_2), and the tertiary alone (ϵ_3).
b. Noting that the resultant of the part-regulations of two circuits together is an addition if power flows from one to the other, and a subtraction otherwise, calculate the per-unit regulations of the primary-secondary (ϵ_{12}), the primary-tertiary (ϵ_{13}), the secondary-tertiary (ϵ_{23}), and the tertiary-secondary (ϵ_{32}).

4-30.

The primary winding of an audio transformer is connected to a source, while its secondary is connected to a speaker that has a resistive impedance of 8 ohms over the frequency range in which it is generally used. The parameters of the transformer equivalent circuit are given below:

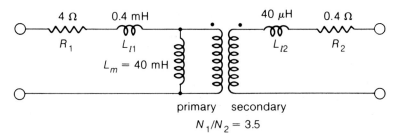

primary   secondary

$N_1/N_2 = 3.5$

   a. It is required to supply a power of 60 watts to the speaker in the mid-frequency range. Neglecting the internal resistance of the source, compute the applied voltage at the source terminals.

   b. Assuming that the same applied voltage is maintained over the whole frequency range, find the frequencies at which the speaker power is reduced to 30 watts.

4-31.

A single-phase, 500-kVA, 2,400:240-V transformer with negligible resistance and 0.05 per-unit equivalent leakage reactance is operated in parallel with another single-phase 250-kVA, 2,400:240-V transformer with negligible resistance and 0.04 per-unit equivalent leakage reactance. These two transformers in parallel operation are to share a load of 750 kVA at 0.8 lagging power factor.

   a. Neglecting the exciting current of the transformers, find the load on each transformer. Which transformer is overloaded and why?

   b. In addition to the data given above, consider the equivalent resistance of the 500-kVA transformer to be 0.01 per unit and that of the 250-kVA transformer to be 0.015 per unit. How would the results change?

5

PRINCIPLES OF ELECTROMECHANICAL ENERGY CONVERSION

The transformer, which we have studied in the last chapter, is an electromagnetic device transmitting electrical energy with a change in the levels of voltage and current from one side to the other. However, it is not an energy-conversion device. In this chapter we shall study the principles of electromechanical energy conversion and their application to simple devices. Electromechanical energy conversion involves the interchange of energy between an electrical and a mechanical system. When the energy is converted from electrical to mechanical form, the device is displaying *motor action. Generator action* involves converting mechanical energy into electrical energy. Electromechanical energy converters embody three essential features: (1) an electric system, (2) a mechanical system, and (3) a coupling field.

Both electric and magnetic fields store energy, and useful mechanical forces can be derived from them. In air or other gas at normal pressure the dielectric strength of the medium restricts the working electric field intensity to about 3×10^6 V/m, and consequently the stored electric energy density to

$$\frac{1}{2}\ \epsilon_0\ E^2 = \frac{1}{2}\frac{10^{-9}}{36\pi}\ (3 \times 10^6)^2 \simeq 40 \text{ J/m}^3$$

where ϵ_0 is the permittivity of free space, given by $10^{-9}/(36\pi)$ or 8.854×10^{-12} F/m, and E is the electric field intensity. This corresponds to a force density of 40 N/m^2. While there is no comparable restriction on magnetic fields, the saturation of ferromagnetic media required to complete the magnetic circuit limits the working magnetic flux density to about 1.6 T, for which the stored magnetic energy density in air is about

$$\frac{1}{2}\frac{B^2}{\mu_0} = \frac{1}{2}\frac{(1.6)^2}{4\pi \times 10^{-7}} \simeq 1 \times 10^6 \text{ J/m}^3$$

where μ_0 is the permeability of free space, and B is the magnetic flux density. As this is nearly 25,000 times as much as for the electric field, almost all industrial electric machines are magnetic in principle and are magnetic-field devices. Because magnetic poles occur in pairs (north and south) and the movement of a conductor through a natural north-south sequence induces an emf that changes direction in accordance with the magnetic polarity (i.e., an alternating emf), the devices are inherently ac machines.

Three basic principles associated with all electromagnetic devices are (1) *induction,* (2) *interaction,* and (3) *alignment.* The essentials for the production of an electromotive force by magnetic means are electric and magnetic circuits, mutually interlinked. The induced emf is given by Faraday's law of induction:

$$e = -\frac{d\lambda}{dt} = -N\frac{d\phi}{dt}$$

which is the same as Equation 3.1.9, stated earlier. The induced emf will be acting in the direction of positive current as shown in Figure 5.0.1(b) with a source (or generator) convention. Sometimes

FIGURE 5.0.1   Circuit conventions: *(a)* load (or sink, or motor) convention (note that the power *into* the circuit is positive when v and i are positive), *(b)* source (or generator) convention (note that the power *delivered* by this circuit to the external circuit is positive when v and i are positive).

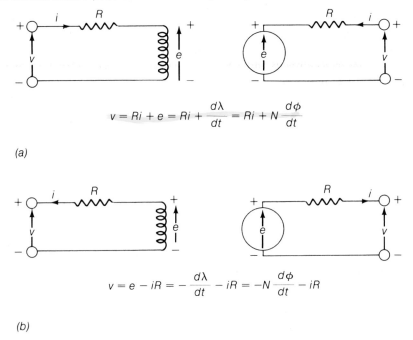

$$v = Ri + e = Ri + \frac{d\lambda}{dt} = Ri + N\frac{d\phi}{dt}$$

(a)

$$v = e - iR = -\frac{d\lambda}{dt} - iR = -N\frac{d\phi}{dt} - iR$$

(b)

it is more convenient to consider the emf as directed in opposition to positive current, as shown in Figure 5.0.1*(a)* with a load (or sink, or motor) convention, in which case

$$e = +\frac{d\lambda}{dt} = +N\frac{d\phi}{dt}$$

which is the same as Equation 4.2.1, stated earlier.

The change of flux linkage in a coil may occur in one of the following three ways:

a. The flux remaining constant, the coil moves through it.
b. The coil remaining stationary with respect to the flux, the flux varies in magnitude with time.
c. The coil may move through a time-varying flux; that is to say, both changes may occur together.

In *(a)*, the flux-cutting rule given by Equation 3.1.12 or 3.1.13 yields the *motional emf* (or the *speed emf*) that is always associated with the conversion of energy between the mechanical and

FIGURE 5.0.2   Principle of interaction.

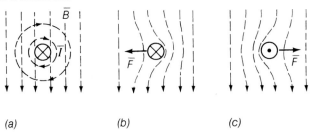

(a)   (b)   (c)

FIGURE 5.0.3   Torque produced by forces caused by interaction of
current-carrying conductors and magnetic fields.

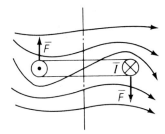

electrical forms. In *(b)*, since there is no motion involved, no energy conversion takes place; Equation 4.2.1 gives the *transformer emf* (or the *pulsational emf*). Both emf components will be present in *(c)*.

Current-carrying conductors, when placed in magnetic fields, experience mechanical force, which can be calculated by Equation 3.1.14 or 3.1.15. Consider the flux density $\overline{B}$ of an undisturbed uniform field, shown in Figure 5.0.2*(a)*, on which the introduction of a current-carrying conductor imposes a corresponding field component, developing the resultant as in Figure 5.0.2*(b)* for the case in which the current is directed into and perpendicular to the plane of the paper, as symbolized by the cross in the figure. In the neighborhood of the conductor, as seen in Figure 5.0.2*(b)*, the resultant flux density is greater than *B* on one side and less than *B* on the other side. Figure 5.0.2*(c)* shows the conditions corresponding to the current's being directed out of and perpendicular to the plane of the paper, as symbolized by the dot. The direction of the mechanical force developed is such that it tends to restore the field to its original undisturbed and uniform configuration, as shown in Figures 5.0.2*(b)* and *(c)*. The force is always in such a direction that the energy stored in the magnetic field is minimized. Figure 5.0.3 shows a one-turn coil in a magnetic field, and illustrates how torque is produced by forces caused by interaction of current-carrying conductors and magnetic fields.

FIGURE 5.0.4   Principle of alignment.

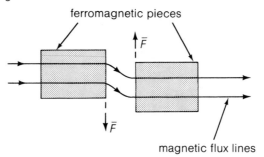

ferromagnetic pieces

magnetic flux lines

Pieces of highly permeable material such as iron situated in an ambient medium of low permeability, such as air, in which a magnetic field is established, experience mechanical forces that tend to align them with the field direction in such a way that the reluctance of the system is minimized. This principle of alignment is illustrated in Figure 5.0.4, showing the direction of forces. The force is always in such a direction that the net magnetic reluctance is reduced, and the magnetic flux path is shortened.

With this general background we shall proceed to evaluate the electromagnetic forces and torques associated with magnetic-field systems.

5.1   FORCES AND TORQUES IN MAGNETIC FIELD SYSTEMS

A method of computing the electromagnetic force, based on the Lorentz force equation, is given by Equations 3.1.14 and 3.1.15. However, this is applicable only when the flux density is known and the geometry of the system is rather simple. In general, the *energy method* presented here is much more convenient. The energy method of determining electromagnetic forces depends on the principle of conservation of energy. Energy can be neither created nor destroyed, even though, within an isolated system, energy may be converted from one form to another form, and transferred from an energy source to an energy sink. The total energy in the system must be a constant. Since we shall be concerned with low frequencies and low velocities (i.e., velocities much less than that of light), it is fair to assume that no energy is radiated into space, no mass is converted into energy, and no dielectric losses exist.

The energy conversion process involves interchange between electrical and mechanical energy via the stored energy in the magnetic field. This stored energy, which can be determined for any configuration of the system, is a *state function* defined solely by the functional relationships between variables and by the final values of these variables. Thus the energy method provides a powerful tool for determining the coupling forces of electromechanics.

FIGURE 5.1.1 Schematic representation of a lossy electromechanical system.

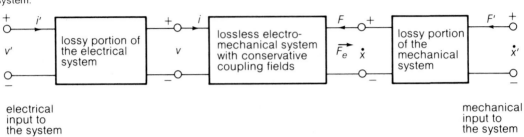

FIGURE 5.1.1 Schematic representation of a lossy electromechanical system.

Figure 5.1.1 shows a schematic representation of a lossy electromechanical system divided into its simpler component parts. Note that all types of dissipation have been excluded to form a *conservative* (or *lossless*) energy-conversion part that can be described by state functions to yield the electromechanical coupling terms in electromechanics. A property of a conservative system is that its energy is a function of its state only and is described by the same independent variables that describe the state. State functions at a given instant of time depend solely on the state of the system at that instant of time and not on past history; they are independent of how the system is brought to that particular state.

For the case of a *sink* of electrical energy, such as an electric motor, the principle of conservation of energy allows one to write:

$$
\begin{array}{l}
\text{Electrical input} \\
\text{energy from source}
\end{array}
=
\begin{array}{l}
\text{Mechanical output} \\
\text{energy to load}
\end{array}
+
\begin{array}{l}
\text{Increase in stored} \\
\text{field energy}
\end{array}
+
\begin{array}{l}
\text{Energy loss} \\
\text{converted to heat}
\end{array}
\qquad (5.1.1)
$$

The energy losses associated with this form of energy conversion are *(a)* the energy loss due to the resistances of the circuits, *(b)* the energy loss due to friction and windage associated with motion, and *(c)* the energy loss associated with the coupling field. Considering the coupling field to be a magnetic field, the field losses are due to hysteresis and eddy-current losses, i.e., the core losses in the magnetic system. Since these losses are usually small, they may be neglected, or their effect may be included in the lossy portion of the electrical system. Then, considering only the conservative (or *lossless*) portion of the system, one has

$$
\begin{array}{l}
\text{Electrical energy from} \\
\text{source minus electrical} \\
\text{system losses}
\end{array}
=
\begin{array}{l}
\text{Mechanical energy to load} \\
\text{plus mechanical system} \\
\text{losses}
\end{array}
+
\begin{array}{l}
\text{Increase in magnetic-} \\
\text{coupling field energy} \\
\text{stored}
\end{array}
\qquad (5.1.2)
$$

or

$$
\begin{array}{l}
\text{Input electrical energy} \\
\text{to the lossless} \\
\text{electromechanical system}
\end{array}
= \text{Mechanical work done} + \text{Increase in stored energy}
\qquad (5.1.3)
$$

FIGURE 5.1.2   A simple electromechanical system.

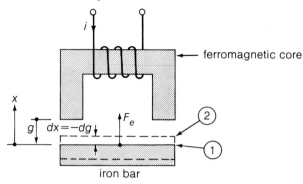

iron bar

The increase in energy stored in the magnetic field is considered here, while neglecting the energy stored in the electric field. In incremental form, in time dt, Equation 5.1.3 can be expressed as

$$dW_e = dW + dW_m \qquad (5.1.4)$$

or, referring to Figure 5.1.1, it may be written as

$$v \, i \, dt = -F \, dx + dW_m \qquad (5.1.5)$$

where $(-F \, dx)$ corresponds to the mechanical output of the lossless electromechanical system, which may also be expressed as $(F_e \, dx)$, in which F_e is the mechanical force of electrical origin due to magnetic field coupling. Then it follows

$$F_e \, dx = v \, i \, dt - dW_m \qquad (5.1.6)$$

where dW_m is the increase in energy stored in the magnetic field, and dx is an arbitrary incremental displacement. Since v, which is the same as the induced emf in the lossless electromechanical system, can be expressed in terms of the flux linkage λ by means of Faraday's law of induction as

$$v = \frac{d\lambda}{dt} \qquad (5.1.7)$$

Equation 5.1.6 can be rewritten as

$$F_e \, dx = i \, d\lambda - dW_m \qquad (5.1.8)$$

In a simple electromechanical system with a singly-excited electrical system and a mechanical system consisting of only one-dimensional motion, such as the one shown in Figure 5.1.2, the independent variable on the electrical side is either the current i or the flux linkage λ. The λ-i characteristic of the nonlinear magnetic system with no core loss is given by a single-valued nonlinear relationship, typically shown in Figure 3.4.1. Thus, if the independent variables are the current i and the coordinate x, then the flux linkage λ is a function of both i and x:

$$\lambda = \lambda(i, x) \tag{5.1.9}$$

in which case $d\lambda$ can be expressed as

$$d\lambda = \frac{\partial \lambda}{\partial i} \, di + \frac{\partial \lambda}{\partial x} \, dx \tag{5.1.10}$$

where di is an arbitrary incremental change in i. Also, since the stored magnetic energy is also a function of i and x, it follows that

$$dW_m = \frac{\partial W_m}{\partial i} \, di + \frac{\partial W_m}{\partial x} \, dx \tag{5.1.11}$$

Substituting Equations 5.1.10 and 5.1.11 in Equation 5.1.8, one gets

$$F_e \, dx = \left(- \frac{\partial W_m}{\partial x} + i \frac{\partial \lambda}{\partial x} \right) dx + \left(- \frac{\partial W_m}{\partial i} + i \frac{\partial \lambda}{\partial i} \right) di \tag{5.1.12}$$

In order that the force F_e be independent of the change in the current i (and λ) during the arbitrary displacement, the coefficient of di in Equation 5.1.12 must be zero. Consequently, the force F_e is always given by

$$F_e = - \frac{\partial W_m(i, x)}{\partial x} + i \frac{\partial \lambda(i, x)}{\partial x} \tag{5.1.13}$$

in which W_m and λ are functions of independent variables i and x. It is possible to express Equation 5.1.13 in a simpler form in terms of magnetic coenergy W_m' based on Equations 3.4.7 and 3.4.8:

$$F_e = + \frac{\partial W_m'(i, x)}{\partial x} \tag{5.1.14}$$

If, on the other hand, the independent variables are chosen as λ and x, it follows that

$$i = i(\lambda, x) \tag{5.1.15}$$

$$di = \frac{\partial i}{\partial \lambda} \, d\lambda + \frac{\partial i}{\partial x} \, dx \tag{5.1.16}$$

TABLE 5.1.1 MECHANICAL FORCE OF ELECTRICAL ORIGIN CAUSED BY THE MAGNETIC COUPLING FIELD

Stored magnetic energy	$W_m = \int_0^\lambda i \, d\lambda$	(3.4.6)
Magnetic coenergy	$W_m' = \int_0^i \lambda \, di$	(3.4.7)
Relation between energy and coenergy	$W_m + W_m' = \lambda i$	(3.4.8)
Conservation of energy principle applied to conservative coupling fields for an arbitrary displacement dx	$F_e \, dx = i \, d\lambda - dW_m$	(5.1.8)

Independent Variables	Electromechanical Coupling Force Evaluated from Stored Magnetic Energy	Electromechanical Coupling Force Evaluated from Magnetic Coenergy
Current i Coordinate x	$F_e = -\dfrac{\partial W_m(i,x)}{\partial x} + i\dfrac{\partial \lambda(i,x)}{\partial x}$ (5.1.13)	$F_e = +\dfrac{\partial W_m'(i,x)}{\partial x}$ (5.1.14)
Flux Linkage λ Coordinate x	$F_e = -\dfrac{\partial W_m(\lambda,x)}{\partial x}$ (5.1.19)	$F_e = \dfrac{\partial W_m'(\lambda,x)}{\partial x} - \lambda\dfrac{\partial i(\lambda,x)}{\partial x}$ (5.1.20)

Note: For the case of a rotational electromechanical system, the force F_e and the linear displacement dx are to be replaced by the torque T_e and the angular displacement $d\theta$, respectively.

$$dW_m = \frac{\partial W_m}{\partial \lambda} d\lambda + \frac{\partial W_m}{\partial x} dx \qquad (5.1.17)$$

$$F_e \, dx = i \, d\lambda - \frac{\partial W_m}{\partial \lambda} d\lambda - \frac{\partial W_m}{\partial x} dx \qquad (5.1.18)$$

or

$$F_e = -\frac{\partial W_m(\lambda, x)}{\partial x} \qquad (5.1.19)$$

Note that W_m in Equation 5.1.19 is a function of independent variables λ and x. Based on Equation 3.4.8, Equation 5.1.19 may also be written as

$$F_e = \frac{\partial W_m'(\lambda, x)}{\partial x} - \lambda \frac{\partial i(\lambda, x)}{\partial x} \qquad (5.1.20)$$

Table 5.1.1 summarizes the relations associated with the mechanical force of electrical origin caused by the magnetic coupling field.

   Although we have considered a simple system that can be described with only one electrical and one mechanical variable, an electromechanical system in general may need several electrical and mechanical variables for its modeling description, in which case the expressions given in Table 5.1.1 need to be generalized. The results of Table 5.1.1 are completely general and independent of any electrical source variations. A similar development can be made for determining mechanical forces due to electric field coupling in an electromechanical system.

   It should be observed that the force F_e is independent of the variations of λ and i during the arbitrary displacement; however, the changes in λ and i follow the functional relationship given by the λ-i characteristic of the magnetic system, which is also the reason that only one of two variables (λ or i) can be treated as independent. Since the stored magnetic energy is a state function, it does not matter how the configuration of the system is arrived at.

   Figure 5.1.3(a) illustrates a differential movement of the operating point in a λ-i diagram corresponding to a differential displacement dx of the iron bar of Figure 5.1.2 made at low speed, i.e., at constant current. The increase in coenergy from position 1 to position 2 can be observed, and Equation 5.1.14 can be applied to evaluate F_e readily (with i treated as a constant), when it is more convenient to express λ as a function of i and x.

   Figure 5.1.3(b) illustrates a differential movement of the operating point in the λ-i diagram corresponding to a differential displacement dx of the iron bar of Figure 5.1.2 made at high speed, i.e., at constant flux linkage (or voltage). The decrease in energy from position 1 to position 2 can be observed, and Equation 5.1.19 can be applied to evaluate F_e readily (with λ treated as a constant), when it is more convenient to express the coil current i as a function of λ and x.

   It must be emphasized that the holding of i or λ as a constant is a mathematical restriction imposed by the selection of independent coordinates, and has nothing to do with the terminal electrical constraints. The generality of the force expressions holds good even when the change from position 1 to position 2 follows a general path, shown in Figure 5.1.3(c), in which neither condition of constant i or λ applies.

   For a linear magnetic system, however, the λ-i characteristic is a straight line, in which case the magnetic energy and coenergy are always equal in magnitude. Equation 3.5.2 applies:

$$W_m = \frac{1}{2} i\lambda = \frac{Li^2}{2} \qquad (3.5.2)$$

from which

$$F_e = \frac{1}{2} i^2 \frac{\partial L}{\partial x} = -\frac{1}{2} \lambda \frac{\partial i}{\partial x} \qquad (5.1.21)$$

follows from the force expressions given in Table 5.1.1.

   It should now be clear that, in order for energy conversion to take place, the electromechanical device must have at least one component capable of storing energy, and this stored energy must be a function of the space variable. While the foregoing analysis is concerned with the force F_e and the linear displacement dx, for the case of a rotational electromechanical system, the torque T_e and the angular displacement $d\theta$ must be introduced. We shall now take up a few examples to illustrate the application of the force or torque expressions.

FIGURE 5.1.3 Energy balance in a nonlinear electromechanical system: *(a)* constant current operation, *(b)* constant flux linkage (or voltage) operation, *(c)* a general case.

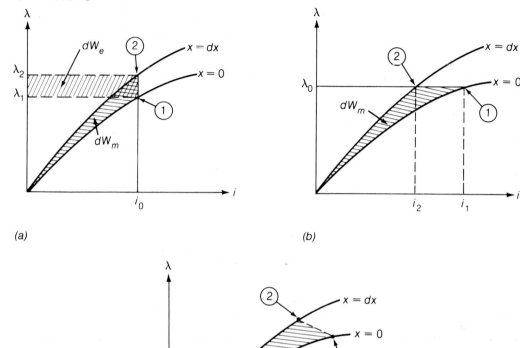

(a)

(b)

(c)

EXAMPLE 5.1.1

The λ-i relationship for an electromechanical system shown in Figure 5.1.2 is given by

$$\lambda = \frac{0.1 i^{1/2}}{g}$$

which holds good between the limits $0 < i < 4$ A and $0.04 < g < 0.10$ m. If the current is maintained at 2A, find the mechanical force (of electrical origin caused by the magnetic coupling field) on the iron bar for $g = 0.06$ m.

SOLUTION

Noting that the λ-i relationship is nonlinear, and that λ is given as a function of the current i and the space variable coordinate g, it is convenient to apply Equation 5.1.14, for which we need to evaluate the magnetic coenergy W_m' of the system by Equation 3.4.7.

$$W_m' = \int_0^i \lambda \, di = \int_0^i \frac{0.1 i^{1/2}}{g} \, di = \frac{0.1}{g} \frac{2}{3} i^{3/2} \text{ Joules}$$

$$F_e = \frac{\partial W_m'(i, g)}{\partial g} = \frac{\partial}{\partial g}\left[\frac{0.1}{g} \frac{2}{3} i^{3/2}\right] = -\frac{0.1}{g^2} \frac{2}{3} i^{3/2}$$

For $g = 0.06$ m and $i = 2$ A, one obtains

$$F_e = -\frac{0.1}{0.06^2} \times \frac{2}{3} \times (2)^{3/2} = -52.37 \text{ newtons}$$

The negative sign indicates that the force F_e acts in such a direction as to decrease the air-gap length g. Note that a positive displacement dx in Figure 5.1.2 corresponds to a reduction dg in the air-gap length, i.e., $dx = -dg$.

This problem may alternatively be solved by expressing i as a function of λ and g, evaluating the magnetic energy W_m, and making use of Equation 5.1.19:

$$i = \left(\frac{\lambda g}{0.1}\right)^2$$

$$W_m = \int_0^\lambda i \, d\lambda = \int_0^\lambda \frac{g^2}{0.1^2} \lambda^2 \, d\lambda = \frac{g^2}{0.1^2} \frac{\lambda^3}{3}$$

$$F_e = -\frac{\partial W_m(\lambda, g)}{\partial g} = -\frac{\partial}{\partial g}\left(\frac{g^2}{0.1^2} \frac{\lambda^3}{3}\right) = -\frac{\lambda^3}{3} \cdot \frac{2g}{0.1^2}$$

When $i = 2$ A and $g = 0.06$ m,

$$\lambda = \frac{0.1 \times 2^{1/2}}{0.06}$$

and

$$F_e = -\frac{0.1^3 \times 2^{3/2}}{0.06^3} \times \frac{1}{3} \times \frac{2 \times 0.06}{0.1^2}$$

$$= -\frac{0.1}{0.06^2} \times \frac{2}{3} \times 2^{3/2} = -52.37 \text{ N}$$

which is the same as the result obtained before. The selection of the energy or coenergy function as a basis for analysis is a matter of convenience, depending upon the initial description of the system and the desired variables in the result.

EXAMPLE 5.1.2

Let the λ-i relationship for the electromechanical system shown in Figure 5.1.2 be given by

$$i = \lambda^{1/2} + 20\lambda(x - 0.1)^2, \quad x < 0.1 \text{ m}$$

Compute the force on the iron bar at $x = 0.05$ m in terms of λ.

SOLUTION

Since the λ-i relationship is nonlinear, and i is given as a function of λ and x, it is convenient to evaluate $W_m(\lambda, x)$ and apply Equation 5.1.19.

$$W_m = \int_0^\lambda i \, d\lambda = \frac{2}{3} \lambda^{3/2} + 10\lambda^2(x - 0.1)^2$$

$$F_e = -\frac{\partial W_m(\lambda, x)}{\partial x} = -20\lambda^2(x - 0.1)$$

At $x = 0.05$ m, $F_e = +\lambda^2$, which shows that the force is proportional to the square of the voltage, assuming of course that the leakage is neglected. The positive sign indicates that the force F_e acts in such a direction as to increase the coordinate x in its positive direction, or to decrease the air-gap length g. In this problem, it is rather inconvenient to express λ as a function of i and x.

EXAMPLE 5.1.3

The magnetic structure shown with dimensions in the following figure is made out of a ferromagnetic material that has negligible reluctance. The rotor is free to rotate about a vertical axis. Neglect leakage and fringing.

a. Obtain an expression for the torque acting on the rotor.
b. Calculate the torque for a current of 1.5 amps and the dimensions given with the figure.
c. If the maximum flux density in the airgap is to be limited to 1.5 Wb/m² because of saturation in the ferromagnetic structure, compute the maximum torque of the device.

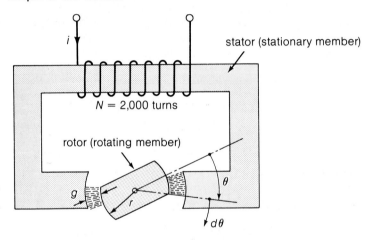

Axial length perpendicular to the plane of the paper = h = 0.05 m
Length of a single airgap = g = 0.004 m
Radius of rotor face = r = 0.04 m

SOLUTION

a. Choosing as independent variables the current i and the space coordinate θ, the expression for the torque is given by

$$T_e = \frac{\partial W_m{}'(i, \theta)}{\partial \theta}$$

Since the air-gap region is linear, the magnetic energy or coenergy density in the air gap is given by Equation 3.4.3:

$$w_m = w_m' = \frac{B_g^2}{2\mu_0} = \frac{\mu_0 H_g^2}{2} \text{ joules/m}^3$$

where B_g is the air-gap flux density, H_g is the air-gap field intensity, and μ_0 is the permeability of free space. Noting that the total length of the air gap is $2g$, the volume of the overall air-gap region is calculated as

$$2gh(r + 0.5g)\,\theta \text{ m}^3$$

where θ is the angle in radians between the stator-pole tip and the adjacent rotor-pole tip as shown in the figure, and $[(r + 0.5g)\,\theta]$ is the mean arc length in the air gap. Then

$$W_m' = \mu_0\, H_g^2\, gh(r + 0.5g)\,\theta \text{ joules}$$

$$T_e = \mu_0\, H_g^2\, gh(r + 0.5g)$$
$$= \frac{B_g^2}{\mu_0}\, gh(r + 0.5g) \text{ newton - meters}$$

The torque acts in such a direction as to align the rotor pole faces with the stator pole faces, in the positive direction of θ as shown. The relationship between the current and the air-gap field intensity H_g is given by

$$Ni = 2g\, H_g$$

or

$$H_g = \frac{Ni}{2g}$$

Making use of the above, the mechanical torque T_e of electrical origin caused by the magnetic coupling field may be expressed as

$$T_e = \frac{\mu_0\, N^2\, i^2}{4g^2}\, gh(r + 0.5g)$$

or

$$T_e = \frac{i^2}{2}\left[\frac{\mu_0\, N^2}{2g}\, h(r + 0.5g)\right]$$

For a linear magnetic system, T_e may also be calculated from Equation 5.1.21 by simply replacing F_e and dx by T_e and $d\theta$, respectively:

$$T_e = \frac{i^2}{2}\frac{dL}{d\theta}$$

For our problem,

$$L = \frac{N^2}{R} = \frac{\mu_0 A N^2}{l}$$

where

A = cross-sectional area of the air-gap region
$\quad = h(r + 0.5\ g)\ \theta,$

and

$l = 2\ g,$ the total air-gap length

The same result for T_e is obtained by substituting and working out the details.
b. The self-inductance L in terms of θ is given by

$$L = \frac{4\pi \times 10^{-7} \times 0.05(0.04 + 0.002) \times 2{,}000^2}{0.008} \theta \text{ henry}$$

or

$$L = 1.32\theta$$

$$T_e = \frac{i^2}{2} \cdot \frac{dL}{d\theta} = \frac{(1.5)^2}{2} \times 1.32 = 1.485 \text{ newton-meters}$$

Note that the torque acts to increase the inductance by pulling on the rotor so as to reduce the reluctance of the magnetic path linking the coil.
    The air-gap flux density B_g corresponding to a current of 1.5 amps is given by

$$B_g = \mu_0 H_g = \frac{\mu_0 Ni}{2g} = \frac{4\pi \times 10^{-7} \times 2{,}000 \times 1.5}{2 \times 0.004} = 0.47 \text{ T}$$

c. Corresponding to $B_{g_{max}}$ of 1.5 Wb/m^2, the current is calculated as

$$i_{max} = \frac{B_g(2g)}{\mu_0 N} = \frac{1.5 \times 0.008}{4\pi \times 10^{-7} \times 2,000} = 4.77 \text{ A}$$

$$T_{e_{max}} = \frac{i^2}{2} \frac{dL}{d\theta} = \frac{4.77^2}{2} \times 1.32 = 15.02 \text{ N·m}$$

This may also be obtained by directly substituting in the torque expression given in terms of B_g.

5.2 SINGLY EXCITED AND MULTIPLY EXCITED MAGNETIC FIELD SYSTEMS

Before we proceed to analyze magnetic field systems excited by more than one electrical circuit, let us consider an elementary *reluctance machine* that is singly excited, carrying only one winding on its stationary member, called the stator. The device is very similar to the one analyzed in Example 5.1.3. But now we would like to consider sinusoidal excitation to the stator winding, while the rotor is free to rotate about an axis normal to the plane of the paper. For notational convenience, the diagram is redrawn in Figure 5.2.1(a), showing an elementary rotating reluctance machine. We shall assume that the reluctances of the stator and rotor iron are negligible; also, we shall neglect the leakage and fringing.

The stator and rotor poles are so shaped that the reluctance varies sinusoidally about a mean value as shown in Figure 5.2.1(b). $\mathbf{R}_d$ is the reluctance of the magnetic system when the rotor is in the direct-axis position ($\theta = 0$), and $\mathbf{R}_q$ is the reluctance when the rotor is in the quadrature-axis position ($\theta = \pi/2$). For each revolution of the rotor, there are two cycles of reluctance. The space variation of inductance is also of double frequency, since the inductance is inversely proportional to the reluctance. The inductance of the stator winding as a function of space coordinate θ measured from the direct axis, as shown in Figure 5.2.1(a) is given by

$$L(\theta) = L_0 + L_2 \cos 2\theta \qquad L = \frac{\lambda}{i} \tag{5.2.1}$$

which is sketched in Figure 5.2.1(c). Let the stator coil excitation be

$$i_s = I_s \sin \omega_s t \tag{5.2.2}$$

We shall investigate the instantaneous and average electromagnetic torques produced because of this sinusoidal excitation whose angular frequency is ω_s.

FIGURE 5.2.1   A singly-excited magnetic field system with a rotor: *(a)* an elementary rotating reluctance machine, *(b)* reluctance variation with rotor position, *(c)* inductance variation with rotor position.

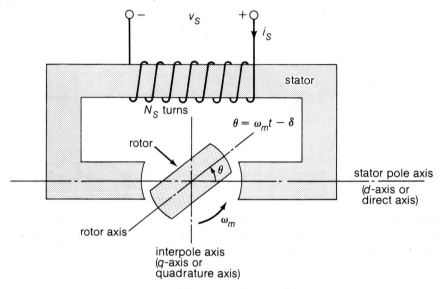

(a)

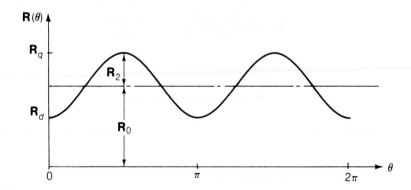

(b)

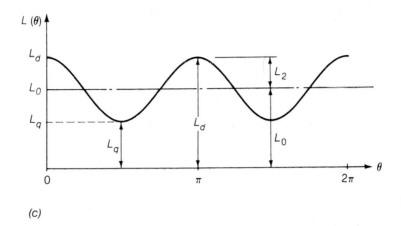

(c)

The electromagnetic torque can be found by Equation 5.1.14 from the coenergy in the magnetic field of the air-gap region, since the independent variables are the current i and the space coordinate θ. Since the air-gap region is linear, Equations 3.5.2 and 5.1.21 apply:

$$W_m' = W_m = \frac{1}{2} L(\theta) \, i_s^2 \qquad (5.2.3)$$

and

$$T_e = \frac{\partial W_m'(i_s, \theta)}{\partial \theta} = \frac{1}{2} i_s^2 \frac{\partial L(\theta)}{\partial \theta} \qquad (5.2.4)$$

Substituting the current and inductance variations, one obtains

$$T_e = -I_s^2 \, L_2 \sin 2\theta \, \sin^2 \omega_s t \qquad (5.2.5)$$

Let the rotor now be allowed to rotate at an angular velocity ω_m, so that at any instant θ is given by

$$\theta = \omega_m t - \delta \qquad (5.2.6)$$

where $\theta = -\delta$ is the angular position of the rotor at $t = 0$, when the current i_s is zero.

Equation 5.2.5 for the instantaneous torque, as a function of time, may be rearranged in terms of ω_m and ω_s by making use of the following trigonometric identities:

$$\sin^2 A = \frac{1 - \cos 2A}{2}$$

and

$$\sin A \cos B = \frac{1}{2} \sin(A + B) + \frac{1}{2} \sin (A - B)$$

The instantaneous electromagnetic torque is then given by

$$
T_e = -\frac{1}{2} I_s^2 L_2 \Big\{ \sin 2(\omega_m t - \delta)
$$
$$
-\frac{1}{2} \sin 2[(\omega_m + \omega_s) t - \delta]
$$
$$
-\frac{1}{2} \sin 2[(\omega_m - \omega_s) t - \delta] \Big\}
$$

(5.2.7)

The above torque expression consists of three sinusoidally time-varying terms of frequencies $2\omega_m$, $2(\omega_m + \omega_s)$ and $2(\omega_m - \omega_s)$. The time-average value of these three sine terms is zero unless, in one of them, the coefficient of t becomes zero. Since $\omega_m \neq 0$, the necessary condition for nonzero time-average torque is then

$$|\omega_m| = |\omega_s|$$

(5.2.8)

corresponding to which

$$(T_e)_{av} = -\frac{1}{4} I_s^2 L_2 \sin 2\delta$$

(5.2.9)

Expressed in terms of L_d and L_q, the maximum and minimum values of inductance as shown in Figure 5.2.1(c), known as the direct-axis inductance and quadrature-axis inductance, respectively,

$$(T_e)_{av} = -\frac{1}{8} I_s^2 (L_d - L_q) \sin 2\delta$$

(5.2.10)

A number of conclusions can be drawn from a closer examination of the preceding analysis. The machine develops an average torque only at one particular speed, given by Equation 5.2.8, for either direction of rotation. This particular speed is known as the *synchronous speed*, when the speed of mechanical rotation in radians per second is equal to the angular frequency of the electrical source. Since the torque in this particular electromechanical energy converter is due to

FIGURE 5.2.2   Variation of the electromagnetic torque developed by a synchronous reluctance machine.

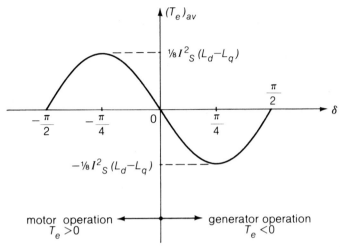

the variation of reluctance with rotor position, the device is known as a *synchronous reluctance machine*. As seen from Equation 5.2.10, the torque is zero if $L_d = L_q$, i.e., if there is no inductance or reluctance variation with rotor position.

Figure 5.2.2 shows the variation of the average electromagnetic torque developed by the machine as a function of the angle δ, which is known as the *torque angle*. The angle δ gives a measure of the torque. For $\delta \not< 0$, $(T_e)_{av} > 0$; that is, the developed torque acts in the direction of rotation, and the machine operates as a motor. This torque tends to maintain the speed of the rotor against friction, windage, and any external load torque applied to the rotor shaft. The maximum torque for motor operation occurs at $\delta = -\pi/4$, and it is known as the *pull-out torque*. Any load requiring a torque greater than the maximum torque results in unstable operation of the machine; the machine pulls out of synchronism and comes to a standstill.

For $\delta > 0$, $(T_e)_{av} < 0$; that is, the torque developed opposes the rotation; an external driving torque drives the machine against the electromagnetically developed torque to maintain the rotor at synchronous speed. The mechanical energy supplied to the system, after meeting the friction and windage losses, is converted to electrical energy; that is to say the machine operates as a generator. However, this can occur only if the stator winding is already connected initially to an ac source, which becomes a sink as the external driving torque is applied and the machine starts to generate. The maximum torque for generator operation occurs at $\delta = \pi/4$. If the driving torque externally applied to the shaft exceeds the sum of the developed torque and that due to friction and windage, then the machine tends to be driven above synchronous speed, and may run away unless the speed of the prime mover is controlled. However, under such conditions, continuous energy conversion ceases. The machine can develop only a certain maximum power, and there is a limit to the rate of energy conversion.

It should be emphasized that the singly excited synchronous reluctance motor cannot start by itself. Also, under excitation from a source of constant amplitude and sinusoidally time-varying voltage, the coil current can be shown to contain a third-harmonic component because of the variation of reluctance with rotor position. Similarly, with sinusoidal current excitation given by Equation 5.2.2, the induced voltage in the stator coil will contain a third-harmonic component. This undesirable feature makes reluctance machines less useful as practical generators and limits their size as motors. However, reluctance machines (modified to develop starting torque) are popularly used to drive electric clocks, record players, and other small mechanisms that need constant speed.

EXAMPLE 5.2.1

In continuation with the analysis of the elementary rotating reluctance machine that is presented above, with the assumed current-source excitation of Equation 5.2.2, find the voltage induced in the stator winding at synchronous speed.

SOLUTION

$$e_s = p \lambda_s = \frac{d}{dt} [L(\theta) \, i_s]$$

where p is the time-derivative operator (d/dt). Noting that L is not a constant, but a function of θ, which in turn is a function of time, we have

$$e_s = \frac{d}{dt} [L_0 + L_2 \cos 2\theta)(i_s)]$$

Applying the product rule for differentiation, one gets

$$e_s = (L_0 + L_2 \cos 2\theta)(p \, i_s) - 2L_2 \, i_s \sin 2\theta (p\theta)$$

The second term on the right-hand side is the speed-voltage term caused by mechanical motion, proportional to the instantaneous speed. Including the current variation with time, one gets

$$e_s = (L_0 + L_2 \cos 2\theta)(\omega_s I_s \cos \omega_s t) \\ - 2L_2(I_s \sin \omega_s t) \sin 2\theta (p\theta)$$

Substituting $\theta = \omega_m t - \delta$; $p\theta = \dfrac{d\theta}{dt} = \omega_m$, and $\omega_m = \omega_s$,

$$e_s = [L_0 + L_2 \cos 2(\omega_s t - \delta)](\omega_s I_s \cos \omega_s t) \\ - 2L_2 \omega_s I_s \sin \omega_s t \sin 2(\omega_s t - \delta)$$

Rearranging with the aid of the following trigonometric identities:

$$\cos A \cos B = \frac{1}{2} \cos (A + B) + \frac{1}{2} \cos (A - B)$$

$$\sin A \sin B = \frac{1}{2} \cos (A - B) - \frac{1}{2} \cos (A + B)$$

one obtains

$$\begin{aligned} e_s = &\; \omega_s L_0 I_s \cos \omega_s t \\ &+ \frac{1}{2} \omega_s L_2 I_s[\cos(3\omega_s t - 2\delta) + \cos(\omega_s t - 2\delta)] \\ &- \omega_s L_2 I_s [\cos(\omega_s t - 2\delta) - \cos(3\omega_s t - 2\delta)] \end{aligned}$$

or

$$\begin{aligned} e_s = &\; \omega_s L_0 I_s \cos \omega_s t - \frac{1}{2} \omega_s L_2 I_s \cos(\omega_s t - 2\delta) \\ &+ \frac{3}{2} \omega_s L_2 I_s \cos(3\omega_s t - 2\delta) \end{aligned}$$

which has a third-harmonic component. The flux will also contain a third-harmonic component.

EXAMPLE 5.2.2

Consider the elementary rotating reluctance machine shown in Figure 5.2.1. The reluctance variation is given by

$$\mathbf{R}(\theta) = \mathbf{R_0} - \mathbf{R_2} \cos 2\theta$$

Let the stator be excited from a voltage source

$$v_s = V_s \sin \omega_s t$$

The resistance of the winding may be neglected.

    a. Develop an expression for the instantaneous electromagnetic torque T_e in terms of the flux ϕ in the magnetic circuit and the reluctance.

    b. Let $\theta = \omega_m t + \theta_0$ where ω_m is the angular velocity of the rotor. Obtain the expression for T_e in terms of ω_m, ω_s, and θ_0.

    c. Determine the necessary condition for nonzero time-average torque, and find the corresponding expression for the average torque $(T_e)_{av}$.

    d. Show that the coil current contains a third-harmonic component when the machine is operating at synchronous speed.

SOLUTION

    a. Treating the flux linkage λ and the coordinate θ as independent variables, the electromagnetic torque is given by

$$T_e = -\frac{\partial W_m(\lambda,\ \theta)}{\partial \theta}$$

Since $W_m = \frac{1}{2}\phi^2\ \mathbf{R}(\theta)$ for a linear system, it follows that

$$T_e = -\frac{1}{2}\ \phi^2\ \frac{d\mathbf{R}(\theta)}{d\theta}$$

Compare the above expressions to those of Equations 5.2.3 and 5.2.4, which show that the torque always acts in such a direction that the resulting rotation increases the inductance, or decreases the reluctance. For the given reluctance variation, one obtains

$$T_e = -\frac{1}{2}\ \phi^2\ \frac{d}{d\theta}(\mathbf{R_0} - \mathbf{R_2}\cos 2\theta)$$

or

$$T_e = -\ \phi^2\ \mathbf{R_2}\sin 2\theta$$

    b. Neglecting the resistance of the electric circuit, the flux in the magnetic circuit is related to the voltage v_s through the relation

$$v_s = N_s\frac{d\phi}{dt}$$

or, ϕ is given by

$$\phi = -\phi_{max} \cos \omega_s t$$

where $\phi_{max} = V_s/N_s \omega_s$, and the constant of integration may be assumed as zero in the steady-state analysis of linear circuits. Substituting the time variation of ϕ and θ in the expression for T_e, one gets

$$T_e = -\phi_{max}^2 \cos^2 \omega_s t \; \mathbf{R_2} \sin 2(\omega_m t + \theta_0)$$

The above may be rearranged with the use of trigonometric identities:

$$\cos^2 A = \frac{1 + \cos 2A}{2}$$

$$\cos A \sin B = \frac{1}{2} [\sin(A + B) - \sin(A - B)]$$

Then,

$$T_e = -\frac{\mathbf{R_2} \phi_{max}^2}{2} \left\{ \sin 2(\omega_m t + \theta_0) + \frac{1}{2} \sin 2[(\omega_m + \omega_s)t \right.$$
$$\left. + \theta_0] + \frac{1}{2} \sin 2[(\omega_m - \omega_s)t + \theta_0] \right\}$$

c. Since

$$(T_e)_{av} = \frac{1}{2\pi} \int_0^{2\pi} T_e \, d\theta$$

the average value of each of the three sinusoidally time-varying terms is zero unless, in one of them, the coefficient of t is zero. Because $\omega_m \neq 0$, the necessary condition for nonzero time-average torque is given by

$$|\omega_m| = |\omega_s|$$

corresponding to which

$$(T_e)_{av} = -\frac{\mathbf{R_2} \phi_{max}^2}{4} \sin 2\theta_0 = -\frac{(\mathbf{R_d} - \mathbf{R_q}) \phi_{max}^2}{8} \sin 2\theta_0 \qquad *$$

d. The current is related to the flux by

$$\phi = N_s i_s / \mathbf{R}$$

For the condition $\omega_m = \omega_s$, i.e., at synchronous speed,

$$i_s = - \frac{\phi_{max}}{N_s} \cos \omega_s t [\mathbf{R_0} - \mathbf{R_2} \cos 2(\omega_s t + \theta_0)]$$

By making use of the identity

$$\cos A \cos B = \frac{1}{2} \cos(A + B) + \frac{1}{2} \cos(A - B),$$

one has

$$i_s = - \frac{\phi_{max} \mathbf{R_0}}{N_s} \cos \omega_s t + \frac{\phi_{max} \mathbf{R_2}}{2N_s} [\cos(\omega_s t + 2\theta_0) + \cos(3\omega_s t + 2\theta_0)]$$

which shows that the coil current contains a third-harmonic component.

Energy conversion implies rotation for machines because power is the product of a torque and a speed, and energy is converted only when the speed is nonzero. The equation of mechanical motion (or the torque equation) describing a rotating electromechanical energy converter is of the form

$$T = J\ddot{\theta} + \alpha \dot{\theta} + K\theta - T_e \tag{5.2.11}$$

where T is the torque applied externally to the shaft, T_e is the electromagnetic torque developed internally and applied to the shaft by the electrical system, J is the rotor moment of inertia, α is the rotor damping factor, K is the torsional spring constant, $\ddot{\theta}$ is $d^2\theta/dt^2$; $\dot{\theta}$ is $d\theta/dt$, and θ is the angular displacement. The power converted from electrical to mechanical form is

$$p_{em} = T_e \dot{\theta} = T_e \omega_m \tag{5.2.12}$$

and the mechanical power flowing into the machine from external mechanical systems attached to the shaft is

$$p_m = T\dot{\theta} = T \omega_m \tag{5.2.13}$$

FIGURE 5.2.3   Torque variation with angle.

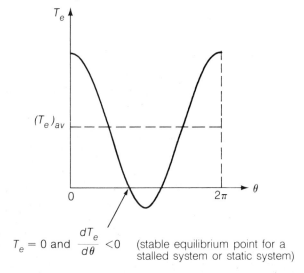

$$T_e = 0 \text{ and } \frac{dT_e}{d\theta} < 0 \quad \text{(stable equilibrium point for a stalled system or static system)}$$

Considering $\dot{\theta}$ to be positive, i.e., for positive rotation, motor action is defined by a conversion of electrical power to mechanical power, i.e., $p_{em} > 0$ and $T_e > 0$; and the generator action is defined by a conversion from mechanical power to electrical power, i.e., $p_{em} < 0$ and $T_e < 0$.

With no externally applied shaft torque T and with zero spring constant K, the condition for *starting* a motor in the positive direction is given by

$$T_e > 0 \text{ for } 0 \le \theta \le 2\pi \tag{5.2.14}$$

When Equation 5.2.14 is not satisfied, the machine may come to rest when $T_e = 0$ and stable equilibrium is achieved when $d\,T_e/d\theta < 0$, as shown in Figure 5.2.3. Since all physical machines have rotor inertia and since kinetic energy is stored in the rotating inertia of the rotor, the condition for successful *running* of a machine is less restrictive than that of Equation 5.2.14 and is given by

$$(T_e)_{av} = \frac{1}{2\pi} \int_0^{2\pi} T_e \, d\theta > 0 \tag{5.2.15}$$

which implies that the *average* torque be greater than zero over a revolution. Thus a machine with the torque-angle characteristic as shown in Figure 5.2.3 will run successfully.

The synchronous reluctance machine of Figure 5.2.1 would work equally well even if the exciting coil were placed on the rotor instead of on the stator. Singly excited devices are usually employed to produce uncontrolled bulk forces in such devices as relays, solenoids, and force actuators. When it is desired to obtain forces proportional to electric signals, and signals proportional to forces and velocities, the devices must be multiply excited (i.e., having two or more paths for excitation and exchange of energy with sources). Permanent magnets may be used as one of the

FIGURE 5.2.4   An elementary doubly-excited magnetic-field system.

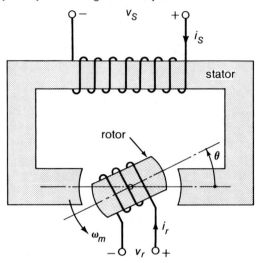

excitation sources. While one excitation path sets the level of the electrical or magnetic field, the other can be made to handle signals. Loudspeakers, torque motors, and tachometers are examples of signal-handling devices, while most of the known types of motors and generators are examples of continuous energy conversion and power handling.

MULTIPLY EXCITED MAGNETIC-FIELD SYSTEMS   Let us consider an elementary multiply excited magnetic system shown in Figure 5.2.4 with two sets of electrical terminals and one mechanical terminal. This model is the same as in Figure 5.2.1 but with the addition of a rotor coil that is connected to its electrical source by means of *carbon brushes* bearing on *slip rings* (or *collector rings*). The flux linkages of the stator and rotor windings can be expressed as functions of the coil currents:

$$\lambda_s = L_{ss}\, i_s + L_{sr}\, i_r \qquad\qquad (5.2.16)$$

$$\lambda_r = L_{sr}\, i_s + L_{rr}\, i_r \qquad\qquad (5.2.17)$$

where L_{ss} and L_{rr} are the self-inductances of the stator and rotor coils respectively, and L_{sr} is the stator-rotor mutual inductance. All these inductances are generally functions of the angle θ between the magnetic axes of the stator and rotor windings. The equations of motion for the electrical side (i.e., the volt-ampere equations) are given by

$$v_s = R_s\, i_s + p\, \lambda_s$$
$$= R_s\, i_s + L_{ss}(p\, i_s) + i_s(p\, L_{ss}) + L_{sr}(p\, i_r) + i_r(p\, L_{sr}) \quad (5.2.18)$$

$$v_r = R_r\, i_r + p\, \lambda_r$$
$$= R_r\, i_r + L_{sr}(p\, i_s) + i_s(p\, L_{sr}) + L_{rr}(p\, i_r) + i_r(p\, L_{rr}) \quad (5.2.19)$$

Since there are two sets of electrical terminals, two volt-ampere equations result, as in Equations 5.2.18 and 5.2.19. Neglecting the reluctances of the stator- and rotor-iron circuits, the electromagnetic torque can be found either from the energy or coenergy stored in the magnetic field of the air-gap region:

$$T_e = -\frac{\partial W_m(\lambda_s,\, \lambda_r,\, \theta)}{\partial \theta} = +\frac{\partial W_m{}'(i_s,\, i_r,\, \theta)}{\partial \theta} \quad (5.2.20)$$

For a linear system, the energy or coenergy stored in a pair of mutually coupled inductors is given by Equation 3.5.14 (also see Problem 3–14).

$$W_m{}'(i_1,\, i_2,\, \theta) = \frac{1}{2} L_{ss}\, i_s{}^2 + L_{sr}\, i_s\, i_r + \frac{1}{2} L_{rr}\, i_r{}^2 \quad (5.2.21)$$

The instantaneous electromagnetic torque is then given by

$$T_e = \frac{i_s{}^2}{2}\frac{d\,L_{ss}}{d\theta} + i_s\, i_r\, \frac{d\,L_{sr}}{d\theta} + \frac{i_r{}^2}{2}\frac{d\,L_{rr}}{d\theta} \quad (5.2.22)$$

The first and third terms on the right-hand side of Equation 5.2.22, involving angular rate of change of self-inductance are the *reluctance-torque* terms; the middle term, involving angular rate of change of mutual inductance, is the torque caused by the interaction of fields produced by the stator and rotor currents. It is this mutual-inductance torque that is most commonly exploited in practical rotating machines. Multiply excited systems with more than two sets of electrical terminals can be handled in a similar manner as for two pairs by assigning additional independent variables to the terminals.

If the self-inductances L_{ss} and L_{rr} are independent of angle θ, the reluctance torque is zero, and the torque is produced only by the mutual term $L_{sr}(\theta)$, as seen from Equation 5.2.22. We shall consider such a case in the following example.

EXAMPLE 5.2.3

Consider an elementary two-pole rotating machine with a uniform (or smooth) air gap as shown in Figure 5.2.5, in which the cylindrical rotor is mounted within the stator made up of a hollow cylinder coaxial with the rotor. The stator and rotor windings are distributed over a number of slots so that their mmf waves can be approximated by space sinusoids. A consequence of such a type of construction is that one can fairly assume that the self-inductances L_{ss} and L_{rr} are constant, but the mutual inductance L_{sr} is given by

$$L_{sr} = L \cos \theta$$

where θ is the angle between the magnetic axes of the stator and rotor windings. Let the currents in the two windings be given by

$$i_s = I_s \cos \omega_s t$$

$$i_r = I_r \cos (\omega_r t + \alpha)$$

and let the rotor rotate at an angular velocity

$$\omega_m = \dot{\theta} \text{ rad/s}$$

such that the position of the rotor at any instant is given by

$$\theta = \omega_m t + \theta_0$$

Assume that the reluctances of the stator and rotor-iron circuits are negligible, and that the stator and rotor are concentric cylinders neglecting the effect of slot openings.

    a. Derive an expression for the instantaneous electromagnetic torque developed by the machine.

    b. Find the necessary condition for the development of an average torque in the machine.

    c. Obtain the expression for the average torque corresponding to the following cases:

       (i) $\omega_s = \omega_r = \omega_m = 0$; $\alpha = 0$

       (ii) $\omega_s = \omega_r$; $\omega_m = 0$

       (iii) $\omega_r = 0$; $\omega_s = \omega_m$; $\alpha = 0$

       (iv) $\omega_m = \omega_s - \omega_r$, where ω_s and ω_r are two different angular frequencies

    d. With the assumed current-source excitations of Part (c), determine the voltages induced in the stator and rotor windings at the corresponding angular velocity ω_m at which an average torque results.

FIGURE 5.2.5 An elementary two-pole rotating machine with a uniform air gap.

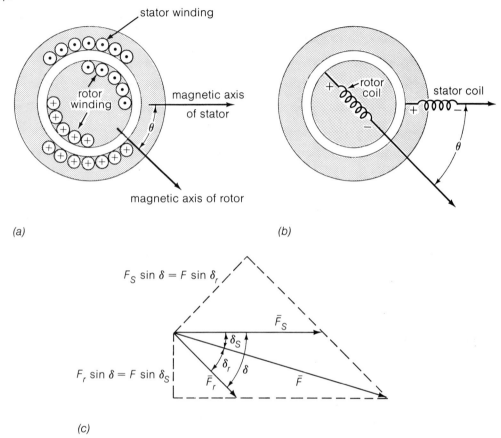

(a)

(b)

$$F_S \sin \delta = F \sin \delta_r$$

$$F_r \sin \delta = F \sin \delta_S$$

(c)

SOLUTION

a. Equations 5.2.16 through 5.2.22 apply. With constant L_{ss} and L_{rr}, Equation 5.2.22 simplifies as

$$T_e = i_s i_r \frac{d L_{sr}}{d\theta} = -i_s i_r L \sin \theta$$

when the variation of L_{sr} as a function of θ is substituted. For the given current variations, the instantaneous electromagnetic torque developed by the machine is given by

$$T_e = -L I_s I_r \cos \omega_s t \cos(\omega_r t + \alpha) \sin(\omega_m t + \theta_0)$$

Using the trigonometric identities, the product of the three trigonometric terms in the above equation may be expressed to yield

$$T_e = \frac{-L\,I_s\,I_r}{4}\left\{\sin[[\omega_m + (\omega_s + \omega_r)]t + \alpha + \theta_0]\right.$$
$$+\ \sin[[\omega_m - (\omega_s + \omega_r)]t - \alpha + \theta_0]$$
$$+\ \sin[[\omega_m + (\omega_s - \omega_r)]t - \alpha + \theta_0]$$
$$\left.+\ \sin[[\omega_m - (\omega_s - \omega_r)]t + \alpha + \theta_0]\right\}$$

b. The average value of each of the sinusoidal terms in the above equation is zero, unless the coefficient of t is zero in that term; that is, the average torque $(T_e)_{av}$ developed by the machine is zero unless

$$\omega_m = \pm(\omega_s \pm \omega_r)$$

which may also be expressed as

$$|\omega_m| = |\omega_s \pm \omega_r|$$

c. (i) The excitations are direct currents I_s and I_r. For the given conditions of $\omega_s = \omega_r = \omega_m = 0$ and $\alpha = 0$,

$$T_e = -L\,I_s\,I_r \sin\theta_0$$

which is a constant. As such

$$(T_e)_{av} = -L\,I_s\,I_r \sin\theta_0$$

The machine operates as a *dc rotary actuator*, developing a constant torque against any displacement θ_0 produced by an external torque applied to the rotor shaft.

(ii) With $\omega_s = \omega_r$, both excitations are alternating currents of the same frequency. For the conditions $\omega_s = \omega_r$ and $\omega_m = 0$

$$T_e = -\frac{L\,I_s\,I_r}{4}[\sin(2\omega_s t + \alpha + \theta_0) + \sin(-2\omega_s t - \alpha + \theta_0)$$
$$+ \sin(-\alpha + \theta_0) + \sin(\alpha + \theta_0)]$$

The machine operates as an *ac rotary actuator,* and the developed torque is fluctuating. The average value of the torque is

$$(T_e)_{av} = -\frac{L\,I_s\,I_r}{2}\sin\theta_0\cos\alpha$$

Note that α becomes zero if the two windings are connected in series, in which case cos α becomes unity.

(iii) With $\omega_r = 0$, the rotor excitation is a direct current I_r. For the conditions $\omega_r = 0$, $\omega_s = \omega_m$, and $\alpha = 0$,

$$T_e = -\frac{L\,I_s\,I_r}{4}[\sin(2\omega_s t + \theta_0) + \sin(\theta_0) + \sin(2\omega_s t + \theta_0)$$
$$+\,\sin(\theta_0)]$$

or

$$T_e = -\frac{L\,I_s\,I_r}{2}[\sin(2\omega_s t + \theta_0) + \sin\theta_0]$$

The machine operates as an idealized *single-phase synchronous machine,* and the instantaneous torque is pulsating. The average value of the torque is

$$(T_e)_{av} = -\frac{L\,I_s\,I_r}{2}\sin\theta_0$$

since the average value of the double-frequency sine term is zero. If the machine is brought up to *synchronous* speed ($\omega_m = \omega_s$), an average unidirectional torque is established, and continuous energy conversion takes place at synchronous speed. Note that the machine is not self-starting, since an average unidirectional torque is not developed at $\omega_m = 0$ with the specified electrical excitations.

(iv) With $\omega_m = \omega_s - \omega_r$, the instantaneous torque is given by

$$T_e = -\frac{L\,I_s\,I_r}{4}[\sin(2\omega_s t + \alpha + \theta_0) + \sin(-2\omega_r t - \alpha + \theta_0)$$
$$+\,\sin(2\omega_s t - 2\omega_r t - \alpha + \theta_0) + \sin(\alpha + \theta_0)]$$

The machine operates as a *single-phase induction machine,* and the instantaneous torque is pulsating. The average value of the torque is

$$(T_e)_{av} = -\frac{L I_s I_r}{4} \sin(\alpha + \theta_0)$$

If the machine is brought up to a speed of $\omega_m = (\omega_s - \omega_r)$, an average unidirectional torque is established, and continuous energy conversion takes place at the *asynchronous* speed of ω_m. Note that the machine is not self-starting, since an average unidirectional torque is not developed at $\omega_m = 0$ with the specified electrical excitations.

The pulsating torque, which may be acceptable in small machines, is in general an undesirable feature in a rotating machine, working either as a generator or a motor, since it may result in speed fluctuation, vibration, noise, and waste of energy. In magnetic-field systems excited by single-phase alternating sources, the torque pulsates while the speed is relatively constant; consequently, pulsating power becomes a feature. This calls for an improvement; in fact, by employing polyphase windings and polyphase sources, constant power is developed in a balanced system.

d. The flux-linkage relations are given by

$$\lambda_s = L_{ss}\, i_s + L_{sr}(\theta)\, i_r = L_{ss}\, i_s + L\, i_r \cos\theta$$

$$\lambda_r = L_{sr}(\theta)\, i_s + L_{rr}\, i_r = L\, i_s \cos\theta + L_{rr}\, i_r$$

in which L_{ss}, L_{rr}, and L are constants. The volt-ampere equations are then given by

$$v_s = R_s\, i_s + p\, \lambda_s$$

$$v_r = R_r\, i_r + p\, \lambda_r$$

where R_s and R_r are the winding resistances, and p is the time-derivative operator d/dt.

Substituting for λ_s and λ_r, and recognizing that θ is a variable and a function of time, one obtains

$$v_s = R_s\, i_s + L_{ss}\, p\, i_s + L \cos\theta(p\, i_r) - L\, i_r \sin\theta(p\theta)$$

$$v_r = R_r\, i_r + L_{rr}\, p\, i_r + L \cos\theta(p\, i_s) - L\, i_s \sin\theta\, (p\theta)$$

where $p\theta$ is the instantaneous speed ω_m. The fourth terms on the right-hand side of the above equations are caused by mechanical motion and are proportional to the instantaneous speed. These are the speed-voltage terms, which are the coupling terms relating the interchange of power between electrical and mechanical systems.

The voltages induced in the stator and rotor windings will now be found for each of the parts in *(c)*.

(i) $i_s = I_s$; $i_r = I_r$; $\omega_m = 0$

so that

$e_s = 0$ and $e_r = 0$.

(ii) $i_s = I_s \cos \omega_s t$; $i_r = I_r \cos(\omega_s t + \alpha)$; $\omega_m = 0$

so that

$e_s = -\omega_s L_{ss} I_s \sin \omega_s t - \omega_s L I_r \cos \theta_0 \sin(\omega_s t + \alpha)$

$e_r = -\omega_s L_{rr} I_r \sin(\omega_s t + \alpha) - \omega_s L I_s \sin \omega_s t \cos \theta_0$

(iii) $i_s = I_s \cos \omega_s t$; $i_r = I_r$; $\omega_m = \omega_s$

so that

$e_s = -\omega_s L_{ss} I_s \sin \omega_s t - \omega_s L I_r \sin(\omega_s t + \theta_0)$

$e_r = -\omega_s L I_s \sin \omega_s t \cos(\omega_s t + \theta_0) - \omega_s L I_s \cos \omega_s t \sin(\omega_s t + \theta_0)$

or

$e_r = -\omega_s L I_s \sin(2\omega_s t + \theta_0)$

Note that the stator current induces a double-frequency voltage in the rotor circuit.

(iv) $i_s = I_s \cos \omega_s t$; $i_r = I_r \cos(\omega_s t + \alpha)$; $\omega_m = \omega_s - \omega_r$

so that

$$e_s = -\omega_s L_{ss} I_s \sin \omega_s t$$
$$- \omega_r L I_r \sin(\omega_r t + \alpha) \cdot \cos[(\omega_s - \omega_r) t + \theta_0]$$
$$- (\omega_s - \omega_r) L I_r \cos(\omega_r t + \alpha) \cdot \sin[(\omega_s - \omega_r) t + \theta_0]$$

or

$$e_s = -\omega_s L_{ss} I_s \sin \omega_s t$$
$$- \frac{\omega_s L I_r}{2} [\sin(\omega_s t + \alpha + \theta_0) - \sin[(-\omega_s + 2\omega_r) t + \alpha - \theta_0]]$$
$$- \omega_r L I_r \sin[(-\omega_s + 2\omega_r) t + \alpha - \theta_0]$$

$$e_r = -\omega_r L_{rr} I_r \sin(\omega_r t + \alpha)$$
$$- \omega_s L I_s \sin \omega_s t \cdot \cos[(\omega_s - \omega_r) t + \theta_0]$$
$$- (\omega_s - \omega_r) L I_s \cos \omega_s t \sin[(\omega_s - \omega_r) t + \theta_0]$$

or

$$e_r = -\omega_r L_{rr} I_r \sin(\omega_r t + \alpha)$$
$$+ \frac{\omega_r L I_s}{2} [\sin[(2\omega_s - \omega_r) t + \theta_0] - \sin(\omega_r t - \theta_0)]$$
$$- \omega_s L I_s \sin[(2\omega_s - \omega_r) t + \theta_0]$$

In Example 5.2.3, we have considered a two-pole rotating machine. When a machine has more than two poles, only a single pair of poles needs to be considered since the electric, magnetic, and mechanical conditions associated with every other pole-pair are repetitions of those for the pole-pair under consideration. The angle subtended by one pair of poles in a P-pole machine (or one cycle of flux distribution) is defined to be 360 *electrical degrees* or 2π *electrical radians*. So the relationship between the mechanical angle θ_m and the angle θ in electrical units is given by

$$\theta = \frac{P}{2} \theta_m \qquad (5.2.23)$$

because there are $P/2$ complete wavelengths (or cycles) in one complete revolution. In view of this, Equation 5.2.20 for the electromagnetic torque needs to be modified as

$$T_e = \frac{\partial W_m'(i_s, i_r, \theta_m)}{\partial \theta_m} = \frac{\partial W_m'(i_s, i_r, \theta)}{\partial \theta} \frac{d\theta}{d\theta_m} \qquad (5.2.24)$$

where T_e is the electromagnetic torque acting in the positive direction of θ_m, and the derivative is to be taken with respect to the actual mechanical angle θ_m because one is dealing here with mechanical variables. Thus, for the Example 5.2.3, if it were a P-pole machine, the expression for torque would be

$$T_e = -\frac{P}{2} L i_s i_r \sin\left(\frac{P}{2} \theta_m\right) \qquad (5.2.25)$$

The negative sign in the above equation means that the electromagnetic torque acts in such a direction as to bring the magnetic fields of the stator and rotor into alignment.

The voltage and torque equations for the idealized elementary machine of Example 5.2.3 with a uniform air gap have been derived on the basis of the *coupled-circuit viewpoint*. These can also be obtained from the *magnetic-field viewpoint,* as shown below. Since the mmf waves of the stator and rotor are considered to be spatial sine waves, they can be represented by the space vectors $\overline{F}_s$ and $\overline{F}_r$ drawn along the magnetic axes of the stator and rotor mmf waves, as in Figure 5.2.5(c), with the phase angle δ (in electrical units) between their magnetic axes. The resultant mmf $\overline{F}$ acting across the air gap is also a sine wave given by the vector sum of $\overline{F}_s$ and $\overline{F}_r$, so that

$$F^2 = F_s^2 + F_r^2 + 2F_s F_r \cos \delta \qquad (5.2.26)$$

where F's are the peak values of the mmf waves. Assuming the air-gap field to be entirely radial, the resultant $\overline{H}$-field is a sinusoidal space wave whose peak value is given by

$$H_{peak} = F/g \qquad (5.2.27)$$

where g is the radial length of the air gap (or the clearance between the rotor and stator), which is considered to be small compared with the radius of either stator or rotor. The currents in machine windings produce magnetic flux in the air gap, and the flux paths are completed through the stator and rotor iron.

The energy or coenergy-density in the air-gap region at a point where the magnetic field intensity is H is given by Equation 3.4.3:

$$w_m = w_m' = \frac{1}{2} \mu_0 H^2 \tag{5.2.28}$$

The average coenergy density obtained by averaging over the volume of the air-gap region is

$$(w_m')_{av} = \frac{\mu_0}{2} \text{ (average value of } H^2) \tag{5.2.29}$$

$$= \frac{\mu_0}{2} \frac{H^2_{peak}}{2} = \frac{\mu_0}{4} \frac{F^2}{g^2}$$

since the average value of the square of a sine wave is one-half of the square of its peak value. The total coenergy for the air-gap region is then given by

$$W_m' = (w_m')_{av} \text{ (volume of air-gap region)} \tag{5.2.30}$$

$$= \frac{\mu_0}{4} \frac{F^2}{g^2} \pi \, Dlg$$

where D is the average diameter at the air gap, and l is the axial length of the machine. Equation 5.2.30 may be rewritten as follows by using Equation 5.2.26:

$$W_m' = \frac{\pi \, Dl \, \mu_0}{4g} (F_s^2 + F_r^2 + 2F_s \, F_r \cos \delta) \tag{5.2.31}$$

The torque in terms of the interacting magnetic fields is obtained by taking the partial derivative of the field coenergy with respect to the angle δ. For a two-pole machine, it is given by

$$T_e = \frac{\partial W_m'}{\partial \delta} = -\frac{\pi \, Dl \, \mu_0}{2g} F_s \, F_r \sin \delta = -K \, F_s \, F_r \sin \delta \tag{5.2.32}$$

in which K is a constant determined by the dimensions of the machine. The torque for a P-pole cylindrical machine with a uniform air gap is then

$$T_e = -\frac{P}{2} K \, F_s \, F_r \sin \delta \tag{5.2.33}$$

Equations 5.2.32 and 5.2.33 show that the torque is proportional to the peak values of the interacting stator and rotor mmfs and also to the sine of the space-phase angle δ (expressed in electrical units) between them. The interpretation of the negative sign is the same as before, in that the fields tend to align themselves by decreasing the displacement angle δ between the fields. If the rotor-mmf axis is fixed relative to the rotor winding, the angle δ between the mmf axes is the same as the angle θ describing the rotor position in Figure 5.2.3.

Equation 5.2.33 shows that it is possible to obtain a constant torque, varying neither with time nor with rotor position, provided that the two mmf waves are of constant amplitude and have constant angular displacement from each other. While it is easy to conceive the two mmf waves having constant amplitudes, the question would then be how to maintain a constant angle between the stator and rotor-mmf axes, if one winding is stationary and the other is rotating. Three possible answers to this question arise: *(i)* if the stator-mmf axis is fixed in space, the rotor-mmf axis must also be fixed in space, even when the rotor winding is physically rotating; *(ii)* if the rotor-mmf axis is fixed relative to the rotor, the stator-mmf axis must rotate at the rotor speed relative to the stationary stator windings; *(iii)* the two mmf axes may rotate at such speeds relative to their windings that they remain stationary with respect to each other.

5.3 ELEMENTARY CONCEPTS OF ROTATING MACHINES

The most widely used electromechanical device is the magnetic field-type rotating machine. The main purpose of most rotating machines is electromechanical energy conversion, i.e., to convert energy between electrical and mechanical systems, either for electric power generation (as in generators or sources) or for the production of mechanical power to do useful tasks (as in motors or sinks). These machines range in size and capacity from small motors that consume only a fraction of a watt to large generators that produce several hundreds of megawatts. In spite of the wide variety of types, sizes, and methods of construction, all these machines operate on the same principle, namely, the tendency of two magnets to align themselves. Thus, an analysis of an idealized structure of one electric machine will provide the essential concepts necessary to understand the operation of most practical electric machines.

The study of electric machines from the coupled-circuit viewpoint is based on the fundamental consideration that machines can be viewed in terms of sets of linear lumped circuits in relative motion. Relative motion exists between the two magnetic members of the electric machine, namely, the stator, which is the stationary member, and the rotor, which is the rotating member. In general, electrical machines consist of two sets of windings (or coils) in which one set of coils can rotate with respect to the other. The mechanical motion between these two sets of coils is generally restricted to one degree of freedom, which, in most instances, is rotary motion. The annular space between the inner surface of the stator and the outer surface of the rotor is known as the air gap. The radial length of the air gap is always kept very small compared to the radial dimension of the outer surface of the rotor in order to produce a large magnetic field for a given current. The two windings in relative motion are known as the *field winding,* which produces the flux density, and the *armature winding,* in which the working emf is induced. The field structure on which the field circuit is located can, in general, be physically situated on either the stationary member (stator) or the rotating member (rotor) of the machine, depending on constructional convenience.

While permanent magnets may be used in small machines as the primary sources of flux, in the majority of machines the field is electromagnetic, and field coils carrying the field current are wound on a magnetic structure. The iron core forming this structure is laminated in order to reduce the field iron losses if the field current is alternating or contains an alternating component.

TABLE 5.3.1 CLASSES OF ROTATING MACHINES

Type of Machine	Stator		Rotor		Typical Examples
	Number of Circuits	Normal type of excitation	Number of Circuits	Normal type of excitation	
	Cylindrical Stator and Rotor Structures				
synchronous machines (single and polyphase)	one or more than one (symmetrical winding)	single or polyphase balanced	one	dc	alternators synchronous motors
polyphase induction machines	more than one (symmetrical winding)	polyphase balanced or unbalanced	more than one (symmetrical winding)	short-circuited	3-phase induction motors, 2-phase servo-motors
commutator machines	one or more than one	dc, single or polyphase	commutated winding	short-circuited	amplidynes, metadynes, Schrage motors, repulsion motors, dc machines
single-phase induction machines	one or two (unsymmetrical winding)	single-phase	more than one (symmetrical winding)	short-circuited	split-phase, capacitor, shaded-pole motors, ac tachometers, synchros
	Saliency on Either Stator or Rotor, but Not Both				
synchronous machines (single and polyphase)	one or more than one (symmetrical winding)	single or polyphase balanced	one (salient rotor)	dc	salient-pole alternators, reluctance motors
Commutator machines	one or more than one (salient stator)	ac or dc	commutated winding	ac or dc	conventional dc machines, universal motors, metadynes, rotary amplifiers

Two different types of construction are used for the field circuit. One type is the *salient-pole* arrangement, in which the field coils are *concentrated* and wound around the protruding poles; this form of construction is used only for machines with direct-current field supply. The air gap around the periphery of the machine is characteristically nonuniform in this arrangement. The second type is the *non-salient-pole* construction for smooth (or uniform) air-gap machines, in which the field coils are *distributed* in slots cut into a cylindrical magnetic structure; this arrangement is commonly used on certain forms of large ac generators known as turbo-alternators.

Table 5.3.1 summarizes the classes of rotating machines that we shall consider in this text.

FIGURE 5.3.1 Schematic diagram of a dc machine. (4 pole) <SALIENT POLE>

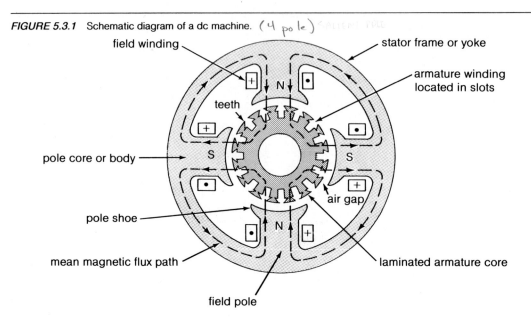

field winding

stator frame or yoke

armature winding located in slots

teeth

pole core or body

pole shoe

air gap

mean magnetic flux path

laminated armature core

field pole

The winding in which the voltage is to be induced is known as the armature winding, and the structure containing it is called an *armature*. In general, the electrical circuit representing the armature of the machine consists of coils distributed in slots cut into a cylindrical magnetic structure. Iron cores are employed for magnetic circuits of both the armature and field in order to provide the most effective flux path. The armature currents are always alternating; under these circumstances, because the armature iron is subjected to a varying magnetic flux, eddy currents will be induced in it. In order to minimize the eddy-current loss, the armature iron is built up of thin laminations. Schematic diagrams showing the flux paths in a four-pole dc machine, in a salient-pole ac machine with six poles, and in a two-pole synchronous machine with a cylindrical rotor are given in Figures 5.3.1, 5.3.2, and 5.3.3 respectively. Photos 5–1 and 5–2 show typical stators of a dc machine and a synchronous machine, respectively, while Plates 5–3 and 5–4 illustrate the construction of salient- and non-salient-pole rotors of synchronous machines.

In general, either the field or the armature can physically rotate. When the armature of the machine is on the rotor, the machine is known as a rotating armature machine. The form of different rotating machines (ac or dc) is governed by external constraints, such as the form of electrical supply connected to the field circuit or the mechanical method of connection to the rotating member. In the case of a dc machine, as shown in Figure 5.3.1, the armature is the rotating member, or rotor, and the field structure is the stationary member, or stator. For most of the ac machines, on the other hand, the armature windings are located on the stator, and the field winding is on the rotor, which may be of salient-pole or non-salient-pole construction, as in Figures 5.3.2 and 5.3.3.

FIGURE 5.3.2 Schematic diagram of a salient-pole ac machine (6 poles)
(synchronous machine).

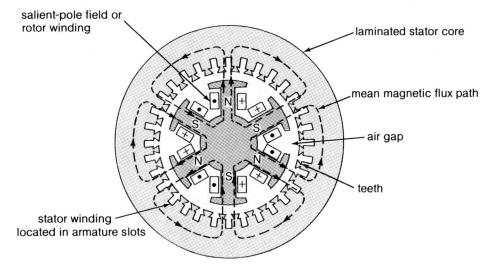

salient-pole field or
rotor winding

laminated stator core

mean magnetic flux path

air gap

teeth

stator winding
located in armature slots

FIGURE 5.3.3 Schematic diagram of a nonsalient-pole synchronous
machine.

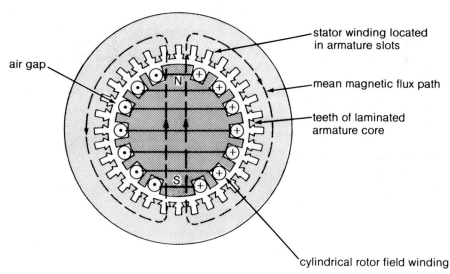

air gap

stator winding located
in armature slots

mean magnetic flux path

teeth of laminated
armature core

cylindrical rotor field winding

PHOTO 5.1   Typical stator of a large dc machine.   Photo courtesy of
Westinghouse Electric Corporation.

PHOTO 5.2   Typical stator of a synchronous machine.   Photo courtesy of
Westinghouse Electric Corporation.

PHOTO 5.3 Eight salient-pole rotor of a synchronous machine. Photo courtesy of Westinghouse Electric Corporation.

PHOTO 5.4 Cutaway view of a two-pole nonsalient rotor of a synchronous generator. Photo courtesy of General Electric Company.

Currents in the machine windings produce magnetic flux in the air gap between the stator and rotor as shown in Figures 5.3.1, 5.3.2, and 5.3.3, while the flux path is completed through the stator and rotor iron. This situation is equivalent to the appearance of magnetic poles on both stator and rotor, the number of such poles depending on the specific winding design. By using the concept of interaction between magnetic fields, it can be shown that electromagnetic torque cannot be obtained by using unequal numbers of rotor and stator poles, and that all rotating machines must have the same number of poles on the stator and on the rotor. In motor and generator action, then, the magnetic fields tend to line up, pole to pole. In the case of a generator, the electromagnetic torque opposes the mechanical torque applied from the prime mover, which is the source of mechanical energy. In a motor, the direction of rotation is determined by the electromagnetic torque, and the speed voltages that are produced act in opposition to the applied voltages. Thus, the electromagnetic torque and rotational voltages, which are essential for electromechanical energy conversion, are produced in both generators and motors. Constructionally, generators and motors of the same type differ only in details necessary for the best adaptation of a machine for its intended use. In general, any machine can be used either as a motor or as a generator for energy conversion.

Besides motoring and generating, there is another possible mode of operation for the machine, known as *braking*. These three possible modes of operation are schematically shown in Figure 5.3.4(a), (b), and (c):

   a. Motoring mode: There is electrical power input and mechanical power output. The electromagnetic torque T_e drives the machine against the load torque T. The input voltage v drives the current into the winding against the generated emf e.
   b. Generating mode: There is mechanical power input and electrical power output. The torque T applied externally to the shaft drives the machine against the electrically developed torque T_e. The generated emf e drives current out of the winding against the terminal voltage v.
   c. Braking mode: The machine has both mechanical and electrical energy input. The total input is dissipated as heat. While the machine is driven by the externally applied torque T, the electromagnetic torque T_e is opposing T, thereby braking the machine. The electric braking of motor drives is achieved by causing the motor to act as a generator, receiving mechanical energy from the moving parts and converting it to electrical energy, which is dissipated in a resistor or pumped back into the power line. Note that the applied voltage v and the generated emf e do not oppose each other.

Many forms of electrical machines operate from a three-phase ac supply, and the armature winding of such a machine consists of a three-phase distributed winding for which the three separate phase windings are wound in an identical manner, displaced by 120 electrical degrees from each other. The phase windings may be connected in wye or delta. It is shown in Section 5.4 that, in general, a rotating field of constant amplitude and sinusoidal space distribution of mmf around the periphery of the stator is produced by a q-phase winding ($q > 1$) located on the stator and excited by balanced q-phase currents when the respective phase windings are wound $2\pi/q$ electrical radians apart in space. The constant amplitude is $q/2$ times the maximum contribution of

FIGURE 5.3.4   Three possible modes of operation of a rotating electrical machine: *(a)* motoring mode, *(b)* generating mode, *(c)* braking mode.

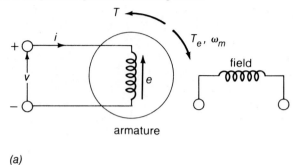

(a)

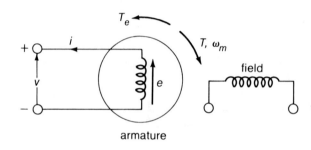

(b)

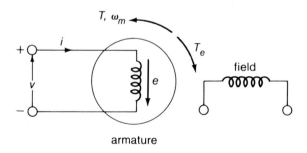

(c)

PHOTO 5.5   Commutator of a dc machine on the rotor shaft.   Photo courtesy of Westinghouse Electric Corporation.

any one phase, and the speed is $\omega = 2\pi f$ electrical radians per second, where f is the frequency of the electrical supply in hertz. For a P-pole machine the rotational speed of the mmf is the synchronous speed given by

$$\omega_m = \frac{2}{P}\,\omega \text{ rad/s} \qquad (5.3.1)$$

or

$$n = \frac{120f}{P} \text{ rpm} \qquad (5.3.2)$$

Thus, a polyphase winding excited by balanced polyphase currents will be seen to produce an effect equivalent to that of spinning a permanent magnet about an axis perpendicular to the magnet, or to that of the rotation of the dc-excited field poles. The armature, in general, may be located either on the stator or on the rotor of the machine; the speed of rotation of the rotating mmf in space is then the algebraic sum of the speed of the mmf relative to the structure and the mechanical speed of the structure itself.

Special mechanical arrangements need to be provided when electrical connections are to be made to the rotating member. Such connections are usually made through carbon brushes bearing on either a *slip ring* or a *commutator*, mounted on, but insulated from, the rotor shaft and rotating with the rotor. A slip ring is a continuous ring, usually made of brass, to which only one electrical connection is made. In order to supply direct current to the field winding located on the rotor of a synchronous machine, for example, two slip rings are provided. A commutator, on the other hand, is an elegant mechanical switch, consisting of a cylinder formed of hard-drawn copper segments separated and insulated from each other by mica. See Photo 5–5 for a picture of a commutator of a dc machine on the rotor shaft.

FIGURE 5.3.5 Schematic representations of a dc machine: *(a)* schematic arrangement, *(b)* circuit representation.

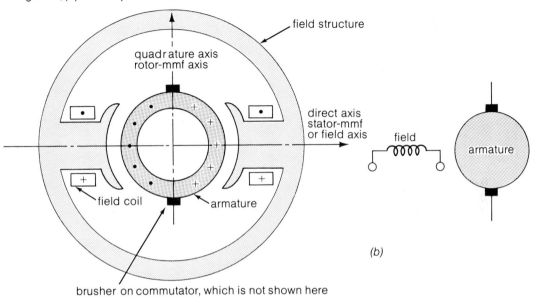

(a)

(b)

brusher on commutator, which is not shown here

Note: The geometrical position of the brushes in an actual machine is approximately 90 electrical degrees from their position in the schematic diagram because of the shape of the end connections to the commutator.

In the conventional dc machine, for example, full-wave rectification of the alternating voltage induced in individual armature coils is achieved by means of a commutator, which makes a unidirectional voltage available to the external circuit through the stationary carbon brushes held against the commutator surface. The armature windings of dc machines are located on the rotor because of the necessity for commutation. The winding connected to the commutator, known as the commutator winding, can be viewed as a pseudostationary winding because it produces a stationary flux when carrying a direct current, as a stationary winding would. The direction of the flux axis is determined by the position of the brushes, and it is shown later in the detailed analysis of commutator action in dc machines that the flux axis corresponds to the brush axis, i.e., the line joining the two brushes. The brushes are so located that commutation (i.e., reversal of current in the commutated coil) occurs when the coil sides are in the neutral zone, midway between the field poles. The axis of the armature mmf is then 90 electrical degrees from the field (or direct) axis of the field poles, i.e., in the quadrature axis. Figure 5.3.5 shows the schematic representations of a dc machine. The commutator is then a device for changing the connections between a rotating closed winding and an external circuit at the instants when the individual coil-generated voltages reverse. Thus, in a dc machine, this enables the output voltage to be constant and unidirectional.

FIGURE 5.3.6 *(a)* Slip-ring action and *(b)* commutator action connections.

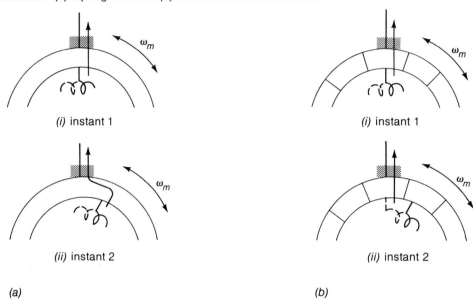

(i) instant 1   *(i)* instant 1

(ii) instant 2   *(ii)* instant 2

(a)   (b)

Because the armature mmf axis is fixed in space because of the switching action of the commutator (even though the closed armature winding on the rotor is rotating), the commutator winding becomes pseudostationary.

The basic difference between the action of slip rings and that of a commutator is that the conducting coil connected to the slip ring is always connected to the brush, regardless of the mechanical speed ω_m of the rotor and the rotor position, while, in the case of the commutator, the conducting coil is conducting current only when it is physically under the commutator brush, i.e., when it is stationary with respect to the commutator brush. This is illustrated in Figure 5.3.6. Then it follows that the frequency of the voltage at the slip-ring brush must be equal to that of the rotor mmf relative to the rotor, i.e., $\omega_r = \pm (\omega_s - \omega_m)$, and that at the commutator brush must be that of the rotor mmf in space, i.e., $\omega_s = \omega_m \pm \omega_r$, where $\omega_s (= 2\pi f_s)$ is the angular speed of the mmf set up when polyphase ac of frequency f hertz is applied to the stator.

As seen from Equation 5.2.33 and the discussion that followed, to maintain torque production when the machine rotates, the relative current-sheet patterns giving rise to the mmfs and the angle δ between their magnetic axes must be maintained. The manner in which this is achieved depends on the form (dc or ac) of the stator and rotor currents, and on the windings in which they circulate.

TABLE 5.3.2   CLASSIFICATION OF ELECTROMAGNETIC DEVICES $(\omega_s = \omega_m \pm \omega_r)$

Air-Gap flux	Electric Supply and Windings		Name of the Device
	Member 1 (primary or stator)	Member 2 (secondary or rotor)	
Fixed-Axis alternating	ac(ω_s) concentrated	ac(ω_s) concentrated	transformer (not an energy-conversion device)
alternating	ac(ω_s) phase	ac($\omega_r = \omega_s - \omega_m$) phase	1-phase induction machine
constant	ac($\omega_s = \omega_m$) phase	dc($\omega_r = 0$) concentrated	1-phase synchronous machine
constant	dc($\omega_s = 0$) concentrated	dc($\omega_r = 0$) commutator	dc commutator machine
alternating	ac(ω_s) distributed	ac(ω_r) commutator	1-phase commutator machine
Rotating-Wave (or traveling-wave, in case of a linear form that the following can take) constant	ac($\omega_s = \omega_m$) polyphase	dc($\omega_r = 0$) concentrated or distributed	polyphase synchronous machine, salient-pole or non-salient-pole type (3-phase machines are most common)
constant	ac(ω_s) polyphase	ac($\omega_r = \omega_s - \omega_m$) polyphase	polyphase induction machine (3-phase machines are most common)

Machine windings are excited so as to develop particular current-sheet and mmf patterns. A winding for the dc excitation of a working flux is usually concentrated, while for an ac flux it is commonly distributed to reduce the leakage. As mentioned earlier and as discussed later in detail, polyphase windings can be arranged to yield, to a close degree of approximation, sinusoidally distributed current sheets and rotating mmfs. Transformers and rotating machines can then be classified depending on *(i)* the kind of electrical supply (dc or ac) to which each of its windings is connected, and *(ii)* the type of connections made between the winding and its supply, i.e., by phase or tapped windings, or by switched connections as in a commutator winding. Table 5.3.2 lists a number of possible combinations, giving the names by which the machines are generally known.

ELEMENTARY SYNCHRONOUS MACHINES   Figure 5.3.7 shows an elementary single-phase, two-pole synchronous machine. In almost all cases, the armature winding of a synchronous machine is on the stator, and the field winding is on the rotor, because it is constructionally advantageous to have the low-power field winding on the rotating member. The field winding is excited by direct current that is supplied by a dc source connected to carbon brushes bearing on slip rings (or collector rings). The armature windings, though distributed in the slots around the inner periphery of the stator in an actual machine, is shown, for simplicity, consisting of a single coil of N turns (see Figure 5.3.8) indicated in cross section by the two coil sides a and $-a$ placed in

FIGURE 5.3.7 Elementary single-phase two-pole synchronous machine.

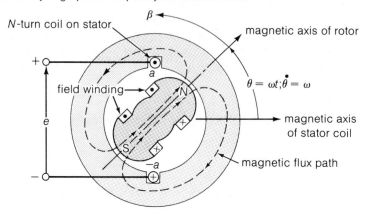

FIGURE 5.3.8 A three-turn single coil of armature winding.

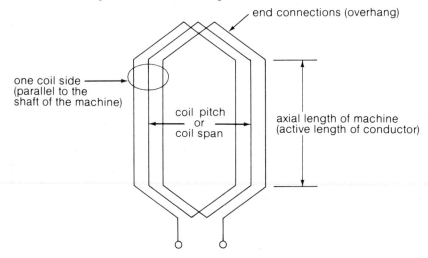

diametrically opposite narrow slots. The conductors forming these coil sides are placed in slots parallel to the machine shaft and are connected in series by means of the end connections as shown in Figure 5.3.8. The coil in Figure 5.3.7 spans 180 electrical degrees (or a complete pole pitch, i.e., the peripheral distance from the center line of a north pole to the center line of an adjacent south pole) and is hence denoted as a *full-pitch* coil. Only a two-pole synchronous machine with salient-pole construction is shown in Figure 5.3.7 for simplicity and convenience; the flux paths are shown by dotted lines.

The space distribution of the radial air-gap flux density around the air-gap periphery can be made to approximate a sinusoidal distribution by properly shaping the rotor-pole faces facing the air gap:

$$B = B_m \cos \beta \tag{5.3.3}$$

where B_m is the peak value at the rotor-pole center, and β is measured in electrical radians from the rotor-pole axis (or the magnetic axis of rotor) as shown in Figure 5.3.7. The air-gap flux per pole is the integral of the flux density over the pole area: for a two-pole machine,

$$\phi = \int_{-\pi/2}^{+\pi/2} B_m \cos \beta \, lr \, d\beta = 2B_m \, lr \tag{5.3.4}$$

and for a *P*-pole machine,

$$\phi = \frac{2}{P} \, 2B_m \, lr \tag{5.3.5}$$

where l is the axial length of the stator, r is the average radius at the air gap, and for a *P*-pole machine, the pole area is $2/P$ times that of a two-pole machine of the same length and diameter.

The flux linkage with the stator coil is $N\phi$ when the rotor poles are in line with the magnetic axis of the stator coil. If the rotor is turned at a constant speed ω by a source of mechanical power connected to its shaft, the flux linkage with the stator coil varies as the cosine of the angle θ between the magnetic axes of the stator coil and rotor:

$$\lambda = N\phi \cos \omega t \tag{5.3.6}$$

where time t is arbitrarily taken as zero when the peak of the flux-density wave coincides with the magnetic axis of the stator coil. By Faraday's law, the voltage induced in the stator coil is given by

$$e = -\frac{d\lambda}{dt} = \omega \, N\phi \sin \omega t - N \frac{d\phi}{dt} \cos \omega t \tag{5.3.7}$$

The minus sign associated with Faraday's law in Equation 5.3.7 implies generator reference directions as explained earlier and as shown in Figure 5.3.7. Considering the right-hand side of

Equation 5.3.7, the first term is a speed voltage caused by the relative motion of the field and the stator coil; the second term is a transformer voltage, which is negligible in most rotating machines under normal steady-state operation because the amplitude of the air-gap flux wave is fairly constant. The induced voltage is then given by the speed voltage itself:

$$e = \omega \, N\phi \, \sin \omega t \qquad (5.3.8)$$

The above equation may alternatively be obtained by the application of the cutting-of-flux concept given by Equation 3.1.13, from which the motional emf is given by the product of B_{coil}, the total active length of conductors l_{eff} in the two coil sides and the linear velocity of the conductor relative to the field, provided that these three are mutually perpendicular. For the case under consideration,

$$e = B_{coil} \, l_{eff} \, v = (B_m \sin \omega t)(2lN)(r \, \omega_m)$$

or

$$e = (B_m \sin \omega t)(2lN)(r \, 2\omega/P) = \omega N \frac{2}{P} 2B_m \, lr \sin \omega t \qquad (5.3.9)$$

which is the same as Equation 5.3.8 along with Equation 5.3.5.

   The resulting coil voltage is thus a time function having the same sinusoidal waveform as the spatial distribution B. The coil voltage passes through a complete cycle for each revolution of the two-pole machine of Figure 5.3.7. So its frequency in hertz is the same as the speed of the rotor in revolutions per second; that is, the electrical frequency is synchronized with the mechanical speed of rotation. Thus a two-pole synchronous machine, under normal steady-state conditions of operation, revolves at 60 rps or 3,600 rpm in order to produce a 60-Hz voltage. However, for a P-pole machine in general, the coil voltage passes through a complete cycle every time a pair of poles sweeps by; i.e., $P/2$ times in each revolution. The frequency of the voltage wave is then given by

$$f = \frac{P}{2} \cdot \frac{n}{60} \text{ hertz} \qquad (5.3.10)$$

where n is the mechanical speed of rotation in rpm. The synchronous speed in terms of the frequency and the number of poles is given by

$$n = \frac{120f}{P} \text{ rpm} \qquad (5.3.11)$$

FIGURE 5.3.9   Elementary single-phase four-pole synchronous machine.

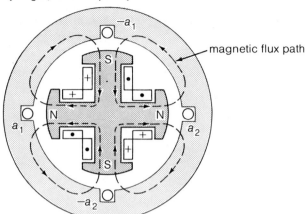

The radian frequency ω of the voltage wave in terms of ω_m, the mechanical speed in radians per second, is given by

$$\omega = \frac{P}{2}\,\omega_m \tag{5.3.12}$$

which is in accordance with Equation 5.2.23. Figure 5.3.9 shows an elementary single-phase synchronous machine with four salient poles; the flux paths are shown by dotted lines. Two complete wavelengths (or cycles) exist in the flux distribution around the periphery, since the field coils are connected such that poles of alternate north and south polarities form. The armature winding now consists of two coils $(a_1, -a_1)$ and $(a_2, -a_2)$ connected in series by their end connections. The span of each coil is one-half wavelength of flux, or 180 electrical degrees for the full-pitch coil. Since the generated voltage now goes through two complete cycles per revolution of the rotor, the frequency is then twice the speed in revolutions per second, consistent with Equation 5.3.10.

The field winding may be concentrated around the salient poles as shown in Figures 5.3.7 and 5.3.9, or distributed in slots around the cylindrical rotor as in Figure 5.3.3 (shown for a two-pole machine). By properly shaping the pole faces in the former case and by appropriately distributing the field winding in the latter case, an approximately sinusoidal field is produced in the air gap. A salient-pole rotor construction is best suited mechanically for hydroelectric generators because hydroelectric turbines operate at relatively low speeds, and a relatively large number of poles is required in order to produce the desired frequency (60 Hz in the United States) in accordance with Equation 5.3.11. Salient-pole construction is also employed for most synchronous motors.

The non-salient-pole (smooth or cylindrical) rotor construction is preferred for turbine-driven alternators (known also as turbo-alternators or turbine generators), which are usually of two or four poles driven by steam turbines or gas turbines, operating best at relatively high speeds. The rotors for such machines may be made either from a single steel forging or from several forgings shrunk together on the shaft.

Going back to Equation 5.3.8, the maximum value of the induced voltage is

$$E_{max} = \omega N\phi = 2\pi f N\phi \qquad (5.3.13)$$

and its rms value is

$$E_{rms} = \frac{2\pi}{\sqrt{2}} f N\phi = 4.44 f N\phi \qquad (5.3.14)$$

which are identical in form to the corresponding emf equations for a transformer. Consistent with the discussion that followed Equations 3.1.12 and 3.1.13, the effect of a time-varying flux in association with stationary transformer windings is the same as that of relative motion of a coil and a constant-amplitude spatial flux-density wave in a rotating machine. The space distribution of flux density is transformed into a time-variation of voltage because of the time element introduced by mechanical rotation. The induced voltage is a single-phase voltage for single-phase synchronous machines of the nature discussed so far. As pointed out at the end of the solution of Example 5.2.3, Part *(c)*, in order to avoid the pulsating torque, one could think of employing polyphase windings and polyphase sources to develop constant power under balanced conditions of operation.

In fact, with very few exceptions, three-phase synchronous machines are most commonly used for power generation; in general, three-phase ac power systems including power generation, transmission, and utilization have become most popular because of their relative economic advantages. Photos 5–6 and 5–7 show typical salient-pole and non-salient-pole three-phase synchronous generators. An elementary three-phase, two-pole synchronous machine with one coil per phase (chosen for simplicity) is shown in Figure 5.3.10*(a)*. The coils are displaced by 120 electrical degrees from each other in space so that three-phase voltages of positive phase sequence *a-b-c*, displaced by 120 electrical degrees from each other in time, could be produced. Figure 5.3.10*(b)* shows an elementary three-phase, four-pole synchronous machine with one slot per pole per phase. There are 12 coil sides or six coils in all; two coils belong to each phase, which may be connected in series either in wye or delta as shown in Figure 5.3.10*(c)* and *(d)*. Equation 5.3.14 can be applied to give the rms voltage per phase when *N* is treated as the total series turns per phase. The coils may also be connected in parallel to increase the current rating of the machine. In actual ac machine windings, instead of concentrated full-pitch windings, distributed fractional-pitch armature windings are commonly used in order to make better use of iron and copper, and to make waveforms of the

PHOTO 5.6   A salient-pole three-phase synchronous generator.   Photo courtesy of Marathon Electric Company.

PHOTO 5.7   A nonsalient-pole three-phase synchronous generator.   Photo courtesy of Electric Machinery Manufacturing Company.

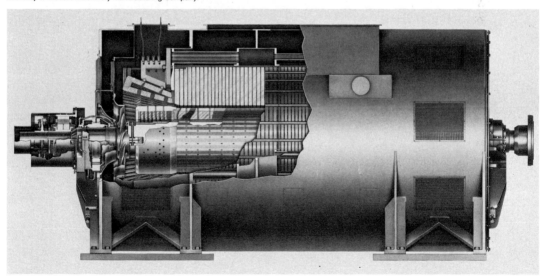

FIGURE 5.3.10 Elementary three-phase synchronous machines:
(a) salient two-pole machine, *(b)* salient four-pole machine, *(c)* phase
windings connected in wye, *(d)* phase windings connected in delta.

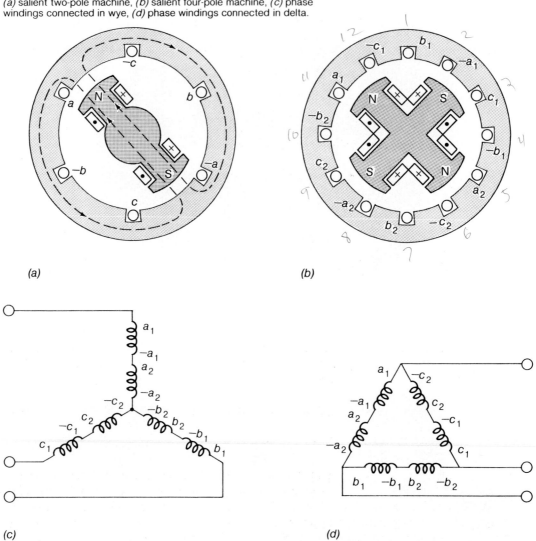

(a)

(b)

(c)

(d)

FIGURE 5.3.11   Simplified two-pole synchronous machine with nonsalient poles.

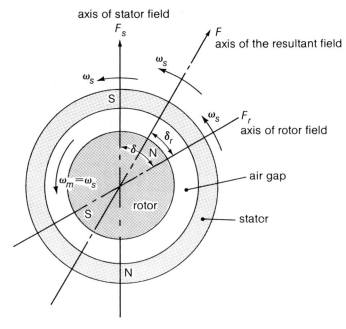

generated voltage (in time) and the armature mmf (in space) as nearly sinusoidal as possible; that is to say, the armature coils of each phase are distributed in a number of slots and the coil span may be shorter than a full pitch. In such cases Equation 5.3.14 will be modified as

$$E_{rms} = 4.44 k_w f N_{ph} \phi \text{ volts per phase} \tag{5.3.15}$$

where k_w is a winding factor (less than unity; usually about 0.85 to 0.95), and N_{ph} is the number of series turns per phase.

The polyphase synchronous machine operates with direct current supplied to the field winding (assumed to be on the rotor, which is usually the case) through two slip rings, and polyphase ac supplied to the armature (assumed to be on the stator). The rotor mmf, which is obtained from a dc source, is stationary with respect to the rotor structure. When carrying balanced polyphase currents, as stated earlier and shown later in Section 5.4, the armature winding produces a magnetic field in the air gap rotating at synchronous speed (Equation 5.3.11 or 5.3.12) as determined by the system frequency and the number of poles in the machine. But the field produced by the dc rotor winding revolves with the rotor. For the production of a steady unidirectional torque, the rotating fields of stator and rotor must be traveling at the same speed, and therefore the rotor must turn precisely at the synchronous speed. Under such conditions, the second possibility for the development of a constant torque suggested in the discussion following Equation 5.2.33 is satisfied. The rotor-mmf axis of the sinusoidally distributed rotor-mmf wave of constant amplitude in the air gap will remain at a constant angle to that of the stator. The resulting physical system is shown in Figure 5.3.11, in which the poles correspond to peak values of gap mmf components produced by the rotor and stator windings. The polyphase machine in which the rotor rotates in synchronism with the rotating mmf wave produced by the stator is called a polyphase synchronous machine. The actual flux-density distribution in the air gap is produced by the resultant mmf, the maximum of which occurs at some angle between the mmf axes. Since the rotor is rotating at the same speed as the resultant flux wave produced in the air gap, no emf is induced in the rotor winding because there is no variation in the associated flux linkage. The essential condition for energy conversion in the cylindrical machine with single stator and rotor windings was expressed by

$$\omega_m = \pm (\omega_s \pm \omega_r) \tag{5.3.16}$$

where the $\pm$ outside the parentheses implied that the direction of rotation was immaterial in such single-winding machines. However, in a machine with polyphase stator winding, this can no longer be the case, since the rotor-mmf axis must rotate in the same direction as that of the stator in order to satisfy the condition for constant torque deduced from Equation 5.2.33. Thus, for machines with polyphase stators, Equation 5.3.16 should be modified as

$$\omega_m = \omega_s \pm \omega_r \tag{5.3.17}$$

Equation 5.2.33 applies for the electromagnetic torque produced by the non-salient-pole (or cylindrical-rotor) machine. The torque can also be expressed in terms of the resultant mmf wave $\overline{F}$ (see Problem 5–14):

$$T_e = -\frac{P}{2} K F_s F_r \sin \delta \tag{5.2.33}$$

or

$$T_e = -\frac{P}{2} K F F_s \sin \delta_s \tag{5.3.18}$$

or

$$T_e = -\frac{P}{2} K F F_r \sin \delta_r \tag{5.3.19}$$

where δ_s and δ_r are the angles, as shown in Figure 5.2.5(c), between $\overline{F}$ and $\overline{F}_s$, and $\overline{F}$ and $\overline{F}_r$, respectively. Alternatively, the torque can be expressed in terms of the resultant flux ϕ per pole produced by the combined effect of the stator and rotor mmfs (see Problem 5–14d):

$$T_e = -\frac{\pi}{2} \left(\frac{P}{2}\right)^2 \phi F_r \sin \delta_r \tag{5.3.20}$$

FIGURE 5.3.12 Torque-angle characteristic curve of a cylindrical-rotor synchronous machine (for a given field current and a fixed terminal voltage, i.e., for a constant resultant air-gap flux).

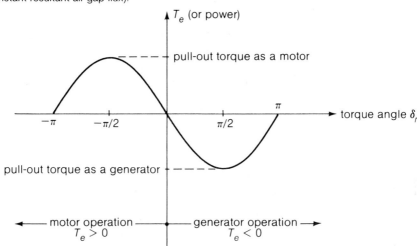

where δ_r is the angle between the rotor-mmf axis and the resultant-flux or mmf axis. When the armature terminals are connected to a balanced polyphase *infinite bus* (which is a high capacity, constant-voltage, constant-frequency system), the resultant air-gap flux ϕ is approximately constant, independent of shaft load; under normal operating conditions the armature-resistance voltage drop is negligible, and the armature-leakage flux is relatively small compared with the resultant air-gap flux ϕ, which is given by Equation 5.3.15 as

$$\phi = \frac{\text{Terminal phase voltage}}{4.44 \; k_w \, f \, N_{ph}} \qquad\qquad (5.3.21)$$

The rotor mmf F_r, determined by the dc field current, is also a constant under normal operating conditions. So, as seen from Equation 5.3.20, any variation in the torque requirements of the load have to be taken care of entirely by variation of the angle δ_r; and that is the reason that δ_r is known as the *torque angle* (or the *load angle*) of a synchronous machine. The effect of salient poles on the torque-angle characteristic is discussed later in detail under the chapter on synchronous machines.

The torque-angle characteristic curve of a cylindrical-rotor synchronous machine is shown in Figure 5.3.12 as a function of the angle δ_r. For $\delta_r < 0$, $T_e > 0$; the developed torque is positive and acts in the direction of rotation; the machine operates as a motor. If, on the other hand, the machine is driven by a prime mover so that δ_r becomes positive, the torque is then negative, and the machine operates as a generator. It should be noted that at standstill, i.e., when $\omega_m = 0$, no

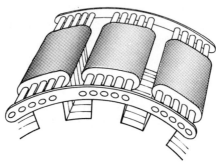

FIGURE 5.3.13   A sketch of damper bars located on the salient-pole
shoes of a synchronous machine.

average unidirectional torque is developed by the synchronous machine, and such a synchronous motor is not capable of self-starting because it has no starting torque. A method must be provided for bringing the machine up to synchronous speed.

Since a synchronous machine operates only at synchronous speed under steady-state conditions, when the load on the synchronous machine is to be changed, the machine cannot operate at synchronous speed during the load transition; the readjustment process is in fact a dynamic one. In the case of a generator connected to an infinite bus, an increase in the electrical output power of the generator is brought about by increasing the mechanical input power supplied by the prime mover; the speed of the rotor increases momentarily during the process, and the axis of the rotor-field mmf advances relative to the axis of the armature mmf and also relative to the axis of the resultant air-gap mmf. The increase in torque angle results in an increase in electrical power output, and the torque angle will continue to increase until the output power is equal to the net input power; the machine will then lock itself into synchronism and continue to rotate at synchronous speed until a further load change takes place. The same general argument can be made for the operation of a synchronous motor, except that in this case an increase in mechanical load decreases the speed of the rotor during the transition period, so that the axis of the rotor-field mmf falls behind that of the stator mmf or the resultant air-gap mmf by the required value of the load angle.

When the load on a synchronous machine changes, the load angle changes from one steady value to another and, during this transition, oscillations in the load angle and consequent associated mechanical oscillations (known as *hunting*) will occur. In order to damp out these oscillations, it is a common practice to provide an additional short-circuited winding (made out of copper or brass bars located in pole-face slots on the pole shoes of a salient-pole machine and connected together at the ends of the machine) known as a *damper winding* (or an *amortisseur winding*), as shown in Figure 5.3.13, on the field structure. This is also useful in getting a synchronous motor started, as explained later. When the machine is operating under steady-state conditions at synchronous speed, however, this winding has no effect because there is no rate of change of flux linkage with this winding and hence no voltage is induced in it. As seen from Figure 5.3.12, when δ_r is $\pm \pi/2$ or $\pm 90°$, the maximum torque (or power), called the *pull-out torque* (or the *pull-out power*), is reached for a fixed terminal voltage and a given field current. If the load requirements exceed this value, the motor slows down because of the excess shaft torque and synchronous-motor action is lost because the rotor and stator fields are no longer stationary with respect to each other. Any load requiring a torque greater than the maximum torque results in unstable operation of the machine; the machine pulls out of synchronism, exhibiting the phenomenon known as *pulling out of step* or losing synchronism. The motor is usually disconnected from the electric supply by the action of automatic circuit breakers, and the machine comes to a standstill. It should be noted here that the pull-out torque can be increased by increasing either the field current or the terminal voltage. In the case of a generator connected to an infinite bus, synchronism will be lost if the torque applied by the prime mover exceeds the maximum generator pull-out torque; the speed will then increase rapidly unless the quick-response governor action comes into play on the prime mover to control the speed.

ELEMENTARY DIRECT-CURRENT MACHINES   A dc machine operates with direct current applied to its field winding, which is generally located on the salient-pole stator of the machine, and direct current applied to a commutator (via the brushes) connected to the armature winding situated inside slots on the cylindrical rotor, as shown schematically in Figures 5.3.1 and 5.3.5. In terms of the discussion that followed Equation 5.2.33 for development of a constant torque, the possibility (*i*) is satisfied; that is, in this case the stator-mmf axis is fixed in space and the rotor-mmf axis is also fixed in space, even when the rotor winding is physically rotating, because of the commutator action, which has been briefly discussed earlier. Thus the dc machine will operate under steady-state conditions, whatever the rotor speed ω_m. The dc motor is capable of producing starting torque. The armature current in the armature winding is alternating; the action of the commutator is to change the armature current from a frequency governed by the mechanical speed of rotation to zero frequency at the commutator brushes connected to the external circuit.

For the case of a dc machine with a flux per pole ϕ, the total flux cut by one conductor in one revolution is given by ϕP, where P is the number of poles of the machine. If the speed of rotation is n rpm, the emf generated in a single conductor will be given by

$$e = \frac{\phi P n}{60} \qquad (5.3.22)$$

For an armature with Z conductors and α parallel paths, the total generated armature emf E_a is given by

$$E_a = \frac{P \phi n Z}{60\alpha} \qquad (5.3.23)$$

Since the angular velocity ω_m is given by $2\pi n/60$, Equation 5.3.23 becomes

$$E_a = \frac{PZ}{2\pi\,\alpha} \phi\,\omega_m = K_a\,\phi\,\omega_m \qquad (5.3.24)$$

where K_a is the design constant given by $(PZ/2\pi\ \alpha)$. This is the speed voltage appearing across the brush terminals in the quadrature axis (see Figure 5.3.5) because of the field excitation producing ϕ in the direct axis. That is why, in the schematic circuit representation of a dc machine, the field axis and the brush axis are shown in quadrature, i.e., perpendicular to each other. The generated voltage as observed from the brushes is the sum of the rectified voltages of all the coils in series between brushes. If the number of coils is sufficiently large, the ripple in the wave-form of the armature voltage (as a function of time) becomes very small, thereby making it direct or constant in magnitude.

The instantaneous electrical power associated with the speed voltage should be equal to the instantaneous mechanical power associated with the electromagnetic torque, the direction of power flow being determined by whether the machine is operating as a motor or generator.

$$T_e\ \omega_m = E_a\ I_a \tag{5.3.25}$$

With the aid of Equation 5.3.23, it follows that

$$T_e = K_a\ \phi\ I_a \tag{5.3.26}$$

which is created by the interaction of the magnetic fields of stator and rotor. As mentioned earlier, if the machine is acting as a generator, this torque opposes rotation; if it is acting as a motor, the electromagnetic torque acts in the direction of rotation.

If there is a large number of conductors on the rotor and a commutated dc source is adopted to excite the rotor, the variation of armature mmf around the periphery is not sinusoidal; in fact, it is very nearly triangular, as discussed later (in Section 6.7). However, this is not important, as the machine is not to be connected to an ac system. For the same reason, it is not necessary for the distribution of stator mmf around the air gap to be sinusoidal. In fact, the air-gap flux distribution for a dc machine usually approximates a flat-topped wave (nearly rectangular) rather than the sine wave found in ac machines. The location of brushes on the commutator arrangement connected to the armature winding ensures that the rotor and stator-mmf axes are at all times at right angles to one another as shown in Figure 5.3.5. The expression for the torque given by Equation 5.2.33 is applicable to the dc machine, provided the constant K is adjusted for the nonsinusoidal mmf distributions and $\sin\delta$ is made equal to unity. Not only are the conditions for constant torque fulfilled but also the condition for maximum torque and hence for maximum energy conversion is satisfied, since $\delta = \pm\ \pi/2$, depending upon whether generator or motor action occurs for a given direction of rotation.

Since both the armature and field circuits carry direct current in the case of a dc machine, they can be connected together either in series or in parallel. The machine is known as a *shunt machine* when the armature and field circuits are connected in parallel; in such a case, the field coils are wound with a large number of turns carrying a relatively small current. When the circuits are connected in series, the machine is known as a *series machine;* the field winding in such a case carries the full armature current and is wound with a smaller number of turns. If the machine is provided with a series winding and a shunt field winding, it is known as a *compound machine;* in such a case, the series field may be connected either *cumulatively,* so that its mmf adds to that

Figure 5.3.14 Field-circuit connections of direct-current machines:
(a) separately excited machine, *(b)* shunt machine, *(c)* series machine,
(d) compound machine.

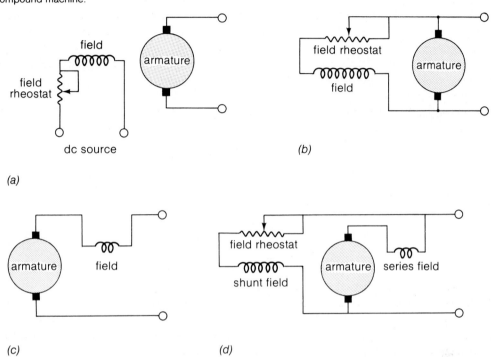

of the shunt field, or *differentially,* so that it opposes. The differential connection is very rarely used. The voltage of both shunt and compound generators (or the speed in case of motors) is controlled over reasonable limits by means of a field rheostat in the shunt field. The machines are said to be *self-excited* when the machine supplies its own excitation of the field windings as in the above-mentioned cases. However, the field windings may be *separately excited* from an external dc source. A small amount of power in the field circuit can control a large amount of power in the armature circuit. The dc generator may then be viewed as a power amplifier. Some possible field-circuit connections of dc machines are shown in Figure 5.3.14.

In the case of self-excited generators, residual magnetism must be present in the ferromagnetic circuit of the machine in order to get the self-excitation process started. For the case of a dc *generator,* the relation between the steady-state generated emf E_a and the terminal voltage V_t is given by

$$V_t = E_a - I_a R_a \qquad (5.3.27)$$

PHOTO 5.8 Cross-sectional view of a typical dc machine.   Photo courtesy of
Westinghouse Electric Corporation.

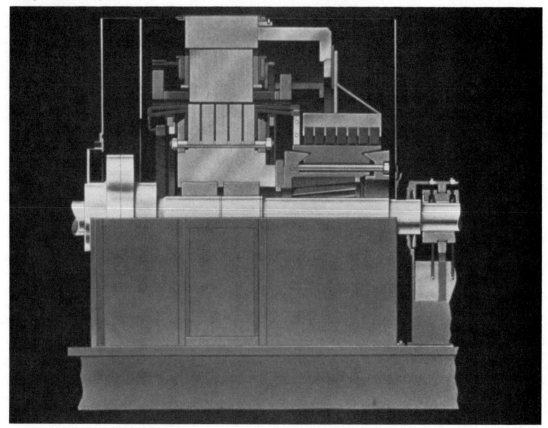

where I_a is the armature current *output* and R_a is the armature circuit resistance. For the case of a dc *motor*, the relationship is given by

$$V_t = E_a + I_a R_a \qquad\qquad (5.3.28)$$

where I_a is now the armature current *input*. Under steady-state conditions, while volt-ampere characteristic curves are of interest for dc generators, speed-torque characteristics are of interest for dc motors. Depending on the method of excitation of the field windings, a wide variety of operating characteristics can be obtained; these make the dc machine very versatile and adaptable for control. Photo 5–8 shows a cross-sectional view of a typical dc machine.

FIGURE 5.3.15 Mmf-axes of an induction machine.

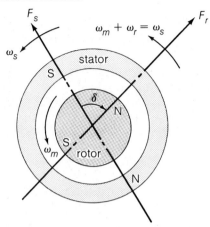

ELEMENTARY INDUCTION MACHINES   In our discussion that followed Equation 5.2.33, the third possible method of producing constant torque was stated to be that of causing the mmf axes of stator and rotor to rotate at such speeds relative to their windings that they remain stationary with respect to each other. If the stator and rotor windings are polyphase and carry polyphase ac, then both the stator- and rotor-mmf axes may be caused to rotate relative to their windings. Such a machine will have polyphase stator ac excitation at ω_s, polyphase rotor ac excitation at ω_r, and the rotor speed ω_m satisfying Equation 5.3.17. Let us consider the rotor speed $\omega_m = \omega_s - \omega_r$ and the same phase sequence of sources. A rotating magnetic field of constant amplitude, rotating at ω_s rad/s relative to the stator is produced because of polyphase stator excitation. Also a rotating magnetic field of constant amplitude, rotating at ω_r rad/s relative to the rotor is produced because of polyphase rotor excitation. The speed of rotation of the rotor magnetic field relative to the stator will be $(\omega_m + \omega_r)$ or ω_s, if the rotor is rotating with a positive speed of rotation ω_m in the direction of rotating fields. In that case, the condition for energy conversion at constant torque is satisfied. Such a situation is diagramatically illustrated in Figure 5.3.15. The machine under these conditions is operating as a double-fed polyphase machine. Normally, in an induction machine with polyphase stator and rotor windings, only a source to excite the stator is employed, and the rotor excitation at the appropriate frequency is induced from the stator winding; that is why the machine is known as an induction machine.

In an induction machine, the stator winding (Photo 5–9) is essentially the same as that of a synchronous machine. Equation 5.3.15 and the considerations leading to it hold good as in the case of a synchronous machine. When excited from a balanced polyphase source, the polyphase winding produces a magnetic field in the air gap rotating at a synchronous speed determined by the number of poles and the applied stator frequency, given by Equation 5.3.11. On the rotor, the winding is electrically closed on itself and very often has no external terminals; the rotor may be one of two types: *(i)* A *wound rotor* has a polyphase winding similar to and wound for the same

PHOTO 5.9   Stator of an induction motor.   Photo courtesy of General Electric Company.

number of poles as the stator winding. The terminals of the rotor winding (wye- or delta-connected in the case of three-phase machines) are brought to insulated slip rings mounted on the shaft. Carbon brushes bearing on these slip rings make the rotor terminals available to the circuitry external to the motor. The rotor winding is usually short-circuited through external resistances that can be varied. *(ii)* A *squirrel-cage rotor* (Photo 5–10) has a winding consisting of conducting bars of copper or aluminum embedded in slots cut in the rotor iron and short-circuited at each end by conducting end rings. The machine with such a rotor is known as a squirrel-cage induction machine, which is the electromagnetic machine most widely used as a motor because of its extreme simplicity and ruggedness. Although the induction motor is the most common of all motors, the induction machine is very rarely used as a generator, because its performance characteristics as a generator have not been found satisfactory for most applications. The induction machine with a wound rotor may also be used as a *frequency changer.*

The polyphase induction motor (Photo 5–11) operates with polyphase ac applied to the primary winding, located usually on the stator of the polyphase machines. Three-phase motors are most popularly used in practice, while two-phase motors are used in control systems. The induction machine has emf (and consequently current) induced in the short-circuited secondary (or rotor) winding by virtue of the primary rotating mmf. Such a machine is then singly excited. The induction machine may be regarded as a generalized transformer in which energy conversion takes place, and electric power is transformed between the stator and rotor along with a change of frequency and a flow of mechanical power.

PHOTO 5.10   A squirrel-cage rotor of a three-phase induction motor.
Photo courtesy of Westinghouse Electric Corporation.

PHOTO 5.11   Totally enclosed fan-cooled 10-hp, three-phase, 60-Hz, 230/
460-V squirrel-cage induction motor.   Photo courtesy of Siemens-Allis Corporation.

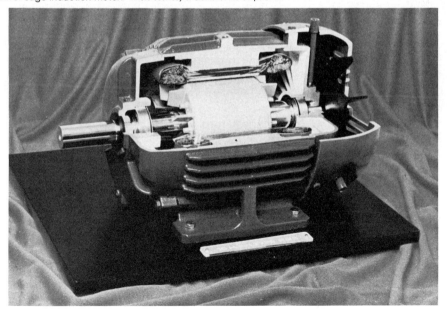

Let us assume that the rotor is turning at a steady speed of n rpm in the same direction as the rotating stator field. Let the synchronous speed of the stator field be n_1 rpm, as given by Equation 5.3.11, corresponding to the applied stator frequency f_s Hz or ω_s rad/s. It is convenient to introduce the concept of a *per-unit slip* **S** given by

$$\mathbf{S} = \frac{\text{Synchronous speed} - \text{Actual rotor speed}}{\text{Synchronous speed}}$$

$$= \frac{n_1 - n}{n_1} = \frac{\omega_s - \omega_m}{\omega_s} \qquad (5.3.29)$$

The rotor is then traveling at a speed $(n_1 - n)$ or $n_1\mathbf{S}$ rpm in the backward direction with respect to the stator field. This relative motion of the flux and rotor conductors induces voltages of frequency $(\mathbf{S}f_s)$, known as the *slip frequency,* in the rotor winding. Thus the induction machine is similar to a transformer in its electrical behavior, but with an additional feature of frequency change. The frequency f_r Hz of the secondary (or rotor) currents is then given by

$$f_r = \frac{\omega_r}{2\pi} = \mathbf{S}f_s \qquad (5.3.30)$$

At standstill, $\omega_m = 0$ so that the slip $\mathbf{S} = 1$ and $f_r = f_s$; that is, the machine then acts as a simple transformer with an air gap and a short-circuited secondary winding. A steady starting torque is produced because the condition for energy conversion at constant torque is satisfied; hence the polyphase induction motor is self-starting. However, at synchronous speed, $\omega_m = \omega_s$ so that the slip $\mathbf{S} = 0$ and $f_r = 0$; no induction takes place because there is no relative motion of the flux and rotor conductors. Thus, at synchronous speed, the value of the secondary mmf will be zero, and no torque will be produced. That is to say that the induction motor cannot run at synchronous speed. The no-load speed of the induction motor is usually of the order of 99.5 percent of synchronous speed so that the no-load per-unit slip is about 0.005, and the full-load per-unit slip is of the order of 0.05. Thus the polyphase induction motor is effectively a constant-speed machine.

An induction machine, connected to a polyphase exciting source on its stator side, can be made to generate (i.e., the power flow would be reversed compared to that of a motor) if its rotor is driven mechanically by an external means at above synchronous speed, so that $\omega_m > \omega_s$ and the slip becomes negative. If the machine is driven mechanically in the direction opposite to its primary rotating mmf, then the slip is greater than unity and the machine acts as a brake. For example, let the machine be operating normally as a loaded motor: if two of the three-phase supply lines to the stator are reversed, the direction of the stator-rotating mmf will reverse (this aspect is explained in detail in Section 5.4); the rotor will then be rotating in the direction opposite to that of the rotating mmf, so that the machine will act as a brake and the speed will rapidly come to zero, at which time the electric supply may be removed from the machine. Such a reversal of two supply lines of the three-phase system is a useful method of rapidly stopping the motor and is generally referred to as *plugging* or *plug-braking*. If, however, the electric supply is not removed at zero speed, the machine will reverse its direction of rotation because of the change of phase

FIGURE 5.3.16 General form of torque-speed curve (or torque-slip characteristic) for a polyphase induction machine.

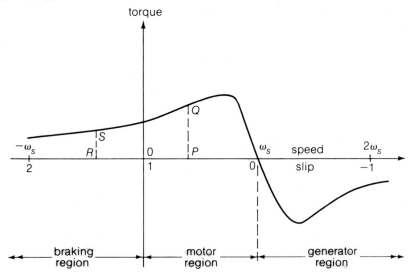

sequence of the supply resulting from the interchange of the two stator leads. The general form of the torque-speed curve (or torque-slip characteristic) for a polyphase induction machine between rotor-speed limits of $-\omega_s \leq \omega_m \leq 2\omega_s$, corresponding to a range of slips of $2 \leq \mathbf{S} \leq -1$, is shown in Figure 5.3.16.

In general, concerning the polyphase operation of induction machines, it should be pointed out that a rotor with any number of phases will develop torque in a stator of the same or any other number of phases, single-winding stators only excepted. The student should be able to reason it out from the necessary condition to be fulfilled for producing the torque. However, the number of poles in the stator and rotor must be the same for torque production, as mentioned earlier. In the case of a wound-rotor induction machine, the rotor must be wound for the same number of poles as the stator; for a squirrel-cage machine, however, an equal number of poles is induced in the rotor. The emfs induced in the squirrel-cage conductors produce currents that circulate through the bars and the end rings, thereby effectively forming a rotor winding with a large number of phases. In machines of small size, sometimes, solid iron rotors as well as rotors with a thin cylinder of conductor (such as copper) enclosing the rotor magnetic core are employed; also, in what are known as composite-rotor induction machines, alternate layers of copper and iron in the form of thin cylinders enclosing the rotor-iron structure are used for high-speed and high-frequency operation.

The torque that exists at any mechanical speed other than synchronous speed is known as an *asynchronous torque.* The induction machine is also known as an *asynchronous machine,* since no torque is produced at synchronous speed and the machine runs at a speed other than synchronous speed; in fact, as a motor, the machine will run only at a speed that is less than synchronous speed with positive slip. The factors influencing the general shape of the torque-speed characteristic (shown in Figure 5.3.16) can be appreciated in terms of the torque equation (5.3.20). Noting that the resultant air-gap flux ϕ is nearly constant when the stator-applied voltage and frequency are constant as seen by Equation 5.3.21, and that the rotor mmf F_r is proportional to the rotor current I_r, the torque may be expressed as

$$T_e = K_1 I_r \sin \delta_r \tag{5.3.31}$$

where K_1 is a constant and δ_r is the angle between the rotor-mmf axis and the resultant-flux or mmf axis. The rotor current I_r is determined by the rotor-induced voltage (proportional to slip) and rotor impedance. Since the slip is small under normal running conditions, as already mentioned, the rotor frequency $f_r = S f_s$ is very low (of the order of 3 Hz in 60-Hz motors for a per-unit slip of 0.05). Hence, in this range the rotor impedance is largely resistive, and the rotor current is very nearly proportional to and in phase with the rotor voltage; that is, the rotor current is very nearly proportional to slip. An approximately linear torque-speed relationship can be observed in the range of low values of slip in Figure 5.3.16. Further, the rotor-leakage reactance being very small compared with rotor resistance, the rotor-mmf wave will lag approximately 90 electrical degrees behind the resultant flux wave, and therefore $\sin \delta_r$ will be approximately equal to unity.

As slip increases, the rotor impedance increases because of the increasing effect of rotor-leakage inductance; the rotor current will then be somewhat less than proportional to slip. The rotor current will lag further behind the induced voltage, and the rotor-mmf wave will lag further behind the resultant-flux wave, so that $\sin \delta_r$ decreases. The torque increases with increasing values of slip up to a point and then decreases, as shown in Figure 5.3.16 for the motor region. The maximum torque that the machine can produce is sometimes referred to as the *breakdown torque,* because it limits the short-time overload capability of the motor. Higher starting torque can be obtained by inserting external resistances in the rotor circuit, as is usually done in the case of wound-rotor induction motors; these resistances can be cut out for the normal running conditions in order to operate the machine with a higher efficiency.

You may recall that we mentioned that a synchronous motor has no starting torque, and it is usually provided with a damper (or amortisseur) winding located in the rotor-pole faces. Such a winding acts as a squirrel-cage winding to make the synchronous motor start as an induction motor and come up almost to synchronous speed, with the dc field winding unexcited; if the load and inertia are not too large, when the field winding is energized from a dc source, the motor will then pull into synchronism and act as a synchronous motor.

So far the discussion on induction machines applies only to machines operating from a polyphase supply. Of particular interest is the single-phase induction machine (Photo 5-12), which is most widely used as a fractional-horsepower ac motor supplying the motive power for all types of equipment at the home, office, and factory. For the sake of simplicity, let us consider a single-phase

PHOTO 5.12   Single-phase squirrel-cage induction motor.   Photo courtesy of
Marathon Electric Company.

induction motor with a squirrel-cage rotor and a stator carrying a single-phase winding, connected to a single-phase ac supply. In such a case as this, the primary mmf cannot be rotating; but in fact, it is pulsating in phase with the variations in the single-phase primary current. However, it can be shown (see Section 5.4) that any pulsating mmf can be resolved in terms of two rotating mmfs of equal magnitude, rotating in synchronism with the supply frequency but in opposite directions. The wave rotating in the same direction as the rotor is known as the *forward-rotating wave,* while the other, rotating in the opposite direction, is called the *backward-rotating wave.* Then the slip, $\mathbf{S}_f$, of the machine with respect to the forward-rotating wave is given by

$$\mathbf{S}_f = \frac{\omega_s - \omega_m}{\omega_s} \qquad (5.3.32)$$

which is the same as Equation 5.3.29. However, the slip $\mathbf{S}_b$ of the machine with respect to the backward-rotating wave will be given by

$$\mathbf{S}_b = \frac{-\omega_s - \omega_m}{\omega_s} = 2 - \mathbf{S}_f \qquad (5.3.33)$$

FIGURE 5.3.17 Approximate shape of the torque-speed curve for a single-phase induction motor.

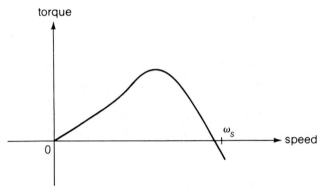

S_f and S_b are known as the forward (or positive-sequence) slip and the backward (or negative-sequence) slip, respectively. Assuming that the two component mmfs exist separately, the frequency and magnitude of the component emfs induced in the rotor by their presence will, in general, be different because S_f is not equal to S_b. Then the machine can be thought of as producing a steady total torque as the algebraic sum of the component torques. At standstill, however, $\omega_m = 0$ and the component torques are equal and opposite so that no starting torque is produced. Thus it is clear that a single-phase induction motor is not capable of self-starting; but it will continue to rotate once started in any direction. In practice, additional means are provided to get the machine started (usually as an asymmetrical two-phase motor) from a single-phase source, and the machine is then run as a single-phase motor. An approximate shape of the torque-speed curve for a single-phase motor can readily be obtained from that of a three-phase machine shown in Figure 5.3.16. Corresponding to a positive slip of $S_f = OP$ in Figure 5.3.16, the positive-sequence torque is PQ; then, corresponding to $S_b = 2 - S_f = OR$, the negative-sequence torque is RS, so that the resultant torque is given by $(PQ - RS)$. This procedure can be repeated for a range of slips $(1 \le S \le 0)$ so as to give the general form of the torque-speed characteristic of a single-phase induction motor, as shown in Figure 5.3.17.

5.4   ROTATING MAGNETIC FIELDS

In this section we set out to show, as stated earlier, that a rotating field of constant amplitude and sinusoidal space distribution of mmf around the periphery of the stator is produced by a three-phase winding located on the stator and excited by balanced three-phase currents when the respective phase windings are wound $2\pi/3$ electrical radians (or 120 electrical degrees) apart in space.

FIGURE 5.4.1   Simple two-pole, three-phase winding arrangement on a stator.

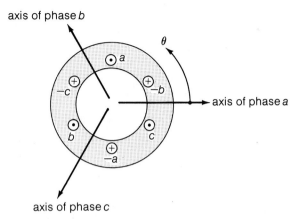

axis of phase b

θ

axis of phase a

axis of phase c

Let us consider a simple two-pole, three-phase winding arrangement on a stator as shown in Figure 5.4.1; the windings of the individual phases are displaced by 120 electrical degrees from each other in space around the air-gap periphery. The reference directions are given for positive phase currents. The concentrated full-pitch coils, shown here for simplicity and convenience, do in fact represent the actual distributed windings producing sinusoidal mmf waves centered on the magnetic axes of the respective phases. Thus these three sinusoidal mmf waves are then displaced by 120 electrical degrees from each other in space. Let a balanced three-phase excitation be applied with phase sequence *a-b-c*:

$$i_a = I \cos \omega_s t$$

$$i_b = I \cos(\omega_s t - 120°)$$ (5.4.1)

$$i_c = I \cos(\omega_s t - 240°)$$

where I is the maximum value of the current, and the time $t = 0$ is arbitrarily chosen when the *a*-phase current is a positive maximum. Each phase current is an ac wave varying in magnitude sinusoidally with time. Hence the corresponding component mmf waves vary sinusoidally with time. The sum of these components yields the resultant mmf.

Analytically, the resultant mmf at any point at an angle θ from the axis of phase *a* is given by

$$F(\theta) = F_a \cos \theta + F_b \cos(\theta - 120°) + F_c \cos(\theta - 240°)$$ (5.4.2)

But the mmf amplitudes vary with time according to the current variations:

$$F_a = F_m \cos \omega_s t$$

$$F_b = F_m \cos(\omega_s t - 120°) \tag{5.4.3}$$

$$F_c = F_m \cos(\omega_s t - 240°)$$

Then, on substitution, it follows that

$$\begin{aligned} F(\theta, t) = {} & F_m \cos \theta \cos \omega_s t \\ & + F_m \cos(\theta - 120°) \cos(\omega_s t - 120°) \\ & + F_m \cos(\theta - 240°) \cos(\omega_s t - 240°) \end{aligned} \tag{5.4.4}$$

By the use of the trigonometric identity

$$\cos \alpha \cos \beta = \frac{1}{2} \cos(\alpha - \beta) + \frac{1}{2} \cos(\alpha + \beta)$$

and noting that the sum of three equal sinusoids displaced in phase by 120° is equal to zero, Equation 5.4.4 can be simplified as

$$F(\theta, t) = \frac{3}{2} F_m \cos(\theta - \omega_s t) \tag{5.4.5}$$

which is the expression for the resultant mmf wave; it has a constant amplitude $(3/2) F_m$, is a sinusoidal function of the angle θ, and it rotates in synchronism with the supply frequency; hence it is called a rotating field. The constant amplitude is 3/2 times the maximum contribution of any one phase F_m. The angular velocity of the wave is $\omega_s = 2\pi f_s$ electrical radians per second, where f_s is the frequency of the electrical supply in hertz. For a P-pole machine, the rotational speed is given by

$$\omega_m = \frac{2}{P} \omega_s \text{ rad/s or } n = \frac{120 f_s}{P} \text{ rpm} \tag{5.4.6}$$

which is the synchronous speed.

The same result may be obtained graphically as shown in Figure 5.4.2. The spatial distribution of the mmf of each phase and that of the resultant mmf (given by the algebraic sum of the three components at any given instant of time) are shown in Figure 5.4.2. Part *(a)* applies for that instant when the *a*-phase current is a positive maximum, while part *(b)* refers to that instant when the *b*-phase current is a positive maximum; the intervening time corresponds to 120 electrical degrees.

FIGURE 5.4.2 Generation of a rotating mmf: *(a)* spatial mmf distribution at the instant in time when the *a*-phase current is a maximum, *(b)* spatial mmf distribution at the instant in time when the *b*-phase current is a maximum.

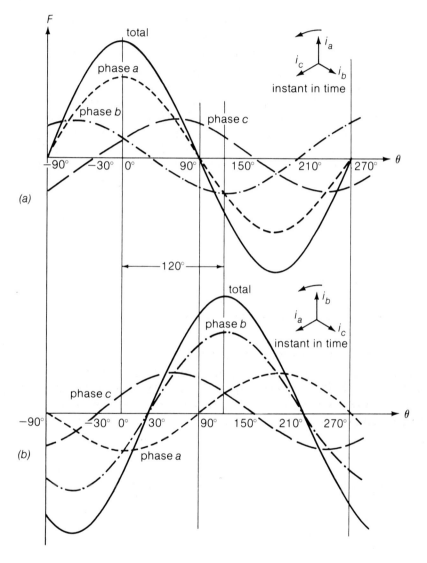

It can be seen from Figure 5.4.2 that, during this time interval, the resultant sinusoidal mmf waveform has traveled (or rotated through) 120 electrical degrees of the periphery of the stator structure carrying the three-phase winding. That is to say, the resultant mmf wave is rotating in synchronism with time variations in current, with its peak amplitude remaining constant at 3/2 times that of the maximum phase value. Note that the peak value of the resultant stator mmf wave coincides with the axis of a particular phase winding when that phase winding carries its peak current. The graphical process may be continued for different instants of time to show that the resultant mmf is in fact rotating in synchronism with the supply frequency.

Although the analysis here is carried out only for a three-phase case, it holds good for any q-phase ($q > 1$; i.e., polyphase) winding excited by balanced q-phase currents when the respective phases are wound $2\pi/q$ electrical radians apart in space. However, in a balanced two-phase case, note that the two phase windings are displaced 90 electrical degrees in space, and the phase currents in the two windings are phase-displaced by 90 electrical degrees in time. The constant amplitude of the resultant rotating mmf can be shown to be $q/2$ times the maximum contribution of any one phase. With the reluctance of the magnetic circuit neglected, the corresponding flux density in the air gap of the machine is then given by

$$B_g = \frac{\mu_0 F}{g} \qquad (5.4.7)$$

where g is the length of the air gap.

PRODUCTION OF ROTATING FIELDS FROM SINGLE-PHASE WINDINGS   In this subsection, we would like to show that a single-phase winding carrying alternating current produces a stationary pulsating flux that can be represented by two counter-rotating fluxes of constant and equal magnitude.

Let us consider a single-phase winding, shown in Figure 5.4.3(a), carrying alternating current $i = I \cos \omega t$. It will produce a flux-density distribution whose axis is fixed along the axis of the winding and that pulsates sinusoidally in magnitude. The flux density along the coil axis is proportional to the current and is given by $B_m \cos \omega t$, where B_m is the peak flux density along the coil axis.

Let the winding be on the stator of a rotating machine with uniform air gap, and let the flux density be sinusoidally distributed around the air gap. Then, the instantaneous flux density at any position θ from the coil axis can be expressed as

$$B(\theta) = (B_m \cos \omega t) \cos \theta \qquad (5.4.8)$$

which may be rewritten as

$$B(\theta) = \frac{B_m}{2} \cos(\theta - \omega t) + \frac{B_m}{2} \cos(\theta + \omega t) \qquad (5.4.9)$$

by the use of the trigonometric identity.

FIGURE 5.4.3   Single-phase winding carrying ac, producing a stationary pulsating flux or equivalent rotating flux components.

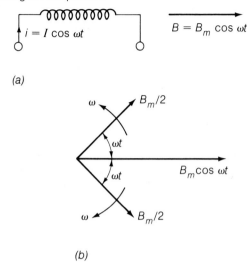

(a)

(b)

The sinusoidal flux-density distribution given by Equation 5.4.8 can be represented by a vector $B_m \cos \omega t$ of pulsating magnitude on the axis of the coil as shown in Figure 5.4.3(b). Alternatively, as suggested by Equation 5.4.9, this stationary pulsating flux-density vector can be represented by two counter-rotating vectors of constant magnitude $B_m/2$, as shown in Figure 5.4.3(b). While Equation 5.4.8 represents a standing space wave varying sinusoidally with time, Equation 5.4.9 represents the two rotating components of constant and equal magnitude, rotating in opposite directions at the same angular velocity given by $d\theta/dt = \omega$. The vertical components of the two rotating vectors in Figure 5.4.3(b) will always cancel, and the horizontal components will yield a sum equal to $B_m \cos \omega t$, the instantaneous value of the pulsating vector.

As indicated earlier, this principle is often used in the analysis of single-phase machines. The two rotating fluxes are considered separately, as if each one represents the rotating flux of a three-phase machine, and the effects are then superimposed. If the system is linear, the principle of superposition holds and yields correct results. However, in a nonlinear system with saturation, one has to be careful in reaching any conclusions since the result is not as obvious as in a linear system.

PHASE SPLITTING   The technique of phase splitting is a means to obtain the two currents in a two-phase winding equal in magnitude and 90° displaced in phase, by using only a single-phase supply. The effect is then the same as if the two identical coils were supplied by a two-phase supply.

FIGURE 5.4.4   Split-phase system: *(a)* schematic diagram of windings for simulating two-phase performance from a single-phase supply, *(b)* phasor diagram for part *(a)*.

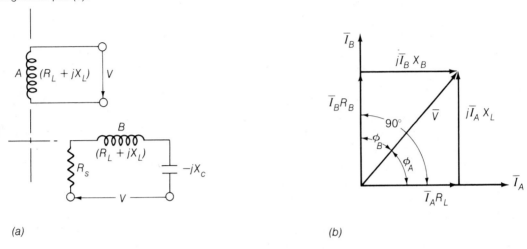

(a)   (b)

Let us consider two windings, as shown in Figure 5.4.4*(a)*, displaced by 90° in space, and connected to the same single-phase supply of voltage V. Let the inductive impedance of each coil be $Z_L = R_L + jX_L$, in which R_L is usually much less than X_L. Let a resistance R_S and a capacitance of reactance X_C be connected in series with coil B, as shown. The total impedance of the circuit B is then given by

$$Z_B = R_B + jX_B = (R_L + R_S) - j(X_C - X_L) \qquad (5.4.10)$$

If the values are so chosen that

$$R_B = X_L \; ; X_B = R_L \qquad (5.4.11)$$

then $|Z_B| = |Z_L|$; which makes the two currents have the same magnitude but with a phase difference of 90°, as seen from the phasor diagram for the split-phase system in Figure 5.4.4*(b)*. Thus a two-phase system is simulated, and hence a rotating flux of constant magnitude is produced from the two-phase system, as discussed earlier.

The system described above (with some modifications) is often used in a single-phase induction motor for the purpose of making it self-starting. If the values used for R_S and X_C are such that the two currents $\bar{I}_A$ and $\bar{I}_B$ are not exactly equal in magnitude, or not exactly 90° out of phase, then the system becomes an unbalanced two-phase system.

5.5 BASIC ASPECTS OF ELECTROMECHANICAL ENERGY CONVERTERS

As one goes deeper into the study of rotating machines, detailed differences and particularly challenging problems emerge from among various machine types. In spite of a number of special problems peculiar to a given class of machines, there is a group of interrelated problems common to all machine types. It is the purpose of this section to touch upon these briefly. These include losses and efficiency, ventilation and cooling, machine ratings, magnetic saturation, leakage and harmonic fluxes, and problems generally associated with machine applications. The matters related to machine ratings, insulation, and allowable temperature rise, as well as the determination of losses, are subjects of standardization by professional organizations such as the Institute of Electrical and Electronics Engineers (IEEE), the National Electrical Manufacturers Association (NEMA), the American National Standards Institute (ANSI), and the International Electrotechnical Commission (IEC).

The limitation on the performance of a machine basically stems from the properties of the materials of which it is made. Much of the great progress made over the years in electric machinery is due to the improvements in the quality and characteristics of steel and insulating materials and to innovative cooling methods. In fact, the progress in cooling methods is the most important single factor that has made it possible to increase power ratings of turboalternators for a given physical size. A steady increase in the horsepower rating from 7.5 to 120 hp, obtained for a given size of an induction-motor frame over a period of the last eight decades, illustrates the effect of continued progress in science and technology.

LOSSES AND EFFICIENCY   Losses are an inevitable part of any energy-transforming device, even though they play essentially no basic role in the energy-conversion process. As in the case of transformers, the efficiency of a machine is given by

$$\text{Efficiency} = \frac{\text{Output}}{\text{Input}} = \frac{\text{Input} - \text{Losses}}{\text{Input}}$$
$$= 1 - \frac{\text{Losses}}{\text{Input}} = \frac{\text{Output}}{\text{Output} + \text{Losses}} \qquad (5.5.1)$$

In view of accuracy and generally high values of efficiency, the efficiency of an electric machine is most commonly determined by the measurement of losses instead of by directly measuring the input and output under load conditions. Many of the definitions and methods of measuring losses and of specifying efficiencies as well as other performance characteristics are standardized by various national and international professional agencies; one should refer to these standards whenever the need arises.

Typical values of full-load efficiencies for rotating machines are about 50 percent for fractional-horsepower motors, 75 percent for 1-hp motors, 90 percent for 100-hp motors, and 98 percent for 100-MVA turboalternators. Rotating machines generally operate quite efficiently except at light loads. As in the case of transformers, maximum efficiency occurs at a load for which variable losses are equal to constant losses. Since most of the practical machines operate at less than full load (about 80–90 percent of full load), thereby making an allowance either for some reserve capacity in emergencies or for some growth in demand, it is a common practice to design machines with maximum efficiency at about 80 percent of full load.

Besides affecting the efficiency and hence the operating cost of the machine, the losses determine the heating of the machine and consequently the rating or power output that can be obtained from a machine without overheating and causing deterioration of insulation over a reasonable period. Also, the current components for supplying the losses and the associated voltage drops affect regulation and other performance characteristics of a machine. In terms of the basic physical phenomena and from the viewpoint of methods of measurement and testing, the losses in a machine are classified as follows, as in the case of transformers, with the addition of mechanical losses:

1. *No-load (or open-circuit) core losses* consist of the hysteresis and eddy-current losses, as discussed in Section 3.2 under "Iron Losses," which occur because of the periodic magnetic reversals in the ferromagnetic parts of the machine. The principal core losses due to the main flux (without accounting for the effect of spatial harmonics in flux density due to slots and due to non-sinusoidal current density) are confined to the armature side of the air gap in dc and synchronous machines. In induction machines, these are largely confined to the stator.

   Additional core losses result from harmonic components of flux (such as slot harmonics), harmonic components of mmf, and end-leakage flux caused in the end portions of the winding. The losses due to the dominating alternating effect in the teeth are known as *tooth losses,* which may be quite pronounced in induction machines. In the case of dc and synchronous machines, the rotational effect is predominant at the surface of the poles, and the losses it causes are known as *surface losses;* in order to minimize these, the iron in the pole shoes may be laminated. While the losses caused by harmonic components of flux do not in general depend on load, the other losses, due to harmonic components of mmf and end-leakage flux, generally depend upon the load and increase with the load on the machine.

   The principal core losses are mostly constant at a given voltage, and the no-load core losses are considered as the constant losses of the machine, since they are nearly independent of load. The additional core losses are usually included into what are known as *stray-load losses* mentioned below in this classification.

2. *Mechanical losses* are *friction and windage losses* caused by friction of bearings, brushes, and windage. While the mechanical losses are generally functions of the machine speed, these are usually assumed to be practically constant for small speed variations. The sum of the mechanical losses and the no-load core losses is usually termed as the *no-load rotational losses,* which are sensibly constant.

3. *Copper losses (or I^2R losses)* are due to the resistance of the machine windings. The principal copper losses consist of the I^2R losses in the windings, where R is taken to be the dc resistance of the winding, usually corrected to 75°C to allow for the effect of the usual temperature rise under normal operation. The effect of the brush-contact resistance in dc machines is conventionally taken into account by assuming a full-load

drop of 2 volts in series with the armature circuit. The prinicpal copper losses, which depend on the square of the load current, vary much more widely than the no-load rotational losses.

Additional copper losses result from the increase in winding resistance caused by the skin effect and the consequent nonuniform current-density distribution in the conductors. Also, the mmf and slot harmonics produce I^2R losses in induction machines and damper windings of synchronous machines. These additional copper losses are generally included with the stray-load losses discussed below.

4. *Stray-load losses,* as mentioned earlier, consist of the losses arising from the nonuniform current distribution in the conductors and the additional core losses produced in the iron by the distortion of the magnetic flux distribution caused by the load currents. It is usually very difficult to determine the stray-load losses accurately; hence, estimates are based on tests, experience, and judgment. While these may vary from 0.5 percent of the output in large machines to 5 percent of the output in medium-size machines, it is customarily assumed to be 1 percent of the output in dc machines.

VENTILATION AND COOLING   In electrical machines, as in any other energy-conversion device, some of the energy is dissipated as heat during operation. The problem of transferring heat without increasing the temperature beyond reasonable limits in different parts of the electrical machine can be very complex. The difficulty generally increases with increasing size, because the surface area from which the heat is to be carried away increases approximately as the square of the dimensions, whereas the heat developed by the losses is roughly proportional to the cube of the dimensions.

Two main types of ventilating systems are: *(i)* the self-ventilating system, in which ventilating devices such as fans are mounted on the machine shaft, and *(ii)* the separate (or independent) ventilating system, in which ventilating devices, independent of the actual machine, are externally driven. Both systems may use either open-circuit or closed-circuit systems. The stator and rotor cores are usually provided with radial ducts for the passage of air or any other cooling medium, such as hydrogen. The flow of the ventilating medium may be made axial in small machines, or radial in large machines, or both in larger machines. The draft of the cooling medium may be induced or forced. Closely associated with the ventilating systems are the types of enclosures commonly employed for the machines: *(i)* open, *(ii)* protected, and *(iii)* totally enclosed. These different types of enclosures are necessitated by considerations of safety and the environment in which the machine is to be used. For example, totally enclosed machines are used in coal mines and chemical plants.

The cooling problem is particularly serious in large turboalternators where economy, mechanical requirements, transportation limitations, and erection and assembly problems demand compactness, especially for the rotor forging. Rather elaborate systems of cooling ducts have to be provided to ensure that the cooling medium removes most effectively the heat arising from the losses. Closed ventilating systems are commonly used for moderate sizes of machines. Here the

TABLE 5.5.1   THERMAL CLASSIFICATION OF ELECTRICAL INSULATING MATERIALS

Class	Maximum Permissible Temperature Rise in °C Beyond the Ambient Temperature of 40°C	Materials
0	50	paper, cotton, silk
A	65	cellulose, phenolic resins
B	90	mica, glass, asbestos with organic binder
F	115	same as above, with suitable binder
H	140	mica, glass, asbestos with silicone binder, silicone resin, teflon
C	>180	mica, porcelain, glass

cooling medium (such as air or hydrogen), after being heated by its passage through the machine, is made to go through a heat exchanger (using air or water as the cooling agent) and then recirculated through the machine. In the case of most modern large turboalternators, water-cooled stators (the conductors have hollow passages through which cooling water circulates in direct contact with copper conductors) and hydrogen-cooled rotors are quite common, while water-cooling is employed in both the stator and rotor in a few cases. Typical cooling arrangements for large turbogenerators are shown in Photos 5–13 and 5–14. Direct-conductor cooling, also known as inner cooling, either by liquid (such as water) or gas (such as hydrogen) under pressure (1 to 5 atmospheres) is the most important single improvement that has made it possible to double the output of a turbine generator for a given physical size, and increase the efficiency of the unit to more than 99 percent. A rating of 1,500 MVA or more for a single unit is quite common these days for machines used for power generation in electric utility systems. The constructional details of machines greatly depend on the type of cooling methods employed. Complex problems may arise in appropriately sealing the bearings when hydrogen under pressure is used as a coolant. In the case of hydroelectric generators, in which vertical mounting (instead of the usual horizontal) is necessitated by the water-turbine construction, the design of thrust bearings may pose serious problems.

MACHINE RATINGS   The temperature rise resulting from the losses in a machine plays an important role in the rating of a machine, since the life of a machine should not be unduly shortened by overheating and consequent deterioration of the materials that are used in electrical machinery for the purpose of insulating current-carrying conductors from the core. Life expectancy of a large industrial electric machine usually ranges from 10 to 50 years or more; but it may be only a matter of minutes in some military or missile applications, while it may be of the order of a few thousand hours in some aircraft and electronic equipment. Various insulating materials employed in electrical machinery are classified according to the maximum allowable temperature rise that can be withstood safely. Table 5.5.1 gives the IEEE classification of electrical insulating materials by tolerance to heat.

PHOTO 5.13 Hydrogen cooling arrangement for a large two-pole turbogenerator. Photo courtesy of General Electric Company.

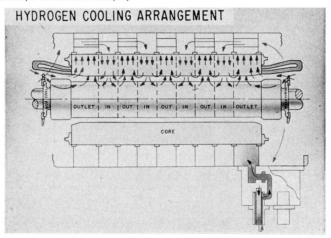

PHOTO 5.14 Ventilating scheme for a large turboalternator. Photo courtesy of Electric Machinery Manufacturing Company.

For the same class of insulation, the permissible rise of temperature depends on the type of enclosure and cooling method used; it is lower for a *general-purpose motor* than for a *special-purpose motor,* in order to allow a greater factor of safety in view of generally unknown service conditions. The general-purpose motor is one of standard rating up to 200 hp (with 450 rpm or more) with standard operating characteristics and mechanical construction for use under normal service conditions without restriction to a particular application or type of application. But, on the other hand, a special-purpose motor is designed either to achieve some particularly desirable operating characteristics or to satisfy a particular mechanical requirement, or both. The ASA (American Standards Association) recommends that a general-purpose motor should be able to operate successfully at 115 percent of its rating continuously as long as the voltage does not vary more than ±10 percent from the rated value and the frequency does not vary more than ±5 percent from the rated value and that the sum of these two percentages is not more than 10 at any one time. *Service factor* is a multiplier, which, applied to the rated output, gives a permissible loading that may be carried continuously under the specified conditions for that service factor. Thus, for general-purpose motors, a service factor of 1.15 is allowed. The NEMA and IEEE standards should be consulted for the allowable temperature rise for different types of machines.

The rating of a machine gives its working capabilities under specified electrical and environmental conditions. The most common machine rating is called the *continuous rating,* which gives the output (in kW for dc generators, kVA at a specified power factor for ac generators, and hp for motors) that can be carried indefinitely without exceeding the specified temperature rise. In contrast to this is the *short-time rating* given to some machines, defining the load that can be carried without overheating for a specified time due to intermittent, periodic, or varying duty. Standard periods for short-time ratings are 5, 15, 30, and 60 minutes. The duty cycle is usually represented by the horsepower-time curve, and a motor rating is normally chosen on the basis of the rms value given by

$$\text{rms hp} = \sqrt{\frac{\Sigma(\text{hp})^2(\text{time})}{\text{running time} + (\text{standstill time}/k)}} \qquad (5.5.2)$$

in which k is a constant accounting for reduced ventilation at standstill, and is given a value of 3 for an open motor. In practice, the result is rounded off to the next higher commercially available motor size. Special features, such as frequent starting or reversal and high torque peaks in the duty cycle, must be given adequate consideration in the selection of a motor. Special-purpose motors with short-time ratings are, in general, designed to have better torque-producing ability than continuously rated motors to produce the same power output; however, they will naturally have a lower thermal capacity.

In general, every machine has a *nameplate* attached to the frame that is inscribed with relevant information regarding the voltage, current, power, power factor, speed, frequency, phases, and allowable temperature rise. Most often, the nameplate rating is the continuous rating, unless otherwise specified. Ac generators and transformers are rated in terms of kVA rather than kW, because their losses (and heating) are approximately determined by the voltage and current, regardless of power factor; the physical size and cost of ac power-system apparatus are roughly proportional to the kVA rating.

MAGNETIC SATURATION   From the study of ferromagnetic materials and their properties in Section 3.2, we have seen that the magnetization *B-H* characteristic is generally linear up to a point (on the knee of the curve), beyond which it is generally nonlinear. When a magnetic material is saturated, the mmf required to produce a given flux is greater than that required under unsaturated conditions. In other words, a given mmf will produce less flux under saturation than under unsaturated conditions. For specified mmfs in the windings of a rotating machine, the fluxes depend not only on the reluctance of the air gaps but also on that of the iron portions of the magnetic circuits that may be saturated. In order fully to utilize the magnetic properties of the iron and optimize the machine design, the machine iron is worked at fairly saturated levels of flux density, such that the normal operating point on open circuit is near the knee of the *open-circuit characteristic* (or the *no-load saturation curve,* which is similar to the magnetization *B-H* characteristic) given by the plot of the open-circuit (no-load) voltage of the generator versus field excitation current, at rated speed of the machine. Basically, it is the characteristic magnetization curve for the particular magnetic-circuit geometry (involving air and iron) of the machine. Saturation may therefore influence the machine performance characteristics to a considerable degree.

As a first crude approximation, one neglects the reluctance of the iron portions of the magnetic circuit, and only air gaps are considered for mmf calculations. As a next step towards refinement, iron portions may be considered to have constant permeability and be included in the calculations. As a further step, depending on the actual working saturated conditions of the machine, permeabilities of the iron may be accounted for. This is rather difficult, analytically, since saturation affects different parts of the machine to varying degrees; in addition, the load-saturation curve is not the same as the open-circuit (no-load) characteristic. However, serious efforts are made to reproduce the magnetic conditions at the air gap correctly by increasing the air-gap length to a sufficient degree to take care of the effects of saturation, and air-gap nonuniformities due to slots and ventilating ducts. Final experimental test results will show the validity of the approach adopted. From an engineering point of view, such an approach may be quite adequate for a number of situations. In other cases, where there is too much discrepancy between the test and computed results, one has to look for more refined methods of calculation.

Several empirical and approximate methods have been suggested to account for the effects of magnetic saturation in machines, and these do work well within certain limits under specified conditions for a class of machines under consideration. Linear mathematical models and circuit techniques will always retain their important proper place for the analysis and understanding of the rotating machines. However, better understanding of the fundamental phenomena (including saturation) will be gained substantially by moving in the right direction towards the determination of flux distribution within a machine at various conditions of operation. But, field-theory methods of analysis are rather too involved.

LEAKAGE AND HARMONIC FLUXES   The flux linking the stator and rotor windings is the mutual flux between the two windings, which is assumed to be sinusoidally distributed in space in the analysis given so far. In real machines, we have fluxes that do not link both the windings, and hence are called *leakage fluxes*. Examples of these include slot-leakage flux due to slots, end-leakage flux due to end connections of the winding, and pole-to-pole leakage in salient-pole structures. The effect of the leakage flux, as in the case of transformers, is accounted for by defining leakage inductances that will cause the inductive voltage drop of fundamental frequency induced in ac windings. The paths that the leakage fluxes take in a practical machine are very complicated; the load and saturation also affect these considerably. Exact methods of calculating leakage inductances are usually too involved to be used, but empirical formulas have been developed for particular situations to yield reasonable solutions.

In addition to the leakage fluxes mentioned above, there are *harmonic fluxes* in the air gap, i.e., the flux-density distribution produced by a winding contains not only a fundamental component but also several other harmonics in space. However, it is possible, as we shall see later in the chapter on machine windings, to reduce harmonic fluxes and even to eliminate one or more undesirable harmonic components completely. Harmonic fluxes, besides causing harmonics in induced emf, produce secondary effects in the energy-conversion process itself, and develop undesirable *parasitic torques* that may become responsible for vibration and noise. In induction motors, for example, the problem may become rather serious: for certain stator and rotor slot numbers, slot-harmonic fluxes may cause the machine to *cog;* the seventh harmonic flux will develop a torque that may keep the machine, when started from standstill, from rising above one-seventh the normal speed and make the machine *crawl.* Any detailed discussion of harmonic fluxes and their effects on the machine performance is outside the scope of this book.

GENERAL NATURE OF MACHINE-APPLICATION PROBLEMS   At the outset, we may state that there are several important, interesting, challenging, and complex engineering problems associated with the design, development, and manufacture of rotating machines, most of which are beyond the scope of this book. We would like to touch upon the general nature of machine-application considerations.

For the case of motors it is the torque-speed characteristics, while for the case of generators it is the volt-ampere or voltage-load characteristics, that are of major importance. One should also know the limits between which these characteristics can be varied and how such variations can be obtained. Further, relevant economic features—such as the efficiency, power factor, relative costs, and the effect of losses on the heating and machine rating—need to be looked into.

The motor is generally supplied with electric power from a constant-voltage source, and it drives a mechanical load whose torque requirements vary with the speed at which it is driven. When the electromagnetic torque that can be supplied by the motor is equal to the mechanical torque that the load can absorb, the steady-state operating speed is determined. The requirements of motor loads generally vary from one application to another: in ordinary hydraulic pumps, the speed remains approximately constant while the load varies; in phonograph turntables, absolutely constant speed is required; in cranes and several traction-type drives, a varying speed characteristic with heavy torques at low speeds and light torques at high speeds is needed; others, such as machine tool drives, may demand constant speeds adjustable over a wide range, while still others may need adjustable varying speed. For any motor application, the starting torque, maximum torque, and running torque characteristics, along with the current requirements, should be looked into.

Similar remarks apply for machines operating as generators. The terminal voltage and power output of a generator are fixed by the characteristics of both the generator and its load. When the electrical load is connected to the generator, an operating point is attained such that the generator gives exactly what the load can take. Normally it is required that the terminal voltage remain substantially constant over a wide range of load. On the other hand, in some applications, the terminal voltage may be required to vary with load in some particular manner to provide greater flexibility and better control.

A generator, even of the size of 1,500 MVA or more, is only one component in a complex modern power system. Similarly, a motor may be only one component of an involved electromechanical system for industrial applications that may demand dynamic controls of great accuracy and rapid response. In such system-related applications, the electromechanical transient behavior of the system as a whole becomes a major consideration. In such cases, then, the emphasis shifts from the steady-state characteristics to those of transient behavior. Thus, it becomes necessary for us to study not only the steady-state but also the transient (or dynamic) analysis and behavior of rotating machinery. It is one thing to study the detailed dynamic behavior of a rotating machine as an individual unit, and it is quite different when the same machine becomes a component in a major system. The degree of detail of representation, the mathematical model, and the simplifying assumptions that can be made depend on the particular problem at hand. This fact underscores the importance of a thorough understanding of the fundamentals and orders of magnitude of the effects, which is essential for the choice of the right model with the available data.

BIBLIOGRAPHY

Fitzgerald, A. E.; Kingsley, C., Jr.; and Kusko, A. *Electric Machinery.* 3d ed. New York: McGraw-Hill, 1971.

Majmudar, H. *Electromechanical Energy Converters.* Boston: Allyn and Bacon, 1965.

Matsch, L. W. *Electromagnetic and Electromechanical Machines.* New York: Intext Educational Publishers, 1972.

Morgan, A. T. *General Theory of Electrical Machines.* London: Heyden & Son, 1979.

Nasar, S. A., and Unnewehr, L. E. *Electromechanics and Electric Machines.* New York: John Wiley & Sons, 1979.

Slemon, G. R., and Straughen, A. *Electric Machines.* Reading, Mass.: Addison Wesley, 1980.

White, D. C., and Woodson, H. H. *Electromechanical Energy Conversion.* New York: John Wiley & Sons, 1959.

Woodson, H. H., and Melcher, J. R. *Electromechanical Dynamics—Part I: Discrete Systems.* New York: John Wiley & Sons, 1968.

PROBLEMS*

5–1.

Table 5.1.1 of the text summarizes the relations associated with the mechanical force of electrical origin caused by the magnetic field coupling. Develop similar relations for determining mechanical forces due to electric field coupling in an electromechanical system. Choose as the independent variables the space coordinate x on the mechanical side, and either the voltage v or the charge q on the electrical side.

*Assume SI units unless otherwise stated.

5-2.

Consider the electromagnetic plunger shown in the figure below:

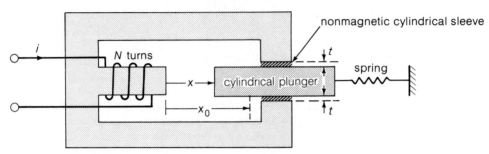

x_0 is the position of the plunger when spring is relaxed.

The λ-i relationship for the normal working range is experimentally found to be

$$\lambda = \frac{K\,i^{2/3}}{x + t}$$

where K is a constant. Determine the electromagnetic force on the plunger by the application of Equations 5.1.13, 5.1.14, 5.1.19, and 5.1.20.

Interpret the significance of the sign that you get in the force expression.

5-3.

In Problem 5-2, neglect the saturation of the core, leakage, and fringing. Neglect also the reluctance of the ferromagnetic circuit. Assuming that the cross-sectional area of the center leg is twice the area of the outer legs, obtain expressions for the inductance of the coil and the electromagnetic force on the plunger.

5-4.

Consider the system shown in Figure 5.1.6 and assume the λ-i characteristic of the system to be linear.

    a. Draw a diagram similar to that of Figure 5.1.7(a), and show that the electrical energy input divides equally between increasing stored energy and doing mechanical work.

    b. Draw a diagram similar to that of Figure 5.1.7(b), and show that the mechanical work done equals the reduction in stored energy.

5-5.

In the magnetic field structure shown in the figure that follows, the rotor core is displaced 1 cm in an axial direction from its correct position. Assuming the air-gap length under the pole shoes to be constant, compute the axial force tending to bring the rotor to its correct position.

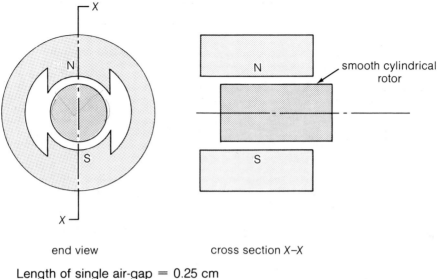

end view   cross section X–X

Length of single air-gap = 0.25 cm
Rotor diameter = 25 cm
Air-gap flux density = 0.8 T
Angle subtended by each pole shoe = 120°

5–6.

For the electromagnet shown in the following figure, the λ-*i* relationship for the normal working range is given by

$$i = a \lambda^2 + b \lambda (x - d)^2$$

where *a* and *b* are constants. Determine the force applied to the plunger by the electrical system.

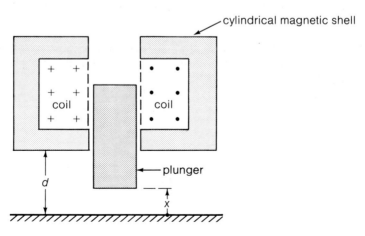

5-7.

A rotating machine of the form shown in Figure 5.2.1 has a coil inductance that can be approximated by

$$L(\theta) = 0.02 - 0.04 \cos 2\theta - 0.03 \cos 4\theta \text{ henry}$$

A current of 5 A (rms) at 60 Hz is passed through the coil, and the rotor is driven at a speed, which can be controlled, of ω_m rad/s.

   a. Find the values of ω_m at which the machine can develop average torque.
   b. At each of the speeds obtained in part *(a)*, determine the maximum value of the average torque and the maximum mechanical power output.

5-8.

Consider an elementary rotating reluctance machine of the type shown in Figure 5.2.1, but with four rotor poles instead of two. The poles are so shaped that the reluctance of the magnetic system is given by

$$\mathbf{R}(\theta) = (4 \times 10^5 - 3 \times 10^5 \cos 4\theta) \text{ A/Wb}$$

The stator coil has 100 turns and negligible resistance. An ac voltage of 110 volts (rms) at 60 Hz is applied to the coil terminals.

   a. Sketch the function of $\mathbf{R}$ versus θ.
   b. Determine the synchronous speed of the rotor and the maximum average torque that the machine can develop.

5-9.

A reluctance motor of the type illustrated in Figure 5.2.1 has a direct-axis reluctance $\mathbf{R}_d$ of 20×10^6 A/Wb and a quadrature-axis reluctance $\mathbf{R}_q$ of 60×10^6 A/Wb. Assume that the reluctance varies sinusoidally with rotor position as in Figure 5.2.1*(b)*. The stator winding has 5,000 turns and is excited by a 230-V, 50-Hz supply. Neglect the winding resistance.

   a. Determine the speed at which the machine can develop an average torque.
   b. Find the maximum average motoring torque and the power developed.
   c. Compute the rms value of the coil current.

5-10.

An electromagnetic relay may be modelled by its lumped-parameter system, as shown in the figure. Consider that there is no externally applied mechanical force. Neglect saturation of the ferromagnetic circuit, which is considered to be infinitely permeable. Ignore leakage and fringing fluxes. Assume the friction force to be linearly proportional to the velocity, and the spring force linearly proportional to the elongation. Let the resistance of the electrical circuit be R and the inductance

$$L(x) = \frac{A}{B + x}$$

where A and B are constants. Obtain the equations of motion for the electrical and mechanical sides of the electromechanical system.

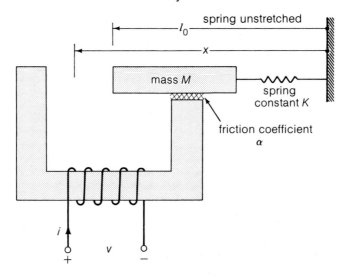

5–11.

Consider an elementary cylindrical-rotor two-phase synchronous machine with uniform air gap. As illustrated in the schematic diagram for Problem 5–11, it is similar to that of Figure 5.2.5 except that there are two identical stator windings in quadrature instead of one winding as shown in that figure. The self-inductance of the rotor winding, known also as the field winding, is a constant given by L_{ff} henrys; the self-inductance of each stator winding is a constant given by $L_{aa} = L_{bb}$; the mutual inductance between the stator windings is zero since they are in space quadrature; the mutual inductance between a stator winding and the rotor winding depends on the angular position of the rotor:

$$L_{af} = L \cos \theta \; ; \; L_{bf} = L \sin \theta$$

where L is the maximum value of the mutual inductance, and θ is the angle between the magnetic axes of the stator a-phase winding and the rotor field winding.

  a. Let the instantaneous currents be i_a, i_b, and i_f in the respective windings. Obtain a general expression for the electromagnetic torque T_e in terms of these currents, angle θ, and L.

b. Let the stator windings carry balanced two-phase currents given by

$$i_a = I_a \cos \omega t \; ; \; i_b = I_a \sin \omega t$$

and the rotor winding be excited by a constant direct current I_f. Let the rotor revolve at synchronous speed so that its instantaneous angular position θ is given by $\theta = \omega t + \delta$. Derive the torque expression under these conditions and describe its nature.

c. For conditions of part *(b)*, neglecting the resistance of the stator windings, obtain the volt-ampere equations at the terminals of stator phases *a* and *b*, and identify the speed-voltage terms.

5–12.

Consider the machine configuration of Problem 5–11. Let the rotor be stationary and constant direct currents I_a, I_b, and I_f be supplied to the windings. Further, let I_a and I_b be equal. If the rotor is now allowed to move, will it rotate continuously or will it tend to come to rest? If the latter, find the value of θ for stable equilibrium.

5–13.

Let us consider an elementary salient-pole, two-phase synchronous machine with non-uniform air gap. The schematic representation is the same as for Problem 5–11. Structurally the stator is similar to that of Figure 5.2.5 except that there are two identical stator windings in quadrature instead of one winding as shown in that figure. The salient-pole

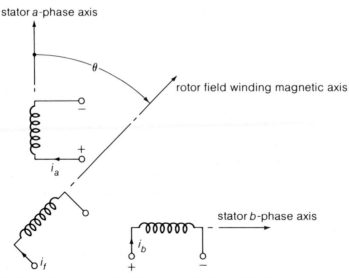

Schematic diagram for Problems 5–11 and 5–13.

rotor with two poles is similar to that of Figure 5.2.4, carrying the field winding connected to slip rings. The inductances are given below:

$$L_{aa} = L_0 + L_2 \cos 2\theta$$

$$L_{bb} = L_0 - L_2 \cos 2\theta$$

$$L_{ab} = L_2 \sin 2\theta$$

$$L_{af} = L \cos \theta$$

$$L_{bf} = L \sin \theta$$

L_{ff} is a constant, independent of θ.

L_0, L_2, and L are positive constants, and θ is the angle between the magnetic axes of the stator a-phase winding and the rotor field winding.

a. Let the stator windings carry balanced two-phase currents given by

$$i_a = I_a \cos \omega t \; ; \; i_b = I_a \sin \omega t$$

and the rotor winding be excited by a constant direct current I_f. Let the rotor revolve at synchronous speed so that its instantaneous angular position θ is given by

$$\theta = \omega t + \delta$$

Derive an expression for the torque under these conditions and describe its nature.
b. Compare it with that of Problem 5–11(b).
c. Explain whether the machine can be operated as a motor and as a generator.
d. Suppose that the field current I_f is brought to zero. Will the machine continue to run?

$\times$ **5–14.**
Consider the analysis leading up to Equation 5.2.33 for the torque of an elementary cylindrical machine with uniform air gap.

a. Express the torque in terms of F, F_s, and δ_s, where δ_s is the angle between $\overline{F}$ and $\overline{F}_s$.
b. Express the torque in terms of F, F_r, and δ_r, where δ_r is the angle between $\overline{F}$ and $\overline{F}_r$.

c. Neglecting magnetic saturation, obtain the torque in terms of B, F_r, and δ_r, where B is the peak value of the resultant flux-density wave.

d. Let ϕ be the resultant flux per pole given by the product of the average value of the flux density over a pole and the pole area. Express the torque in terms of ϕ, F_r, and δ_r, where ϕ is the resultant flux produced by the combined effect of the stator and rotor mmfs.

5–15.

An electromagnetic structure is characterized by the following inductances:

$$L_{11} = L_{22} = 4 + 2 \cos 2\theta$$

$$L_{12} = L_{21} = 2 + \cos \theta$$

Neglecting the resistances of the windings, find the torque as a function of θ when both windings are connected to the same ac voltage source such that

$$v_1 = v_2 = 220\sqrt{2} \sin 314t$$

5–16.

By using the concept of interaction between magnetic fields, show that electromagnetic torque cannot be obtained by using a four-pole rotor in a two-pole stator.

5–17.

Is a reluctance torque produced in the following devices, when the coils carry direct current?

    a. Salient-pole stator carrying a coil, and salient-pole rotor.
    b. Salient-pole rotor carrying a coil, and salient-pole stator.
    c. Salient-pole stator carrying a coil, and cylindrical rotor.
    d. Salient-pole rotor carrying a coil, and cylindrical stator.
    e. Cylindrical stator carrying a coil, and cylindrical rotor.
    f. Cylindrical rotor carrying a coil, and cylindrical stator.
    g. Cylindrical stator carrying a coil, and salient-pole rotor.
    h. Cylindrical rotor carrying a coil, and salient-pole stator.

5–18.

Consider an elementary doubly-excited magnetic-field system, such as the one shown in Figure 5.2.4. The stator and rotor inductances are given by

$$L_{ss} = 0.75 + 0.25 \cos 2\theta$$

$$L_{rr} = 0.45 + 0.15 \cos 2\theta$$

$$L_{sr} = 0.8 \cos \theta$$

Let both windings carry a constant current of 1 A.

    a. Calculate the torque when $\theta = 45°$.
    b. If the rotor moves slowly from $\theta = 90°$ to $\theta = 0°$, find
       (i) the work done.
       (ii) the change in stored energy.
       (iii) the electrical input.
    c. If the rotor rotates at a speed of 200 rad/s with constant winding currents of 1 A, determine the generated emfs e_s and e_r at the instant when $\theta = 45°$.

X 5-19.

An elementary two-pole rotating machine with uniform air gap, as shown in Figure 5.2.5, has a stator-winding self-inductance L_{ss} of 50 mH, a rotor-winding self-inductance L_{rr} of 50 mH, and a maximum mutual inductance L of 45 mH. If the stator were excited from a 60-Hz source, and the rotor from a 25-Hz source, at what speed or speeds would the machine be capable of converting energy?

X 5-20.

A rotating electrical machine with uniform air gap has a cylindrical-rotor winding with inductance $L_2 = 1$ H and a stator winding with inductance $L_1 = 3$ H. The maximum mutual inductance is 2 H and varies sinusoidally with the angle θ between the winding axes. Resistances of the windings are negligible. Compute the mean torque if the stator current is 10 A (rms), the rotor is short-circuited, and the angle between the winding axes is 45°.

X 5-21.

Consider an elementary three-phase, four-pole alternator with a wye-connected armature winding, consisting of full-pitch concentrated coils, as shown in Figures 5.3.10(b) and (c). Each phase coil has three turns, and all the turns in any one phase are connected in series. The flux per pole, sinusoidally distributed in space, is 0.1 Wb. The rotor is driven at 1,800 rpm.

    a. Calculate the generated rms voltage in each phase.
    b. If a voltmeter is connected across the two line terminals, what would it read?
    c. For the *a-b-c* phase sequence, taking $t = 0$ at the instant when the flux linkages with *a*-phase are a maximum, express the three phase voltages as functions of time. Draw a corresponding phasor diagram of these voltages with the *a*-phase voltage as reference. Represent the line-to-line voltages on the phasor diagram drawn above, and obtain the time functions of the same.

X 5-22.

A wye-connected, three-phase, 50-Hz, six-pole synchronous alternator develops a voltage of 1,000 volts (rms) between the lines, when the rotor dc field current is 3 A. If this alternator is to generate 60-Hz voltages, find the new synchronous speed, and calculate the new terminal voltage for the same field current.

5-23.

Determine the synchronous speed in rpm and the useful torque in newton-meters of a 200-hp, 60-Hz, six-pole synchronous motor operating at its rated full load.

5-24.

From a three-phase, 60-Hz system, through a motor-generator set consisting of two directly coupled synchronous machines, electrical power is supplied to a three-phase, 50-Hz system.

    a. Determine the minimum number of poles that the motor may have.
    b. Determine the minimum number of poles that the generator may have.
    c. With the number of poles decided, find the speed in rpm at which the motor-generator set will operate.

5-25.

A dc machine, operating as a generator, develops 400 volts at its armature terminals, corresponding to a field current of 4 amps, when the rotor is driven at 1,200 rpm and the armature current is zero.

    a. If the machine produces 20 kW of electromagnetic power, find the corresponding armature current and the electromagnetic torque produced.
    b. If the machine is operated as a motor supplied from a 400-volt dc supply, and if the motor is delivering 30 hp to the mechanical load, determine *(i)* the current taken from the supply, *(ii)* the speed of the motor, and *(iii)* the electromagnetic torque, if the field current is maintained at 4 A. Assume that the armature circuit resistance is negligible and the mechanical losses are negligible.
    c. If this machine is operated from a 440-V dc supply, determine the speed at which this motor will run and the current taken from the supply, assuming that there is no mechanical load, no friction and windage losses, and the field current is maintained at 4 A.

5-26.

Consider the operation of a dc shunt motor that is affected by each of the following changes in its operating conditions. Explain the corresponding approximate changes in the armature current and speed of the machine.

    a. The field current is doubled, with the armature terminal voltage and load torque remaining the same.
    b. The armature terminal voltage is halved, with the field current and load torque remaining the same.
    c. The field current as well as the armature terminal voltage are halved, with the horsepower output remaining the same.

d. The armature terminal voltage is halved, with the field current and horsepower output remaining the same.

e. The armature terminal voltage is halved, with the field current remaining the same, and the load torque varies as the square of the speed.

5–27.

A three-phase, 50-Hz induction motor has a full-load speed of 700 rpm, and a no-load speed of 740 rpm.

a. Determine the number of poles of the machine.

b. Find the slip and the rotor frequency at full load.

c. What is the speed of rotation of the rotor field at full load
  (i) with respect to the rotor?
  (ii) with respect to the stator?

5–28.

A three-phase, 60-Hz induction motor runs at almost 1,800 rpm at no-load, and 1,710 rpm at full load.

a. How many poles has the motor?

b. What is the per-unit slip at full load?

c. What is the frequency of rotor voltages at full load?

d. At full load, find the speed of
  (i) the rotor field with respect to the rotor.
  (ii) the rotor field with respect to the stator.
  (iii) the rotor field with respect to the stator field.

5–29.

Consider a three-phase induction motor with a normal torque-speed characteristic. Neglecting the effects of stator resistance and leakage reactance, discuss the approximate effect on the characteristic if *(i)* the applied voltage and frequency are halved, and *(ii)* only the applied voltage is halved, but the frequency is at its normal value.

5–30.

A four-pole, three-phase, wound-rotor induction machine is to be used as a variable frequency supply. The frequency of the supply connected to the stator is 60 Hz.

a. Let the rotor be driven at 3,600 rpm in either direction by an auxiliary synchronous motor. If the slip-ring voltage is 20 V when the rotor is at standstill, what frequencies and voltages can be available at the slip rings?

b. If the slip-ring voltage is 400 V when the rotor frequency is 120 Hz, at what speed must the rotor be driven in order to give 150 Hz at the slip-ring terminals, and what will be the slip-ring voltage in this case?

5-31.

A three-phase wound-rotor induction machine with its shaft rigidly coupled to the shaft of a three-phase synchronous motor is used to change balanced 60-Hz voltages to other frequencies at the wound-rotor terminals brought out through slip rings. Both the machines are electrically connected to the same balanced three-phase, 60-Hz source. Let the synchronous motor, which has four poles, drive the interconnecting shaft in the clockwise direction, and let the eight-pole balanced three-phase stator winding of the induction machine produce a counterclockwise rotating field, i.e., in the direction opposite to that of the synchronous motor. Determine the frequency of the rotor voltages of the induction machine.

5-32.

For a balanced two-phase stator supplied by balanced two-phase currents, carry out the steps leading up to an equation such as Equation 5.4.5 for the rotating mmf wave.

5-33.

A split-phase induction motor has two stator windings of impedance $(5 + j50)\,\Omega$, displaced by 90 electrical degrees. Compute the necessary resistance and capacitance in series with one winding in order to obtain a simulated balanced two-phase current system when both windings are supplied from a 60-Hz, 110-V single-phase supply.

5-34.

Consider a rectangular coil $PQRS$ (shown in the figure) moving in the x-direction in the xy-plane with a velocity U m/s in a traveling-wave magnetic field distribution given by the flux-density function

$$B = B_m \cos(\beta x - \omega t) \text{ Wb/m}^2$$

directed in the z-direction perpendicular to the plane of the coil. Determine the total induced voltage, identifying the transformer-emf and speed-emf components *(i)* by the method of relative velocity, and *(ii)* by the application of Faraday's law of induction.

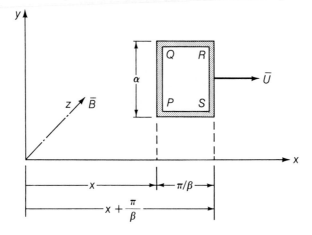

5-35.

A certain 10-hp, 230-V motor has a rotational loss of 600 W, a stator copper loss of 350 W, a rotor copper loss of 350 W, and a stray load loss of 50 W. It is not stated whether the motor is an induction, synchronous, or dc machine.

    a. Calculate the full-load efficiency.

    b. Compute the efficiency at half-full load, assuming that the stray-load and rotational losses do not change with load.

5-36.

A synchronous motor operates continuously on the following duty cycle: 50 hp for 8 minutes, 100 hp for 8 minutes, 150 hp for 10 minutes, 120 hp for 20 minutes, and no-load for 14 minutes. Specify the continuous-rated hp of the motor required.

6

MACHINE WINDINGS

In an electromagnetic rotating machine, energy conversion takes place through the medium of the magnetic field, an emf being induced in any coil that experiences a change of flux linkage. In the case of a practical machine, there are usually several coils, connected in series or in series-parallel circuits to form a winding. Machine windings are the means by which theory is translated into practice, and their importance warrants a short chapter. These windings may also be viewed as a unifying link among different machines.

What is the best way of placing the coils to collect their emfs in the most effective manner? The stationary windings of a transformer present no problem, as it is apparent that they can be readily wound in close proximity on the iron core to reduce the leakage, as discussed earlier. The situation is rather different when it comes to rotating machines. In fact, the study of armature windings is rather specialized, and there are many design considerations that lead to a choice among different possibilities for a given situation. Since these cannot be dealt with fully in this book, a simplified treatment of winding problems is presented to give practical support to theoretical analyses.

6.1 *BASIC WINDING ARRANGEMENTS IN ROTATING MACHINES*

Machine windings can be classified as either *field windings* or *armature windings*. Armature windings may further be subdivided as *dc armature windings* (also known as commutator windings) and *ac armature windings*. Such windings are generally formed by wire-wound coils of one or more *turns* (see Figure 5.3.8) placed in slots arranged to form either *single-layer* or *double-layer windings*.

In a single-layer winding one side of a coil, known as a *coil side,* occupies the whole of one slot, whereas in a double-layer winding there are two separate coil sides in any one slot. Most polyphase ac and dc armature windings are double-layer windings with multiturn coils. A double-layer winding, in general, has a lower leakage reactance and produces a better waveform than the corresponding single-layer winding.

When one coil side is under the influence of a north pole and the other coil side is under the influence of a south pole, the emfs induced are in opposite directions; but a back-to-back connection is made so as to add these two emfs around the coil circuit. In such a case, the coil is known as a *full-pitch coil,* with the *coil span* equal to *one pole pitch*. The coils are sometimes *chorded (short-pitched);* i.e., the coil sides are not exactly one pole pitch apart, and the coil span is somewhat shorter than a pole pitch. The coils are laid in succession, i.e., *distributed,* around the periphery of the machine, and for mechanical reasons they are insulated and embedded in *slots* that may be *open, semiclosed,* or *closed.* When the slot opening is the same as the slot width, the slots are said to be open; when the slot opening is less than the slot width, the slots are said to be semiclosed; and when there is no slot opening at all, the slots are said to be closed. Semiclosed and closed slots are normally used in induction motors, while open slots are most frequently used in large synchronous machines. The reactances are affected by the shape and relative dimensions of the slots.

The separate coils of a winding can be interconnected in a variety of ways, and it is really in the interconnections that windings differ; however, the common features remain that, in progressing from one coil to another in series, the emfs are additive, and any time-phase displacement

FIGURE 6.1.1   Methods of interconnection of armature coils: *(a)* lap, *(b)* wave.

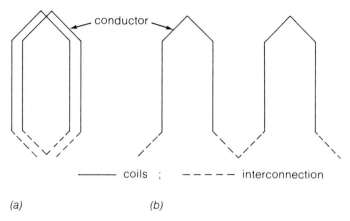

coils ;   – – – – – interconnection

(a)   (b)

between the coil emfs is usually kept small by suitable choice of connection. The two most general methods of interconnection are illustrated in Figure 6.1.1: namely, the *lap* and *wave* connection. The dc machine armature windings are always of the closed type of double-layer lap or waveform connected to a *commutator*. The commutator itself, as mentioned earlier, consists of many copper segments insulated from and running parallel to one another, and clamped around the periphery of an insulated cylinder mounted on the machine shaft. Each segment is extended to a coil junction, so that there are as many segments as there are coils.

With reference to Figure 6.1.2, let us get ourselves familiarized with the usual armature winding terms and symbols:

conductor: the active length of wire or strip in the slot along the axial length of the machine core. The conductor may be laminated and transposed (i.e., occupying different physical locations) along its length to reduce eddy currents.

coil: two conductors separated by a pole pitch or nearly so, connected in series to form a single-turn coil; more conductors placed virtually in the same position magnetically can be formed into multiturn coils. Figure 5.3.10 shows a three-turn lap coil with six conductors.

coil side: one-half of the coil, lying in one slot, the coil being either single- or multiturn; only one coil side per slot for a single-layer winding, while there are at least two coil sides in any one slot for a double-layer winding. A commutator winding may have several coil sides in one slot. Figure 6.1.2 shows one slot arrangement with two-turn coils and with three coil-sides per layer, in which case there are 12 conductors per slot insulated from one another. All the conductors in one layer in the slot are usually taped together to share a common insulation to earth or the iron core.

pitch: There are various pitches of interest as shown in Figure 6.1.2. The coil-pitch or coil-span (also known as slot span of the coil) and the pole pitch have already been referred to. In terms of the dc armature winding, the *back pitch* y_b (also known as coil

FIGURE 6.1.2   Armature winding terms and symbols.

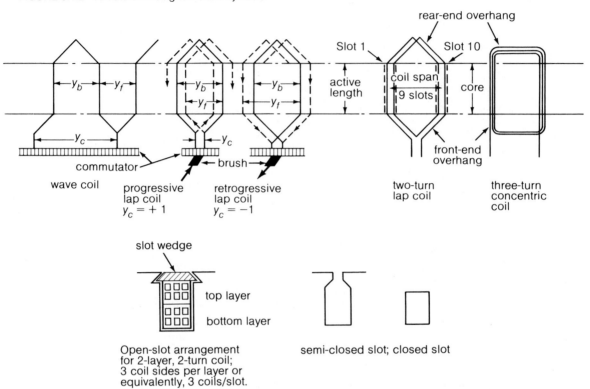

pitch), the *front pitch* y_f, and the *commutator pitch* y_c (also known as coil-end pitch) can be seen from Figure 6.1.2. These pitches or distances are usually measured or specified in terms of the commutator segments between the end connections of a coil. Thus, for a single commutator lap winding, y_c is ± 1, the positive sign indicating a progression to the right when tracing through the winding, and the negative sign indicating a retrogression to the left. In all cases, the commutator pitch is equal to the algebraic sum of the front pitch and the back pitch of the coil. The commutator pitch for a *multiplex* lap winding of degree m will be m commutator segments.

With this background, we shall move on to discuss briefly various classes of machine windings. See Photos 6–1 and 6–2 for a laminated (stranded) armature-coil construction and the details of open-slot arrangements.

PHOTO 6.1   Typical laminated armature coil consisting of several strands
of copper wire in parallel.   Photo courtesy of Westinghouse Electric Corporation.

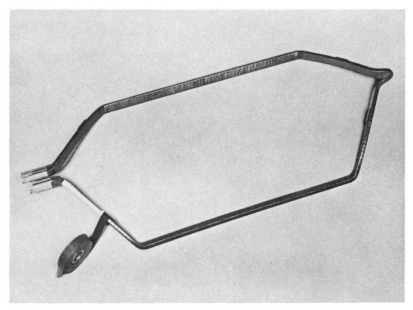

PHOTO 6.2   Cross sections of open-slot arrangements with insulation for
double-layer windings.   Photo courtesy of Westinghouse Electric Corporation.

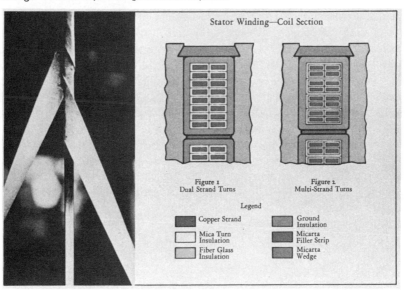

FIGURE 6.2.1   Concentrated coil on a salient pole: *(a)* schematic arrangement, *(b)* realistic arrangement.

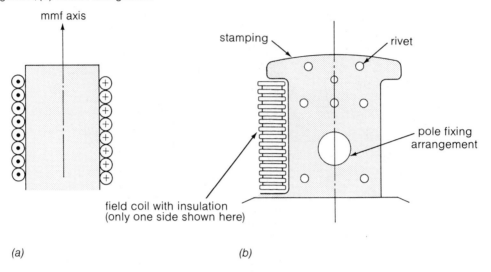

(a)   (b)

6.2 DC FIELD WINDINGS

Synchronous machines and dc machines require dc-excited field windings. Two different forms of construction are used for the field circuit: *(i)* salient-pole arrangement with a concentrated winding, used for either dc or ac machines, and *(ii)* distributed arrangement on the nonsalient (cylindrical) rotors of ac machines. Figure 6.2.1 shows a concentrated coil on a salient pole. In order to obtain the current pattern indicated, the turns of the concentrated coil are to be connected in series; it does not even matter in which order the coils are connected, except perhaps from the viewpoint of constructional convenience. Photo 6–3 shows the dc machine field windings wound on a pole, while Photo 6–4 illustrates the salient-pole field arrangement for a synchronous machine.

For high-speed rotors of turbogenerators that are usually run at either 3,600 or 1,800 rpm (for 60-Hz systems), the centrifugal forces are too great to be sustained by a salient-pole construction, so that the field winding is distributed into slots as shown in Figure 5.3.3 for a two-pole machine. A distributed two-pole winding with six coils (12 conductors) is shown in Figure 6.2.2 in a simple schematic fashion to illustrate the principle; 1–1′, 2–2′, etc., are connected to form coils 1, 2, etc., respectively. The required current distribution is produced by simply connecting the six coils in series. The winding is called a concentric winding, because all six coils are arranged in a concentric fashion. Concentric windings are most commonly employed for the field windings of synchronous machines, if the field winding is on the rotor.

PHOTO 6.3   DC machine field windings wound-on-pole along with pole-face compensating winding bars.   Photo courtesy of Westinghouse Electric Corporation.

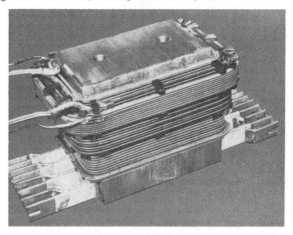

PHOTO 6.4   Salient-pole field arrangement being manufactured for a synchronous machine.   Photo courtesy of Westinghouse Electric Corporation.

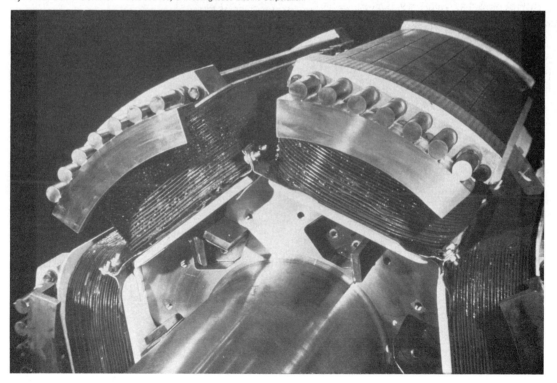

FIGURE 6.2.2   Distributed winding. *(a)* Single-layer distributed winding (concentric), this arrangement most commonly used for nonsalient pole synchronous machine rotor field windings; *(b)* single-layer distributed winding with 180° coil span; *(c)* double-layer distributed winding (concentric); *(d)* distributed double-layer winding with 180° coil span.

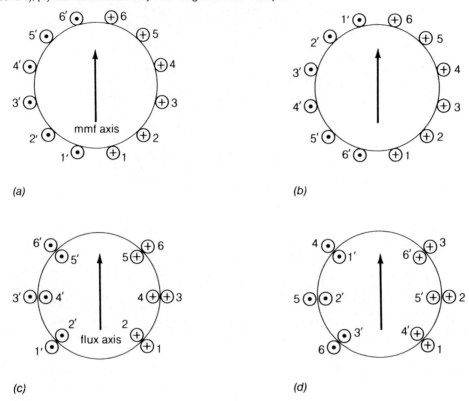

The winding arrangement may easily be modified to other schemes such as those shown in Figure 6.2.2*(b)*, *(c)*, and *(d)*, while still yielding the mmf axis in the same position.

Liquid helium-cooled superconducting rotor field windings are feasible, and these are used in large cryogenerators or superconductive turboalternators.

6.3 DC ARMATURE WINDINGS

As mentioned already, these windings (also known as commutator windings) are always of the closed continuous type of double-layer lap or waveform. The simplex lap winding has as many parallel paths as there are poles, while for the simplex wave windings, there are always two parallel paths. Photo 6–5 shows a dc armature being wound.

PHOTO 6.5 DC armature being wound. Photo courtesy of Westinghouse Electric Corporation.

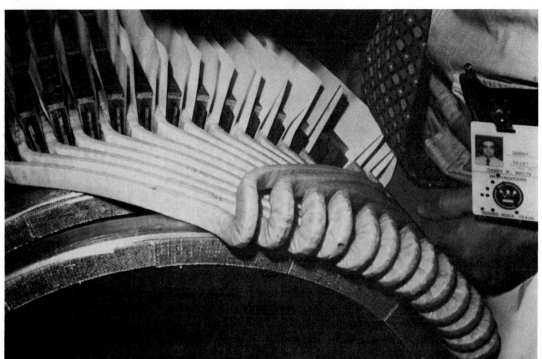

Before proceeding further, let us try to justify a statement made earlier in the text that a rotating commutator winding is called a pseudostationary winding that produces a stationary flux when carrying a direct current. Consider a six-coil closed continuous two-pole winding connected to a six-segment commutator, as shown in Figure 6.3.1(a), producing the current distribution illustrated when a current is passed into the winding through the brushes located on the commutator. Note that the flux axis is along the brush axis as shown in the figure.

Next, let us consider an instant when the commutator and the winding have rotated by 60°, but with the same position of brushes, as shown in Figure 6.3.1(b). It can be seen that coils 3 and 6 have moved from one parallel circuit to the other; but the current directions in them are changed and so also their physical position, such that the resultant spatial current distribution and the position of the flux axis remain the same as in Figure 6.3.1(a). By considering successive 60-degree movements, it can be seen that the resultant total mmf (or flux) is a constant in a direction fixed by the brush axis, irrespective of the position of the individual coils. Hence, the commutator winding is viewed as a pseudostationary winding.

FIGURE 6.3.1   Commutator winding as a pseudo-stationary winding.
(a) Six-coil closed continuous two-pole winding connected to a commutator,
corresponding current distribution and the mmf axis; *(b)* winding in
(a) rotated by 60°, but with the same position of brushes, corresponding
current distribution and the mmf axis; *(c)* winding in *(a)* with brushes
rotated by 60°, corresponding current distribution and the mmf axis.

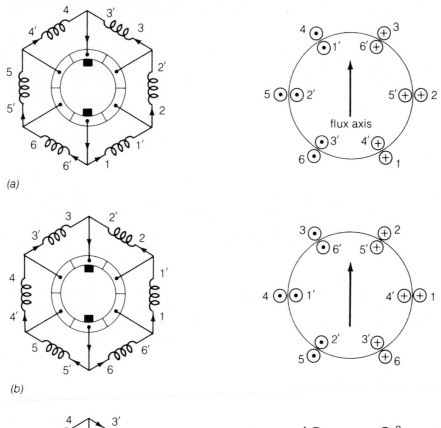

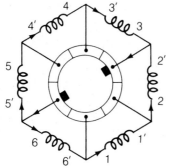 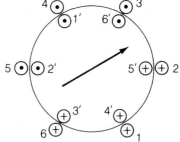

FIGURE 6.3.2 Developed diagram of a four-pole, two-layer simplex lap dc armature winding with 20 slots and 20 single-turn coils.

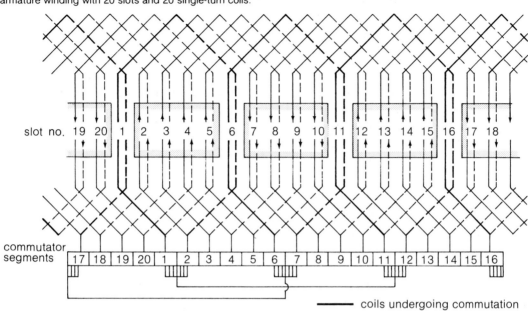

——— coils undergoing commutation

In a practical machine with a large number of coils, intervals much less than 60° can be considered and the same conclusions reached. In practice, however, the axis of flux oscillates forward and backward by an angle equal to that between adjacent commutator segments. By keeping the winding as in Figure 6.3.1(a), but moving the position of brushes by 60°, as shown in Figure 6.3.1(c), it can be seen that the flux axis has also been moved by 60°. Thus it is obvious that the flux axis corresponds to the brush axis.

LAP WINDINGS   A simplex lap winding is one in which the coil sides are placed in slots approximately one pole pitch apart, the coil leads are connected to adjacent commutator segments, and the number of parallel paths is equal to the number of poles. This is best illustrated by considering an example.

Let us consider a four-pole, two-layer, simplex lap winding with 20 slots, and let the number of conductors per coil side be one (i.e., single-turn coils) for the sake of simplicity. The number of coils as well as the number of commutator segments is 20, which is the same as the number of slots. Instead of showing a polar diagram of the winding located in the slots of an armature of circular cross section, a developed diagram is shown in Figure 6.3.2 for better clarity and understanding.

The slots per pole being $20/4 = 5$ in our case, the following pitches are selected in terms of the slots or slot pitches:

$$y_b = 5 \; ; y_f = -4 \; ; y_c = y_b + y_f = +1 \qquad\qquad (6.3.1)$$

The above choice of $y_c = +1$ makes the winding progressive, instead of retrogressive. The development of the winding is rather straightfoward, and four parallel paths through the armature can be observed in Figure 6.3.2. The brushes on the commutator are located as nearly as possible on the center of the poles so that the coil undergoing commutation (i.e., reversal of current directions) is situated in the interpolar region. Corresponding to the four parallel armature paths, the lap winding has to have four brush sets, two of which will have the same polarity. The polarity of the brushes may be assigned here arbitrarily, because it is definitely determined only if the pole polarities and direction of rotation are known.

The generated voltages in each of the parallel paths of a lap winding should be equal if the machine is geometrically and magnetically symmetrical. However, in practice, since these voltages are not exactly the same and hence cause circulating currents through the armature and brushes, most machines with lap windings are provided with *equalizers* (copper straps of large cross-sectional area) joining points on the winding that are 360 electrical degrees apart. The number of equalizer bars is given by the number of slots per pole pair, i.e., 10 in our example; each bar joins two coils together. Equalization prevents excessive brush currents and helps commutation.

WAVE WINDINGS   A simplex wave winding is one in which the coil sides are placed in slots approximately one pole pitch apart, the coil leads are connected to commutator bars approximately two pole pitches apart, and the number of parallel paths in the armature is always two. This is again best illustrated by considering an example. The commutator pitch y_c for a simplex wave winding in terms of commutator segments is given by

$$y_c = y_b + y_f = \frac{K \pm 1}{\text{Pole pairs}} \qquad\qquad (6.3.2)$$

where K is the number of commutator segments. A simplex wave winding can be obtained only if y_c is an integer. In some cases where y_c is not an integer, a so-called dummy coil is introduced into the winding to maintain the mechanical symmetry of the armature, and electrically the coil is left on open-circuit.

Let us consider a four-pole, two-layer, simplex wave winding with 19 slots, and let the number of conductors per coil side be one (i.e., single-turn coils) for simplicity. The number of coils as well as the number of commutator segments is 19, the same as the number of slots. A developed winding diagram is shown in Figure 6.3.3. The commutator pitch y_c is given by $(19 \pm 1)/2 = 9$

FIGURE 6.3.3 Developed diagram of a four-pole, two-layer simplex wave dc armature winding with 19 slots and 19 single-turn coils. (Note: Only two brush sets are really necessary.)

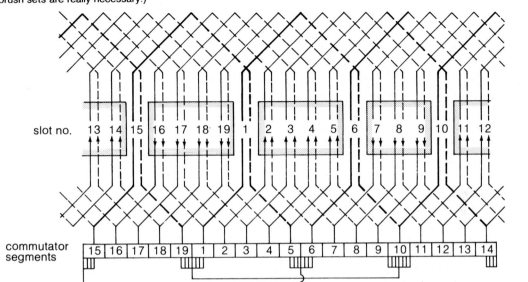

or 10; choosing $y_c = 9$, $y_b = 5$, and $y_f = 4$, the winding arrangement is completed as shown. Only two parallel paths through the armature can be observed in Figure 6.3.3. Only two brush sets are really necessary; however, in large machines the same number of brush sets as poles is usually employed.

Regarding the location of the brushes, as seen in Figures 6.3.2 and 6.3.3, the brushes lie physically on the center lines of the poles; however, considering the active parts of the conductors to which the brushes are connected, they lie on the interpolar (quadrature) axis, which is 90 electrical degrees away from the direct-pole axis. In view of the greater number of parallel paths available, lap windings are normally used for dc machines with large current-carrying capacity. In practice, there are usually more than two coil sides in any one slot, so that the number of commutator segments is greater than the number of armature slots. For a given output voltage, such a practice leads to reduced voltage between adjacent commutator segments, which in turn reduces the risk of flashover between segments. Rarely used are the multiplex windings that have two or more times (depending on the degree of multiplicity) as many parallel paths for the same number of poles as the simplex windings.

6.4 AC ARMATURE WINDINGS

When associated with a heteropolar magnetic field, a q-phase ac armature winding develops emfs
in all of its phases; when the phases are wound $2\pi/q$ electrical radians apart in space, these emfs
are normally equal in magnitude and displaced in time-phase relationship by $2\pi/q$ electrical radians, except in the case of a two-phase winding when a displacement of $\pi/2$ electrical radians
applies for both space and time. The winding is composed of conductors in slots distributed around
the periphery of the air gap, connected together at the ends, and grouped to form separate phase
windings. Polyphase windings are usually double-layer, with the two layers arranged one above
the other in the slot. The single-layer polyphase winding is used in the wound rotors of small
induction motors, but seldom used in the stators of these motors. Polyphase windings can be of
lap or wave type; lap windings are most commonly used while wave windings are employed mainly
in wound rotors of medium-size and large induction motors. Photo 6–6 shows an induction-motor
stator being wound.

The ac armature windings can be considered in terms of several separate sections, known as *phase belts;* a three-phase, two-layer winding can be completed with either three phase belts per pole pair, i.e., with 120 electrical-degree spread, or six phase belts per pole pair, i.e., with 60 electrical-degree spread. The more common arrangement is to use 60° spread for three-phase windings. It may be noted here that a three-phase, 60° spread winding is effectively a six-phase winding. Two-phase machines normally have double-layer windings with four phase belts per pole pair, i.e., 90° spread.

A developed sectional diagram for a three-phase, two-pole, two-layer, full-pitch lap ac armature winding with two *slots per pole per phase* (spp), i.e., a total of 3 × 2 × 2 = 12 slots, is shown in Figure 6.4.1*(a)* with 60° spread and in Figure 6.4.1*(b)* with 120° spread. Notice the winding arrangement in Figure 6.4.1*(a)* along with the instantaneous directions of current flow that make the phase sequence come out as *A-B-C.* The coil groups assigned to each phase, shown in Figure 6.4.1*(c)* with detailed coil interconnections, are connected either in series or in parallel according to the voltage to be produced in the case of a generator, or the voltage impressed in the case of a motor. The phase end leads are brought out for wye or delta connection.

All the coils shown in Figure 6.4.1 are full-pitch coils. Short-pitch (chorded) windings are quite often used in polyphase armature windings. The effect of using chorded coils is to reduce the length of the coil-end connections (thereby saving a little copper) and also to reduce significantly, as we will see later, the magnitude of certain harmonics in the waveform of the generated emf and mmf. In most ac machines, the coil pitch is generally between two-thirds and one full pole pitch, and the coils are usually short-pitched as a standard practice. Figure 6.4.2 shows the flattened layout of a three-phase, two-pole, two-layer, *fractional-pitch* (5/6 pitch coils) lap ac armature winding with spp (number of slots per pole phase) = 2 and 60° spread.

It can be seen from Figure 6.4.1*(a)* that the coil sides occupying the top and bottom layers of each slot belong to the same phase for the case of full-pitch windings. As is typical of fractional-pitch windings, shown in Figure 6.4.2, the coil sides occupying the top and bottom layers of some slots are of different phases. Individual phase groups are still displaced by 120 electrical degrees from the groups in other phases so that three-phase voltages are produced.

The ac armature windings may be classified as *integral-slot windings* and *fractional-slot windings,* depending on whether spp is an integer or an improper fraction. The fractional slot windings offer great flexibility of design and manufacture, and allow a wider latitude in the choice of the number of slots that can be used for the armature punchings. These are not considered here.

Squirrel-cage windings used for the rotors of most induction motors have solid, uninsulated bars in the slots and are connected on each end of the rotor by a short-circuiting ring, as mentioned already. In small rotors of induction motors up to about 50 hp, the bars and rings are usually die-cast of aluminum, while in large motors the bars are usually made of copper.

FIGURE 6.4.1   A flattened layout of a three-phase, two-pole, two-layer, full-pitch lap ac armature winding with two slots per pole per phase. *(a)* 60-electrical-degree phase spreading, *(b)* 120-electrical-degree phase spreading, *(c)* coil interconnections shown in detail for part *(a)*. (Note: The signs indicate the instantaneous directions of current flow, in or out of the plane of paper.)

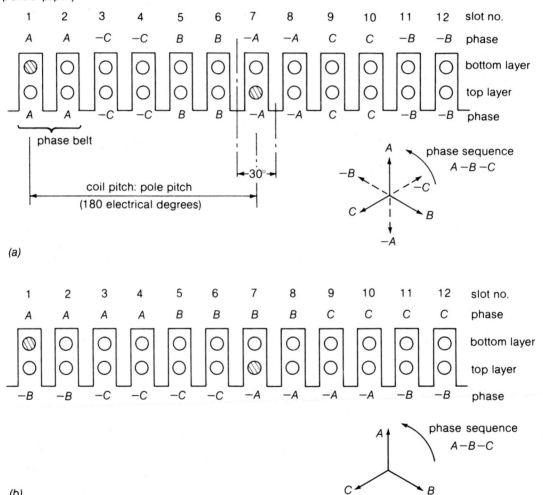

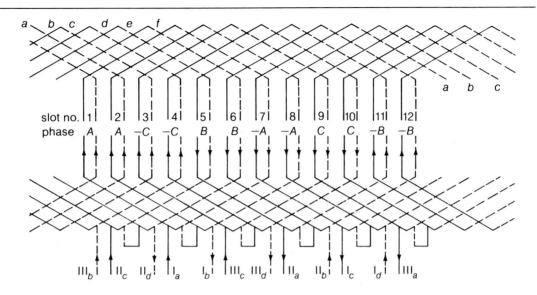

series connection of coil groups:

$$I_b - I_d \; ; \; II_b - II_d \; ; \; III_b - III_d$$

parallel connections of coil groups:

$$I_a - I_d, \; I_b - I_c \; ; \; II_a - II_d, \; II_b - II_c \; ;$$
$$III_a - III_d, \; III_b - III_c$$

(c)

FIGURE 6.4.2 A flattened layout of a three-phase, two-pole, two-layer, fractional-pitch (5/6 pitch coils) lap ac armature winding with spp = 2 and 60° spread.

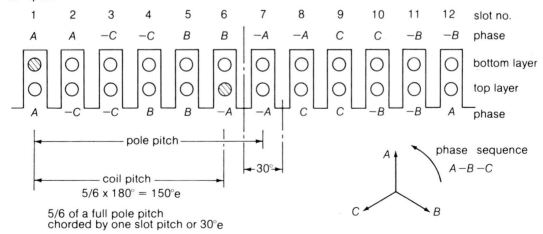

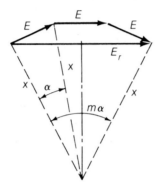

FIGURE 6.5.1   Coil voltage phasors and phasor sums to calculate the fundamental distribution factor.

6.5 WINDING FACTORS

There are two factors to be discussed in this section: *(i)* the distribution factor, also known as the breadth factor; and *(ii)* the pitch factor, also known as the coil-span or coil-pitch factor.

DISTRIBUTION FACTOR   The coils of an ac armature winding of a practical machine are distributed in space so that the induced voltages in the coils are not in phase, but are displaced from each other by the slot angle α. The effect of distributing the winding is taken into account by introducing the *distribution factor* k_d, which is defined as the ratio of the actual resultant emf E_r given by the phasor sum to the arithmetic sum of the individual coil emfs, each given by E. Thus, for an ac armature winding with m coils per pole per phase, the distribution factor is given by

$$k_d = E_r / m E \tag{6.5.1}$$

For a particular arrangement with $m = 3$, illustrated in Figure 6.5.1, it follows that

$$E_r = 2 \times \sin (m\alpha/2) \; ; \; E = 2 \times \sin (\alpha/2) \tag{6.5.2}$$

and

$$k_d = \frac{\sin (m\alpha/2)}{m \sin (\alpha/2)} \tag{6.5.3}$$

where α is measured in electrical degrees. For common, three-phase, two-layer windings with 60° spread, it follows that $m\alpha/2$ is 30 electrical degrees so that

$$k_d = \frac{\sin(\pi/6)}{m\sin(\pi/6m)} = \frac{0.5}{m\sin(\pi/6m)} \qquad (6.5.4)$$

Equations 6.5.3 and 6.5.4 give the fundamental distribution factor, and apply for integral slot windings where m is the number of slots per phase belt in which the winding is distributed.

The above discussion assumes a sinusoidal-generated emf. If the emf contains harmonics, then the distribution factor is different for each harmonic. Since the phase difference between the nth harmonic of adjacent coil voltages is $n\alpha$, the distribution factor for the nth harmonic is given by

$$k_{dn} = \frac{\sin(nm\alpha/2)}{m\sin(n\alpha/2)} \qquad (6.5.5)$$

With $n = 1$, i.e., for the fundamental, Equation 6.5.5 reduces to that of Equation 6.5.3. If the number of series coils in a phase winding is sufficiently large, then α becomes rather small, and an approximation can be made as follows:

$$k_{dn} \simeq \frac{\sin(nm\alpha/2)}{nm\alpha/2} \qquad (6.5.6)$$

in which α is expressed in electrical radians.

Distributing the winding will result in a marked reduction in the harmonic content of the flux-density distribution and consequently the generated voltage, as shown later. If it is assumed that the generated voltages are sinusoidal, the output voltage from a distributed winding of N turns will be equal in magnitude to that of a concentrated winding of (k_dN) turns. That is why it is said that the effective number of turns of the distributed winding is (k_dN).

The fundamental distribution factor for double-layer, integral-slot windings of common three-phase machines with 60° spread can be found by the application of Equation 6.5.4. For values of m, i.e., number of coils per phase belt, ranging from 1 to 6, k_{d1} varies from 1.0 to 0.956. Thus, distributed windings lower the fundamental generated voltage to some extent, since the voltage expression given by Equation 5.3.14 is to be multiplied by k_{d1}.

Equation 6.5.6 shows that k_{dn} decreases continually as n, the order of the harmonic, increases; it follows that, if α can be made very small, the higher harmonics could be virtually eliminated. This is achieved practically by *skewing* the slots by a full slot pitch, in which case some part of the conductor occupies every angular position over a slot pitch, so that it is equivalent to an infinite number of infinitesimally short lengths displaced by infinitesimally small angles. The winding can then be said to be uniformly distributed so that Equation 6.5.6 can be applied. Skewing introduces constructional difficulties, but is often used in induction motors to smooth out the variations in air-gap permeance and obtain a more uniform torque and a quieter motor. If the slots are not skewed by a full slot pitch, it becomes necessary to introduce another factor called the *skew factor*, which has to be applied to Equation 6.5.5 to account for the effect of skewing.

Let us now consider an example to show the effect of distributing the winding and skewing the slots.

EXAMPLE 6.5.1

A machine has nine slots per pole.

a. Calculate the fundamental distribution factor for the following cases:
   (i) One winding distributed in all the slots.
   (ii) One winding using only the first two-thirds of the slots per pole, i.e., 120° groups.
   (iii) Three equal windings placed sequentially in 60° groups.
b. Considering the case *(iii)* above, determine the distribution factors for the 5th, 17th, and 19th harmonics for straight slots as well as for slots skewed by one slot pitch.

SOLUTION

a. In each case, the slot angle $\alpha = 180/9 = 20$ electrical degrees. The values of m, the number of slots in a group, are 9, 6, and 3, respectively, for cases *(i)*, *(ii)*, and *(iii)*. Equation 6.5.5—or, approximately, Equation 6.5.6—can be applied with $n = 1$ for the fundamental.

$$\text{(i)} \quad k_{d1} = \frac{\sin(9 \times 20°/2)}{9 \sin(20°/2)} = 0.64, \text{ from Equation 6.5.5}$$

or approximately, from Equation 6.5.6

$$k_{d1} = \frac{\sin(\pi/2)}{\pi/2} = 0.637$$

$$\text{(ii)} \quad k_{d1} = \frac{\sin(6 \times 20°/2)}{6 \sin(20°/2)} = 0.831$$

or approximately $k_{d1} = \dfrac{\sin(\pi/3)}{\pi/3} = 0.827$

$$\text{(iii)} \quad k_{d1} = \frac{\sin(3 \times 20°/2)}{3 \sin(20°/2)} = 0.960$$

or approximately $k_{d1} = \dfrac{\sin(\pi/6)}{\pi/6} = 0.955$

These results illustrate one of the reasons for using three or more separate windings or phases. Dividing the winding into six groups per pole pair instead of three increases the voltage, and hence the power output for the same number of conductors, by 15 percent from a factor of 0.831 to one of 0.96. As stated earlier, these six groups could form a six-phase winding, or three groups of two sections each to form a three-phase winding as shown in Figure 6.4.1*(a)*. This example further illustrates why only two-thirds of the slots are usually employed for a single-phase winding. The additional slots would only increase the output by $(9 \times 0.64)/(6 \times 0.831) = 1.15$ times, while the increase in the amount of copper would be 50 percent.

For a simple two-phase winding, this slotting arrangement is not suitable, because an exact displacement of 90° is not possible with an odd number of slots. Considering eight slots per pole as an example, the distribution factor for two equal windings would then be

$$\frac{\sin[4 \times 180°/(8 \times 2)]}{4 \sin[180°/(8 \times 2)]} = 0.907$$

which is not quite as good as for the three-phase winding with 60° groups.

b. $k_{d5} = \dfrac{\sin(3 \times 5 \times 20°/2)}{3 \sin(5 \times 20°/2)} = 0.216$, from Equation 6.5.5,

or for skewed slots,

$$k_{d5} = \frac{\sin 150°}{150 \times \pi/180} = 0.191, \text{ from Equation 6.5.6.}$$

$$k_{d17} = \frac{\sin(3 \times 17 \times 20°/2)}{3 \sin(17 \times 20°/2)} = 0.959,$$

or for skewed slots,

$$k_{d17} = \frac{\sin 510°}{510 \times \pi/180} = \frac{\sin 150°}{510 \times \pi/180} = 0.056.$$

$$k_{d19} = \frac{\sin(3 \times 19 \times 20°/2)}{3 \sin(19 \times 20°/2)} = 0.959,$$

or for skewed slots,

$$k_{d19} = \frac{\sin 570°}{570 \times \pi/180} = \frac{\sin 210°}{570 \times \pi/180} = -0.05.$$

With straight slots, some harmonics are reduced; but unfortunately the pronounced tooth harmonics caused by armature slotting are not reduced. For example, the 17th and 19th harmonics arise with nine slots per pole. With skewed slots, however, there is a progressive reduction in the higher harmonics as seen from the above example. In order to take full advantage, the degree of skewing should be equivalent to a full slot pitch.

PITCH FACTOR   If the individual coils of a winding are short-pitched, that is to say, if the coil span is less than a pole pitch, the coil emf is reduced, and the *pitch factor* k_p is defined as

$$k_p = \frac{\text{emf of short-pitched coil}}{\text{emf of full-pitched coil}} \tag{6.5.7}$$

This is equivalent to the ratio of the phasor sum of coil-side emfs to the arithmetic sum of coil-side emfs. When the coil pitch is less than a pole pitch (180 electrical degrees), the flux linking the coil and hence the induced voltage is less than that for a full-pitch coil. Considering the situation shown in Figure 6.5.2, the coil pitch ρ in electrical degrees of the fractional-pitch coil is less than 180 electrical degrees; and the coil is located in a sinusoidal flux-density field wave. The ratio of the flux linking with the short-pitched coil to the flux that would have linked with a full-pitch coil is given by

$$\frac{\int_{-\rho/2}^{+\rho/2} \cos \theta \, d\theta}{\int_{-\pi/2}^{+\pi/2} \cos \theta \, d\theta} = \frac{2\int_0^{\rho/2} \cos \theta \, d\theta}{2\int_0^{\pi/2} \cos \theta \, d\theta} = \frac{[\sin \theta]_0^{\rho/2}}{[\sin \theta]_0^{\pi/2}} = \sin \frac{\rho}{2} \tag{6.5.8}$$

where θ is in electrical degrees. The expression obtained in Equation 6.5.8 also gives the pitch factor since the emfs are simply proportional to the flux linkages. Thus, we have

$$k_p = \sin \frac{\rho}{2} = \sin \left\{ \frac{\text{coil span in electrical degrees}}{2} \right\} \tag{6.5.9}$$

If the field is nonsinusoidal, containing odd harmonics (note that even harmonics need not be considered in ac machines because of the symmetry), the magnitude of the pitch factor for the nth harmonic is given by

$$k_{pn} = \sin \frac{n\rho}{2} \tag{6.5.10}$$

FIGURE 6.5.2   Fractional-pitch coil in a sinusoidal field.

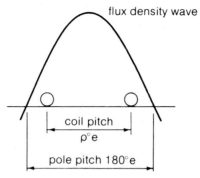

flux density wave

coil pitch
$\rho°$e

pole pitch 180°e

For odd harmonics, the sign and magnitude of the pitch factor are given in terms of the short pitch or pitch deficiency $(180 - \rho)$ by

$$k_{pn} = \cos \frac{n(180 - \rho)}{2} \qquad (6.5.11)$$

The effect of pitch on the phase of any harmonic in a fractional-pitch winding, as compared with its phase in a full-pitch winding, depends on the order of the harmonic. All odd harmonics are either not changed in phase or reversed. The magnitudes of the harmonics in fractional-pitch windings, as compared with their magnitudes in a full-pitch winding having the same number of turns, are given in Table 6.5.1.

TABLE 6.5.1   MAGNITUDES OF HARMONICS IN FRACTIONAL-PITCH WINDINGS

Pitch	Harmonic				
	1	3	5	7	11
2/3	0.866	0.000	0.866	0.866	0.866
4/5	0.951	0.588	0.000	0.588	0.951
5/6	0.966	0.707	0.259	0.259	0.966
6/7	0.975	0.782	0.434	0.000	0.782

The voltage waveform can be improved by using short-pitched coils. Any harmonic can be eliminated from the voltage generated in an ac armature coil by choosing a pitch that makes the pitch factor zero for that harmonic. To eliminate the nth harmonic,

$$k_{pn} = \sin\left(\frac{n\rho}{2}\right) = 0, \rho = \frac{2k\pi}{n} \qquad (6.5.12)$$

where k is any integer. For example, to eliminate the fifth harmonic, coil pitches of 2/5, 4/5, 6/5, etc., can be used. However, since any departure from full pitch reduces the fundamental by an amount that increases progressively with the departure of the pitch from unity, in practice the fractional pitch that is nearest to unity, in this case 4/5, is used. Pitches greater than unity require more copper for the coil-end connections than shortened pitches and possess no compensating advantage. The seventh harmonic is totally eliminated (chorded out) with the choice of 6/7 pitch. A 5/6 pitch is particularly satisfactory for reducing harmonics as it cuts significantly both the fifth and seventh harmonics.

It can be shown that third harmonic voltages and multiples of third harmonic such as the 9th, 15th, etc., caused by corresponding harmonic components in the flux-density wave do not appear in the line-to-line voltages of normally designed either wye- or delta-connected three-phase machines. Some difficulty may occasionally be caused by such triplex harmonics in the line-to-neutral voltages of wye-connected machines, in which case 2/3 pitch may be used. Usually the most significant harmonics to be minimized by the use of fractional-pitch windings are the fifth and seventh. Higher harmonics than the ninth are so small that little attention is required except in rare cases.

For single-layer windings, only the distribution factor is effective. Even though individual coils may be short-pitched or over-pitched, the coils are full-pitched on the average because the coil sides of a phase under neighboring poles occupy the same group of slots relative to the poles.

EXAMPLE 6.5.2

An armature of an ac machine has 18 slots per pole. It is wound with a two-layer, three-phase winding consisting of single-turn coils. Each coil spans 15 slots. Calculate the pitch factors for the fundamental k_{p1}, for the fifth harmonic k_{p5}, and for the seventh harmonic k_{p7}.

SOLUTION

The slot angle $\alpha = 180/18 = 10$ electrical degrees. The coil span in electrical degrees is $\rho = 15 \times 10 = 150°\text{e}$. The coil is short-pitched by $180 - \rho = 180 - 150 = 30$ electrical degrees. From Equation 6.5.11 it follows that

$$k_{p1} = \cos \frac{30°}{2} = 0.966$$

$$k_{p5} = \cos \frac{5 \times 30°}{2} = 0.259$$

$$k_{p7} = \cos \frac{7 \times 30°}{2} = -0.259$$

The signs are associated with the relative phase angles of the harmonics in the resultant emf wave. The signs are usually not of practical significance so that Equation 6.5.10 can, as well, be applied.

FIGURE 6.6.1   Space harmonics in the flux-density wave. (Note: Only third-harmonic is shown for simplicity.)

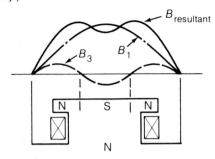

6.6 EMF PRODUCED BY AN ARMATURE WINDING

The time variation of emf for a single conductor corresponds to the spatial variation of air-gap flux density. This is not purely sinusoidal, and often contains significant lower-order odd harmonics. By suitable winding design through distributing and chording, the harmonics could be appreciably reduced while the fundamental is only reduced by a little. Consequently, the waveform of the generated emf approaches more nearly a pure sine shape. The winding factors (discussed in Section 6.5) express mathematically the per-unit reduction of the fundamental as well as of each harmonic, which takes place as a result of distribution and chording. In practice, the harmonics are often reduced to negligible proportions, while the fundamental is only reduced by about 10 percent. The same winding factors also apply to the space distribution of mmf, and so reduce the mmf space harmonics.

For an ac machine, considering only the fundamental, Equation 5.3.14 has been developed for the induced voltage. The rms-generated voltage per phase for a concentrated winding having N_{ph} turns in series per phase is given by

$$E_1 = 4.44 \, f \, N_{ph} \, \phi_1 \text{ rms volts per phase} \qquad (6.6.1)$$

where subscript 1 denotes the fundamental, f the frequency, and ϕ_1 the fundamental field flux per pole. When both the fundamental distribution factor and pitch factor for the distributed, fractional-pitch winding are applied, the rms voltage per phase is modified to

$$E_1 = 4.44 \, k_{w1} \, f \, N_{ph} \, \phi_1 \text{ rms volts per phase} \qquad (6.6.2)$$

which is the same as Equation 5.3.18, and in which k_{w1} is the winding factor for the fundamental given by

$$k_{w1} = k_{d1} \cdot k_{p1} \qquad (6.6.3)$$

which is slightly less than unity for a polyphase machine.

Time harmonics in the emf are produced by space harmonics in the flux-density wave. Figure 6.6.1 shows the flux distribution consisting of a fundamental and an inphase third harmonic. It can be imagined as being produced by three harmonic poles superimposed on each fundamental pole as shown. Harmonic voltages can be calculated if the field form is analyzed into its fundamental and harmonic flux densities given by

$$B = B_1 \sin\theta + B_3 \sin 3\theta + B_5 \sin 5\theta$$
$$+ B_7 \sin 7\theta + \ldots \qquad (6.6.4)$$

For the harmonics, the frequency of flux changes is increased three, five, seven, etc., times depending on the order of the harmonic, which tends to bring about an increase of voltage; but corresponding reductions in harmonic pole-pitch area and flux components, as seen in Figure 6.6.1, tend to offset the increase of voltage. Thus, harmonic voltages can be obtained by reducing the fundamental voltage in the ratio of the corresponding flux density and winding factors. That is to say, for the nth harmonic,

$$E_n \text{ per phase } = E_1(B_n/B_1)(k_{dn}/k_{d1})(k_{pn}/k_{p1}) \qquad (6.6.5)$$

or equivalently

$$E_n \text{ per phase } = 4.44\ k_{dn}\ k_{pn}\ k_{fn}\ f\ N_{ph}\ \phi_1 \qquad (6.6.6)$$

where ϕ_1 is the fundamental flux per pole, and $k_{fn} = (B_n/B_1)$ is a coefficient, depending on the nature of the surfaces facing the air gap, corresponding to the nth harmonic.

The ratio of the voltage magnitude of any harmonic to the fundamental is then given by

$$\frac{E_n}{E_1} = \frac{k_{dn}k_{pn}k_{fn}}{k_{d1}k_{p1}} \qquad (6.6.7)$$

For normal designs and machine proportions, the ratios (k_{dn}/k_{d1}), (k_{pn}/k_{p1}), and the coefficient k_{fn} are all much less than unity, so that their product results in a very small quantity; and hence the harmonics in the terminal voltage may usually be neglected under normal conditions.

The rms phase value of the complete voltage wave is given by

$$E_{ph} = \sqrt{E_1{}^2 + E_3{}^2 + E_5{}^2 + \ldots}$$
$$= E_1\sqrt{1 + (E_3{}^2/E_1{}^2) + (E_5{}^2/E_1{}^2) + \ldots} \qquad (6.6.8)$$

Since the value of the radical $\sqrt{1 + (E_3{}^2/E_1{}^2) + (E_5{}^2/E_1{}^2) + \ldots}$ is very nearly unity, the rms phase value of the complete voltage is very nearly the same as the rms value of the fundamental alone, when the harmonics are relatively small in comparison with the fundamental. Thus, for practical purposes, the rms value of the generated voltage per phase is given by

$$E_{ph} = 4.44\ k_{d1}\ k_{p1}\ f\ N_{ph}\ \phi_1 \text{ rms volts} \qquad (6.6.9)$$

FIGURE 6.6.2 A closed continuous two-pole commutator winding. (Note: The commutator brushes are shown, for convenience, making direct contact with the armature winding.)

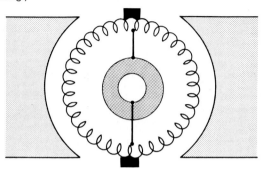

which is the same as Equation 5.3.15 given earlier. The rms line-to-line voltage for balanced wye connection is given by $\sqrt{3}$ times the phase value, but excluding the third harmonic, if in fact the harmonic components are considered. This is because, for the wye connection, the line voltage is obtained from the difference of the phase voltages, inphase components cancel, so that the line voltage is free from third and multiples of third (triplex) harmonics.

Let us next consider the case of a conventional dc machine with a closed continuous commutator winding on its armature. Equations 5.3.22 to 5.3.24 have already been obtained for the generated emf. The same can also be derived from a different point of view, which brings out a better understanding of the interrelationship between the ac and dc machine windings.

The induced voltages in a commutator winding are inherently alternating, and between any two points the resultant can be obtained by vectorial summation. Winding factors that have been discussed may also be applied; but the slot angle is usually small such that Equation 6.5.3 with the denominator replaced by $m\alpha/2$ in electrical radians is used for the distribution factor, and the chording factor may usually be neglected (i.e., pitch factor of unity is taken) since the chording angle is small. Let us consider two tappings opposite to each other on a two-pole commutator winding shown in Figure 6.6.2. Assuming a sinusoidal distribution of flux, the vectorial summation is proportional to the diameter, while the arithmetic sum is proportional to one-half of the periphery; that is to say,

$$k_d = \frac{\text{Chord}}{\text{Arc}} = \frac{2}{\pi} \qquad (6.6.10)$$

The above may also be obtained by the application of Equation 6.5.3, with the approximation made for the denominator:

$$k_d = \frac{\sin(\pi/2)}{\pi/2} = \frac{2}{\pi}$$

If the diametrical tappings are connected to two slip rings as shown in Figure 6.6.2, for the position indicated, the instantaneous ac voltage will be at its maximum value, and the rms value of the slip-ring voltage will be 0.707 times of that maximum value. If the winding is connected to a commutator on which two brushes are located on the quadrature axis, as shown schematically in Figure 6.6.2, the voltage across the brushes will be that maximum voltage at a substantially constant value as a function of time. In actual practice, however, there is a small superimposed high-frequency ripple because of the finite number of commutator bars and slots. Thus we see that when a commutator winding rotates in a magnetic field, a unidirectional emf is obtained whose magnitude depends on the brush position; further, the magnitude of the emf is the same as the instantaneous emf that would be obtained from a slip-ring winding when the rotor position with the tappings corresponds with the commutator brush axis.

As stated already, for a dc machine, a sinusoidal flux waveform is not needed. The flux density is arranged to be as high as possible over the pole pitch in order to get the maximum output voltage; of course, other considerations, such as commutation, need to be looked into. This leads to a field form that approaches a rectangular shape. In view of the large number of space harmonics in such a field form, the vectorial addition, though feasible, becomes rather complicated. Hence, a simpler method is usually employed for calculating the generated emf. The average voltage per conductor is calculated on the basis of the average flux density, since the dc voltage (which is an average value) is of interest. This, when multiplied by the number of conductors in series, gives the total dc voltage across the brushes.

In a conventional dc machine, the brush axis is fixed relative to the stator. If, however, there is continuous relative motion between the poles and brushes, the voltage at the brushes will in fact be alternating. This principle is made use of in ac commutator machines.

EXAMPLE 6.6.1

The field form of a wye-connected, three-phase, 60-Hz, 10-pole alternator has a spatial flux-density distribution given by

$$B = \sin \theta + 0.3 \sin 3\theta + 0.2 \sin 5\theta \text{ tesla}$$

The machine has 180 slots, lap-wound with double-layer, three-turn coils in 60° groups. Each coil spans 15 slots. The armature diameter is 1.5 m and the core length is 0.50 m. Determine the following:

   a. Instantaneous emf per conductor.
   b. Instantaneous emf per coil.
   c. RMS phase and line-to-line voltages.

SOLUTION

a. Surface area per pole pitch = (pole pitch)(axial length of the core)

$$= \frac{\pi \times 1.50}{10} \times 0.50 = 0.2357 \text{ m}^2$$

Fundamental flux per pole = $\phi_1 = B_{1 \text{ av}} \times$ area

$$= 1 \times (2/\pi) \times 0.2357 = 0.15 \text{ Wb}$$

where $2/\pi$ is the factor for the average of a sinusoidal wave with an amplitude of unit magnitude.

The maximum or peak value of fundamental voltage per conductor is given by $\pi f \phi$, as seen from Equation 5.3.13, in which N is the number of turns of a coil, or one-half of the number of conductors; i.e., $2N$ is the number of conductors per coil.

Thus, peak value of fundamental conductor voltage = $\pi \times 60 \times 0.15$ = 28.3 V. The peak harmonic conductor voltages are proportional to their corresponding peak flux densities: for the third harmonic, 0.3×28.3 = 8.49 V; for the fifth harmonic, $0.2 \times 28.3 = 5.66$ V. The instantaneous conductor voltage is then given by

$$(28.3 \sin \omega t + 8.49 \sin 3\omega t + 5.66 \sin 5\omega t) \text{ volts}$$

where $\omega = 2\pi \times 60 = 377$ rad/s, and $t = 0$ when the voltage is zero, i.e., t is measured from the point of zero voltage.

b. The number of slots and the coil span are the same as in Example 6.5.2. The pitch factors for the fundamental and fifth harmonic are calculated in that example:

$$k_{p1} = \cos\left(\frac{30°}{2}\right) = 0.966$$

$$k_{p3} = \cos\left(\frac{3 \times 30°}{2}\right) = 0.707$$

$$k_{p5} = \cos\left(\frac{5 \times 30°}{2}\right) = 0.259$$

Since there are six conductors in a three-turn coil, it follows that the peak fundamental coil voltage = $6 \times 28.3 \times 0.966 = 164$V, the peak third harmonic coil voltage = $6 \times 8.49 \times 0.707 = 36$V, the peak fifth harmonic coil voltage = $6 \times 5.66 \times 0.259 = 8.8$V. The instantaneous coil voltage is then given by

$$(164 \sin \omega t + 36 \sin 3\omega t + 8.8 \sin 5\omega t) \text{ volts}$$

c. The distribution factors can be calculated from Equation 6.5.5 for each harmonic. The slot angle α being $180/18 = 10$ electrical degrees, and the number of slots per group m being $18/3 = 6$ for $60°$ grouping,

$$k_{d1} = \frac{\sin(60°/2)}{6\sin(60°/2)} = \frac{0.5}{6 \times 0.087} = 0.958$$

$$k_{d3} = \frac{\sin(180°/2)}{6\sin(30°/2)} = \frac{1}{6 \times 0.259} = 0.644$$

$$k_{d5} = \frac{\sin(300°/2)}{6\sin(50°/2)} = \frac{0.5}{6 \times 0.423} = 0.197$$

Total number of coils per phase $= 180/3 = 60$, since two coil sides occupy each slot in a double-layer winding. The rms phase emf may now be calculated:

$$\text{rms fundamental phase emf} = (164/\sqrt{2}) \times 60 \times 0.958$$
$$= 6{,}666.7 \text{ V}$$

$$\text{rms third harmonic phase emf} = (36/\sqrt{2}) \times 60 \times 0.644$$
$$= 983.8 \text{ V}$$

$$\text{rms fifth harmonic phase emf} = (8.8/\sqrt{2}) \times 60 \times 0.197$$
$$= 73.6 \text{ V}$$

These can also be calculated by the direct application of Equation 6.6.6, in which N_{ph} is the number of series turns per phase, given by 180 in our example because of the 60 three-turn coils per phase.

The rms phase voltage is then given by Equation 6.6.8:

$$E_{ph} = \sqrt{6{,}666.7^2 + 983.8^2 + 73.6^2} = 6{,}739.3 \text{ V}$$

$$\text{The rms line-to-line voltage} = \sqrt{3}\sqrt{6{,}666.7^2 + 73.6^2}$$
$$= \sqrt{3} \times 6{,}667.1 = 11{,}547.4 \text{ V}$$

Note that the third-harmonic component is excluded in the calculation of the line-to-line voltage.

EXAMPLE 6.6.2

The armature of a four-pole dc machine has a simplex-lap wound commutator winding with 120 two-turn coils. If the flux per pole is 0.02 Wb, calculate the dc voltage appearing across the brushes located on the quadrature axis, when the machine is running at 1,800 rpm.

SOLUTION

This can be solved by the direct application of Equation 5.3.23:

$$E_a = \frac{P \phi n Z}{a \times 60}$$

For our example,

$$P = 4; \phi = 0.02; n = 1,800$$
$$Z = \text{number of conductors} = 120 \times 2 \times 2 = 480$$
$$a = \text{number of parallel paths} = 4, \text{ for the simplex-lap}$$
$$\text{winding (same as number of poles)}$$

Therefore,

$$E_a = \frac{4 \times 0.02 \times 1,800 \times 480}{4 \times 60} = 288 \text{ V dc}$$

6.7 MMF PRODUCED BY WINDINGS

Having discussed the formation of windings, we shall examine in this section the mmf and flux-density distributions produced by windings. A knowledge of the flux-density distribution waveforms is important, because they determine the waveforms of generated emfs in the machine and also determine the winding inductances as well as the variation of these as functions of rotor position.

Most of the analysis in this book is based on inductances and the coupled-circuit viewpoint. The general expressions for emf and torque, based on self- and mutual inductances of the machine windings, have already been derived. Emf and torque calculations can, of course, be performed directly in terms of the magnetic fields, as we have briefly seen earlier. However, the concepts of self- and mutual inductances make it possible to express the magnetic flux linkages in terms of the inductances and the winding currents; such an approach is quite convenient as the subsequent work can be performed by the familiar circuit analysis, and this in fact becomes the basis for the

FIGURE 6.7.1   A two-pole cylindrical machine with a stator and rotor, having a single coil of N turns on the rotor.

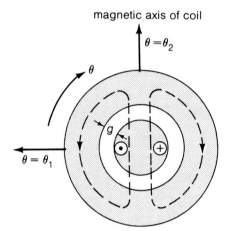

generalized theory of electrical machines. While harmonics can be taken into account with the use of the inductance method, saturation effects, if present, cannot easily be taken care of. Based on the mmf and flux-density distributions produced by windings, it is possible to develop expressions for the winding inductances in cylindrical and salient-pole machines by evaluating the appropriate flux linkages per ampere of the appropriate current.

CYLINDRICAL MACHINE WITH A SINGLE COIL   Let us start by considering a simple case of a two-pole machine with a cylindrical stator and rotor having a single coil of N turns on the rotor, as shown in Figure 6.7.1, carrying a current of I amperes. The flux produced by the coil, with the directions of current as shown, divides itself into two parallel paths as illustrated by the dotted flux loops. The mmf applied to each of these parallel paths is NI ampere-turns, and Ampere's circuital law can be applied around the closed path in order to obtain the magnetic field intensity and hence the magnetic flux density, recognizing that each flux loop contains two air gaps in series. In other words, the mmf at each pole is $NI/2$ ampere-turns per pole. We shall neglect the reluctance of the iron compared with that of the air gaps, and consequently assume that all the applied mmf acts across the air gaps alone. With the convention that mmf at the air gap is positive if acting radially outwards, and vice versa, one obtains the mmf distribution, F_θ, as a function of the position around the air gap, as shown in Figure 6.7.2 in a developed form that is laid out flat. The flux density at any position θ is given by $B_\theta = \mu_0 H_\theta = \mu_0 F_\theta/g$, where g is the single radial air-gap length. If the air gap is constant at all values of θ, the flux density B_θ for any position θ is proportional to the mmf, and hence it follows that Figure 6.7.2 will also represent the flux-density distribution as a function of θ to a different scale.

FIGURE 6.7.2 The mmf and flux-density distribution corresponding to Figure 6.7.1 in a flattened (developed) layout.

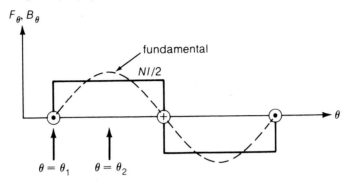

The wave of mmf distribution is shown in Figure 6.7.2 by the steplike distribution of amplitude $\pm NI/2$; on the basis of narrow slot openings, the mmf wave jumps abruptly by NI in crossing from one side to the other of the coil. The rectangular waveshape of the mmf shown in Figure 6.7.2 can be analyzed by Fourier analysis into harmonic components. If θ is measured from zero position θ_1, one obtains

$$F_\theta = \frac{4F}{\pi} \sum_{n \text{ odd}} \frac{\sin(n\theta)}{n} \qquad (6.7.1)$$

where $F = NI/2$.

If θ is measured from zero position θ_2, i.e., the magnetic axis of the coil, one obtains

$$F_\theta = \frac{4F}{\pi} \sum_{n \text{ odd}} \pm \frac{\cos(n\theta)}{n} \qquad (6.7.2)$$

where $F = NI/2$, and the sign alternates between $+$ and $-$, starting with the positive sign for the fundamental. Explicitly,

$$F_\theta = \frac{4F}{\pi} [\cos \theta - \frac{1}{3} \cos 3\theta + \frac{1}{5} \cos 5\theta - \dots]$$

The fundamental is sketched in Figure 6.7.2, the amplitude of which is given by $[(4/\pi)NI/2]$. The peak of the fundamental sinusoidal space wave is aligned with the magnetic axis of the coil.

FIGURE 6.7.3   A two-pole cylindrical machine with uniformly distributed winding of ten conductors in *(a)*, and its equivalent continuous conductor arrangement in *(b)*.

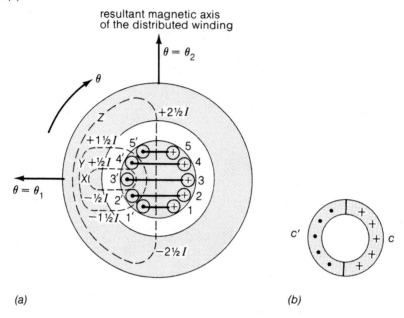

(a)   (b)

DISTRIBUTED WINDING   Let us next consider the case of a two-pole cylindrical machine with a uniformly distributed winding, consisting of 10 conductors as shown in Figure 6.7.3*(a)*, each conductor carrying a current of I amperes in the directions indicated. It is assumed that pairs of conductors are connected to form coils as shown, 1–1′, 2–2′, etc. Considering the three flux loops sketched on either side of the magnetic axis of the distributed winding, the mmf applied to the flux loop X is I ampere-turns, and that applied to the flux loop Y is $3I$ ampere-turns, while the mmf applied to the flux loop Z is $5I$ ampere-turns. Noting that, at any position of the air gap between two adjacent conductors, the mmf remains at the same constant value, the mmf at various parts of the air gap as a function of θ can be determined by inspection. The mmf distribution showing the variation of mmf around the air gap is given in Figure 6.7.4 by a series of steps, each of height I ampere-conductors in the slot, in a flattened (developed) layout. If the radial air gap is a constant, the value of the flux density at any position θ is simply proportional to the mmf, and as such the same waveshape of Figure 6.7.4 also represents the flux-density distribution to a different scale.

FIGURE 6.7.4 The mmf and flux-density distribution corresponding to Figure 6.7.3 in a flattened (developed) layout.

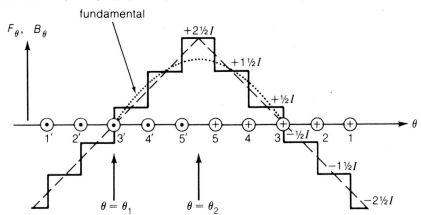

It can be seen that the distributed winding produces a closer approximation to a sinusoidal mmf wave than the concentrated coil of Figure 6.7.1. Thus, distributing the winding results in a marked reduction in the harmonic content of the flux-density distribution, and hence causes the induced voltage to be more nearly sinusoidal. The resultant space-fundamental component of the mmf wave of a distributed winding is less than the sum of the fundamental components of the individual coils, because the magnetic axes of the individual coils are not aligned with the resultant.

If the conductors 1 to 5 and 1' to 5' are replaced by continuous conductors C and C', as shown in Figure 6.7.3(b), carrying a total current of $5I$ amperes, then the mmf will increase and decrease linearly around the air gap as shown by the dotted triangular waveform in Figure 6.7.4. This is equivalent to the assumption that the winding is made up of an infinite number of conductors Z with an infinitesimally small circumferential width, each carrying $(1/Z)$ of the total current. In practice, since a much larger number of conductors than five is normally used, the triangular waveform with linear variation can be taken as a reasonable approximation.

The triangular waveshape of the mmf shown in Figure 6.7.4 can be analyzed by Fourier analysis into harmonic components. If θ is measured from zero position θ_1, one obtains

$$F_\theta = \frac{8F}{\pi^2} \sum_{n \text{ odd}} \left[\pm \frac{\sin(n\theta)}{n^2} \right] \tag{6.7.3}$$

where $F = NI/2$, N being 5 in Figure 6.7.4. Equivalently,

$$F_\theta = \frac{4F}{\pi} \sum_{n \text{ odd}} [\pm k_n \sin(n\theta)] \tag{6.7.4}$$

FIGURE 6.7.5   A two-pole cylindrical machine with partially distributed
winding over a portion of the rotor.

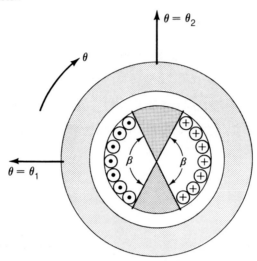

where $k_n = 2/n^2\pi$. The sign alternates between $+$ and $-$, starting with the positive sign for the
fundamental. If θ is measured from zero position θ_2, i.e., the resultant magnetic axis of the distributed winding, one obtains

$$F_\theta = \frac{8F}{\pi^2} \sum_{n\ odd} \frac{\cos(n\theta)}{n^2} = \frac{4F}{\pi} \sum_{n\ odd} k_n \cos(n\theta) \qquad (6.7.5)$$

where F and k_n are the same as given before.

The fundamental is sketched in Figure 6.7.4, the amplitude of which is given by $8F/\pi^2$. The
peak of the fundamental sinusoidal space wave is aligned with the resultant magnetic axis of the
distributed winding.

PARTIALLY DISTRIBUTED WINDING   Instead of a distributed winding fully covering the
rotor surface as in Figure 6.7.3, consider a distributed rotor winding in which the conductors,
carrying currents as shown in Figure 6.7.5, are distributed over an angle β. The case of $\beta = 180$
electrical degrees corresponds to that of Figure 6.7.3, whereas $\beta = 0$ represents the single coil of
Figure 6.7.1. For the case of three-phase windings, each phase occupies one-third of the air-gap
periphery, and for the usual 60° groupings β is equal to 60 electrical degrees.

FIGURE 6.7.6 The mmf and flux-density distribution corresponding to
Figure 6.7.5 in a flattened (developed) layout.

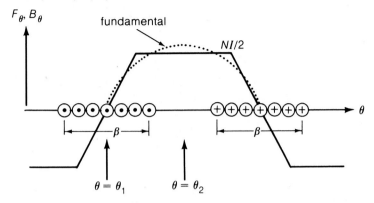

By following the procedure indicated for the previous cases, it should be obvious that the mmf
and flux-density distributions will be as shown in Figure 6.7.6. Note that the flat tops with constant
mmf correspond to the parts of the periphery where there are no conductors, and a straight-line
approximation is used over the rest. The trapezoidal waveshape of the mmf shown in Figure 6.7.6
can be analyzed by Fourier analysis into harmonic components. If θ is measured from zero position
θ_1, one obtains

$$F_\theta = \frac{4F}{\pi} \sum_{n \text{ odd}} k_n \sin(n\theta) \tag{6.7.6}$$

where $F = NI/2$ and

$$k_n = \frac{\sin(n\beta/2)}{(n^2\beta/2)}$$

If θ is measured from the zero position θ_2, i.e., the resultant magnetic axis of the distributed winding, one obtains

$$F_\theta = \frac{4F}{\pi} \sum_{n \text{ odd}} [\pm k_n \cos(n\theta)] \tag{6.7.7}$$

where F and k_n are the same as given for Equation 6.7.6; the sign alternates between $+$ and $-$,
starting with the positive sign for the fundamental.

The magnitude of any harmonic component of mmf produced by the distributed winding is nk_n times that produced by the concentrated winding, as seen from Equations 6.7.1 and 6.7.6, or Equations 6.7.2 and 6.7.7. Thus

$$nk_n = \frac{\sin(n\beta/2)}{n\beta/2} = k_{dn} \qquad (6.7.8)$$

is called the distribution factor for the nth harmonic, which is the same as given by Equation 6.5.6 earlier. The distributed winding of N turns is equivalent to a concentrated winding of $(k_{dn}N)$ turns, as far as the nth harmonic is concerned.

The fundamental with the amplitude of $(4Fk_1/\pi)$ is aligned with its peak occurring on the resultant magnetic axis of the distributed winding. In general, for a P-pole machine with a distributed winding of N_{ph} series turns per phase, the amplitude of the space-fundamental mmf wave is given by

$$F_{a1} = \frac{4}{\pi} k_1 \frac{N_{ph}}{P} i_a \qquad (6.7.9)$$

where i_a is the current in phase a; the factor $4/\pi$ arises from the Fourier-series analysis of the rectangular mmf wave of a concentrated full-pitch coil, and k_1 accounts for the distribution of the winding. The effects of space harmonics in ac machines, as we have studied earlier, can be made small through the use of distributed windings with fractional-pitch coils.

The above space-fundamental component of the mmf wave produced by current in the distributed phase-a winding is equivalent to the mmf wave produced by a finely divided sinusoidally distributed current sheet placed on the outer periphery of the rotor (or inner periphery of the stator if the winding is located in the slots of the stator). The mmf is a standing wave whose fundamental spatial distribution around the periphery is sinusoidal, described by $\cos\theta$ or $\sin\theta$ depending upon the zero-position chosen for θ. As stated before, its peak is along the magnetic axis of phase-a winding, and its peak amplitude is proportional to the instantaneous current i_a as given by Equation 6.7.9. If the phase current i_a is a function of time given by $i_a = I \cos \omega t$, the time maximum of the peak will be given by

$$F_m = \frac{4}{\pi} k_1 \frac{N_{ph}}{P} I \qquad (6.7.10)$$

The effect of balanced three-phase currents in all the three phases has been mentioned in Section 5.4.

Thus far, we have seen that the waveshapes of the mmf and flux-density distributions produced by windings around the air-gap periphery can be analyzed by Fourier analysis into harmonic components:

$$F_\theta = \Sigma F_n \sin(n\theta) \text{ or } \Sigma F_n \cos(n\theta) \qquad (6.7.11)$$

$$B_\theta = \Sigma B_n \sin(n\theta) \text{ or } \Sigma B_n \cos(n\theta) \qquad (6.7.12)$$

FIGURE 6.7.7   A two-pole machine with cylindrical rotor and salient-pole stator fitted with concentrated stator winding.

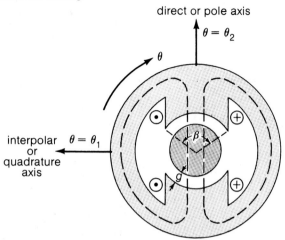

depending on the reference chosen for $\theta = 0$; $B_n = \mu_0 F_n/g$. Note that there are no even harmonics because of the particular symmetry of the waveforms, i.e., $F(\theta + \pi) = -F(\theta)$. Discussions are often restricted only to the fundamental sinusoidal component of mmf or flux-density distribution because:

(i) This is quite often a reasonable mathematical approximation, which also happens to be very convenient for the analysis.

(ii) By suitably distributing and chording the winding, the effect of harmonics can be reduced significantly.

(iii) If, in fact, the approximation is not good enough, the effects of each harmonic can be included by following the same general principles of analysis developed for the fundamental.

Thus, in introductory machine analyses, it is often sufficiently accurate to neglect all but the fundamental components.

CONCENTRATED COILS ON SALIENT POLES   Consider a two-pole machine with a cylindrical rotor and a salient-pole stator, as shown in Figure 6.7.7, with concentrated stator coils of N total number of turns on salient poles, carrying a current of I amperes. The flattened layout of the stator in a developed form is given in Figure 6.7.8 along with the corresponding mmf wave-

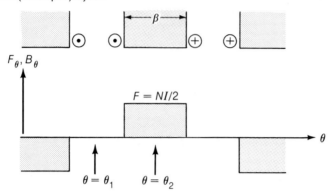

FIGURE 6.7.8 The mmf and flux-density distribution corresponding to
Figure 6.7.7 in a flattened (developed) layout.

shape. Maximum mmf of constant value ($NI/2$) occurs over the polar region, and no mmf exists in the interpolar region. If the air gap in the polar region is a constant, the flux density under the pole is proportional to the mmf, and hence the waveform of Figure 6.7.8 also gives the waveshape of the flux-density distribution to a different scale; the maximum constant value of B is given by $\mu_0 F/g$, where F is ($NI/2$) and g is the radial air-gap length under the pole.

The Fourier analysis of the waveform leads to the following results. With zero position of θ as θ_1,

$$F_\theta = \frac{4F}{\pi} \sum_{n \text{ odd}} [\pm\, k_n \sin(n\theta)] \tag{6.7.13}$$

where $k_n = [\sin(n\beta/2)]/n$ and β is the angle subtended by the pole-face at the center, as shown in Figure 6.7.7, and the signs alternate starting with the positive sign for the fundamental. With zero position of θ as θ_2,

$$F_\theta = \frac{4F}{\pi} \sum_{n \text{ odd}} k_n \cos(n\theta) \tag{6.7.14}$$

In the case of salient-pole ac machines, where the sinusoidal nature of the flux-density distribution around the air-gap periphery is desirable as in synchronous machines, the body of the pole structure is usually fitted with a wider poleshoe, which is so shaped that the air gap is shortest under the pole center and increases in radial length toward the pole tips. This is usually referred to as *pole-chamfering*. Because the reluctance of the air gap under the poleshoe is not uniform, the flux-density distribution will not be the same as the mmf distribution.

FIGURE 6.7.9 A two-pole machine with salient-pole stator and cylindrical rotor carrying a distributed winding. *(a)* Rotor-mmf axis in line with the stator-pole axis, *(b)* rotor-mmf axis in quadrature to the stator-pole axis.

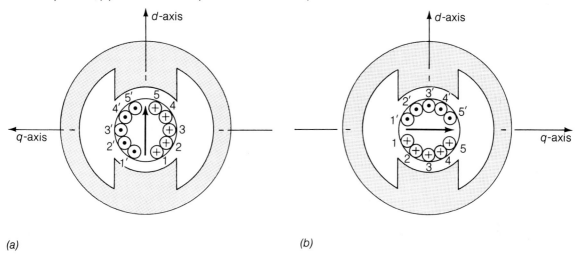

(a)   *(b)*

DISTRIBUTED WINDING ON THE ROTOR WITH SALIENCY ON STATOR   Let us next consider a two-pole machine with saliency on its stator and a cylindrical rotor with a distributed winding of 10 conductors, as shown in Figure 6.7.9, carrying currents as indicated. Two different positions of the rotor-mmf axis with respect to the stator are considered: *(a)* in line with the stator-pole axis, and *(b)* at right angles to the stator-pole axis. Figure 6.7.10 shows the corresponding developed diagrams for the mmf and flux-density waveforms. The mmf waveshape is triangular, as discussed previously, assuming linear variation instead of steps as in the case of a much larger number of conductors than five. But the flux density around the air-gap periphery is not now proportional to the mmf throughout, as the air-gap length is not a constant.

For a given position of the rotor, a rough sketch of the flux-density variation can be drawn by first locating a triangular curve corresponding to the uniform air gap equal to that under the poles, and then reducing the value of the flux density in those portions where the air gap is larger than the uniform value. Dotted curves in Figure 6.7.10 illustrate the approximate waveshapes of the flux-density distribution for the two positions of the rotor considered in Figure 6.7.9. It is seen that the flux-density waveform is now different for different positions of the rotor with respect to the stator. Even though the situation looks rather complicated because of the variable air gap, reasonable approximations can be made to calculate the flux-density distribution in terms of two mutually perpendicular components in the polar (direct) and interpolar (quadrature) axes. The physical structure and symmetry of the salient-pole arrangement suggest that it may be more

FIGURE 6.7.10 The mmf and flux-density distributions corresponding to Figure 6.7.9 *(a)* for the case when the rotor-mmf axis is in line with the stator-pole axis, *(b)* for the case when the rotor-mmf axis is in quadrature to the stator-pole axis.

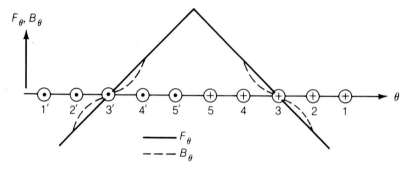

(a)

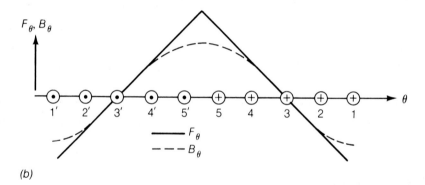

(b)

amenable to mathematical analysis if quantities are resolved into two orthogonal components along the direct and quadrature axes, which correspond to the physical polar and interpolar axes on the machine, that are always 90 electrical degrees away from each other. In fact, such a technique does work well, as we see later on in the analysis of a salient-pole synchronous machine.

BIBLIOGRAPHY

Daniels, A. R. *Introduction to Electrical Machines.* London: Macmillan, 1976.

Garik, M. L., and Whipple, C. C. *Alternating-Current Machines.* 2d ed. New York: D. Van Nostrand, 1961.

Hindmarsh, J. *Electrical Machines and Their Applications.* 2d ed. Oxford: Pergamon Press, 1970.

Kloeffler, R. G.; Kerchner, R. M.; and Brenneman, J. L. *Direct-Current Machinery.* New York: Macmillan, 1948.

Lawrence, R. R., and Richard, H. E. *Principles of Alternating-Current Machinery.* 4th ed. New York: McGraw-Hill, 1953.

Morgan, A. T. *General Theory of Electrical Machines.* London: Heyden & Son, 1979.

Sarma, M. S. *Synchronous Machines.* New York: Gordon and Breach Science Publishers, 1979.

Say, M. G. *Alternating Current Machines.* New York: John Wiley & Sons, Halsted Press, 1976.

Siskind, C. S. *Direct-Current Machinery.* New York: McGraw-Hill, 1952.

Whipple, C. C. *Electric Machinery, Volume 1: Fundamentals and D. C. Machines.* New York: D. Van Nostrand, 1946.

PROBLEMS

6-1.

Draw a developed winding diagram for a two-pole, double-layer, dc armature winding in 16 slots for simplex-lap progressive connection. Indicate the position of the brushes and the poles. You may assume single-turn coils for simplicity.

6-2.

Draw a developed view of the dc commutator simplex-lap winding with the following particulars:

    Number of slots: 27
    Number of poles: 6
    Total number of coils: 54
    Number of coils per slot: 2
    $y_c = +1$; $y_b = 8$; slot span $= 4$

Determine the number of equalizer bars needed for 100 percent equalization, and give their corresponding tappings.

6-3.

Draw a developed winding diagram of a four-pole, two-layer, simplex-wave dc armature winding with 17 slots and 17 single-turn coils. Locate the position of the brushes and the poles.

6-4.

Draw a developed view of the dc commutator simplex-wave winding with the following particulars:

    Number of slots: 32
    Number of poles: 6
    Total number of coils: 64
    Number of coils per slot: 2
    $y_c = 21$ (retrogressive); $y_b = 10$; slot span $= 5$

6-5.

Calculate the armature current per path of a four-pole, 50-kW, 250-volt dc generator if the commutator winding is *(a)* simplex wave, and *(b)* simplex lap. You may neglect the shunt field current.

6-6.

A six-pole dc machine is required to generate an emf of 260 volts when rotating at 450 rpm. The flux per pole is not to exceed 0.02 Wb. If the armature has 74 slots, determine the particulars of a suitable wave winding.

6-7.

Draw a flattened layout of a two-phase, two-pole, two-layer, full-pitch lap ac armature winding in 12 slots with 90° spread.

6-8.

Show that, if entry points displaced by 120 mechanical degrees are to be made to the phases of a three-phase machine with a double-layer winding of 60 electrical degree spread with $(6K + 2)$ poles, where K is an integer (i.e., 2, 8, 14, 20, . . . poles), they must be made in the order of the phase sequence *A-B-C;* and that for a machine with $(6K + 4)$ poles (i.e., 4, 10, 16, 22, . . . poles), they must be in the order of reversed phase sequence, *A-C-B;* and for a machine with $6K$ poles (i.e., 6, 12, 18, . . . poles), it is never possible to make entry points displaced by 120 mechanical degrees.

6-9.

Develop a flattened layout of a three-phase, four-pole, two-layer, lap ac armature winding with 60 electrical degree spread in 24 slots *(a)* with full-pitch coils, and *(b)* with 5/6 fractional-pitch coils. Show the detailed coil interconnections.

Calculate the winding distribution factor and pitch factor for each of the above cases for the fundamental.

6-10.

Draw a flattened layout of a three-phase, four-pole, two-layer, lap ac armature winding with 60 electrical degree spread in 36 stator slots *(a)* with full-pitch coils, and *(b)* with fractional-pitch coils short-pitched by *(i)* one slot, *(ii)* two slots, and *(iii)* three slots.

Determine the winding distribution factor and pitch factor for each of the above cases for the fundamental and the fifth and seventh harmonics.

6-11.

A 72-slot stator is to be wound for three-phase operation. If the coil pitch is six slots, derive a suitable form for the winding arrangement as an eight-pole winding, and draw the developed view showing the position of the phase belts in each layer of the winding *(a)* for 60 electrical degree spread, and *(b)* for 120 electrical degree spread.

Calculate the winding distribution and pitch factors for the fundamental and the fifth and seventh harmonics for each of the above cases.

6-12.

A three-phase, wye-connected, 60-Hz, six-pole alternator has a fundamental flux per pole of 0.02 Wb. The stator has 90 slots with four conductors per slot, and each stator coil spans 12 slots. If the flux in the air gap contains a 30 percent third harmonic and a 20 percent fifth harmonic with respect to the fundamental, determine the rms values of the phase and line-to-line induced emfs.

6-13.

A three-phase, four-pole, star-connected synchronous machine, driven at 1,800 rpm, has a 48-slot stator carrying a 60° spread double-layer winding with 5/6 fractional-pitch coils. There are 10 conductors in each slot. If the maximum value of the sinusoidally distributed field flux is 0.1 Wb, calculate the fundamental distribution and pitch factors, and the phase as well as line rms values of the generated emf.

6–14.

The field form of a salient-pole synchronous machine has been analyzed for the relative magnitudes of the harmonics. The ratios of the harmonics to the fundamental are given below:

$$k_{f1} = 1.0; \; k_{f3} = 0.061; \; k_{f5} = 0.004; \; k_{f7} = 0.035$$

The double-layer, wye-connected armature winding of the machine has three spp, with a fractional pitch of 8/9.

Determine the effect of the harmonics on the rms values of (a) the line-to-neutral voltage, and (b) the line-to-line voltage.

6–15.

A six-pole, double-layer dc armature winding in 28 slots has five turns per coil. If the field flux is 0.025 Wb per pole and the speed of the rotor is 1,200 rpm, find the value of the induced emf when the winding is (a) lap-connected, (b) wave-connected.

6–16.

A four-pole, dc series motor has a lap-connected, two-layer armature winding with a total of 400 conductors. Calculate the gross torque developed for a flux per pole of 0.02 Wb and an armature current of 50 A.

6–17.

A three-phase, 60-Hz, six-pole, wye-connected ac generator has 972 conductors distributed in 54 slots. The coils are short-pitched by one slot. Determine the fundamental-frequency line voltage on no-load (i.e., open circuit) when the fundamental flux per pole is 0.01 Wb.

If the rms current per conductor is 100 A, calculate the peak value of the fundamental and the fifth and seventh harmonic components of the armature mmf wave in ampere-turns per pole.

PART 2

STEADY-STATE THEORY AND PERFORMANCE

7

INDUCTION MACHINES

The object of this chapter is to analyze the polyphase induction machines, introduced in Chapter 5, in sufficient detail for developing equivalent-circuit models for prediction of their steady-state performance. In order to correspond to the number of phases in practical power systems, the polyphase induction motors used in industrial applications are almost without exception three-phase. The stator winding is connected to the ac source and the rotor winding is short-circuited, as in the case of *squirrel-cage* machines, or closed through external resistances, as in the case of *wound-rotor* machines. The cage-type machines are also known as *brushless* machines, and the wound-rotor machines are also called *slip-ring* machines.

The polyphase induction motor was invented by Nikola Tesla in 1886. It is the most widely used, lowest-cost industrial motor. It is estimated that more than 50 million of these motors are now in use in American industry, totalling some 150 million horsepower, and that each normal year's production adds about a million motors more. This is exclusive of the single-phase fractional-horsepower induction motors that are generally employed for operating domestic appliances such as electric fans, refrigerators, washing machines, and farm appliances. The production of these single-phase types in American industry is of the order of 20 million annually.

The polyphase induction motor requires no means for its excitation other than the ac source. Because the induction motor has no inherent means for producing its excitation, it requires reactive power and draws a lagging current. While the power factor at rated load may generally be above 0.8, it is quite low at light loads. In order to limit the reactive power, the magnetizing reactance must be high, and therefore the air gap must be kept relatively short compared to that of a synchronous motor of the same size and rating, except for the case of a small motor. The minimum length of the air gap used in an induction motor is essentially determined by mechanical considerations, and also by such factors as noise and magnetic losses in the tooth faces.

We recall from Chapter 5 that in a polyphase induction motor, ac is supplied to the stator winding directly and to the closed rotor winding by induction from the stator. The balanced polyphase stator and rotor currents produce stator- and rotor-component mmf waves of constant amplitude that rotate in the air gap at synchronous speed and are therefore stationary with respect to each other regardless of the mechanical speed of the rotor. A resultant air-gap flux-density wave is produced as a consequence of these mmfs; the interaction of the flux wave and the rotor-mmf wave gives rise to torque. At all speeds, other than synchronous speed, the conditions necessary for the production of a steady torque are fulfilled. Since no torque is produced at synchronous speed and the induction motor runs below synchronous speed, the induction machines are also known as *asynchronous machines*.

7.1 CONSTRUCTIONAL FEATURES OF POLYPHASE INDUCTION MACHINES

The essential features constitute the following: a laminated stator core carrying a polyphase winding; a laminated rotor core carrying either a squirrel-cage or a polyphase winding, the latter with shaft-mounted slip rings; a stiff shaft to maintain a necessarily short air gap; a frame for the stator housing to carry the bearings; and a cooling system. See Photos 7–1 and 7–2 for typical cross-sectional views of medium-size and small-size three-phase squirrel-cage induction motors.

PHOTO 7.1   A cross-sectional view of a typical medium-size three-phase squirrel-cage induction motor.   Photo courtesy of General Electric Company.

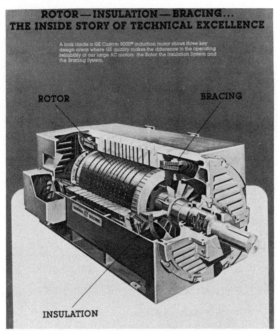

PHOTO 7.2   Cross-sectional view of a drip-proof, fan-cooled, 10-hp, three-phase, 60 Hz, 230/460-V squirrel-cage induction motor.   Photo courtesy of Marathon Electric Company.

PHOTO 7.3 Wound stator of a medium-size three-phase induction motor.
Photo courtesy of Westinghouse Electric Corporation.

The stator core is made up of laminations usually 0.3 to 0.6 mm thick; slightly thicker laminations may be used for small motors or in cases where low core loss is not so important. The laminations are insulated from one another either by an oxide coating produced by heat treatment or by a varnish coating. The built-up laminations are held by flanges in the yoke. Ventilating ducts are provided along the length of the core, spaced every 5 or 7 cm by the use of spacers placed between laminations. The rotor core is built up of laminations of the same material, such as sheet metal, that is used for the stator core, but thicker laminations can be used because of the lower frequencies of the rotor flux. The lamination is of one piece in small motors, while in large motors the laminations are usually segmented and dovetailed to a center spider. Ventilating ducts, when provided for the stator core, are also provided in the rotor core in equal number. In order to force cooling air through the machine, fan blades are usually provided on the ends of the rotor core.

Double-layer windings are most frequently used for the stator windings of the polyphase motors because of the greater ease of their manufacture, assembly, and repair, and also because all coils are alike. The windings are distributed and are generally short-pitched. The pitch and distribution are effective in limiting the magnitudes of the space harmonics that occur in the air-gap flux. Wherever possible, integral-slot windings (for which the number of slots per pole per phase is an integer) are used; however, to be able to utilize standard punchings in as wide a variety of machines as possible, it is often convenient for the manufacturer to use fractional-slot windings. See Photo 7–3 for a view of a wound stator of a medium-size three-phase induction motor.

Since the air-gap reluctance is a variable at different points on the stator as the rotor turns, and since this pulsating reluctance results in pulsating exciting current, irregular torque, increased tooth losses, and noise, the general tendency is to use a large number of slots in order to reduce

PHOTO 7.4 Squirrel-cage rotor of a medium-size three-phase induction motor. Photo courtesy of General Electric Company.

the effect of variable air-gap reluctance. On the other hand, a large number of slots results in narrow teeth and increased manufacturing cost. By making the number of slots on the rotor different from that on the stator and by skewing the rotor slots, it is possible to obtain more uniform reluctance. Partly closed slots on either or both the stator and the rotor help to make the air-gap reluctance more uniform. However, such slots complicate the problem of winding, since the coils must be fed through the narrow slot, turn by turn, or if copper bars are used, as in the case of squirrel-cage rotors, they must be inserted from the ends of the rotor. While both open and partly closed slots are widely used for the stator, partly closed slots are almost always used for the rotor. Once the number of slots on the stator is fixed, a suitable (different) number of rotor slots is chosen in order to avoid magnetic locking and obtain a smooth torque-speed characteristic.

As for the squirrel-cage rotor windings, copper, brass, or aluminum bars that are short-circuited on the ends by end rings are used as the rotor conductors. The bars may be welded, brazed, or bolted to the end rings; or, after building the rotor core, aluminum alloy bars may be cast into the slots with the end rings as integral parts. No insulation is necessary between the bars and the laminated rotor core. For the case of wound-rotor machines, the same types of windings that are used for stators can be used for rotors. The number of poles on the stator and rotor must be the same for torque production, while operation is only slightly affected by a change in the number of phases provided on the rotor winding as long as it is greater than one. While three-phase rotors are commonly used for both three- and two-phase motors, they are also used sometimes in single-phase motors. Bar, strap, or wire is used for rotor windings, the last being employed in cases where many turns are needed. Use of a large number of rotor turns increases the secondary voltage and decreases the current flowing through the slip rings. The value of the resistance to be used across the slip rings for starting or speed control is influenced by the secondary voltage and current; also, the insulation to be provided is determined by the secondary voltage. See Photos 7–4 and 7–5 for typical views of a squirrel-cage rotor and wound rotor of three-phase induction machines.

PHOTO 7.5 Wound rotor of a large three-phase induction motor. Photo courtesy of Westinghouse Electric Corporation.

The windings, air gap, and slot details have to be so selected that the exciting current and machine reactances conform to the desired performance. The air-gap length should be made as small as possible in order to reduce the magnetizing current necessary to set up the air-gap flux, but too small an air gap may increase motor noise and tooth-face losses, and may even prevent the motor from accelerating to its rated speed.

7.2   EQUIVALENT CIRCUIT OF A POLYPHASE INDUCTION MACHINE

A review of the material presented under "Elementary Induction Machines" in Section 5.3 is very helpful at this stage to recall the principle of operation of polyphase induction machines.

As stated in that section, the induction machine may be regarded as a generalized transformer in which energy conversion takes place, and electric power is transferred between the stator and rotor along with a change of frequency and a flow of mechanical power. At standstill, however, the machine acts as a simple transformer with an air gap and a short-circuited secondary winding; the frequency of the rotor-induced emf is the same as the stator frequency at standstill. At any value of the slip under balanced steady-state operation, the rotor current reacts on the stator winding at the stator frequency because the rotating magnetic fields due to the stator and rotor are stationary with respect to each other.

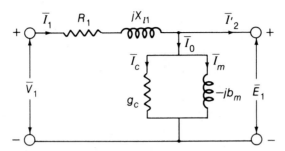

The induction machine may then be viewed as a transformer with an air gap, having a variable resistance in the secondary; the stator of the induction machine corresponds to the transformer primary, and the rotor corresponds to the secondary. For analysis of the balanced steady-state, it is sufficient to proceed on a per-phase basis with the use of phasor concepts, and hence an equivalent circuit on a per-phase basis is now developed. Only machines with symmetrical polyphase windings excited by balanced polyphase voltages are considered.. As in other discussions of polyphase devices, it is convenient to think of three-phase machines as wye-connected, so that currents are always line values and voltages are always line-to-neutral values.

The resultant air-gap flux is produced by the combined mmfs of the stator and rotor currents. For the sake of conceptual and analytical convenience, the total flux is divided into a mutual flux linking both the stator and rotor, and leakage fluxes, which are further subdivided into components such as slot-leakage flux, tooth-top leakage flux, and coil-end-connection leakage flux. An appropriate leakage reactance component may be assigned to each leakage flux component.

Considering the conditions in the stator, the synchronously rotating air-gap flux wave generates balanced polyphase counter emfs in the phases of the stator. The volt-ampere equation for the phase under consideration *in phasor notation* is given by

$$\overline{V}_1 = \overline{E}_1 + \overline{I}_1(R_1 + jX_{l1}) \tag{7.2.1}$$

where $\overline{V}_1$ is the stator terminal voltage, $\overline{E}_1$ is the counter emf generated by the resultant air-gap flux, $\overline{I}_1$ is the stator current, R_1 is the stator effective resistance, and X_{l1} is the stator-leakage reactance.

As in the case of a transformer, the stator (primary) current can be resolved into two components: a load component $\overline{I}_2'$ that produces an mmf that exactly counteracts the mmf of the rotor current, and an exciting component $\overline{I}_0$ (required to create the resultant air-gap flux) that may be resolved into a core-loss component $\overline{I}_c$ in phase with $\overline{E}_1$ and a magnetizing component $\overline{I}_m$ lagging $\overline{E}_1$ by 90°. A shunt branch formed by the core-loss conductance g_c and magnetizing susceptance b_m in parallel, connected across $\overline{E}_1$, will account for the exciting current in the equivalent circuit, as shown in Figure 7.2.1 along with the positive directions for the case of a motor.

Thus far, the equivalent circuit representing the stator phenomena is exactly like that of the transformer primary. Because of the air gap, however, the value of the magnetizing reactance

FIGURE 7.2.2   Slip-frequency equivalent circuit for a rotor phase of a
polyphase induction motor.

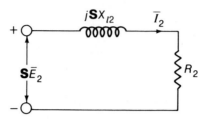

tends to be relatively low compared with that of a transformer; the leakage reactance is larger in proportion to the magnetizing reactance than in transformers. In order to complete the equivalent circuit, the effects of the rotor must be incorporated. This is done by referring the rotor quantities to the stator.

THE ROTOR EQUIVALENT CIRCUIT   Since the frequency of the rotor voltages and currents is the slip frequency, the magnitude of the voltage induced in the rotor circuit is proportional to the slip, as may be verified readily by the voltage equation (4.2.10); also, in terms of the standstill per-phase rotor leakage reactance X_{l2}, the leakage reactance at a slip S will be given by SX_{l2}. With R_2 as the per-phase resistance of the rotor, the slip-frequency equivalent circuit for a rotor phase is shown in Figure 7.2.2, in which E_2 is the per-phase voltage induced in the rotor at standstill. The rotor current I_2 is given by

$$I_2 = \frac{S\,E_2}{\sqrt{R_2{}^2 + (S\,X_{l2})^2}} \qquad (7.2.2)$$

which may be rewritten as

$$I_2 = \frac{E_2}{\sqrt{\left(\dfrac{R_2}{S}\right)^2 + X_{l2}{}^2}} \qquad (7.2.3)$$

resulting in the alternate form of the per-phase rotor equivalent circuit shown in Figure 7.2.3.

DEVELOPMENT OF THE COMPLETE PER-PHASE EQUIVALENT CIRCUIT   All rotor electrical phenomena, when viewed from the stator, become stator-frequency phenomena, because the stator winding simply sees the mmf and flux waves traveling at synchronous speed. Returning to the analogy of a transformer, and considering that the rotor is coupled to the stator just as the secondary of a transformer is coupled to its primary, we may draw the equivalent circuit as shown in Figure 7.2.4.

FIGURE 7.2.3   Alternate form of the per-phase rotor equivalent circuit.

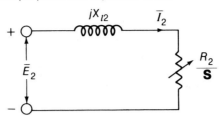

FIGURE 7.2.4   Per-phase equivalent coupled circuit of a polyphase induction motor.

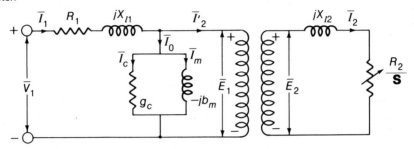

The rotor quantities are now to be referred to the stator. Including the effect of the stator and rotor winding distributions, the ratio of the stator to the rotor voltage is given by

$$\frac{E_1}{E_2} = \frac{k_{w1} N_1}{k_{w2} N_2} \qquad (7.2.4)$$

where k_{w1} and k_{w2} are the winding factors of the stator and rotor windings respectively; N_1 and N_2 are the series-connected turns per phase of the stator and rotor windings respectively. For a cage-type rotor, however, k_{w2} is equal to unity, and the number of turns per phase N_2 is given by $(1/2)(P/2)$, where P is the number of poles.

If m_1 and m_2 are the numbers of phases on the stator and rotor, respectively, the rotor volt-amperes per phase referred to the stator must be the same as the original rotor volt-amperes:

$$m_1 E_2' I_2' = m_2 E_2 I_2 \qquad (7.2.5)$$

where the primed quantities are the rotor quantities referred onto the stator side. For the wound rotor, m_2 is usually equal to m_1; and for the cage-type rotor, the number of phases m_2 is equal to the number of rotor bars per pole-pair. The rotor current I_2', referred to the stator, is given by

$$I_2' = \frac{m_2 \, k_{w2} \, N_2}{m_1 \, k_{w1} \, N_1} I_2 \qquad (7.2.6)$$

By making use of Equation 7.2.6, E_2' can be solved from Equation 7.2.5 and be seen equal to E_1:

$$E_2' = \frac{k_{w1} N_1}{k_{w2} N_2} E_2 = E_1 \tag{7.2.7}$$

Noting that the rotor I^2R losses must be invariant, i.e.,

$$m_1 (I_2')^2 R_2' = m_2 I_2^2 R_2 \tag{7.2.8}$$

it follows with the aid of Equation 7.2.6 that the rotor resistance per phase referred to the stator is given by

$$R_2' = \frac{m_1}{m_2} \left(\frac{k_{w1} N_1}{k_{w2} N_2} \right)^2 R_2 \tag{7.2.9}$$

For the wound rotor, R_2 is the per-phase resistance of the rotor winding. For the cage rotor, R_2 is approximately the resistance of one bar.

The rotor leakage reactance X_{l2}', referred to the stator, is similarly given by

$$X_{l2}' = \frac{m_1}{m_2} \left(\frac{k_{w1} N_1}{k_{w2} N_2} \right)^2 X_{l2} \tag{7.2.10}$$

which leads to the result that the magnetic energy stored in the standstill rotor leakage reactance remains unchanged:

$$\frac{1}{2} m_1 (X_{l2}'/2\pi f_s)(I_2')^2 = \frac{1}{2} m_2 (X_{l2}/2\pi f_s) I_2^2 \tag{7.2.11}$$

where f_s is the stator frequency.

Having thus referred the rotor quantities to the stator, we may now draw the per-phase equivalent circuit of the polyphase induction motor as shown in Figure 7.2.5(a). The combined effect of the shaft load and rotor resistance appears as a reflected resistance $R_2'/\mathbf{S}$, which is a function of slip and therefore of the mechanical load. The quantity $R_2'/\mathbf{S}$ may conveniently be split into two parts:

$$\frac{R_2'}{\mathbf{S}} = R_2' + R_2' \frac{(1 - \mathbf{S})}{\mathbf{S}} \tag{7.2.12}$$

and the equivalent circuit may be redrawn as in Figure 7.2.5(b). R_2' is the per-phase standstill rotor resistance referred to the stator, and $R_2'[(1 - \mathbf{S})/\mathbf{S}]$ is a dynamic resistance that depends on the rotor speed and corresponds to the load on the motor. (See the discussion after Equation 7.4.2.)

When power aspects are to be emphasized, the equivalent circuit is frequently redrawn as in Figure 7.2.5(c), in which the shunt conductance g_c is omitted, and the core losses may eventually be included in efficiency calculations along with the friction, windage, and stray load losses.

FIGURE 7.2.5 Per-phase equivalent circuits of a polyphase induction
motor, referred to the stator.

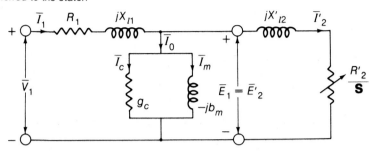

(a)

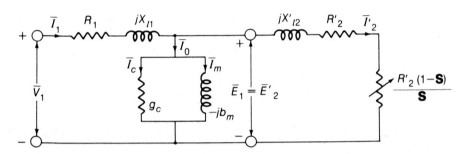

(b)

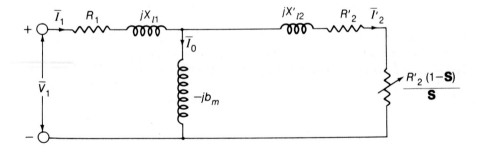

(c)

FIGURE 7.3.1   Per-phase equivalent circuits of a polyphase induction motor corresponding to the no-load test conditions (S $\simeq$ 0).

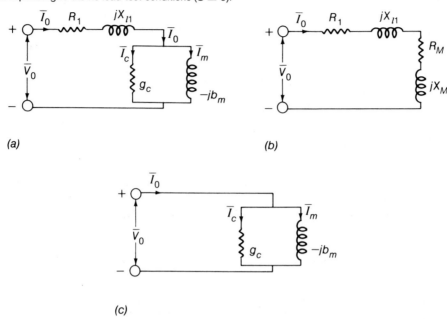

(a)   (b)

(c)

In static-transformer theory, as you may recall, analysis of the equivalent circuit is often simplified by either neglecting the exciting shunt branch entirely or adopting the approximation of moving it out directly to the primary terminals. For the induction machine, however, such approximations may not be permissible under normal running conditions because the presence of the air gap leads to a much higher exciting current (30 to 50 percent of full-load current) and relatively higher leakage reactances.

7.3   EQUIVALENT CIRCUIT FROM TEST DATA

The parameters of the equivalent circuit of the induction machine can be obtained from the *no-load* and *blocked-rotor* tests. These tests correspond to the no-load and short-circuit tests on the transformer.

NO-LOAD TEST   Rated balanced voltage at rated frequency is applied to the stator and the motor is allowed to run on no-load. Input power, voltage, and current are measured; these are reduced to *per-phase* values, denoted by P_0, V_0, and I_0, respectively. When the machine runs on no-load, the slip is close to zero, and the circuit to the right of the shunt branch in Figure 7.2.5(a) is taken to be an open circuit. Thus the equivalent circuit corresponding to the no-load test conditions is given in Figure 7.3.1(a) or equivalently in Figure 7.3.1(b), in which the shunt branch is replaced by an equivalent series impedance to facilitate the evaluation of the motor parameters.

FIGURE 7.3.2 Per-phase equivalent circuits of a polyphase induction
motor corresponding to the blocked-rotor test conditions (S = 1).

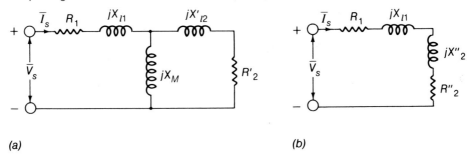

(a)   (b)

The conductance g_c in Figure 7.3.1(a) takes into account not only the stator core losses but also the windage and friction losses. Because of the relatively low value of rotor frequency, the rotor core loss is practically negligible at no-load. From Figure 7.3.1(b), it follows that

$$R_0 = R_1 + R_M = P_0/I_0^2 \qquad (7.3.1)$$

$$Z_0 = V_0/I_0 \qquad (7.3.2)$$

$$X_0 = X_{l1} + X_M = \sqrt{Z_0^2 - R_0^2} \qquad (7.3.3)$$

$$\text{No-load power factor} = \cos\phi_0 = P_0/(V_0 I_0) \qquad (7.3.4)$$

in which R_1 is the stator resistance per phase, and the series resistance $R_M << X_M$, and $X_M \simeq 1/b_m$. The rotational losses given by the sum of the friction, windage, and core losses are found on a per-phase basis by subtracting the stator copper loss from the no-load power input:

$$P_{r0} = P_0 - I_0^2 R_1 \qquad (7.3.5)$$

Sometimes, an approximate per-phase equivalent circuit, shown in Figure 7.3.1(c), is used for the no-load test conditions, when the calculation of shunt-branch parameters becomes much simpler as in the case of a transformer.

BLOCKED-ROTOR TEST In this test, the rotor of the induction motor is blocked so that the slip is equal to unity, and a reduced voltage is applied to the machine stator terminals so that the rated current flows through the stator windings. The input power, voltage, and current values are recorded and reduced to *per-phase values,* denoted respectively by P_s, V_s, and I_s. The iron losses are assumed to be negligible in this test. The equivalent circuit corresponding to the blocked-rotor-test conditions is then given by Figure 7.3.2(a) or equivalently by Figure 7.3.2(b). If the shunt branch of the circuit shown in Figure 7.3.2(a) can be considered to be absent, as in the case of a transformer short-circuit test condition, the calculations become much simpler because X_2'' and R_2'' will be equal to X_{l2}' and R_2', respectively. From Figure 7.3.2(b), it follows that

$$R_{eq} = R_1 + R_2'' = P_s/I_s^2 \tag{7.3.6}$$

$$Z_{eq} = V_s/I_s \tag{7.3.7}$$

$$X_{eq} = X_{l1} + X_2'' = \sqrt{Z_{eq}^2 - R_{eq}^2} \tag{7.3.8}$$

$$\cos \phi_s = P_s/(V_s\, I_s) \tag{7.3.9}$$

For a more complete discussion of tests on induction motors, one may refer to published test codes and procedures (see the bibliography at the end of the chapter), in which the empirical proportions for stator and rotor leakage reactances are listed for three-phase induction motors by their class. When the classification of the motor is not known, it is assumed that

$$X_{l1} = X_{l2}' = 0.5\, X_{eq} \tag{7.3.10}$$

The magnetizing reactance X_M can now be evaluated from Equation 7.3.3.

The value of R_2' requires a closer approximation than that of X_{l2}' because, in the running range, $(R_2'/S) \gg (X_{l1} + X_{l2}')$ and it has a correspondingly greater effect on the performance of the motor within that range. From the equivalent circuits of Figure 7.3.2(a) and (b), it follows that

$$R_2'' + jX_2'' = \frac{(R_2' + jX_{l2}')\, jX_M}{R_2' + j(X_{l2}' + X_M)} \tag{7.3.11}$$

Equating the real parts of both sides, it can be shown that

$$R_2'' = \frac{R_2'\, X_M^2}{(R_2')^2 + (X_{l2}' + X_M)^2} \tag{7.3.12}$$

Since $R_2' \ll (X_{l2}' + X_M)$, Equation 7.3.12 can be approximated as

$$R_2'' \simeq \frac{R_2'\, X_M^2}{(X_{l2}' + X_M)^2} \tag{7.3.13}$$

Substituting from Equation 7.3.6 that $R_2'' = R_{eq} - R_1$, one gets

$$R_2' = (R_{eq} - R_1)\frac{(X_{l2}' + X_M)^2}{X_M^2} \tag{7.3.14}$$

If X_M is much larger than X_{l2}', R_2' will be nearly equal to $(R_{eq} - R_1)$ or R_2''; otherwise, R_2' will be somewhat larger than R_2''.

EXAMPLE 7.3.1

The results of the no-load and blocked-rotor tests on a three-phase, wye-connected, 10-hp, 440-V, 14-A, 60-Hz, 8-pole induction motor with a single squirrel-cage rotor are given below:

> No-load test: line-to-line voltage = 440 V
>   total input power = 350 W
>   line current = 6 A

> Blocked-rotor test: line-to-line voltage = 95 V
>   total input power = 900 W
>   line current = 14 A

The dc resistance of the stator is measured immediately after the blocked-rotor test, and an average value of 0.75 ohm per phase has been obtained. Calculate the parameters of the equivalent circuit shown in Figure 7.2.5*(c)*, and also compute the no-load rotational losses.

SOLUTION

From the no-load test data and Equations 7.3.1 to 7.3.3, one gets

$$R_0 = \frac{350}{3} \times \frac{1}{6^2} = 3.24 \text{ ohms}$$

$$Z_0 = \frac{440}{\sqrt{3}} \times \frac{1}{6} = 42.34 \text{ ohms}$$

$$X_0 = \sqrt{(42.34)^2 - (3.24)^2} = 42.22 \text{ ohms}$$

From the blocked-rotor test data and Equations 7.3.6 to 7.3.8, one gets

$$R_{eq} = \frac{900}{3} \times \frac{1}{(14)^2} = 1.53 \text{ ohms}$$

$$Z_{eq} = \frac{95}{\sqrt{3}} \times \frac{1}{(14)} = 3.92 \text{ ohms}$$

$$X_{eq} = \sqrt{(3.92)^2 - (1.53)^2} = 3.61 \text{ ohms}$$

Taking $X_{l1} = X_{l2}' = 0.5 \times 3.61 = 1.805$ ohms, we get from Equation 7.3.3:

$$X_M = X_0 - X_{l1} = 42.22 - 1.805 = 40.415 \text{ ohms}$$

From Equation 7.3.14, one gets

$$R_2' = (1.53 - 0.75)\left(\frac{1.805 + 40.415}{40.415}\right)^2$$

$$= 0.78 \times \left(\frac{42.22}{40.415}\right)^2 = 0.851 \text{ ohm}$$

The per-phase equivalent circuit is given by the following:

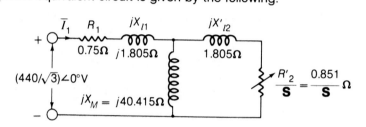

The no-load rotational losses are given by Equation 7.3.5:

$$3\left[\frac{350}{3} - (6^2 \times 0.75)\right] = 350 - (3 \times 6^2 \times 0.75) = 269 \text{ W}$$

Knowing the equivalent-circuit constants, one can compute the machine performance.

7.4 POLYPHASE INDUCTION MACHINE PERFORMANCE

Some of the important steady-state performance characteristics of a polyphase induction motor include the variations of current, speed, losses as the load-torque requirements change, the starting torque, and the maximum torque. Performance calculations can be made from the equivalent circuit. All calculations may be made on a *per-phase basis*, assuming balanced operation of the machine, and the total quantities may be obtained by using an appropriate multiplying factor.

The equivalent circuit of Figure 7.2.5(c), redrawn here for convenience in Figure 7.4.1, is usually employed for the analysis. The core losses, most of which occur in the stator, as well as friction,

FIGURE 7.4.1 Per-phase equivalent circuit of a polyphase induction motor used for performance calculations.

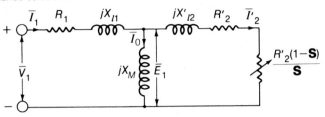

FIGURE 7.4.2 Power flow in an induction motor.

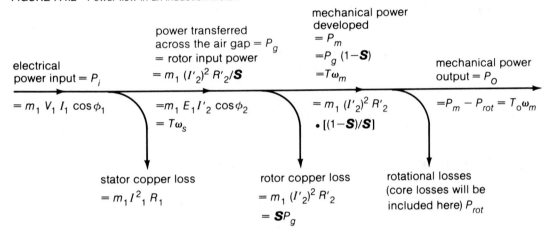

windage, and stray-load losses will be included in efficiency calculations. The power-flow diagram for an induction motor is given in Figure 7.4.2, in which m_1 is the number of stator phases, ϕ_1 is the power-factor angle between $\overline{V}_1$ and $\overline{I}_1$, ϕ_2 is the power-factor angle between $\overline{E}_1$ and $\overline{I}_2'$, T is the internal electromagnetic torque developed, ω_s is the synchronous angular velocity in mechanical radians per second, and ω_m is the actual mechanical rotor speed given by $\omega_s(1 - S)$.

The total power P_g transferred across the air gap from the stator is the difference between the electrical power input P_i and the stator copper loss. P_g is thus the total rotor input power, which is dissipated in the resistance R_2'/S of each phase so that

$$P_g = m_1(I_2')^2 \, R_2'/S = T\omega_s \qquad (7.4.1)$$

where T is the internal electromagnetic torque developed by the machine, and ω_s is the synchronous angular velocity in mechanical radians per second. Subtracting the total rotor copper loss, which is $m_1(I_2')^2 R_2'$ or $\mathbf{S}P_g$, from the above P_g, one gets the internal mechanical power developed:

$$P_m = P_g(1 - \mathbf{S}) = T\omega_m = m_1(I_2')^2 R_2' \frac{(1 - \mathbf{S})}{\mathbf{S}} \qquad (7.4.2)$$

This is the power that is absorbed by a resistance of $\left[R_2' \cdot \dfrac{(1 - \mathbf{S})}{\mathbf{S}} \right]$, which corresponds to the load. That is the reason that the resistance term $R_2'/\mathbf{S}$ has been split into two terms as in Equation 7.2.12 and incorporated in the equivalent circuit of Figure 7.4.1. From Equation 7.4.2, one can see that, of the total power delivered to the rotor, the fraction $(1 - \mathbf{S})$ is converted to mechanical power and the fraction $\mathbf{S}$ is dissipated as rotor copper loss. One can conclude then that an induction motor operating at high values of slip will be an inefficient device.

The total rotational losses including the core losses may be subtracted from P_m to obtain the mechanical power output P_0 that is available in mechanical form at the shaft for useful work:

$$P_0 = P_m - P_{rot} = T_0 \, \omega_m \qquad (7.4.3)$$

The per-unit efficiency of the induction motor is then given by

$$\eta = P_0/P_i \qquad (7.4.4)$$

Let us now illustrate the above procedure and the analysis of the equivalent circuit by taking up the following example.

EXAMPLE 7.4.1

The parameters of the equivalent circuit, given in Figure 7.4.1, for a three-phase, wye-connected, 220-V, 10-hp, 60-Hz, 6-pole induction motor are given below in ohms per phase referred to the stator:

$$R_1 = 0.3 \; ; R_2' = 0.15$$

$$X_{l1} = 0.5 \; ; X_{l2}' = 0.2 \; ; X_M = 15$$

The total friction, windage, and core losses may be assumed to be constant at 400 W, independent of load. For a per-unit slip of 0.02, when the motor is operated at rated voltage and frequency, calculate the stator input current, power factor at the stator terminals, rotor speed, output power, output torque, and efficiency.

SOLUTION

From the equivalent circuit of Figure 7.4.1, the total impedance per phase as viewed from the stator input terminals is given by

$$Z_t = R_1 + jX_{l1} + \frac{jX_M\left(\dfrac{R_2'}{S} + jX_{l2'}\right)}{\dfrac{R_2'}{S} + j(X_M + X_{l2'})}$$

$$= 0.3 + j0.5 + \frac{j15(7.5 + j0.2)}{7.5 + j(15 + 0.2)}$$

$$= (0.3 + j0.5) + (5.87 + j3.10)$$

$$= 6.17 + j3.60 = 7.14\angle30.26° \text{ ohms}$$

Phase voltage $= 220/\sqrt{3} = 127$ V

Stator input current $= 127/7.14 = 17.79$ A

Power factor $= \cos 30.26° = 0.864$

Synchronous speed $= \dfrac{120 \times 60}{6} = 1{,}200$ rpm

Rotor speed $= (1 - 0.02)\, 1{,}200 = 1{,}176$ rpm

Total input power $= \sqrt{3} \times 220 \times 17.79 \times 0.864$
$$= 5{,}856.8 \text{ W}$$

Stator copper loss $= 3 \times 17.79^2 \times 0.3 = 284.8$ W

Power transferred across the air gap $= P_g$
$$= 5{,}856.8 - 284.8$$
$$= 5{,}572 \text{ W}$$

This can be obtained alternatively as follows:

$$P_g = m_1(I_2')^2\, R_2'/S = m_1 I_1^2 R_f$$

where R_f is the real part of the parallel combination of jX_m and $\left(\dfrac{R_2'}{S} + jX_{l2}'\right)$. Thus,

$$P_g = 3 \times 17.79^2 \times 5.87 = 5,573 \text{ W}$$

The internal mechanical power developed $= P_g(1 - S)$

$$= 0.98 \times 5,572$$

$$= 5,460 \text{ W}$$

Total mechanical power output $= 5,460 - 400 = 5,060 \text{ W}$

or $\dfrac{5,060}{745.7} = 6.8 \text{ hp}$

$$\text{Total output torque} = \text{output power}/\omega_m = \frac{\text{output power}}{(1 - S)\,\omega_s}$$

Since $\omega_s = \dfrac{4\pi f}{\text{poles}} = \dfrac{4\pi \times 60}{6} = 40\pi = 125.7$ mech. rad/s, it follows that

$$\text{Total output torque} = \frac{5,060}{(0.98)(125.7)} = 41.08 \text{ N·m}$$

$$\text{Efficiency} = \frac{5,060}{5,856.8} = 0.864 \text{ or } 86.4 \text{ percent}$$

The efficiency may alternatively be calculated from the losses:

$$\text{Total stator copper loss} = 284.8 \text{ W}$$

$$\text{Rotor copper loss} = m_1(I_2')^2\, R_2' = SP_g$$

$$= (0.02)(5,572) = 111.4 \text{ W}$$

$$\text{Friction, windage, and core losses} = 400 \text{ W}$$

$$\text{Total losses} = 284.8 + 111.4 + 400 = 796.2 \text{ W}$$

FIGURE 7.4.3 Typical induction-motor characteristics.

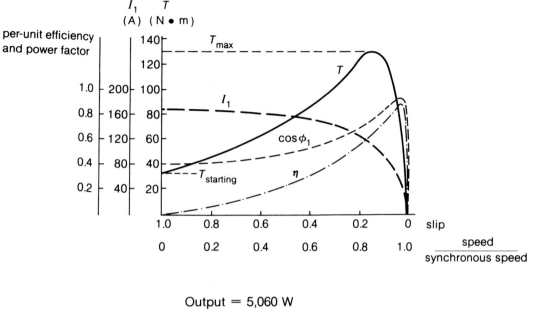

$$\text{Output} = 5{,}060 \text{ W}$$

$$\text{Input} = 5{,}060 + 796.2 = 5{,}856.2 \text{ W}$$

$$\text{Efficiency} = 1 - \frac{\text{Losses}}{\text{Input}} = 1 - \frac{796.2}{5{,}856.2}$$

$$= 1 - 0.136 = 0.864$$

By repeating these calculations for other values of slip, ranging from 0 to 1, using a procedure similar to that given in the previous example, the complete performance characteristics of the induction motor can be determined from its equivalent circuit. Typical induction-motor characteristics are shown in Figure 7.4.3.

We shall now develop an expression for torque as a function of slip and other equivalent-circuit parameters, because the torque-slip characteristic is one of the most important aspects of the induction motor. Recalling Equation 7.4.1 and the equivalent circuit of Fig-

FIGURE 7.4.4   Another form of per-phase equivalent circuit of a
polyphase induction motor shown in Figure 7.4.1.

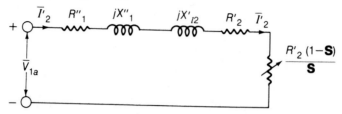

ure 7.4.1, we shall try to obtain an expression for I_2'. Towards that end, let us redraw the equivalent circuit as in Figure 7.4.4. By the application of Thévenin's theorem, one has from Figures 7.4.1 and 7.4.4 the following:

$$\overline{V}_{1a} = \overline{V}_1 - \overline{I}_0(R_1 + jX_{l1}) = \overline{V}_1 \frac{jX_M}{R_1 + j(X_{l1} + X_M)} \qquad (7.4.5)$$

$$R_1'' + jX_1'' = \frac{(R_1 + jX_{l1}) jX_M}{R_1 + j(X_{l1} + X_M)} \qquad (7.4.6)$$

$$I_2' = \frac{V_{1a}}{\sqrt{[R_1'' + (R_2'/S)]^2 + (X_1'' + X_{l2}')^2}} \qquad (7.4.7)$$

$$T = \frac{1}{\omega_s} \frac{m_1 V_{1a}^2 (R_2'/S)}{[R_1'' + (R_2'/S)]^2 + (X_1'' + X_{l2}')^2} \qquad (7.4.8)$$

From neglecting the stator resistance in Equation 7.4.5, negligible error results for most induction motors. If X_M of the equivalent circuit shown in Figure 7.4.1 is sufficiently large that the shunt branch need not be considered, calculations become much simpler; R_1'' and X_1'' will then be equal to R_1 and X_{l1}, respectively; also, V_{1a} will then be equal to V_1.

The general shape of the torque-speed or torque-slip characteristic is shown in Figure 5.3.16, in which the motor region ($0 < S \leq 1$), the generator region ($S < 0$), and the braking region ($S > 1$) are included for completeness. The performance of an induction motor may be characterized by such factors as efficiency, power factor, starting torque, starting current, pull-out (maximum) torque, and maximum internal power developed. Starting conditions are those corresponding to $S = 1$.

The maximum internal, or breakdown, torque T_{max} occurs when the power delivered to (R_2'/S) in Figure 7.4.4 is a maximum. Applying the familiar impedance-matching principle of the circuit theory, this power will be a maximum when the impedance (R_2'/S) equals the magnitude of the impedance between that and the constant voltage V_{1a}. That is to say, the maximum occurs at a value of slip $S_{max\ T}$ for which the following condition is satisfied:

$$\frac{R_2'}{S_{max\ T}} = \sqrt{(R_1'')^2 + (X_1'' + X_{l2}')^2} \qquad (7.4.9)$$

The above result may also be obtained by differentiating Equation 7.4.8 with respect to **S,** or more conveniently with respect to ($R_2'/$**S**), and setting the result equal to zero. This is left out as a desirable exercise for the student. The slip corresponding to maximum torque, $S_{max\ T}$, is thus given by

$$S_{max\ T} = \frac{R_2'}{\sqrt{(R_1'')^2 + (X_1'' + X_{l2}')^2}} \qquad (7.4.10)$$

and the corresponding maximum torque from Equation 7.4.8 comes out as

$$T_{max} = \frac{1}{\omega_s} \frac{0.5 m_1 V_{1a}^2}{R_1'' + \sqrt{(R_1'')^2 + (X_1'' + X_{l2}')^2}} \qquad (7.4.11)$$

which can be verified by the reader. Equation 7.4.11 shows that the maximum torque is independent of the rotor resistance. However, the slip corresponding to maximum torque is directly proportional to the rotor resistance R_2', as seen from Equation 7.4.10. Thus, when the rotor resistance is increased by inserting external resistance in the rotor of a wound-rotor induction motor, the maximum internal torque is unaffected, but the speed or slip at which it occurs is increased. This is shown in Figure 7.4.5. Also, note that the maximum torque and maximum power do not occur at the same speed. The student is encouraged to reason out why it is so.

A conventional induction motor with a squirrel-cage rotor has about a 5 percent drop in speed from no-load to full load, and is thus substantially a constant-speed motor. By employing a wound-rotor motor and inserting external resistance in the rotor circuit, speed variation may be obtained; but poorer efficiency will result. Variation of starting torque (at **S** = 1) with rotor-circuit resistance can also be seen from Figure 7.4.5. As stated in Section 5.3, higher starting torque can be obtained by inserting external resistances in the rotor circuit, and cutting them out eventually for the normal running conditions in order to operate the machine with a higher efficiency. By making the rotor-circuit resistance sufficiently large, it may be possible to make the torque-slip relationship almost linear for the slip range of 0 to 1; two-phase servomotors (used as output actuators in feedback control systems) are usually designed with very high rotor resistance so as to ensure a negative slope for the torque-speed characteristic over the entire operating range.

FIGURE 7.4.5 Effect of changing rotor-circuit resistance on the torque-slip characteristic of a polyphase induction motor.

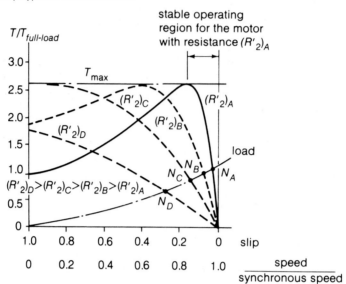

The resistances R_1 and R_2' are kept very small in order to reduce the copper losses and thereby increase the efficiency; the motor would be most compatible with a load running at the highest possible speed with a low value of slip. From Equations 7.4.8 and 7.4.11, it can be seen that the leakage reactance $(X_1'' + X_{l2}')$ must be low to assure good starting torque as well as adequate maximum torque. Power factor can be improved by decreasing the leakage reactances and increasing the magnetizing reactance. However, since the starting current of the motor is essentially limited by the leakage reactances, these should not be reduced below a certain value. An attempt to decrease the core losses beyond a limit by reducing the working flux density would result in an increase in the copper losses for a given load, because the torque (determined by the load) is dependent on the product of the flux density and I_2'.

Since the stator resistance is quite low and has only a negligible influence, one may set $R_1 = R_1'' = 0$, in which case, from Equations 7.4.8 and 7.4.11, it can be shown that

$$\frac{T}{T_{max}} = \frac{2}{(S/S_{max\ T}) + (S_{max\ T}/S)}$$

(7.4.12)

where S and $S_{max\ T}$ are the slips corresponding to T and T_{max}, respectively.

EXAMPLE 7.4.2

For the motor of Example 7.4.1, compute the following:

    a. The load component I_2' of the stator current, the internal torque T, and the internal power P_m for a slip of 0.02.
    b. The maximum internal torque and the corresponding slip as well as speed.
    c. The internal starting torque and the corresponding stator-load current I_2'.

SOLUTION

Let us first reduce the equivalent circuit of Figure 7.4.1 to its Thévenin-equivalent form shown in Figure 7.4.4. With the aid of Equations 7.4.5 and 7.4.6, one obtains

$$\overline{V}_{1a} = \overline{V}_1 \frac{jX_M}{R_1 + j(X_{l1} + X_M)} \simeq \overline{V}_1 \frac{X_M}{X_{l1} + X_M}$$

$$= \frac{220}{\sqrt{3}} \frac{15}{0.5 + 15} \angle 0° = 122.9 \angle 0° \text{ V}$$

$$R_1'' + jX_1'' = \frac{(0.3 + j0.5)\, j15}{0.3 + j(0.5 + 15)} = 0.281 + j0.489$$

a. Corresponding to a slip of 0.02, $\dfrac{R_2'}{S} = \dfrac{0.15}{0.02} = 7.5$. From Equation 7.4.7, one gets

$$I_2' = \frac{122.9}{\sqrt{7.8^2 + 0.689^2}} = \frac{122.9}{7.83} = 15.7 \text{ A}$$

The internal torque T can be calculated either from Equation 7.4.1 or Equation 7.4.8:

$$T = \frac{1}{125.7}(3)(15.7)^2(7.5) = \frac{5{,}546}{125.7} = 44.12 \text{ N·m}$$

From Equation 7.4.2, the internal mechanical power is calculated as

$$P_m = (3)(15.7)^2(7.5)(0.98) = 5{,}435 \text{ W}$$

which is also the same as $T\omega_m = T\omega_s(1 - S) = (44.12 \times 125.7 \times 0.98)$. In Example 7.4.1, this is calculated to be 5,460 W; the small discrepancy is due to errors involved in the calculations and approximations made.

b. From Equation 7.4.10, it follows that

$$S_{max\ T} = \frac{0.15}{\sqrt{0.281^2 + 0.689^2}} = \frac{0.15}{0.744} = 0.202$$

or the corresponding speed at T_{max} is $(1 - 0.202)(1,200) = 958$ rpm. From Equation 7.4.11, the maximum torque can be calculated:

$$T_{max} = \frac{1}{125.7} \frac{(0.5)(3)(122.9)^2}{0.281 + \sqrt{0.281^2 + 0.689^2}}$$

$$= \frac{1}{125.7} \frac{(0.5)(3)(122.9)^2}{0.281 + 0.744} = 175.8\ N \cdot m$$

c. Assuming the rotor-circuit resistance to be constant, with $S = 1$ at starting, from Equations 7.4.7 and 7.4.8, one gets the following:

$$I_2'{}_{start} = \frac{122.9}{\sqrt{(0.281 + 0.15)^2 + (0.489 + 0.2)^2}} = \frac{122.9}{0.813}$$

$$= 151.2\ A$$

$$T_{start} = \frac{1}{125.7}(3)(151.2)^2(0.15) = 81.8\ N \cdot m$$

For such applications as fans and blowers, a motor needs to develop only moderate starting torque. However, some loads such as conveyors require high starting torque to overcome high static torque and load inertia. The starting torque is sometimes made equal to the maximum torque, by choosing the rotor-circuit resistance at starting to be

$$R_2'{}_{start} = \sqrt{(R_1'')^2 + (X_1'' + X_{12}')^2} \qquad (7.4.13)$$

which can easily be obtained from Equation 7.4.9 with $S_{max\ T} = 1$.

The limitations of the equivalent circuit and the conditions under which the circuit-parameters are obtained need to be borne in mind, when investigations are to be carried out over a wide speed range, as in motor-starting problems. Saturation under heavy inrush currents associated with starting, or saturation due to large currents corresponding to the maximum torque conditions, may have a significant effect in reducing the motor reactances. Moreover, the current distribution in the rotor conductors and the rotor resistance may vary significantly over a wide speed range. By simulating the proposed operating conditions as closely as possible, and determining the equivalent-circuit parameters, errors may be kept to a minimum in predicting the machine performance.

7.5 *SPEED CONTROL OF POLYPHASE INDUCTION MOTORS*

The induction motor is used in numerous applications because of its simplicity and ruggedness. Although a good number of industrial drives run at substantially constant speed, there are quite a few applications in which variable speed becomes a necessity. The speed-control capability is essential in such applications as conveyors, hoists, and elevators. Since the induction motor is essentially a constant-speed machine, much thought has been and is being exercised to vary its speed continuously over a wide range of operating conditions through easy and efficient means. The methods of speed control are discussed here only briefly. The interested reader may consult the bibliography given at the end of the chapter.

The appropriate equation to be examined, based on Equation 5.3.29, is

$$n = (1 - \mathbf{S}) \, n_1 = (1 - \mathbf{S}) \, 120 f_s / P \qquad (7.5.1)$$

where n is the actual speed of the machine in rpm, $\mathbf{S}$ is the per-unit slip, f_s is the supply frequency in Hz, P is the number of poles, and n_1 is the synchronous speed in rpm. Equation 7.5.1 suggests that the speed of the induction motor can be varied by varying either the slip or the synchronous speed, which in turn can be varied by changing either the number of poles or the supply frequency. Any method of speed control that depends on the variation of slip is inherently inefficient because the efficiency of the induction motor is approximately proportional to $(1 - \mathbf{S})$. On the other hand, if the supply frequency is constant, varying the number of poles results only in discrete and stepped variation in the speed of the motor. Indeed, all methods of speed control require a sacrifice in performance, cost, and simplicity to a greater or lesser extent, and these disadvantages must be weighed carefully against the advantages offered.

POLE-CHANGING METHOD   If a machine is provided with two stator windings arranged for different numbers of poles, and preferably with a squirrel-cage rotor, so that no change of connections is needed on the secondary, two synchronous speeds become available. It is possible with special connections to have one winding that can be reconnected simply to yield two or even three different numbers of poles. Only discrete changes in the motor speed can be obtained by this technique. Many ingenious methods of varying the number of poles by means of pole-amplitude modulation and phase-modulated pole-changing have been developed recently.

If the changes in speed are accomplished without change in air-gap flux density, the motor develops the same maximum torque for any speed setting, in which case the system is said to have a constant-torque drive. On the other hand, if the changes in speed are made in such a way that the air-gap flux changes with different stator connections, so that the flux is inversely proportional to the speed setting, the system is said to have a constant-horsepower drive.

Another method known as concatenation, formerly used for changing the motor speed, employs two or more separate wound-rotor motors, all mechanically connected to the same shaft, or through gears. If the stator of one motor, with P_1 poles, is connected to the supply line, and its rotor slip rings are connected to the stator of a second motor, with P_2 poles, whose rotor winding (squirrel-cage or wound-rotor type) is short-circuited, the common shaft will be driven at a speed corresponding to $(P_1 + P_2)$ poles. The student should be able to reason it out.

FIGURE 7.5.1 Solid-state control of induction motor.

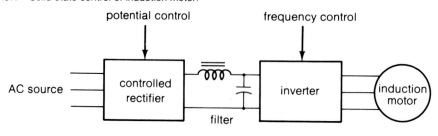

FIGURE 7.5.2 Speed control by changing the slip by means of line-voltage control.

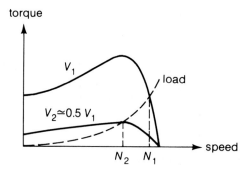

VARIABLE-FREQUENCY METHOD   If it is practicable to vary the supply frequency, then the synchronous speed of the machine can be varied. However, in order to maintain approximately constant air-gap flux density, the line voltage should also be varied directly with the frequency; in such a case, the maximum torque remains very nearly constant. The inherent difficulty in the application of this method has been the lack of an effective and economical source of adjustable frequency. One method is to employ a wound-rotor induction machine as a frequency changer. The advent of solid-state devices with relatively large power ratings has made it possible to use solid-state frequency converters.

One possible combination of power semiconductor converters that would yield the desired variable-frequency, variable-potential excitation from a fixed-frequency, fixed-potential ac supply source is shown in the open-loop system of Figure 7.5.1. For a regenerative system, the single controlled rectifier must be replaced by a dual converter, which is a combination of a rectifier and an inverter that is able to accept the negative current and thereby return power to the supply system.

VARIABLE LINE-VOLTAGE METHOD   The internal torque developed by an induction motor is proportional to the square of the voltage applied to its stator terminals, as seen from Equation 7.4.8 and as shown in Figure 7.5.2 by the two torque-speed characteristics. If the load has the

FIGURE 7.5.3   A closed-loop speed-control system.

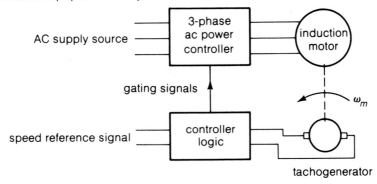

torque-speed characteristic indicated by the dashed line, the speed will be reduced from N_1 to N_2, while the voltage is changed from V_1 to V_2. If the voltage can be varied continuously from V_1 to V_2, the motor speed can also be varied continuously from N_1 to N_2 for the given load.

By employing an ac power controller, the fundamental component of the voltage applied to the motor can be controlled from zero to the value of the supply. A simple closed-loop speed control system is shown in Figure 7.5.3.

VARIABLE ROTOR-RESISTANCE METHOD   The effect of changing rotor-circuit resistance on the torque-slip characteristic of a polyphase wound-rotor induction motor is illustrated in Figure 7.4.5. If the load has the torque-speed curve shown by the dashed line, the speeds corresponding to each of the values of the rotor-circuit resistance are N_A, N_B, N_C, and N_D. By continuous variation of the rotor-circuit resistance, continuous variation of speed is possible.

The main disadvantage of both line-voltage and rotor-resistance control are low efficiency at reduced speeds and poor speed regulation with respect to change in load. You may recall that per-unit speed regulation is the drop in speed from no-load to full-load expressed as a ratio of the full-load base (nominal, rated, or nameplate) speed.

ROTOR-SLIP-FREQUENCY CONTROL   Without sacrificing efficiency at low-speed operation and without affecting the speed by load variation, one can control the induction-motor speed with the availability of power semiconductor converters. Equation 7.4.2 may be expressed as

$$T = \frac{1}{\omega_m} m_1 (I_2')^2 R_2' \frac{1-S}{S} = m_1 (I_2')^2 R_2'/\omega_r \qquad (7.5.2)$$

since

$$\omega_r = S\omega_s = \omega_s - \omega_m = \frac{\omega_r}{S} - \omega_m \qquad (7.5.3)$$

FIGURE 7.5.4   Basic schemes for induction-motor speed control by
auxiliary devices.

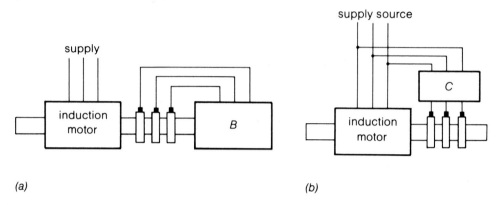

(a)   (b)

(where all ω's are expressed in mechanical radians per second) and a rearrangement yields

$$\frac{1 - S}{S\omega_m} = \frac{1}{\omega_r}$$

(7.5.4)

Equation 7.5.2 shows that, if the rotor frequency ω_r is held constant, the internal torque per rotor ampere will also be a constant; moreover, if ω_r is kept at a lower value, then the torque per rotor ampere will be large. Since ω_s is given by $(\omega_r + \omega_m)$, if ω_r in a control system is determined by a constant reference signal Ω_r, and if the required value of the stator frequency ω_s can be computed from this reference signal and a motor-speed command signal Ω_m, then an exact value of motor speed ω_m, equal to the command signal, can be obtained. Such control systems have been designed to control the speed of induction motors.

ROTOR-SLIP ENERGY RECOVERY METHOD   Considering the power flow in an induction motor as illustrated in Figure 7.4.2, the fraction S of the power transferred across the air gap is transformed by electromagnetic induction to electric power in the rotor circuits. If the rotor circuits are short-circuited, this slip-frequency electric power is wasted as rotor copper loss, and hence operation at reduced speeds is inherently inefficient. The effect of the external resistors introduced into the rotor circuits of a wound-rotor induction motor, in the variable rotor-resistance method of speed control, is to produce slip-frequency voltages that oppose the emfs induced in the rotor windings. If the energy that would be dissipated in the external resistors in such a rheostatic speed-control method could instead be recovered and returned to the ac source, the overall efficiency of the speed-control system would be increased.

Numerous schemes have been tried for recovering the slip-frequency electric power and controlling the slip by means of auxiliary devices. Some of them are rather complicated in their details. However, all of them comprise a means for injecting adjustable voltages of slip frequency into the rotor circuits of a wound-rotor induction motor. The basic schemes are illustrated in Figure 7.5.4: in part (a), the slip rings of the induction motor are connected to auxiliary apparatus,

FIGURE 7.5.5   Rotor-slip energy recovery method of speed control.

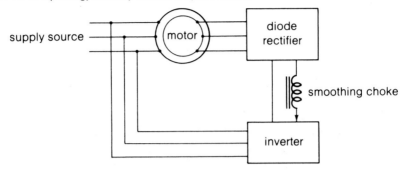

represented by the box *B*, in which the slip-frequency electric power is converted to mechanical power and added on to the shaft power developed by the main induction motor that is being controlled; in part *(b)*, the slip rings of the induction motor are connected to auxiliary frequency-changing devices, represented by the box *C*, in which the slip-frequency electric power is converted to electric power at line frequency and returned to the supply. By adjusting the magnitude and phase of the slip-frequency emfs of the auxiliary devices, the speed and power factor of the main induction motor can be controlled. The auxiliary equipment may consist of rotating machines, adjustable-ratio transformers, or solid-state frequency-converting devices as shown schematically in Figure 7.5.5, in which an inverter is inserted in the rotor circuit. The diode rectifier connected to the three-phase rotor terminals receives the slip-frequency electric power; the smoothed output of the rectifier is then applied to the dc terminals of an inverter operating at the frequency of the supply source, and the three-phase power from the ac terminals of the inverter is fed back to the induction-motor stator source. Such a system permits closed-loop speed control of the motor with increased overall efficiency.

Other schemes for induction-motor speed control by auxiliary devices include concatenation, the brush-shifting Schrage motor, the Kramer control, and the Scherbius control, for the details of which, one may refer to the bibliography given at the end of the chapter. In spite of the advances made in speed-control schemes, researchers are still actively looking for a more satisfactory, economical, and efficient method of speed control of induction motors, particularly for the squirrel-cage type.

7.6 STARTING METHODS FOR POLYPHASE INDUCTION MOTORS

Where high starting torques are required, a wound-rotor induction motor, with external resistances inserted in its rotor circuits, can be used. The starting current can be reduced and high values of starting torque per ampere of starting current can be obtained with *rotor-resistance starting*. The external resistances are generally cut out in steps as the machine runs up to speed.

For cage-rotor machines, the problem is to keep down the starting current, while maintaining adequate starting torque. The input current, for example, may be no more than six times the full-load current, while the starting torque may be about 1.5 times the full-load torque. Depending on the capacity of the available supply system, *direct-on-line starting* may be suitable only for relatively small machines, up to 10-hp rating. Other starting methods include *reduced-voltage starting* by means of *wye-delta starting, autotransformer starting,* or *stator-impedance starting.*

For employing the wye-delta starting method, the machine is designed for delta operation and is connected in wye during the starting period. Because the impedance between line terminals for wye connection is three times that for delta connection, for the same line voltage, the line current at standstill for wye connection is reduced to one-third of the value for delta connection. Since the phase voltage is reduced by a factor of $\sqrt{3}$ during starting, it follows that the starting torque will be one-third of the normal. For autotransformer starting, the setting of the autotransformer can be predetermined to limit the starting current to any desired value. An autotransformer, which reduces the voltage applied to the motor to x times the normal voltage, will reduce the starting current in the supply system as well as the starting torque of the motor to x^2 times the normal values.

While employing the wye-delta starting or the autotransformer starting, if all three line switches are opened simultaneously during the changeover from starting to the normal running condition, the air-gap flux and the speed will decrease during this period, and the transient current surge can be high when the switches are reclosed. Thus, such methods of starting do not necessarily reduce the peak value of starting current, but should reduce the time duration of this current. However, the current surge during switching can be reduced by introducing transition impedances between the starting and motor terminals to maintain the continuity of the current during changeover.

Stator-impedance starting may be employed if the starting-torque requirement is not severe. Series resistances (or impedances) are inserted in the lines in order to limit the starting current. These resistances are shorted out when the motor gains speed. The method has the obvious disadvantage of being inefficient because of the extra losses in the external resistances.

Other methods of starting, such as *part-winding starting* and *multicircuit starting* are sometimes used, in which the motor may be connected asymmetrically during the starting period. In some stator-impedance starting methods, a variable impedance is inserted in only one supply line to the machine during the run-up period.

EXAMPLE 7.6.1

An induction motor, for which the starting current is six times the full-load current and the per-unit full-load slip is 0.04, is to be provided with an autotransformer starter. If the minimum starting torque must be 0.3 times the full-load torque, determine the required tapping on the transformer, and the per-unit supply-source line current at starting.

SOLUTION

From Equation 7.4.1, with slip at starting equal to 1, we have

$$\frac{T_s}{T_{fl}} = \left(\frac{I_s}{I_{fl}}\right)^2 s_{fl}$$

where subscripts *s* and *fl* correspond to starting and full-load conditions.
   Substituting values, $0.3 = (I_s/I_{fl})^2 \times 0.04$, from which

$$I_s/I_{fl} = \sqrt{0.3/0.04} = 2.74$$

or

$$I_s = 2.74 \text{ per-unit, with } I_{fl} \text{ taken as one per-unit}$$

The applied voltage to the motor must reduce the current from six per-unit to 2.74 per-unit; i.e., the tapping must be such as to reduce the voltage to 2.74/6 = 0.456 of the full value. The supply-source line current will then be 2.74 × 0.456 = 1.25 per-unit, where 0.456 is the secondary-to-primary turns ratio.

DEEP-BAR AND DOUBLE-SQUIRREL-CAGE ROTORS   Recall that, at standstill, the rotor frequency is the same as the stator frequency; as the motor accelerates, the rotor frequency decreases to a low value, of the order of 2 or 3 Hz at full load in a typical 60-Hz motor. By using suitable shapes and arrangements for rotor bars, it is possible to design squirrel-cage rotors so that their effective resistance at 60 Hz is several times their resistance at 2 or 3 Hz. Advantage is taken of the inductive effect of the slot-leakage flux on the current distribution in the rotor bars.

   Rotor bars embedded in deep slots, whose depth is two or three times greater than the slot width as shown in Figure 7.6.1*(a)*, provide a high effective resistance and a large torque at starting. Because of the skin effect, the current has a tendency to concentrate at the top of the bars at starting, when the frequency of the rotor currents is high. Under normal running conditions with low slips, because the frequency of rotor currents is much lower, skin effect is negligible and the current tends to distribute almost uniformly throughout the entire rotor-bar cross section; the rotor resistance becomes lower, leading to a higher efficiency.

FIGURE 7.6.1   Forms of a slot for induction-motor rotors with high starting torque.

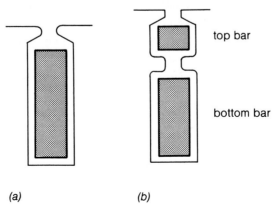

top bar

bottom bar

(a)   (b)

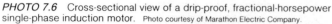

PHOTO 7.6   Cross-sectional view of a drip-proof, fractional-horsepower single-phase induction motor.   Photo courtesy of Marathon Electric Company.

An alternative to the deep-bar rotor is the double-cage arrangement illustrated in Figure 7.6.1*(b)* for attaining higher starting torque and better running efficiency. The squirrel-cage winding consists of two layers of bars short-circuited by end rings. The inner cage, made up of low-resistance bottom bars, is deeply embedded in iron, while the outer cage has relatively high-resistance bars situated close to the inner stator surface. At starting, because of skin effect, the influence of the outer cage dominates, thereby producing a high starting torque. During the normal running period, the current penetrates to full depth into the lower cage because of insignificant skin effect, resulting in an efficient steady-state operation. The inductance of the lower bars is greater than that of the upper ones because of the general nature of the slot-leakage field, and the difference in inductance can be made quite large by properly proportioning the slot dimensions, particularly the constriction in the slot between the two layers of bars.

7.7 SINGLE-PHASE INDUCTION MOTORS

For reasons of simplicity and cost, single-phase power supply is universally preferred for fractional-horsepower motors and is also widely used for motors up to about 5 hp. Single-phase induction motors are usually two- or four-pole, rated at 2 hp or less, while slower and larger motors may be manufactured for special purposes. These are most widely used in domestic appliances and in a very large number of low-power drives in industry. The single-phase induction machine resembles a small, three-phase, squirrel-cage motor except for the fact that, when it is running at full speed, only a single winding on the stator is usually excited. See Photo 7–6 for a cross-sectional view of a fractional-horsepower, single-phase induction motor.

The single-phase stator winding is distributed in slots so as to produce an approximately sinusoidal space distribution of mmf. As discussed in Section 5.3, such a motor inherently has no starting torque, and as indicated in Section 5.4, it must be started by means of an auxiliary wind-

FIGURE 7.7.1   Single-phase induction motor—elements of cross-field theory *(a)* at standstill, *(b)* during rotation.

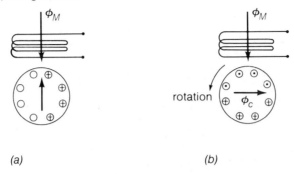

(a)   (b)

ing, displaced in phase position from the main winding, or some similar device. Once started by auxiliary means, the motor will continue to run. Thus, nearly all single-phase induction motors are actually two-phase motors, with the main winding in the direct axis adapted to carry most or all of the current in operation, and an auxiliary winding in the quadrature axis with a different number of turns adapted to provide the needed starting torque.

Since the power input in a single-phase circuit pulsates at twice the line frequency, all single-phase motors will have a double-frequency torque component, which causes slight oscillations in rotor speed and imparts vibration to the motor supports. Means must be provided to prevent this vibration from causing objectionable noise.

Exciting the stator with an alternating emf causes the alternating flux along the main axis, as shown in Figure 7.7.1*(a)*, designated by ϕ_M. The stator-mmf wave is stationary in space, but pulsates in magnitude, the stator-field strength alternating in polarity and varying sinusoidally with time. Currents are induced in the squirrel-cage-rotor conductors by transformer action, these currents being in such a direction as to produce an mmf opposing the stator mmf, as illustrated in Figure 7.7.1*(a)*. The axis of the rotor-mmf wave coincides with that of the stator field, and the torque angle is zero; hence, no starting torque is produced. At standstill, the motor is merely a single-phase static transformer with a short-circuited secondary. However, when the rotor is made to revolve, there is, in addition to the transformer voltage, a speed voltage generated in the rotor by virtue of its rotation in the stationary stator field. The direction of rotational voltages in the rotor conductors being as shown in Figure 7.7.1*(b)*, a component rotor current and a component rotor-mmf wave (whose axis is displaced 90 electrical degrees from the stator axis) are produced. The torque angle for this component of the rotor mmf being 90°, a torque is obtained. Further analysis by what is known as the *cross-field theory*[1] shows that this torque in fact acts in the direction of rotation and that the necessary conditions for the continued production of torque are satisfied.

[1]For a discussion of the cross-field theory, see A. F. Puchstein, T. C. Lloyd, and A. G. Conrad, *Alternating-Current Machines,* 3rd ed. (New York: John Wiley & Sons, 1954), chap. 30; and C. G. Veinott, *Fractional- and Subfractional-Horsepower Electric Motors* (New York: McGraw-Hill, 1970), chap. 2.

FIGURE 7.7.2 *(a)* Equivalent circuit of a single-phase induction motor based on the revolving-field theory, *(b)* Torque-speed characteristics of a single-phase induction motor based on the revolving-field theory.

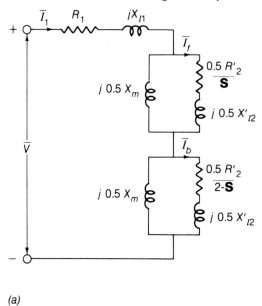

(a)

(b)

The other viewpoint adopted in explaining the operation of the single-phase motor is based on the conditions already established for polyphase motors and is known as the *revolving-field theory*. While both viewpoints have their own advantages, there is little choice between them for computational purposes; the revolving-field point of view, already introduced in Section 5.3, is followed hereafter as a natural extension of the theory applied to the polyphase induction motor. As stated in Section 5.3 and shown in Section 5.4, a stationary pulsating field can be represented by two counterrotating fields of constant magnitude. The equivalent circuit of a single-phase induction motor, then, consists of the series connection of a forward-rotating field equivalent circuit and a backward-rotating field equivalent circuit. Each circuit is similar to that of a three-phase machine, but in the backward-rotating field circuit the parameter S is replaced by $(2 - S)$, as shown in Figure 7.7.2(a). The forward and backward torques are calculated from the two parts of the equivalent circuit, and the total torque is given by the algebraic sum of the two. As shown in Figure 7.7.2 (b), the torque-speed characteristic of a single-phase induction motor is thus obtained as the sum of the two curves, one corresponding to the forward-rotating field and the other corresponding to the backward-rotating field.

The slip S_f of the rotor with respect to the forward-rotating field is given by

$$S_f = S = \frac{n_s - n}{n_s} = 1 - \frac{n}{n_s} \tag{7.7.1}$$

where n_s is the synchronous speed and n is the actual rotor speed. The slip S_b of the rotor with respect to the backward-rotating field is given by

$$S_b = \frac{n_s - (-n)}{n_s} = 1 + \frac{n}{n_s} = 2 - S \tag{7.7.2}$$

Since the amplitude of the rotating fields is one-half of the alternating flux, as seen from Equation 5.4.9, the total magnetizing and leakage reactances of the motor can be divided equally so as to correspond to the forward and backward rotating fields. In the equivalent circuit shown in Figure 7.7.2(a), R_1 and X_{l1} are respectively the resistance and leakage reactance of the main winding, X_m is the magnetizing reactance, and R_2' and X_{l2}' are the standstill values of the rotor resistance and leakage reactance referred to the main stator winding by the use of the appropriate turns ratio. The core loss, which is omitted here, can be accounted for later as if it were a rotational loss. The resultant torque of a single-phase induction motor can be expressed as

$$T_e = \frac{I_f^2(1 - S)}{\omega_m S} R_2' - \frac{I_b^2(1 - S)}{\omega_m(2 - S)} R_2' \tag{7.7.3}$$

The following example illustrates the usefulness of the equivalent circuit in evaluating the motor performance.

EXAMPLE 7.7.1

A 1/4-hp, 230-V, 60-Hz, 4-pole single-phase induction motor has the following parameters and losses:

$$R_1 = 10 \text{ ohms} ; X_{l1} = X_{l2}' = 12.5 \text{ ohms}$$

$$R_2' = 11.5 \text{ ohms} ; X_m = 250 \text{ ohms}$$

Core loss at 230 V is 35 W

Friction and windage loss is 10 W

For a slip of 0.05, determine the stator current, power factor, developed power, shaft-output power, speed, torque, and efficiency when the motor is running as a single-phase motor at rated voltage and frequency with its starting winding open.

SOLUTION

From the given data applied to the equivalent circuit of Figure 7.7.2*(a)*

$$\frac{0.5R_2'}{S} = \frac{11.5}{2 \times 0.05} = 115 \ \Omega$$

$$\frac{0.5R_2'}{2 - S} = \frac{11.5}{2(2 - 0.05)} = 2.95 \ \Omega$$

$$j0.5X_m = j125 \ \Omega$$

and

$$j0.5X_{l2}' = j6.25 \ \Omega$$

For the forward-field circuit, the impedance

$$Z_f = \frac{(115 + j6.25) \, j125}{115 + j131.25} = 59 + j57.65 = R_f + jX_f$$

and for the backward-field circuit, the impedance

$$Z_b = \frac{(2.95 + j6.25) \, j125}{2.95 + j131.25} = 2.67 + j6.01 = R_b + jX_b$$

The total series impedance Z_e is given by

$$Z_e = Z_1 + Z_f + Z_b$$
$$= (10 + j12.5) + (59 + j57.65) + (2.67 + j6.01)$$
$$= 71.67 + j76.16 = 104.6\angle46.74°$$

Input stator current $\bar{I}_1 = \dfrac{230}{104.6\angle46.74°} = 2.2\angle-46.74°$ A

Power factor $= \cos 46.74° = 0.685$ lagging

Developed power $P_d = [I_1{}^2\, R_f](1 - S)$
$$+ [I_1{}^2\, R_b][1 - (2 - S)]$$

$$= I_1{}^2(R_f - R_b)(1 - S)$$

$$= (2.2)^2(59 - 2.67)(1 - 0.05)$$

$$= 259 \text{ W}$$

Shaft-output power $P_0 = P_d - P_{rot} - P_{core}$
$$= 259 - 10 - 35$$
$$= 214 \text{ W or } 0.287 \text{ hp}$$

Speed $= (1 - S)$(synchronous speed)
$$= 0.95 \times \frac{120 \times 60}{4}$$

$$= 1{,}710 \text{ rpm or } 179 \text{ rad/s}$$

Torque $= \dfrac{214}{179} = 1.2$ N·m

Efficiency $= \dfrac{\text{Output}}{\text{Input}} = \dfrac{214}{230 \times 2.2 \times 0.685} = \dfrac{214}{346.6}$

$$= 0.6174 \text{ or } 61.74 \text{ percent}$$

FIGURE 7.7.3   Equivalent circuit of a single-phase induction motor *(a)* no-load conditions, *(b)* blocked-rotor conditions.

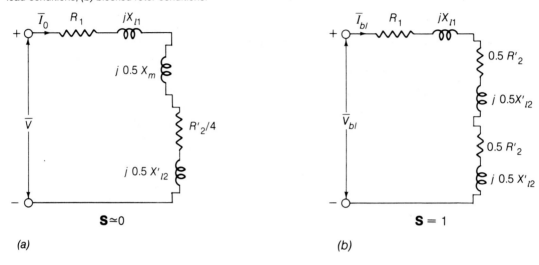

(a)   (b)

The parameters of the equivalent circuit of a single-phase induction motor may be approximately determined from the no-load and blocked-rotor tests similar to those made on a polyphase induction motor. These tests are usually made with the auxiliary winding kept open, except for the capacitor-run motor (see Section 7.8). The no-load test is conducted by running the motor without load at rated voltage and rated frequency. Since the no-load slip is small, the equivalent circuit can be considered to be as in Figure 7.7.3(a), while for the blocked-rotor-test conditions, i.e., for **S** = 1, Figure 7.7.3(b) shows the approximate equivalent circuit neglecting the magnetizing current. Thus, on no-load one has

$$Z_0 = (R_1 + jX_{l1}) + j0.5X_m + (0.25R_2' + j0.5X_{l2}') \qquad (7.7.4)$$

and under blocked-rotor conditions

$$Z_{bl} = (R_1 + jX_{l1}) + (R_2' + jX_{l2}') \qquad (7.7.5)$$

Assuming that $X_{l1} = X_{l2}'$, and measuring the stator resistance R_1, the equivalent-circuit parameters can all be determined.

FIGURE 7.8.1 Split-phase motor: *(a)* schematic diagram, *(b)* phasor diagram at starting, *(c)* typical torque-speed (or slip) characteristic.

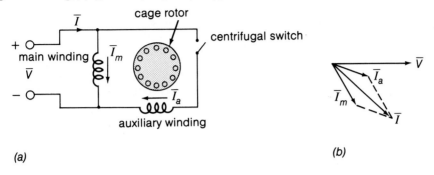

(a)

(b)

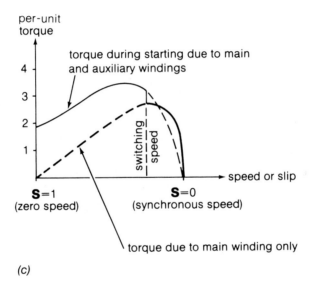

(c)

7.8 STARTING METHODS FOR SINGLE-PHASE INDUCTION MOTORS

The various forms of a single-phase induction motor are grouped into four principal types, depending on the method of starting:

1. *Split-phase* or *resistance-split-phase motors:* Split-phase motors have two stator windings, a main winding and an auxiliary winding, with their axes displaced 90 electrical degrees in space. With this type of motor schematically represented in Figure 7.8.1*(a)*, the auxiliary winding has a higher resistance to reactance ratio than the main winding, so that the two currents are out of phase, as indicated in the

phasor diagram of Figure 7.8.1*(b)*. The motor is equivalent to an unbalanced two-phase motor. The rotating stator field produced by the unbalanced two-phase winding currents causes the motor to start. The auxiliary winding is disconnected by a centrifugal switch or relay when the motor comes up to about 75 percent of the synchronous speed. The torque-speed characteristic of the split-phase motor is of the form shown in Figure 7.8.1*(c)*. A higher starting torque can be developed by a split-phase motor by inserting a series resistance in the starting auxiliary winding. A similar effect may also be obtained by inserting a series inductive reactance in the main winding; this additional reactance is short-circuited when the motor builds up speed.

2. *Capacitor motors:* These have a capacitor in series with the auxiliary winding. There are three varieties: capacitor-start, capacitor run, and permanent-split-capacitor motors. As implied, the first two use a centrifugal switch or relay to open the circuit or reduce the size of the starting capacitor when the motor comes up to speed. A two-value-capacitor motor, with one value for starting and one for running, can be designed for optimum starting and running performance; the starting capacitor is disconnected after the motor starts. The relevant schematic diagrams and torque-speed characteristics are shown in Figures 7.8.2, 7.8.3, and 7.8.4. Motors in which the auxiliary winding and the capacitor are not cut out during the normal running conditions operate, in effect, as unbalanced two-phase induction motors.

3. *Shaded-pole motors:* These are the least expensive of the fractional-horsepower motors, generally rated up to 1/20 hp. They have salient stator poles, with one-coil-per-pole main windings. The auxiliary winding consists of one (or rarely two) short-circuited copper straps wound on a portion of the pole and displaced from the center of each pole, as shown in Figure 7.8.5*(a)*. Such a motor with shading bands is known as the shaded-pole motor. Induced currents in the shading coil cause the flux in the shaded portion of the pole to lag (in time) the flux in the other portion. The result is then like a rotating field moving in the direction from the unshaded to the shaded portion of the pole. A low starting torque is produced, and a typical torque-speed characteristic is shown in Figure 7.8.5*(b)*. The efficiency of shaded-pole motors is rather low.

4. *Repulsion-induction motors:* Before low-cost capacitors became available, repulsion-start and induction-run machines were the most widely used type of single-phase motor in the range of 1/3 to 5 hp. These have distributed rotor windings connected to a commutator (as in the case of a dc machine) with short-circuited brushes, and have a distributed single-phase stator winding in the direct axis only. Repulsion-start motors have a centrifugal device that short-circuits all the commutator segments when the motor comes up to speed and also lifts the brushes off the commutator. The permanently short-circuited rotor brushes are displaced from the direct axis by an angle less than 90 electrical degrees, so that a rotor current is induced by transformer action. The flux produced by this current, which is fixed in position, acts with the

FIGURE 7.8.2   Capacitor-start motor: *(a)* schematic diagram, *(b)* phasor diagram at starting, *(c)* typical torque-speed characteristic.

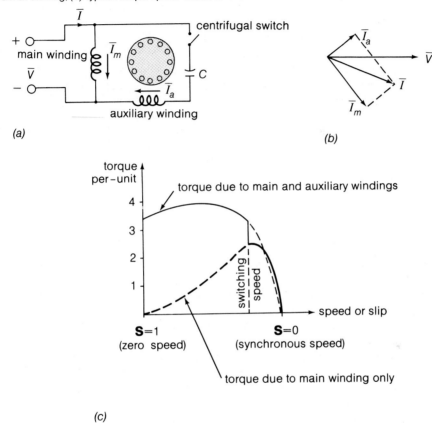

(a)

(b)

(c)

FIGURE 7.8.3   Permanent-split-capacitor motor: *(a)* schematic diagram, *(b)* typical torque-speed characteristic.

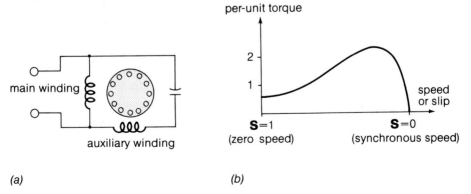

(a)

(b)

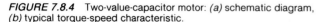

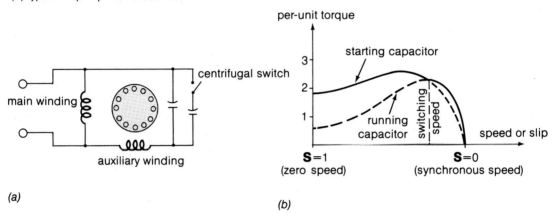

FIGURE 7.8.4 Two-value-capacitor motor: *(a)* schematic diagram, *(b)* typical torque-speed characteristic.

(a)

(b)

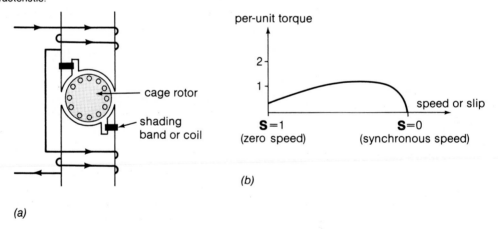

FIGURE 7.8.5 A shaded-pole motor and its typical torque-speed characteristic: *(a)* schematic diagram, *(b)* typical torque-speed characteristic.

(b)

(a)

main winding flux to create a torque. To avoid the centrifugal mechanism, repulsion-induction motors that have in addition a deeply buried, low-resistance squirrel cage in the rotor in order to limit the no-load speed to a little above synchronism, have also been used widely. These machines fall under the general category of ac commutator machines; however, these have been replaced by capacitor motors.

BIBLIOGRAPHY

Alger, P. L. *Induction Machines—Their Behavior and Uses.* 2d ed. New York: Gordon and Breach, 1970.

American National Standards Institute. *ANSI Standard No. C.56.20–1954: Test Code for Induction Motors.* New York, 1954.

Del Toro, V. *Electromechanical Devices for Energy Conversion and Control Systems.* Englewood Cliffs, N. J.: Prentice-Hall, 1968.

Fitzgerald, A. E.; Kingsley, C. Jr.; and Kusko, A. *Electrical Machinery.* 3d ed. New York: McGraw-Hill, 1971.

Hindmarsh, J. *Electrical Machines and Their Applications.* 2d ed. New York: Pergamon, 1970.

Institute of Electrical and Electronics Engineers. *Standard No. 112–1978: Test Procedures for Polyphase Induction Motors and Generators.* New York, 1978.

Knowlton, A. E., ed. *Standard Handbook for Electrical Engineers.* 8th ed. New York: McGraw-Hill, 1949.

Matsch, L. W. *Electromagnetic and Electromechanical Machines.* New York: Intext, 1972.

Nasar, S. A., and Unnewehr, L. E. *Electromechanics and Electric Machines.* New York: John Wiley & Sons, 1979.

National Electrical Manufacturers Association. *Publication No. MG1–1972: Motors and Generators.* New York, 1972.

———. *Publication No. MG2–1951: Standards for Fractional Horsepower Motors.* New York, 1951.

Puchstein, A. F.; Lloyd, T. C.; and Conrad, A. G. *Alternating-Current Machines.* 3d ed. New York: John Wiley & Sons, 1954.

Slemon, G. R., and Straughen, A. *Electric Machines.* Reading, Mass.: Addison-Wesley, 1980.

Veinott, C. G. *Fractional- and Subfractional-Horsepower Electric Motors.* New York: McGraw-Hill, 1970.

PROBLEMS

7-1.

A balanced three-phase 60-Hz voltage is applied to a three-phase, two-pole induction motor. Corresponding to a per-unit slip of 0.05, determine the following:

a. The speed of the rotating stator-magnetic field relative to the stator winding.
b. The speed of the rotor field relative to the rotor winding.
c. The speed of the rotor field relative to the stator winding.
d. The speed of the rotor field relative to the stator field.
e. The frequency of the rotor currents.

Neglecting stator resistance, leakage reactance, and all losses, if the stator-to-rotor turns ratio is 2:1 and the applied voltage is 100 V, find the rotor-induced emf at standstill and at 0.05 slip.

7-2.

The double-layer stator winding of a cage-type, four-pole, three-phase induction motor has 24 turns per phase distributed in 36 slots. The winding is chorded by one slot pitch. Calculate the factor by which the rotor standstill resistance and leakage reactance are to be multiplied for referring them to the stator, if the number of rotor bars is 28.

7-3.

No-load and blocked-rotor tests are conducted on a three-phase wye-connected induction motor with the following results: The line-to-line voltage, line current, and total input power for the no-load test are 220 V, 20 A, and 1,000 W, and for the blocked-rotor test, 30 V, 50 A, and 1,500 W. The stator resistance as measured on a dc test is 0.1 ohm per phase. Determine the parameters of the equivalent circuit shown in Figure 7.2.5(c), and also compute the no-load rotational losses.

7–4.

A three-phase, 5-hp, 220-V, six-pole, 60-Hz induction motor runs at a slip of 0.025 at full load. Rotational and stray-load losses at full load are 5 percent of the output power. Calculate the power transferred across the air gap, the rotor copper loss at full load, and the electromagnetic torque at full load in newton-meters.

7–5.

The power transferred across the air gap of a two-pole induction motor is 24 kW. If the electromagnetic power developed is 22 kW, find the slip. Calculate the output torque, if the rotational loss at this slip is 400 W.

7–6.

The stator and rotor of a three-phase, 440-V, 15-hp, 60-Hz, eight-pole wound-rotor induction motor are both connected in wye and have the following parameters per phase:

$$R_1 = 0.5 \text{ ohm} ; X_M = 1.25 \text{ ohm}$$

$$R_2 = 0.1 \text{ ohm} ; X_{l2} = 0.2 \text{ ohm}$$

The magnetizing impedance is 40 ohms and the core-loss impedance is 360 ohms, both referred to the stator. The ratio of effective stator turns to effective rotor turns is 2.5. The friction and windage losses are given to be 200 W, and the stray-load loss is estimated as 100 W. Using the equivalent circuit of Figure 7.2.5(a), calculate the following for a slip of 0.05, when the motor is operated at rated voltage and frequency applied to the stator, with the rotor slip rings short-circuited: stator input current, power factor at the stator terminals, current in the rotor winding, output power, output torque, and efficiency.

7–7.

a. Starting from Equations 7.4.8 and 7.4.11, show that

$$\frac{T}{T_{max}} = \frac{1 + \sqrt{Q^2 + 1}}{1 + \frac{1}{2} \sqrt{Q^2 + 1} \left(\dfrac{S}{S_{max\,T}} + \dfrac{S_{max\,T}}{S} \right)}$$

where $Q = \dfrac{X_1'' + X_{l2}'}{R_1''}$

With an infinite value of Q, i.e., $R_1'' = 0$ (negligible stator resistance), show that Equation 7.4.12 results.

b. Show that the ratio of the stator-load current I_2' to that at maximum torque I_2' $_{max\ T}$ is given by

$$\frac{I_2'}{I_2'\ _{max\ T}} = \sqrt{\frac{(1 + \sqrt{1 + Q^2})^2 + Q^2}{(1 + \frac{S_{max\ T}}{S} \sqrt{1 + Q^2})^2 + Q^2}}$$

where $Q = \dfrac{X_1'' + X_{l2}'}{R_1''}$.

Obtain the expression for the limiting case when $Q \to \infty$.

7–8.

Considering only the rotor equivalent circuit shown in Figure 7.2.2 or 7.2.3, find

    a. R_2 for which the developed torque would be a maximum.
    b. the slip corresponding to the maximum torque.
    c. the maximum torque.
    d. R_2 for the maximum starting torque.

7–9.

A three-phase induction motor, operating at its rated voltage and frequency, develops a starting torque of 1.6 times the full-load torque, and a maximum torque of 2 times the full-load torque. Neglecting stator resistance and rotational losses, assuming constant rotor resistance, determine the slip at maximum torque, and the slip at full load.

7–10.

A three-phase, wye-connected, 400-V, 4-pole, 60-Hz induction motor has primary leakage impedance of $(1 + j2)\ \Omega$ and secondary leakage impedance referred to the primary at standstill of $(1 + j2)\ \Omega$. The magnetizing impedance is $j40\ \Omega$ and the core-loss impedance is $400\ \Omega$.

Use the *T*-equivalent circuit of Figure 7.2.5(a) for parts (a), (b), and (c):

    a. Calculate the input current and power (i) on a no-load test ($S \simeq 0$) at rated voltage, and (ii) on a blocked-rotor test ($S = 1$) at rated voltage.
    b. Corresponding to a slip of 0.05, compute the input current, torque, output power, and efficiency.
    c. Determine the starting torque and current, the maximum torque and the corresponding slip, the maximum output power and the corresponding slip.

Use the approximate equivalent circuit (for the following parts) obtained by transferring the shunt core loss/magnetizing branch to the input terminals:

d. Repeat part *(b)*.
e. Calculate the primary current, torque, mechanical power input and electrical power output, when the machine is driven as an induction generator with a slip of -0.05.
f. Compute the primary current and the braking torque at the instant of plugging (i.e., reversal of the phase sequence), if the slip immediately before plugging is 0.05.

7–11.

A 500-hp wye-connected wound-rotor induction motor, when operated at rated voltage and frequency, develops its rated full-load output at a slip of 0.02; maximum torque of 2 times the full-load torque at a slip of 0.06, with a referred rotor current of 3 times that at full load; and 1.2 times the full-load torque at a slip of 0.2, with a referred rotor current of 4 times that at full load. Rotational and stray-load losses may be neglected. If the rotor-circuit resistance in all phases is increased to 5 times the original resistance, determine the following:

a. The slip at which the motor will develop the same full-load torque.
b. The total rotor-circuit copper loss at full-load torque.
c. The horsepower output at full-load torque.
d. The slip at maximum torque.
e. The rotor current at maximum torque.
f. The starting torque.
g. The rotor current at starting.

7–12.

The parameters of the per-phase equivalent circuit (shown in Figure 7.4.1) of a three-phase, 600-V, 60-Hz, four-pole, wye-connected wound-rotor induction motor are given below:

$$R_1 = 0.75 \text{ ohm} ; X_{f1} = X_{f2}' = 2.0 \text{ ohms}$$
$$R_2' = 0.80 \text{ ohm} ; X_M = 50 \text{ ohms}$$

The core losses may be neglected.

a. Find the slip at which the maximum developed torque occurs.
b. Calculate the value of the maximum torque developed.
c. What is the range of speed for stable operation of the motor?
d. Determine the starting torque.
e. Compute the per-phase referred value of the additional resistance that must be inserted in the rotor circuit in order to obtain the maximum torque at starting.

7-13.

A squirrel-cage induction motor operates at a slip of 0.05 at full load. The rotor current at starting is five times the rotor current at full load. Neglecting stator resistance, rotational and stray-load losses, assuming constant rotor resistance, calculate the starting torque and the maximum torque in per-unit of full-load torque, as well as the slip at which the maximum torque occurs.

7-14.

Using the approximate equivalent circuit in which the shunt branch is moved to the stator-input terminals, show that the rotor current, torque, and electromagnetic power of a poly-phase induction motor vary almost directly as the slip, for small values of slip.

7-15.

A three-phase, 50-hp, 440-V, 60-Hz, four-pole wound-rotor induction motor operates at a slip of 0.03 at full load with its slip rings short-circuited. The motor is capable of developing a maximum torque of two times the full-load torque at rated voltage and frequency. The rotor resistance per phase referred to the stator is 0.1 ohm. Neglect the stator resistance, rotational and stray-load losses. Find the rotor copper loss at full load, and the speed at maximum torque. Compute the value of per-phase rotor resistance (referred to the stator) that must be added in series to produce a starting torque equal to the maximum torque.

7-16.

A three-phase, 220-V, 60-Hz, four-pole, wye-connected induction motor has a per-phase stator resistance of 0.5 ohm. The no-load and blocked-rotor test data on the motor are given below:

No-load test:	line-to-line voltage = 220 V
	total input power = 600 W, of which 200 W is the friction
	and windage loss
Blocked-rotor test:	line current = 3 A
	line-to-line voltage = 35 V
	total input power = 720 W
	line current = 15 A

a. Calculate the parameters of the equivalent circuit shown in Figure 7.2.5(c).
b. Compute the output power, output torque, and efficiency if the machine runs as a motor with a slip of 0.05.
c. Determine the slip at which the maximum torque is developed, and obtain the value of the maximum torque developed.

7–17.

A three-phase, 2,200-V, 60-Hz, delta-connected squirrel-cage induction motor, when started at full rated voltage, takes a starting current of 693 A from the line and develops a starting torque of 6,250 N·m.

    a. Calculate the ratio of a starting compensator (i.e., an autotransformer starter) such that the current supplied by the 2,200-V line is 300 A. The impedance and exciting current of the compensator may be neglected. Compute the starting torque with the starting compensator.

    b. If a wye-delta starting method is employed, find the starting current and the starting torque.

7–18.

A three-phase, four-pole, 220-V, 60-Hz induction machine with a per-phase resistance of 0.5 ohm is operating at rated voltage as a generator at a slip of -0.04, delivering 12 A of line current and a total output of 4,000 W. The constant losses from a no-load run as a motor are given to be 220 W, of which 70 W represent friction and windage losses. Calculate the efficiency of the induction generator.

7–19.

A 2,200-V, 1,000-hp, three-phase, 60-Hz, 16-pole, wye-connected, wound-rotor induction motor is connected to a 2,200-V, three-phase, 60-Hz bus that is supplied by synchronous generators. The parameters of the per-phase equivalent circuit of Figure 7.4.1 are given below:

$$R_1 = 0.1 \text{ ohm} = R_2' \; ; X_{l1} = 0.625 \text{ ohm} = X_{l2}' \; ;$$
$$X_M = 20 \text{ ohms}$$

If the machine is driven at a speed of 459 rpm to act as a generator of real power, find the rotor current referred to the stator and the real and reactive power output of the induction machine.

7–20.

A three-phase, 440-V, 60-Hz, four-pole induction motor operates at a slip of 0.025 at full load with its rotor circuit short-circuited. This motor is to be operated on a 50-Hz supply so that the air-gap flux wave has the same amplitude at the same torque as on 60 Hz. Determine the 50-Hz applied voltage, and the slip at which the motor will develop a torque equal to its 60-Hz full-load value.

7–21.

The rotor of a wound-rotor induction motor is rewound with twice the number of its original turns, with a cross-sectional area of the conductor in each turn of one-half the original value. Determine the ratio of the following in the rewound motor to the corresponding original quantities: *(a)* full-load current, *(b)* actual rotor resistance, and *(c)* rotor resistance referred to the stator.

    Repeat the problem if the original rotor were rewound with the same number as the original number of turns, but with one-half the original cross-section of the conductor. Neglect the changes in the leakage flux.

7-22.

A wound-rotor induction machine driven by a dc motor, whose speed can be controlled, is operated as a frequency changer. The three-phase stator winding of the induction machine is excited from a 60-Hz supply, while the variable-frequency three-phase power is taken out of the slip rings. The output frequency range is to be 120 to 420 Hz; the maximum speed is not to exceed 3,000 rpm; and the maximum power output at 420 Hz is to be 70 kW at 0.8 power factor. Assuming that the maximum-speed condition determines the machine size, neglecting exciting current, losses, and voltage drops in the induction machine, calculate (a) the minimum number of poles for the induction machine, (b) the corresponding minimum and maximum speeds, (c) the kVA rating of the induction-machine stator winding, and (d) the horsepower rating of the dc machine.

7-23.

A 1/4-hp, 110-V, 60-Hz, four-pole capacitor-start single-phase induction motor has the following parameters and losses:

$$R_1 = 2\ \Omega;\ X_{l1} = 2.8\ \Omega;\ X_{l2}' = 2\ \Omega;\ R_2' = 4\ \Omega;\ X_M = 70\ \Omega$$

$$\text{Core loss at 110 V} = 25\ \text{W ; Friction and windage} = 12\ \text{W}$$

For a slip of 0.05, compute the input current, power factor, power output, speed, torque, and efficiency when the motor is running at rated voltage and rated frequency with its starting winding open.

7-24.

The data obtained from the no-load and blocked-rotor tests conducted on a 110-V, single-phase induction motor are given below:

No-load test:	input voltage = 110 V
	input current = 3.7 A
	input power = 50 W
Blocked-rotor test:	input voltage = 50 V
	input current = 5.6 A

Taking the stator resistance to be 2.0 Ω, friction and windage loss to be 7 W, and assuming $X_{l1} = X_{l2}'$, determine the parameters of the double-revolving-field equivalent circuit.

7-25.

The impedances of the main and auxiliary windings of a 1/3-hp, 120-V, 60-Hz, capacitor-start motor are given as

$$Z_m = (4.6 + j3.8)\ \text{ohms and}\ Z_a = (9.6 + j3.6)\ \text{ohms}$$

Determine the value of the starting capacitance that will cause the main and auxiliary winding currents to be in quadrature at starting.

8

SYNCHRONOUS MACHINES

Large ac power networks operating at a constant frequency of 60 Hz in the United States (50 Hz in Europe) rely almost exclusively on synchronous generators for the generation of electrical energy, and may also have synchronous compensators or condensers at key points for reactive power control. Generators are the largest single-unit electric machines in production, having power ratings in the range of 1,500 MVA; and it is expected that machines of several thousand MVA will come into use in future decades. Private, stand-by, and peak-load plants with diesel or gas-turbine prime-movers also have alternators. Non-land-based synchronous plants can be found on oil rigs, on large aircraft with hydraulically driven alternators operating at 400 Hz, and on ships for variable-frequency supply to synchronous propeller motors. Synchronous motors provide constant-speed industrial drives with the possibility of power-factor correction; but these are not often built in small ratings, for which the induction motor is cheaper.

Analytical methods of examining the steady-state performance of synchronous machines are developed in this chapter. While consideration is initially given to cylindrical-rotor machines, the effects of salient poles are discussed later.

8.1 CONSTRUCTIONAL FEATURES OF SYNCHRONOUS MACHINES

The two basic parts of a synchronous machine are the magnetic field structure, carrying a dc-excited winding, and the armature, often having a three-phase winding in which the ac emf is generated. The use of a rotating dc-field system is almost universal, because it permits the ac windings to be placed on the stator, where they are more conveniently braced against electromagnetic forces and insulated for high voltage. Thus, almost all modern synchronous machines have stationary armatures and rotating field structures.

The *rotor* may be constructed with *salient poles,* or be *cylindrical,* i.e., *round* with no polar projections. The prime-mover speed has a profound influence on the constructional form, and in all large units the limiting feature is the centrifugal force on the rotor. Steam- or gas-turbine-driven machines, known as *turbogenerators (turbo-alternators* or *turbine-generators),* are run at high speed, and usually have two- or four-pole round (cylindrical) rotors of solid forged steel, with a diameter limited to about 1.2 m and an axial length of several meters. See Photo 8–1 for a cross-sectional view of a large turbogenerator. Generators driven by water power, known as *hydroelectric (hydro* or *water-wheel) generators,* are built for a wide range of comparatively low turbine speeds, and are axially short, but have large diameters to accommodate the many salient poles on their rotors. Photo 8–2 shows a cutaway view of a hydroelectric generator. *Diesel-electric generators* are also low-speed machines with salient-pole rotors. The construction of synchronous machines may vary from one to another in several aspects. While the water-wheel-driven hydroelectric generators are commonly of the vertical type with vertically mounted salient-pole rotors, most of the others, including turbine-generators, are of the horizontal type. The constructional details also depend greatly on the type of cooling methods employed. The rotor-end windings or overhangs must be properly secured by the use of end-bells or retaining rings. The feasibility of using *superconducting windings*[1] on the rotor is being investigated for large turbogenerators.

---

[1]Jeffries, M. J., et al., "Prospects for Superconductive Generators in the Electric Utility Industry," *IEEE Transactions on Power Apparatus and Systems,* pp. 1659–69, September/October 1973; Smith, J. L., "Superconductors in Large Synchronous Machines," EPRI Research Project Report prepared at M.I.T., June 1975.

PHOTO 8.1 Cross-sectional view of a large turbogenerator with water-cooled stator and hydrogen-cooled rotor. Photo courtesy of Westinghouse Electric Corporation.

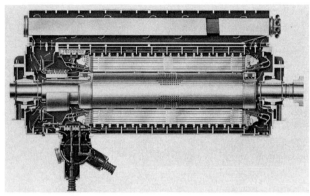

PHOTO 8.2 Cutaway view of a 718/826-MVA, 84-pole, 85.7-rpm, 15-kV hydroelectric generator. Photo courtesy of Canadian General Electric Company.

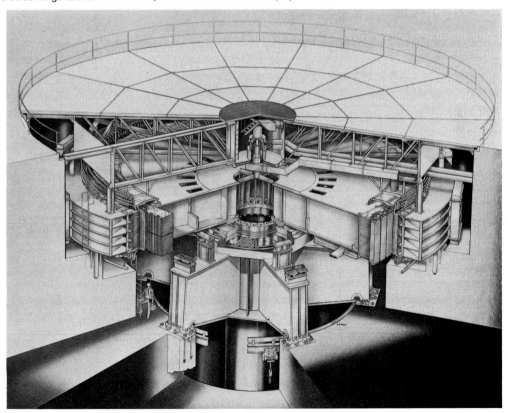

PHOTO 8.3   Brushless exciter assembly of a two-pole turbogenerator.
Photo courtesy of Electric Machinery Manufacturing Company.

The *stator* of a synchronous machine is essentially similar to that of a polyphase induction machine. The stator core consists of punchings (often built up of segmented sectors) of high-quality laminations having slot-embedded double-layer lap windings. Almost all synchronous generators are three-phase wye-connected machines with a 60° phase spread. The end-windings must be securely braced against movement under the impact of short-circuit electromagnetic forces.

Many salient-pole synchronous machines are commonly equipped with *damper (amortisseur) windings,* which consist usually of a set of copper or brass bars set in pole-face slots and connected together at the ends of the machine. (See Figure 5.3.13.) Amortisseur windings have some effect on stability, and serve several useful functions such as the starting of synchronous motors as induction motors, damping rotor oscillations, reducing overvoltages under certain short-circuit conditions, and aiding in synchronizing the machine. The cylindrical rotor of a turbine-generator may be considered as equivalent to an amortisseur of infinitely many circuits.

The dc winding on the rotating field structure may be connected to an external dc source through slip rings and brushes. The dc *excitation* may be provided by a self-excited dc generator, known as the *exciter,* mounted on the same shaft as the rotor of the synchronous machine. In slow-speed machines with large ratings, such as hydrogenerators, the exciter may not be self-excited; rather, a self-excited or a permanent-magnet type *pilot exciter* may be employed to activate the exciter. Because of the maintenance problems associated with direct-coupled dc generators beyond a certain rating, an alternative form of excitation is provided sometimes by silicon diodes and thyristors. The two types of *solid-state excitation* systems are *(a) static systems* having stationary diodes or thyristors that supply the excitation current to the rotor through brushes and slip rings, and *(b) brushless systems* (Photo 8–3) having shaft-mounted rectifiers rotating with the rotor, thereby eliminating the need for brushes and slip rings.

The severe electric and magnetic loadings in a synchronous machine produce heat that must be properly dissipated. The manner in which the active parts of a machine are cooled determines its overall physical size and structure. As stated earlier, in Section 5.7, the *cooling* problem is particularly serious in large turboalternators, where economy, mechanical requirements, transportation limitations, and erection and assembly problems demand compactness, especially for the rotor forging. *Direct-conductor-cooling* methods, using hydrogen or water as coolant, are employed both for the stator and the rotor of large turbogenerators. Such innovative cooling techniques have made it possible to achieve two-pole turbogenerator ratings of 2,000 MVA in a single unit. Superconductive cryo-turbogenerators in large sizes with liquid-helium cooled superconducting field windings should be realizable soon through research and development.[2]

Improved performance is continually sought through better mechanical as well as electromagnetic design for mechanical and electrical stability of the machine. Increased unit ratings bring difficult technological problems not only in the active part of the generator but also in other parts, particularly the end region. Innovative end-region configurations are being used to overcome the associated problems. New concepts, research, development, testing, and design studies continue to open up possibilities permitting further increases in unit ratings with high reliability at a reasonable cost.

8.2 EQUIVALENT CIRCUIT OF A SYNCHRONOUS MACHINE

A review of the material presented under "Elementary Synchronous Machines" in Section 5.3 is very helpful at this stage to recall the principle of operation of synchronous machines.

To start with, for the sake of simplicity, let us consider an unsaturated cylindrical-rotor synchronous machine. Effects of saliency and of magnetic saturation are considered later. Since we are concerned for the present with only the steady-state behavior of the machine, circuit parameters of the field and damper windings need not be considered. However, the effect of the field winding is taken care of by the flux produced by the dc field excitation and the ac voltage generated by the field flux in the armature circuit. As for the armature winding, under balanced conditions of operation, we shall analyze it on a *per-phase* basis. The armature winding obviously has a resistance; the *effective resistance, R_a,* of the armature per phase, which takes into account the effects of operating temperature and alternating currents (causing skin effect), is about 1.6 times the dc resistance measured at room temperature. The leakage reactance associated with the armature winding is caused by the leakage fluxes (caused by the currents in the conductors) that link the armature conductors only, but not the field winding. The leakage reactance may, for convenience, be divided into such components as end-connection leakage reactance and slot-leakage reactance. Let the total *leakage reactance* per phase of the armature winding be denoted by X_l, which includes the effects of not only the leakage across the armature slots and around the coil ends but also those associated with the space-harmonic fields present in the actual armature-mmf wave.

---

[2]Abegg, K., "The Growth of Turbogenerators," *Philosophical Transactions of the Royal Society of London,* vol. 275, no. 1248, pp. 51–68, 1973.

The resultant air-gap flux, $\overline{\phi}_r$, in the machine can be considered as the space-phasor sum of the component fluxes, $\overline{\phi}_f$ and $\overline{\phi}_{ar}$, produced respectively by the field and armature-reaction mmfs, caused by the field and armature currents respectively. From the viewpoint of the armature windings, these fluxes manifest themselves as generated emfs. Thus, let $\overline{E}_f$ be the ac voltage (known as the *excitation voltage*) generated by the field flux, and $\overline{E}_{ar}$ be the voltage generated by the armature-reaction flux. These are proportional to the field and armature currents respectively, each one lagging the flux generating it by 90°. Note that the armature-reaction flux $\overline{\phi}_{ar}$ is in phase with the armature current $\overline{I}_a$ producing it, and hence the armature-reaction emf $\overline{E}_{ar}$ lags the armature current by 90°. The resultant air-gap voltage $\overline{E}_r$ is then given by the phasor sum of the voltages $\overline{E}_f$ and $\overline{E}_{ar}$. Observe that the voltage $\overline{E}_r$ is then lagging behind the flux $\overline{\phi}_r$ by 90°. These combined space and time phasor relationships are shown in Figure 8.2.1 with rms values of the quantities involved, for the case of a generator and motor, with the armature current lagging behind the resultant air-gap voltage. The effect of armature reaction can be considered as that of an inductive reactance X_ϕ accounting for the component voltage generated by the space-fundamental flux created by the armature reaction. This reactance is known as the *magnetizing reactance* or *armature-reaction reactance*. The following may then be written relating the voltage phasors:

$$\overline{E}_f + \overline{E}_{ar} = \overline{E}_f - j\overline{I}_a X_\phi = \overline{E}_r \qquad (8.2.1)$$

The air-gap voltage $\overline{E}_r$ differs from the terminal voltage $\overline{V}_t$ by the voltage drops due to the armature resistance and leakage reactance. Thus, Figure 8.2.2 shows the equivalent circuits with all per-phase quantities, for the case of a cylindrical-rotor synchronous machine working either as a generator or a motor. The sum of the armature leakage reactance and armature-reaction reactance is known as the *synchronous reactance:*

$$X_s = X_\phi + X_l \qquad (8.2.2)$$

and $(R_a + jX_s)$ is called the *synchronous impedance* Z_s. Thus, the equivalent circuit for an unsaturated cylindrical-rotor synchronous machine under balanced polyphase conditions reduces to that shown in Figure 8.2.2(b), in which the machine is represented on a per-phase basis by its excitation voltage $\overline{E}_f$ in series with the synchronous impedance. The excitation voltage takes into account the flux produced by the field current, while the synchronous reactance takes into account all the flux produced by the balanced polyphase armature currents. For an unsaturated cylindrical-rotor machine at constant frequency, note that the synchronous reactance is a constant, and the excitation voltage is proportional to the field current. Figure 8.2.3 shows the corresponding combined space and time phasor diagrams of a cylindrical-rotor synchronous machine working either as a generator or a motor, for the case of a lagging power factor, with the armature current lagging behind the terminal voltage by an angle ϕ.

FIGURE 8.2.1 Phasor relationships among the fluxes and the corresponding voltages with the armature current lagging the resultant air-gap voltage. *(a)* Generator. (Field poles lead the resultant air-gap flux wave. The electromagnetic torque on the rotor acts in opposition to the rotation.) *(b)* Motor. (Field poles lag the resultant air-gap flux wave. The electromagnetic torque acts in the direction of rotation.)

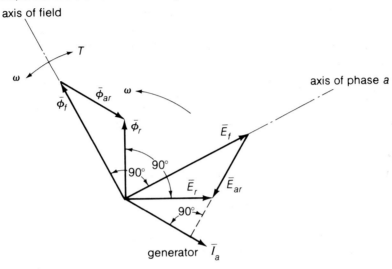

(a)

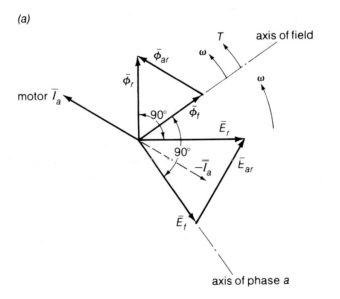

(b)

FIGURE 8.2.2 Per-phase equivalent circuits of a cylindrical-rotor synchronous machine.

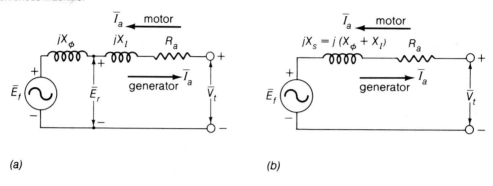

(a)   (b)

FIGURE 8.2.3 Phasor diagrams for a cylindrical-rotor synchronous machine, for the case of a lagging power factor, with the armature current lagging behind the terminal voltage. (a) Generator, (b) motor.

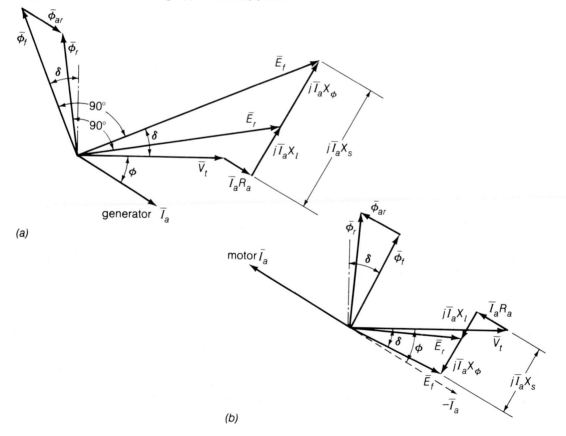

(a)

(b)

FIGURE 8.2.4 Four possible cases of operation of a round-rotor synchronous machine, with negligible armature resistance. *(a)* Overexcited generator (power factor lagging) P > 0; Q > 0; δ > 0; *(b)* underexcited generator (power factor leading) P > 0; Q < 0; δ > 0; *(c)* overexcited motor (power factor leading) P < 0; Q > 0; δ < 0; *(d)* underexcited motor (power factor lagging) P < 0; Q < 0; δ < 0.

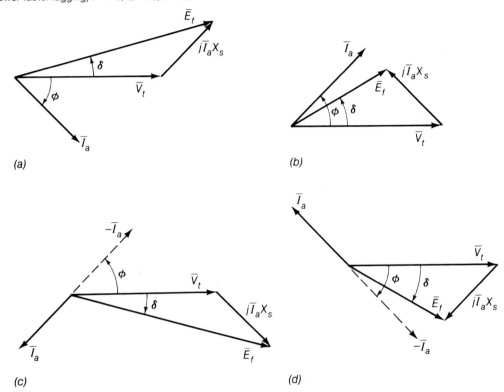

In all but small machines, the armature resistance is usually neglected except insofar as its effect on losses (hence the efficiency) and heating is concerned. With this simplification, Figure 8.2.4 shows the *four* possible cases of operation of a round-rotor synchronous machine, in which the following relation holds good:

$$\overline{V}_t + j\overline{I}_a X_s = \overline{E}_f \tag{8.2.3}$$

Observe that the motor armature current is taken in the opposite direction to that of the generator. The machine is said to be *overexcited* when the magnitude of the excitation voltage exceeds that of the terminal voltage; otherwise, it is said to be *underexcited*. The angle δ between the excitation voltage $\overline{E}_f$ and the terminal voltage $\overline{V}_t$ is known as the *torque angle* or the *power angle* of the synchronous machine. The power-angle performance characteristics are discussed later, in Section 8.4.

The voltage regulation of a synchronous generator at a given load, power factor, and rated speed is defined as

$$\text{Per-unit voltage regulation} = \frac{E_f - V_t}{V_t} \qquad (8.2.4)$$

where V_t is the terminal voltage on load, and E_f is the no-load terminal voltage at rated speed when the load is removed without changing the field current.

EXAMPLE 8.2.1

The per-phase synchronous reactance of a three-phase, wye-connected, 2.5-MVA, 6.6-kV, 60-Hz turboalternator is 10 ohms. The armature resistance may be neglected. Calculate the voltage regulation when the generator is operating at full load (a) with 0.8 power-factor lagging, and (b) with 0.8 power-factor leading. Neglect saturation.

SOLUTION

$$\text{Per-phase terminal voltage } V_t = \frac{6.6 \times 1{,}000}{\sqrt{3}} = 3{,}811 \text{ V}$$

Full-load per-phase armature current I_a
$$= \frac{2.5 \times 10^6}{\sqrt{3} \times 6.6 \times 1{,}000} = 218.7 \text{ A}$$

a. Referring to Figure 8.2.4(a) for the case of an overexcited generator operating at 0.8 power-factor lagging, and applying Equation 8.2.3, we have

$$\overline{E}_f = 3{,}811 + j218.7(0.8 - j0.6)(10) = 5{,}414\angle\tan^{-1} 0.3415$$

$$\text{Per-unit voltage regulation} = \frac{5{,}414 - 3{,}811}{3{,}811} = 0.42$$

b. Referring to Figure 8.2.4(b) for the case of an underexcited generator operating at 0.8 power-factor leading, and applying Equation 8.2.3, we have

$$\overline{E}_f = 3{,}811 + j218.7(0.8 + j0.6)(10) = 3{,}050\angle\tan^{-1} 0.7$$

$$\text{Per-unit voltage regulation} = \frac{3{,}050 - 3{,}811}{3{,}811} = -0.2$$

As seen above, the voltage regulation for a synchronous generator may become negative.

FIGURE 8.3.1 Open-circuit test: *(a)* diagram of connections, *(b)* open-circuit characteristic (OCC).

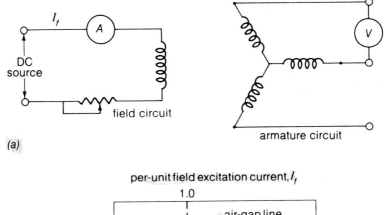

(a)

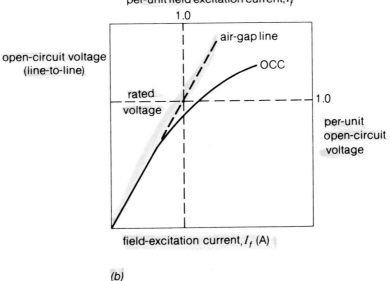

(b)

8.3 OPEN-CIRCUIT AND SHORT-CIRCUIT CHARACTERISTICS

For inclusion of saturation effects and determination of the appropriate machine parameters, two basic characteristic curves corresponding to open-circuit and short-circuit are considered for a synchronous machine. The *open-circuit characteristic* (OCC or the *saturation curve*) is a plot of the armature terminal voltage (usually line-to-line) on open-circuit as a function of the field-excitation current, when the machine is running at rated synchronous speed. See Figure 8.3.1*(a)* for a diagram of connections for the open-circuit test. The curve, indicated as OCC in Figure 8.3.1*(b)*, would be linear as represented by the air-gap line, if it were not for the magnetic saturation of the iron. The OCC is often plotted in per-unit terms, as shown by the alternate set of scales in Figure 8.3.1(b), where one per-unit voltage is the rated voltage and one per-unit field current is the excitation corresponding to the rated voltage on the air-gap line. On the basis of the present choice, the per-unit representation is such as to make the air-gap lines of all synchronous machines identical. Essentially, the OCC gives the relation between the space-fundamental component of the air-gap flux and the mmf of the magnetic circuit, when the field winding is the only mmf source.

The open-circuit characteristic is usually determined experimentally [see Figure 8.3.1(a)] by driving the synchronous machine mechanically at synchronous speed with its armature terminals on open-circuit and reading the terminal voltage corresponding to various values of dc field excitation current. The voltage may be stepped down by means of instrument potential transformers for measurement. From the no-load test, the no-load rotational losses (friction, windage, and core losses) can be determined. The friction and windage losses at synchronous speed are constant, while the core loss is a function of the flux in the machine at no-load, which in turn is proportional to the open-circuit voltage.

For the short-circuit test, the armature terminals are short-circuited through a current-measuring circuit, which may be an instrument current transformer with an ammeter in its secondary. See Figure 8.3.2(a) for a diagram of connections, in which current transformers are not shown. By driving the machine at rated synchronous speed, measurements of armature short-circuit current are made for different values of dc field excitation current; the field current is gradually increased until the armature current is about 1.5 to 2 times the rated current. The plot of short-circuit armature current versus the field current is known as the *short-circuit characteristic* indicated by SCC in Figure 8.3.2, and is practically linear because the magnetic-circuit iron is unsaturated. By drawing the phasor diagram (which is left as an exercise to the student) corresponding to short-circuit conditions, it can be seen that the armature current $\overline{I}_a$ lags the excitation voltage $\overline{E}_f$ by very nearly 90° (because the resistance is much smaller than the synchronous reactance); consequently the magnetic axes of the armature reaction and field are very nearly in line, but with the field and armature mmfs nearly opposing each other. The resultant air-gap flux is only about 0.15 of its normal-voltage value, and hence, the machine is operating in an unsaturated condition.

By representing the open-circuit and short-circuit characteristics on the same graph, as in Figure 8.3.3, the *unsaturated synchronous impedance* per phase is obtained by

$$Z_{s(ag)} = \frac{ot}{\sqrt{3}\ o'd}\ \text{ohms per phase} \tag{8.3.1}$$

where the subscript *(ag)* indicates air-gap line conditions, *ot* is the line-to-line terminal voltage (in volts) of the wye-connected synchronous machine, and *o'd* is the short-circuit armature line current in amperes corresponding to the dc field excitation current *oa*. By expressing the voltage and current in per-unit, the synchronous impedance may be expressed in per-unit.

FIGURE 8.3.2 Short-circuit test: *(a)* diagram of connections, *(b)* short-circuit characteristic (SCC).

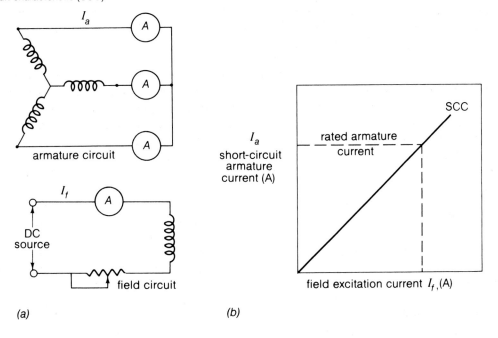

(a)

(b)

FIGURE 8.3.3 Open-circuit and short-circuit characteristics of a synchronous machine.

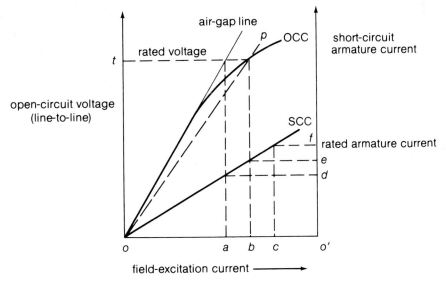

For operation at or near rated terminal voltage, it is sometimes assumed that the machine is equivalent to an unsaturated one whose saturation characteristic is a straight line through the origin and the rated voltage point on the OCC, as shown by the dashed line *op* in Figure 8.3.3. In this approximation, the *saturated synchronous impedance* at rated voltage is given by

$$Z_s = \frac{ot}{\sqrt{3}\ o'e} \text{ ohms per phase} \qquad (8.3.2)$$

This method of handling the effects of saturation usually gives fairly satisfactory results if great accuracy is not required. Since the armature resistance is usually much smaller than the synchronous reactance, the value of the synchronous impedance is often taken as the synchronous reactance itself.

The *short-circuit ratio* is defined as

$$\text{SCR} = \frac{ob}{oc} \qquad (8.3.3)$$

where *ob* is the field current required to generate rated voltage at rated speed on open-circuit, and *oc* is the field current required to produce rated armature current under a sustained three-phase short circuit. It can be shown that the short-circuit ratio is the reciprocal of the per-unit value of the saturated synchronous impedance at rated voltage. Saturation has the effect of increasing the short-circuit ratio.

EXAMPLE 8.3.1

The following data are obtained from the open-circuit and short-circuit characteristics of a three-phase, wye-connected, 135-MVA, 13.8-kV, 60-Hz hydrogen-cooled turbine generator with negligible armature resistance:

From the open-circuit characteristic:		
Line-to-line voltage	13.8 kV	17.3 kV
Field current	550 A	1,208 A
From the air-gap line:		
Line-to-line voltage	13.8 kV	
Field current	480 A	
From the short-circuit characteristic:		
Armature current	3,900 A	5,648 A
Field current	550 A	797 A

Compute *(a)* the unsaturated synchronous reactance, *(b)* the saturated synchronous reactance at rated voltage, *(c)* the short-circuit ratio, *(d)* the estimated field current for rated voltage and rated current at 0.8 power-factor lagging, and *(e)* the voltage regulation.

SOLUTION

a. The field current of 480 A required for rated line-to-line voltage of 13.8 kV on the air-gap line produces a short-circuit armature line current of 480 × 3,900/550 = 3,404 A. Hence, the unsaturated synchronous reactance is given by

$$X_{s(ag)} = Z_{s(ag)} = \frac{13.8 \times 1,000}{\sqrt{3} \times 3,404} = 2.34 \text{ ohms per phase}$$

In per-unit terms with the machine rating as a base, since the rated armature current is 135 × 10⁶/(√3 × 13,800) = 5,648 A and 3,404 A = 3,404/5,648 = 0.6 pu,

$$X_{s(ag)} = Z_{s(ag)} = \frac{1}{0.6} = 1.67 \text{ per unit}$$

b. The field current of 550 A produces rated voltage on the open-circuit characteristic and a short-circuit armature current of 3,900 A, from which the saturated synchronous reactance at rated voltage is given by

$$X_s = Z_s = \frac{13.8 \times 1,000}{\sqrt{3} \times 3,900} = 2.04 \text{ ohms per phase}$$

In per-unit terms, since 3,900 A = $\frac{3,900}{5,648}$ = 0.69 pu,

$$X_s = Z_s = \frac{1}{0.69} = 1.45 \text{ per unit}$$

c. The short-circuit ratio as given by Equation 8.3.3 is

$$\text{SCR} = \frac{550}{797} = 0.69$$

which is the same as the reciprocal of the per-unit saturated synchronous reactance at rated voltage.

d. We shall assume that the machine is equivalent to an unsaturated one whose saturation characteristic is a straight line through the origin and the rated voltage point on the OCC, as shown by the dashed line *op* in Figure 8.3.3.
Applying Equation 8.2.3, we get

$$\overline{E}_f = \frac{13,800}{\sqrt{3}} + j(5,648)(0.8 - j0.6)\, 2.04$$

$$= 7,968 + 6,913 + j9,218 = 14,881 + j9,218$$

$$= 17,505\angle\tan^{-1} 0.62$$

and the line-to-line magnitude of the induced voltage is

$$\sqrt{3} \times 17,505 = 30,319 \text{ V}$$

The field current required to produce this voltage on the line *op* in Figure 8.3.3 is

$$\frac{30,319 \times 550}{13,800} = 1,208 \text{ A}$$

e. The field current of 1,208 A produces a no-load line-to-line voltage of 17,300 V on the OCC. Hence the voltage regulation is

$$\text{Per-unit regulation} = \frac{17,300 - 13,800}{13,800} = 0.254$$

8.4 POWER-ANGLE AND OTHER PERFORMANCE CHARACTERISTICS

The real and reactive power delivered by a synchronous generator or received by a synchronous motor can be expressed in terms of the terminal voltage V_t, generated voltage E_f, synchronous impedance Z_s, and the power angle or torque angle δ. Referring to Figure 8.2.4, it is convenient to adopt a notation which makes the real power P and the reactive power Q delivered by an over-excited generator positive. Accordingly, the generator action corresponds to positive values of δ, while the motor action corresponds to negative values of δ. This is equivalent to omitting the negative sign in Equation 5.3.20, of course with the understanding that the electromagnetic torque acts in the direction to bring the interacting fields into alignment. With the adopted notation it follows that $P > 0$ for generator operation, while $P < 0$ for motor operation. Further, positive Q means delivering inductive vars for a generator action or receiving inductive vars for a motor action; negative Q means delivering capacitive vars for a generator action or receiving capacitive vars for a motor action. It may be observed from Figure 8.2.4 that the power factor is lagging when P and Q are of the same sign, and leading when P and Q are of opposite signs.

The complex power output of the generator in volt-amperes per phase is given by

$$S = P + jQ = \overline{V_t}\,\overline{I_a}^* \tag{8.4.1}$$

where $\overline{V_t}$ is the terminal voltage per phase, $\overline{I_a}$ is the armature current per phase, and the * indicates complex conjugate. Referring to Figure 8.2.4(a), in which the effect of armature resistance has been neglected, and taking the terminal voltage as reference, we have the terminal voltage,

$$\overline{V_t} = V_t + j0 \tag{8.4.2}$$

the excitation voltage or the generated voltage,

$$\overline{E_f} = E_f\,(\cos\delta + j\sin\delta) \tag{8.4.3}$$

the armature current,

$$\overline{I_a} = \frac{\overline{E_f} - \overline{V_t}}{jX_s} = \frac{(E_f\cos\delta - V_t) + jE_f\sin\delta}{jX_s} \tag{8.4.4}$$

where X_s is the synchronous reactance per phase.

$$\begin{aligned}
\overline{I_a}^* &= \frac{(E_f\cos\delta - V_t) - jE_f\sin\delta}{-jX_s} \\
&= \frac{E_f\sin\delta}{X_s} + j\,\frac{E_f\cos\delta - V_t}{X_s}
\end{aligned} \tag{8.4.5}$$

Therefore

$$P = \frac{V_t\,E_f\sin\delta}{X_s} \tag{8.4.0}$$

and

$$Q = \frac{V_t\,E_f\cos\delta - V_t^2}{X_s} \tag{8.4.7}$$

Equations 8.4.6 and 8.4.7 hold good for a cylindrical-rotor synchronous generator with negligible armature resistance. To obtain the total power for a three-phase generator, Equations 8.4.6 and 8.4.7 should be multiplied by 3 when the voltages are line-to-neutral. However, if the line-to-line magnitudes are used for the voltages, these equations give the total three-phase power.

FIGURE 8.4.1 Steady-state power-angle or torque-angle characteristic of a cylindrical-rotor synchronous machine (with negligible armature resistance).

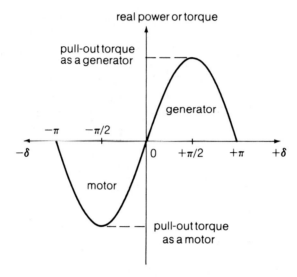

The power-angle or torque-angle characteristic of a cylindrical-rotor synchronous machine is shown in Figure 8.4.1, neglecting the effect of armature resistance. The maximum real power output per phase of the generator for a given terminal voltage and a given excitation voltage is

$$P_{max} = \frac{V_t\, E_f}{X_s} \qquad\qquad (8.4.8)$$

Any further increase in the prime-mover input to the generator causes the real-power output to decrease, and the excess power goes into accelerating the generator, thereby increasing its speed and causing it to pull out of synchronism. Hence the *steady-state stability limit* is reached when $\delta = \pi/2$. For normal steady operating conditions, the power angle or torque angle is well under 90°. The maximum torque or *pull-out torque* per phase that a round-rotor synchronous motor can develop for a *gradually applied* load is

$$T_{max} = \frac{P_{max}}{\omega_m} = \frac{P_{max}}{2\pi(n_s/60)} \qquad\qquad (8.4.9)$$

where n_s is the synchronous speed in rpm. The pull-out torque of a synchronous motor, according to the American Standard Definitions of Electrical Terms, is the maximum sustained torque that the motor will develop at synchronous speed for one minute, with rated voltage applied at rated frequency and with normal excitation.

FIGURE 8.4.2 Phasor diagram showing the effect of increasing the excitation of a synchronous generator, when the real power, frequency, and terminal voltage are constant.

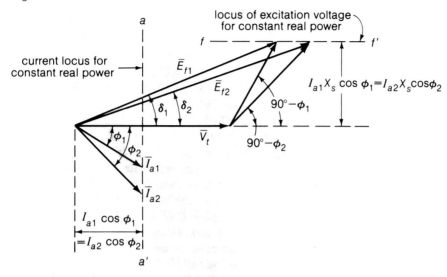

In the steady-state theory of the synchronous machine, with known terminal bus voltage V_t and a given synchronous reactance X_s, the operating variables are six in number, given by P, Q, δ, ϕ, I_a, and E_f. One can only write four independent equations relating these six variables, and hence the synchronous machine is said to have two degrees of freedom. The selection of any two, such as ϕ and I_a, P and Q, δ and E_f, determines the operating point and establishes the other four quantities.

The principal steady-state operating characteristics are the interrelations among the terminal voltage, field current, armature current, real power, reactive power, torque angle, power factor, and efficiency. These can be computed for application studies by means of phasor diagrams, such as those shown in Figure 8.2.4, corresponding to various conditions of operation. Let us, for example, consider the loci of the excitation voltage and armature current for constant real power and variable excitation of a synchronous generator operating at constant terminal voltage and frequency. Neglecting armature resistance and for a given synchronous reactance, Figure 8.4.2 shows the phasor diagram corresponding to two different values of excitation, when the real power, frequency, and terminal voltage are constant. For the real-power output of the generator, $P = V_t I_a \cos \phi$, to be a constant, with constant terminal voltage V_t, $(I_a \cos \phi)$ must be a constant. Thus, the locus of the current for constant real power is given by the vertical line aa' while the locus of the excitation voltage is given by the horizontal line ff'. Observe that the variation in the power-factor angle ϕ is quite significant, whereas the variation in the torque angle δ is only slight.

FIGURE 8.4.3 Phasor diagram showing the effect of increasing the prime-mover input when the field current, terminal voltage, and frequency are constant.

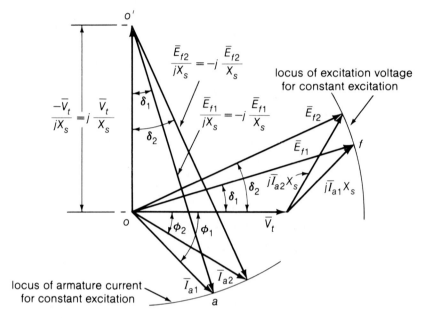

Next, let us consider the loci of the excitation voltage and armature current for constant excitation and variable real-power output of a synchronous generator operating at constant terminal voltage and frequency. Figure 8.4.3 shows the phasor diagram corresponding to two different values of the prime-mover input, when the field current, terminal voltage, and frequency are constant. The locus of excitation voltage for constant excitation is the circle with o as center and the line-segment of as radius, while the locus of the armature current is given by the circle with o' as center and the line-segment $o'a$ as radius, since, neglecting armature resistance,

$$\overline{V}_t + j\overline{I}_a X_s = \overline{E}_f \tag{8.4.10}$$

or

$$\overline{I}_a = \frac{\overline{E}_f}{jX_s} - \frac{\overline{V}_t}{jX_s} = j\frac{\overline{V}_t}{X_s} - j\frac{\overline{E}_f}{X_s} \tag{8.4.11}$$

in which V_t and X_s are constant, and E_f is also a constant for constant excitation. Observe that an increase in the real-power output ($V_t I_a \cos \phi$) obtained by an increase in the prime-mover input is associated with a decrease in the reactive-power output ($V_t I_a \sin \phi$).

FIGURE 8.4.4 Generator compounding curves.

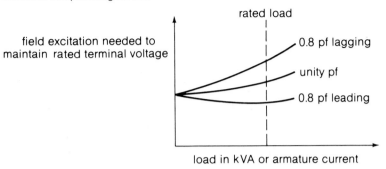

FIGURE 8.4.5 Synchronous motor V-curves.

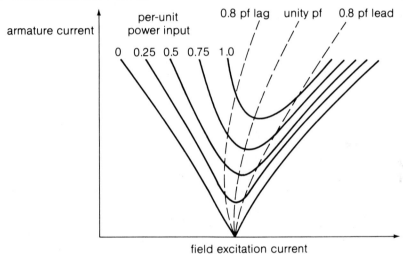

A *generator-compounding curve* is a curve showing the field current required to maintain rated terminal voltage as the constant-power-factor load is varied. Three compounding curves at constant power factors of unity, 0.8 power-factor lagging, and 0.8 power-factor leading are shown in Figure 8.4.4.

For synchronous-motor operation, the power factor as well as the armature current can be controlled by adjusting the field excitation. A *synchronous-motor-V-curve* is a curve showing the relation between the armature current and field current at a constant terminal voltage and with a constant shaft load. Figure 8.4.5 shows a family of V-curves, which are so named because of their characteristic V-shape. It may be observed that, for constant power output, the armature current is a minimum at unity power factor and increases as power factor decreases. The loci of

FIGURE 8.4.6 Synchronous-generator constant-field-current volt-ampere characteristic curves.

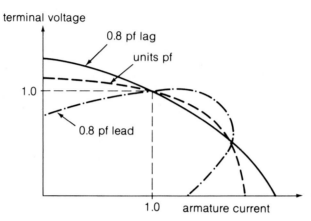

constant power factors are shown by the dashed lines, which are known as *synchronous-motor compounding curves.* If the effect of armature resistance is neglected, the motor and generator compounding curves would be identical except that the lagging and leading-power-factor curves would be interchanged.

A *synchronous condenser* is a synchronous motor running idle with an overexcited field, which may be used in a power system solely for power-factor correction or for control of reactive-power flow. Such a machine in a larger size may be more economical than static capacitors. The V-curve for zero mechanical power in Figure 8.4.5 approximately represents the synchronous-condenser operation. When rotational losses and armature resistance are neglected, the real power of a synchronous condenser is zero with an associated zero torque angle, and the reactive-power output is given by Equation 8.4.7 with cos δ equal to unity.

An electrical system supplied by only one synchronous generator is known as an *isolated system,* for which the concept of an infinite bus having constant voltage and constant frequency does not apply, as there are no other parallel synchronous generators to compensate for changes in field excitation and in prime-mover output in order to maintain constant terminal voltage and frequency. If the generator is driven at a constant speed, thereby maintaining constant frequency, and the field current is increased, the terminal voltage increases; this effort in general is associated with an increase in the real and reactive-power output to an isolated system. Similarly, an increase in prime-mover output, at constant field excitation, generally produces an increase in frequency, terminal voltage, and real as well as reactive power. Figure 8.4.6 shows the *generator constant-field-current volt-ampere characteristics* for three constant power factors and three different values of constant field current. In each case, the field current chosen is the value required to yield rated terminal voltage at rated armature current, as given by the compounding curves of Figure 8.4.4.

FIGURE 8.4.7 Synchronous machine operating modes.

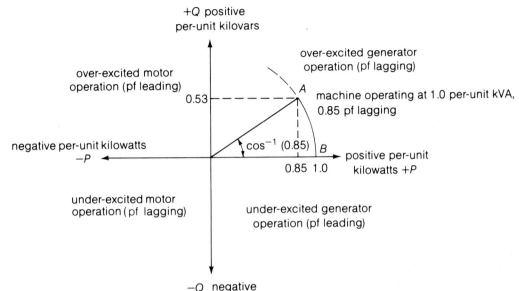

Synchronous generators are usually rated in terms of the maximum kVA or MVA load at a specific voltage and power factor such as 0.8, 0.85, or 0.9 lagging, which they can deliver continuously without overheating. Let us now take a look at the generator capabilities within which the machine may be operated, and the effects of field excitation on the individual machine as well as on the system of which that machine is a part. As mentioned earlier, a synchronous machine can be operated in any quadrant of Figure 8.4.7 in which point A is plotted corresponding to the nameplate rated generator conditions for an assumed typical 0.85 power-factor lagging. Note that in Figure 8.4.7 the positive kilovars $(+Q)$ are taken along the positive y-axis, and hence the lagging-power-factor condition corresponding to an overexcited generator is shown in the first quadrant, on the upper right-hand side. We shall limit our discussion primarily for generator operation and therefore restrict our attention to the right-hand or positive-kilowatt side $(+P)$ of Figure 8.4.7. Since it is usual for generators to be suitable for delivering rated kilovolt-amperes at unity power factor, corresponding to point B in Figure 8.4.7, we may draw the arc AB with its center at 0 and radius equal to rated armature current.

Starting with rated terminal voltage $\overline{V}_t$ and rated armature current $\overline{I}_a$, Figure 8.4.8 gives a typical phasor diagram corresponding to a lagging power-factor angle of ϕ for the case of an overexcited cylindrical-rotor generator with negligible armature resistance. $\overline{E}_f$ is then the internal (generated) voltage corresponding to the rated terminal conditions. It would also be equal to the terminal voltage if full load were removed without making any change in field current.

FIGURE 8.4.8   Typical phasor diagram of an overexcited round-rotor generator with negligible armature resistance.

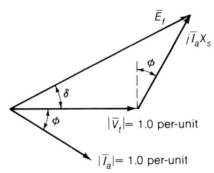

$|\bar{V}_t| = 1.0$ per-unit

$|\bar{I}_a| = 1.0$ per-unit

Dividing the three sides of the phasor triangle by X_s and reorienting the diagram to suit the kilowatt-kilovar coordinates of Figure 8.4.9, triangle OAC is constructed, in which OA represents the rated armature current, OC corresponds to $(1/X_s)$ or the short-circuit ratio, and CA represents the rated full-load field current, E_f/X_s. Then with C as center and CA as radius, the arc AD is drawn, representing the locus of the rated field current, thereby closing off the top of the area within which the machine may be operated. Thus the output of the synchronous generator is limited by *field heating* from D to A, and by *armature heating* from A to B. The vertical intercept OD gives the maximum permissible per-unit kilovars to scale, which is a measure of the capability as a synchronous condenser.

Corresponding to the underexcited generator operation in the fourth quadrant, since this is a region of low field current, one might be tempted to extend the armature current arc AB all the way around to E. But this could be wrong from the viewpoints of system stability as well as localized heating in the machine iron. Modern generators can be operated successfully in the underexcited region down to a line such as HJ in Figure 8.4.9, where the point H is typically at 60 percent rated kVA with zero power-factor leading and the point J is typically at rated kVA with 0.95 power-factor leading. The line HJ will then represent the *end-region heating limit,* which applies to steam-turbine generators in particular, while it is not a limitation of water-wheel generators because of their generally different construction. The reason for the problem of end-region heating stems from the armature reaction end-leakage flux at both ends of the stator.

If the machine under consideration is connected to an infinitely large system through negligibly small impedance, then the *stability limit* may be represented by a straight line CF corresponding to a δ of 90°. Operation along the arc EK in Figure 8.4.9 is impossible as the machine will not remain in synchronism with the system, even if it did not exceed any heating limitation. Since the machine operates in practice through impedance representing transformers, lines, and the paralleled value of the impedances of all the other machines on the system, the resultant external impedance is typically about 0.2 to 0.4 per unit based on the individual machine rating. The effect of this external impedance X_e is to bend upward the line CF to some position such as CG. Thus the part of the boundary dictated by the steady-state stability can be established. With no voltage regulator, the curve CG of Figure 8.4.9 will be bent further upward.

FIGURE 8.4.9 Synchronous generator capability limits.

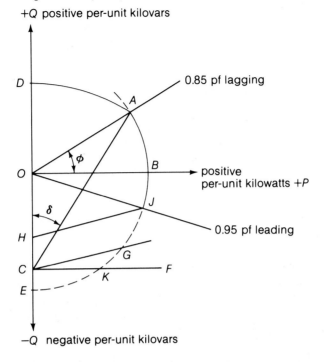

$$OC = \frac{V_t}{X_s} = \frac{1.0}{X_s} = \text{SCR (typically 0.8)}$$

$$OA = \frac{I_a X_s}{X_s} = I_a = 1.0 \text{ per-unit}$$

$$CA = \frac{E_f}{X_s} = I_{fd}$$

DA: field-current limit

AJ: armature-current limit

JH: armature end-region heating limit

CK: steady-state stability limit (with $X_e = 0$ and voltage regulator)

CG: steady-state stability limit (with $X_e > 0$ and voltage regulator)

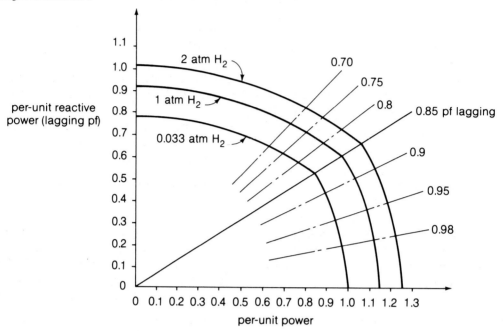

FIGURE 8.4.10 Reactive-power capability curves for a hydrogen-cooled turbine generator, 0.85 pf lagging, 0.8 SCR, with rated kVA at 0.04 atm hydrogen as base kVA.

The permissible generator-operating region is bounded by the limits determined by the field heating, the armature heating, the armature end-region heating, and the steady-state stability. It can be seen from Figure 8.4.9 that an increase of the generator load, without corresponding adjustment of the field current, pushes the power factor of the machine toward the leading (underexcited) region, and an increase of excitation causes the generator power factor to lag more. The more the initial load power factor lags, the greater is the total load the machine can carry before running into one of the limits. Figure 8.4.10 shows a typical set of reactive-power capability curves for a large hydrogen-cooled turbine-generator, along with the effect of increased hydrogen pressure on allowable machine loadings.

The *efficiency* of a synchronous generator at a specified power output and power factor is determined by the ratio of the output to the input; the input power is given by adding the machine losses to the power output. The efficiency is conventionally computed in accordance with a set of rules agreed upon by the ANSI. The losses included in the computation are the following:

a. Armature winding copper loss for all the phases, calculated after correcting the dc resistance of each phase for an appropriate allowable temperature rise depending on the class of insulation used.

b. Field copper loss, based on the field current and measured field-winding dc resistance, corrected for temperature like armature resistance. Note that the losses in field rheostats used to adjust the generated voltage are not charged to the synchronous machine.

c. Core loss, which is read from the open-circuit core-loss curve at a voltage equal to the internal voltage behind the resistance of the machine.

d. Friction and windage loss.

e. Stray load losses, which account for the fact that the effective ac resistance of the armature is greater than the dc resistance because of the skin effect, and for the losses caused by the armature leakage flux.

f. Exciter loss, only if the exciter is an integral component of the alternator, i.e., shares a common shaft or is permanently coupled. Nonintegral exciter losses are not charged to the alternator.

EXAMPLE 8.4.1

A synchronous generator with negligible resistance and a synchronous reactance of 1.0 per unit is connected to an infinite bus of 1.0 per-unit voltage through two parallel three-phase transmission circuits, each having negligible resistance and a reactance of 0.6 per unit, including step-up and step-down transformers at each end, as shown below. The generator rating is chosen as the base for expressing the per-unit values.

a. Under steady-state operation, compute the generator terminal voltage, the excitation voltage, the real-power output, and the reactive power delivered to the infinite bus, when the machine is delivering rated current at unity power factor at its terminals.

b. Let the throttle of the prime mover be adjusted so that there is no real-power transfer between the generator and the infinite bus. Determine the generator terminal and excitation voltages when the generator-field current is adjusted for delivering 0.5 per unit lagging reactive kVA to the infinite bus.

c. Let the system be returned to the operating conditions of part *(a)*, and let one of the two parallel transmission circuits be disconnected by tripping the circuit breakers at its ends, while the generator excitation is kept constant, as in part *(a)*. Calculate the maximum power transfer possible under these conditions, and check whether the generator will remain in synchronism.

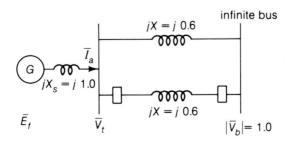

SOLUTION

a. With unity power factor at the generator terminals and a generator current I_a of 1.0 per unit, the phasor diagram is sketched below:

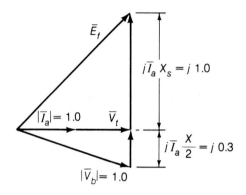

Note that the equivalent reactance of the two transmission circuits in parallel is 0.3 per unit.

$$\text{Generator terminal voltage } V_t = \sqrt{(1.0)^2 - (0.3)^2} = \sqrt{0.91}$$
$$= 0.954 \text{ per unit}$$

$$\text{Generator excitation voltage } E_f = \sqrt{(0.954)^2 + (1.0)^2}$$
$$= \sqrt{1.91} = 1.382 \text{ per unit}$$

$$\text{Real power output} = P = V_t I_a \cos \phi = 0.954 \text{ per unit}$$

Reactive power at infinite bus is 0.3 per unit, with the current $\overline{I}_a$ leading the infinite-bus voltage $\overline{V}_b$.

b. The phasor diagram corresponding to the conditions stated in the problem is given below:

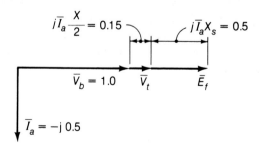

c. When one of the transmission circuits is disconnected, the transmission-circuit reactance is 0.6 pu, and the total reactance including the synchronous reactance is then 1.6 pu. With E_f of 1.382 as in (a), the maximum power transfer possible is

$$P_{max} = \frac{E_f V_b}{(X + X_s)} = \frac{(1.382)(1.0)}{1.6} = 0.864 \text{ pu}$$

which is less than the original power of 0.954 in part (a).

The generator overspeeds and loses synchronism, unless quick-response excitation system and governor action come into play.

EXAMPLE 8.4.2

A 1,000-hp, 2,300-V, wye-connected, three-phase, 60-Hz, 20-pole synchronous motor, for which cylindrical-rotor theory may be used and all losses may be neglected, has a synchronous reactance of 5.00 ohms per phase.

a. The motor is operated from an infinite bus supplying rated voltage at rated frequency, and its field excitation is adjusted so that the power factor is unity when the shaft load is such as to require an input of 750 kW. Compute the maximum torque that the motor can deliver, if the shaft load is slowly increased with the field excitation held constant.

b. Instead of the infinite bus as in part (a), let the power to the motor be supplied by a 1,000-kVA, 2,300-V, wye-connected, three-phase, 60-Hz synchronous generator whose synchronous reactance is also 5.00 ohms per phase. The generator is driven at rated speed, and the field excitations of the generator and motor are so adjusted that the motor absorbs 750 kW at unity power factor and rated terminal voltage. If the field excitations of both machines are then held constant and the mechanical load on the synchronous motor is gradually increased, compute the maximum motor torque under the conditions. Also determine the armature current, terminal voltage, and power factor at the terminals corresponding to this maximum load.

c. Calculate the maximum motor torque if, instead of remaining constant as in part (b), the field currents of the generator and motor are gradually increased so as to always maintain rated terminal voltage and unity power factor while the shaft load is increased.

SOLUTION

The solution neglects reluctance torque since cylindrical-rotor theory is applied.

a. The equivalent circuit and the corresponding phasor diagram for the given conditions are shown below, with subscript m attached to the motor quantities:

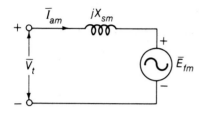 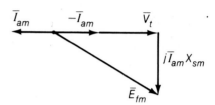

Rated voltage per phase $V_t = \dfrac{2,300}{\sqrt{3}} = 1,328$ V line-to-neutral

Current per phase $I_{am} = \dfrac{750 \times 10^3}{\sqrt{3} \times 2,300 \times 1.0} = 188.3$ A

$I_{am} X_{sm} = 188.3 \times 5 = 941.5$ V

$E_{fm} = \sqrt{(1,328)^2 + (941.5)^2} = 1,628$ V

$P_{max} = \dfrac{E_{fm} V_t}{X_{sm}} = \dfrac{(1,628)(1,328)}{5} = 432,397$ watts/phase

or 1,297,191 watts for three phases

or $\dfrac{1,297.2}{1,000 \times 0.746} = 1.74$ per unit

Synchronous speed $= \dfrac{120 \times 60}{20} = 360$ rpm or 6 rps

$\omega_s = 2\pi \times 6 = 37.7$ rad/s

$T_{max} = \dfrac{1,297.2 \times 10^3}{37.7} = 34,408$ newton-meters

b. With the synchronous generator as the power source, the equivalent circuit and the corresponding phasor diagram for the given conditions are shown below, with subscript *g* attached to the generator quantities:

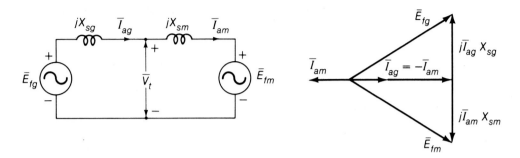

$$E_{fg} = E_{fm} = 1{,}628 \text{ V}$$

$$P_{max} = \frac{E_{fg} E_{fm}}{X_{sg} + X_{sm}} = \frac{(1{,}628)(1{,}628)}{10} = 265 \text{ kW per phase}$$

or 795 kW for three phases

or $\dfrac{795}{746} = 1.07$ per unit.

$$T_{max} = \frac{795 \times 10^3}{37.7} = 21{,}088 \text{ N·m}$$

If a load torque greater than this were applied to the motor shaft, synchronism would be lost; the motor would stall, the generator would tend to overspeed, and the circuit would be opened by circuit-breaker action.

Corresponding to the maximum load, the angle between $\overline{E}_{fg}$ and $\overline{E}_{fm}$ is 90°; from the phasor diagram it follows that

$$V_t = \frac{E_{fg}}{\sqrt{2}} = \frac{1{,}628}{\sqrt{2}} = 1{,}151.3 \text{ V line-to-neutral}$$

or 1,994 V line-to-line

$$I_{ag} X_{sg} = 1{,}151.3 \text{ or } I_{ag} = \frac{1{,}151.3}{5} = 230 \text{ A}$$

The power factor is unity at the terminals.

c. $V_t = 1{,}328$ V and the angle between $\overline{E}_{fg}$ and $\overline{E}_{fm}$ is 90°. Hence it follows that

$$E_{fg} = E_{fm} = (1{,}328)\sqrt{2} = 1{,}878 \text{ V}$$

$$P_{max} = \frac{E_{fg}\,E_{fm}}{X_{sg} + X_{sm}} = \frac{(1{,}878)(1{,}878)}{10} = 352.7 \text{ kW/ph.}$$

or 1,058 kW for three phases.

$$T_{max} = \frac{1{,}058 \times 10^3}{37.7} = 28{,}064 \text{ N·m}$$

EXAMPLE 8.4.3

A three-phase, 350-kVA, 3,300-V, 60-Hz, six-pole, wye-connected synchronous machine, for which cylindrical rotor theory may be applied and all losses may be neglected, has a synchronous reactance of 10.0 ohms per phase.

a. When it operates as a generator delivering full load at rated voltage and a lagging power factor of 0.9, construct the circle diagram and hence determine the excitation voltage and the load angle.
b. Compute the current and power factor when the machine is motoring at maximum torque.
c. Determine the current and torque when the machine is operating as a motor at 0.6 power-factor leading.
d. With the excitation voltage adjusted to the same value as the terminal voltage, find the power factor when the machine is running as a generator delivering 80 A at rated voltage.

SOLUTION

a. Full-load current $= \dfrac{350 \times 10^3}{\sqrt{3} \times 3{,}300} = 61.24$ A

(Voltage per phase)/$X_s = \dfrac{3{,}300}{\sqrt{3}} \times \dfrac{1}{10} = 190.5$ A

$\cos^{-1} 0.9 = 25.9°$

In the diagram shown here for this example, following the construction of Figure 8.4.9, *OC* represents 190.5 A, while *OA* represents 61.24 A at an angle of 25.9° from the horizontal.

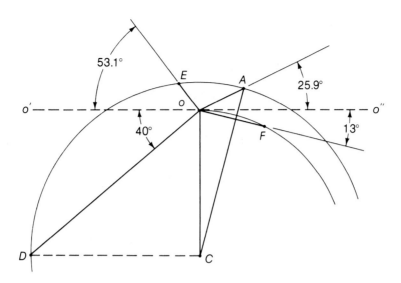

The excitation voltage is proportional to the closing phasor *CA:*

$$226 \times 10 \times \sqrt{3} = 3{,}914.3 \text{ V}$$

The load angle δ is given by the angle *OCA:* 14°.

b. As seen from Figure 8.4.7, the motoring operation occurs in second and third quadrants. The maximum torque corresponds to a torque angle of 90°. With *C* as center and *CA* as radius draw a circle and let angle *OCD* be 90°, in which case *OD* gives the corresponding current of 293 A and the *lagging* power-factor angle is given by the angle *O'OD* of 40°. The power factor is then 0.766 lagging.

c. Draw *OE* at an angle of $\cos^{-1} 0.6 = 53.1°$ from the horizontal as shown. The corresponding current is given by *OE:* 45 A. The power is then given by $\sqrt{3}$ × 3,300 × 45 × 0.6 = 154,321 watts. With six poles at 60 Hz, the synchronous speed is 1,200 rpm or 20 rps. Hence

$$T = \frac{154{,}321}{2\pi \times 20} = 1{,}227.55 \text{ newton-meters}$$

d. With *C* as center and *CO* as radius, construct a circle and draw *OF* equal to 80 A. The machine is operating as a generator with a *leading* power-factor angle *O''OF,* equal to 13°. The power factor is then 0.974 leading.

All the above answers could have been calculated analytically; but the graphical solution is quicker, although less accurate. The student is encouraged to solve the problem analytically and compare the results.

8.5   EFFECTS OF SALIENCY: TWO-REACTANCE THEORY OF SALIENT-POLE SYNCHRONOUS MACHINES

Because of saliency, the reactance measured at the terminals of a salient-pole synchronous machine varies as a function of the rotor position, while this is not so in a cylindrical-rotor machine. The effects of saliency can be taken into account by the two-reaction theory proposed by Blondel and extended by Doherty, Nickle, Park and others.[3] The armature current $\overline{I}_a$ is resolved into two components: $\overline{I}_d$ in time quadrature with, and $\overline{I}_q$ in time phase with the excitation voltage $\overline{E}_f$, as shown in Figure 8.5.1(a) for an unsaturated salient-pole generator operating at a lagging power factor. The component $\overline{I}_d$ of the armature current produces a space-fundamental component armature-reaction flux $\overline{\phi}_{ad}$ along the axes of the field poles, while the component $\overline{I}_q$ produces a space-fundamental component armature-reaction flux $\overline{\phi}_{aq}$ in space quadrature with the field poles. Direct-axis mmfs act on the main magnetic circuit with their magnetic effect centered on the axes of the field poles, whereas the quadrature-axis quantities have their magnetic effect centered on the interpolar axes. The armature-reaction flux $\overline{\phi}_{ar}$, for the case of an unsaturated machine, is given by the space-phasor sum of the components $\overline{\phi}_{ad}$ and $\overline{\phi}_{aq}$. The resultant flux $\overline{\phi}_r$, in the machine is given by the space-phasor sum of $\overline{\phi}_{ar}$ and the main-field flux $\overline{\phi}_f$, as in Figure 8.5.1(a). The angle ψ between $\overline{E}_f$ and $\overline{I}_a$ is known as the *internal power-factor angle*, and β is the space angle between the fundamentals of the ϕ_f and ϕ_r waves.

Similar to the magnetizing reactance X_ϕ of the cylindrical-rotor theory, the inductive effects of the direct- and quadrature-axis armature-reaction flux waves can be accounted for by the direct- and quadrature-axis magnetizing reactances $X_{\phi d}$ and $X_{\phi q}$, respectively. The *direct- and quadrature-axis synchronous reactances* are then given by

$$X_d = X_l + X_{\phi d} \qquad (8.5.1)$$

$$X_q = X_l + X_{\phi q} \qquad (8.5.2)$$

where X_l is the armature leakage reactance, assumed to be the same for both the direct- and quadrature-axis currents. Following the analysis presented in Section 8.2, one has the excitation voltage $\overline{E}_f$ equal to the phasor sum of the terminal voltage $\overline{V}_t$, the armature-resistance drop $\overline{I}_a R_a$, and the synchronous-reactance drops ($j\overline{I}_d X_d + j\overline{I}_q X_q$):

$$\overline{E}_f = \overline{V}_t + \overline{I}_a R_a + j\overline{I}_d X_d + j\overline{I}_q X_q \qquad (8.5.3)$$

This is shown in Figure 8.5.1(b), in which δ is the torque angle or the power angle, ϕ is the power-factor angle, and ($\phi + \delta$) is the internal power factor angle ψ.

---

[3]Blondel, A., "The two-reaction method for study of oscillatory phenomena in coupled alternators," *Revue générale de l'électricité*, vol. 13, pp. 235–51, 515–31, February/March 1923; Doherty, R. E., and C. A. Nickle, "Synchronous Machines I: An extension of Blondel's two-reaction theory," *AIEE Transactions*, vol. 45, pp. 927–42, 1926; Park, R. H., "Two-reaction Theory of Synchronous Machines—Part I," *AIEE Transactions*, vol. 48, pp. 716–27, 1929.

FIGURE 8.5.1 Phasor diagrams of an unsaturated salient-pole synchronous generator operating at a lagging power factor.

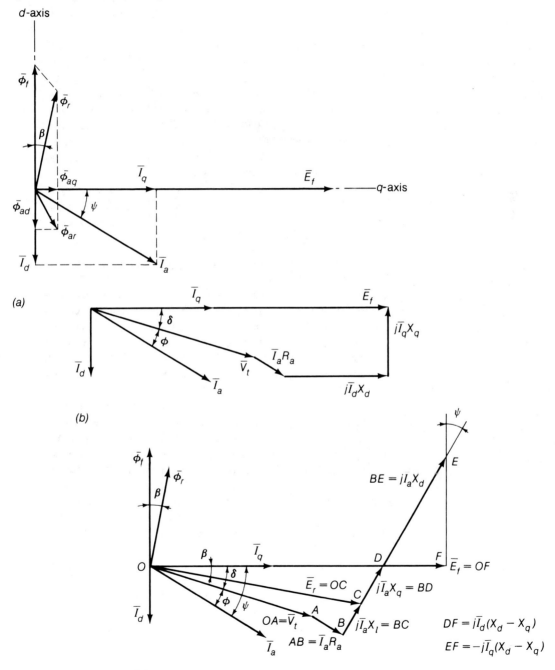

(a)

(b)

$$DF = j\bar{I}_d(X_d - X_q)$$
$$EF = -j\bar{I}_q(X_d - X_q)$$

(c)

Because of the greater reluctance of the air gap in the quadrature axis, X_q is less than X_d and is usually of the order of 0.6 to 0.7 times X_d. The general range of X_d is anywhere from 0.6 to 2.2 per unit, while X_q typically has the range of 0.4 to 1.4 per unit with the machine kVA rating as the base. Even though turbogenerators are essentially cylindrical-rotor machines, a small salient-pole effect is present because of the effect of the rotor slots on the quadrature-axis reluctance.

Once the internal power-factor angle ψ or $(\phi + \delta)$ is known, it is straightforward to draw Figure 8.5.1(b) by resolving the armature current into its d- and q-axis components. However, it is the external power-factor angle ϕ at the machine terminals that is usually known explicitly, rather than the internal power-factor angle. Neglecting the armature resistance, it can be shown that

$$\tan \delta = \frac{I_a X_q \cos \phi}{V_t + I_a X_q \sin \phi} \qquad (8.5.4)$$

Figure 8.5.1(c) shows the construction of the diagram for finding $\overline{E}_f$ from the given values of V_t, I_a, ϕ, R_a, X_d, and X_q, while satisfying Equation 8.5.3. The phasor sum $(\overline{V}_t + \overline{I}_a R_a + j\overline{I}_a X_q)$ given by OD determines the angular position of the q-axis. Hence, the d-axis, which is perpendicular to the q-axis, can be located. By adding DF equal to $j\overline{I}_d(X_d - X_q)$ to OD, one can get the excitation voltage $\overline{E}_f$ for the chosen terminal conditions. Note that in Figure 8.5.1(c) DE is an extension of BD, DF is an extension of OD, and EF is perpendicular to OF. Triangle DEF will vanish if $X_d = X_q$, as in the case of a cylindrical-rotor machine.

Let us next consider how the saliency affects the power-angle characteristic of a synchronous machine. Neglecting the armature resistance, the real power output or the developed power P is given by $V_t I_a \cos \phi$. From Figure 8.5.1(b) it follows that

$$I_q X_q = V_t \sin \delta \qquad (8.5.5)$$

$$I_d X_d = E_f - V_t \cos \delta \qquad (8.5.6)$$

$$I_d = I_a \sin(\phi + \delta) \qquad (8.5.7)$$

$$I_q = I_a \cos(\phi + \delta) \qquad (8.5.8)$$

Solving for $(I_a \cos \phi)$, one obtains

$$I_a \cos \phi = \frac{E_f}{X_d} \sin \delta + \frac{V_t}{2X_q} \sin 2\delta - \frac{V_t}{2X_d} \sin 2\delta \qquad (8.5.9)$$

and

$$P = V_t I_a \cos \phi = \frac{V_t E_f}{X_d} \sin \delta$$
$$+ \frac{1}{2} V_t^2 \left(\frac{1}{X_q} - \frac{1}{X_d} \right) \sin 2\delta \qquad (8.5.10)$$

FIGURE 8.5.2   Steady-state power-angle characteristic of a salient-pole synchronous machine (with negligible armature resistance).

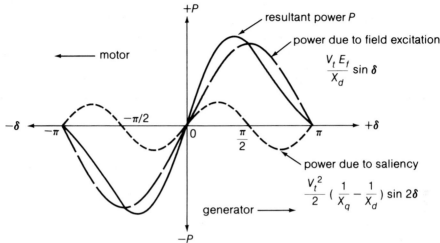

Figure 8.5.2 shows the power-angle characteristic of a salient-pole synchronous machine given by Equation 8.5.10. The resulting power is composed of two terms: one due to field excitation given by the first term of Equation 8.5.10, and the other due to saliency given by the second term of Equation 8.5.10. If $X_d = X_q$, Equation 8.5.10 reduces to Equation 8.4.6. Because of the reluctance torque, for equal voltages and equal values of X_d, a salient-pole machine develops a given torque at a smaller value of δ, and the maximum torque that can be developed is somewhat greater. Hence a salient-pole machine is said to be *stiffer* than one with a cylindrical rotor. Equation 8.5.10 is applicable both for generator and motor operation, with δ being positive for the former and negative for the latter.

8.6 DETERMINATION OF REACTANCES BY TEST DATA

Referring to Figure 8.5.1(b), the terminal current $\overline{I}_a$ may be expressed as follows while taking the terminal voltage $\overline{V}_t$ as reference:

$$\overline{I}_a = \frac{E_f - V_t \cos \delta}{X_d} e^{j[\delta-(\pi/2)]} + \frac{V_t \sin \delta}{X_q} e^{j\delta} \qquad (8.6.1)$$

which may be rewritten as

$$\overline{I}_a = V_t \frac{X_d + X_q}{2X_d X_q} e^{j\pi/2} + V_t \frac{X_d - X_q}{2X_d X_q} e^{j(2\delta - \pi/2)}$$
$$+ \frac{E_f}{X_d} e^{j(\delta - \pi/2)} \qquad (8.6.2)$$

Thus the salient-pole synchronous machine delivers three components of current to the bus. The second term on the right-hand side of Equation 8.6.2 vanishes for the case of a round-rotor synchronous machine, as X_d equals X_q. Even for no excitation (i.e., $E_f = 0$), in the case of a salient-pole synchronous machine, it can be seen that the current has an active component. With sufficient excitation, the current may be made equal to zero; for example, at no-load, for $\delta = 0$, with $E_f = V_t$, I_a becomes zero. Further, it may be seen that, for $E_f = 0$,

$$I_a = I_{max} = \frac{V_t}{X_q}, \text{ for } \delta = \frac{\pi}{2} \qquad (8.6.3)$$

and

$$I_a = I_{min} = \frac{V_t}{X_d}, \text{ for } \delta = 0 \qquad (8.6.4)$$

The current wave is an amplitude-modulated wave, provided δ is varied slowly.

The above analysis suggests a method of determining the direct- and quadrature-axis steady-state, or synchronous, reactances of a synchronous machine by a test known as the *slip test*. The machine is unexcited and balanced voltages are applied at the armature terminals. The rotor is driven at a speed differing slightly from synchronous speed, which is easily calculated from the frequency of the applied voltages and the number of poles of the machine. The armature currents are then modulated at slip frequency by the machine, having maximum amplitude when the quadrature axis is in line with the mmf wave and minimum amplitude when the direct axis aligns with the mmf wave. The armature voltages are also usually modulated at slip frequency because of impedances in the supply lines, the amplitude being greatest when the current is smallest and vice versa. Such variations of voltage and current are illustrated in the oscillograms of Figure 8.6.1. The maximum and minimum values of the voltage and current can also be read on a voltmeter and an ammeter, provided the slip is small. The field winding should be kept open in the slip test so that the slip-frequency current is not induced in it.

The direct-axis synchronous reactance X_d may now be calculated from the ratio of maximum voltage to minimum current, i.e., the ratio of applied volts per phase to the armature amperes per phase for the direct-axis position. The quadrature-axis synchronous reactance X_q is given by the ratio of minimum voltage to maximum current, i.e., the ratio of applied volts per phase to the armature amperes per phase for the quadrature-axis position. For the test to be successful, the slip must be sufficiently small.

The open-circuit and short-circuit tests discussed in Section 8.3 can also be used to determine the direct-axis armature reactance X_d from measurements of the steady-state armature open-circuit voltage and short-circuit current.

FIGURE 8.6.1 Slip-test oscillogram *(a)* armature voltage variation, *(b)* armature current variation.

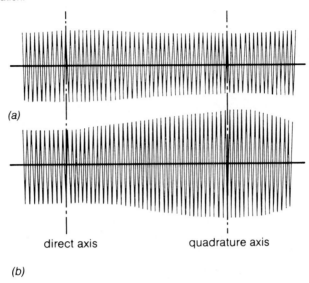

(a)

direct axis   quadrature axis

(b)

8.7 EFFECT OF SATURATION ON REACTANCES AND REGULATION

While both the synchronous reactance X_s of a round-rotor machine and X_d of a salient-pole machine are affected by saturation, the reactance X_q is usually assumed to be unaffected by saturation. Equations 8.3.1 and 8.3.2 based on open-circuit and short-circuit characteristics yield, respectively, the unsaturated and saturated values of the synchronous reactance, as explained in Section 8.3. However, closer approximations of the saturated synchronous reactance can be obtained for round-rotor as well as salient-pole machines from their zero-power-factor and open-circuit characteristics.

The *zero-power-factor rated-current saturation curve* is a graph of armature terminal voltage versus field current, with the armature current held constant at its rated full-load value and lagging power-factor angle constant at or near 90°. Figure 8.7.1 shows the no-load and zero-power-factor rated-current saturation curves and the *Potier triangle ABC* or *A'B'C'* (so named after its inventor). With the assumptions that the leakage reactance is a constant, the armature resistance is negligibly small, and the open-circuit characteristic is also the load saturation curve, the zero-power-factor characteristic would be a curve of exactly the same shape as the no-load saturation curve shifted vertically downward by an amount equal to the leakage-reactance voltage drop and horizontally to the right by an amount equal to the armature-reactance mmf. *O'A'* in Figure 8.7.1 is parallel to *OA*, which is a part of the air-gap line. The *Potier reactance X_p*, as determined from

FIGURE 8.7.1 No-load and zero-power-factor rated-current saturation curves of a synchronous machine and the Potier triangle. (Note: $\overline{E}_p$ in general is a phasor combination of $\overline{V}_t$ and $jX_p\overline{I}_a$ as shown in Figure 8.7.2.)

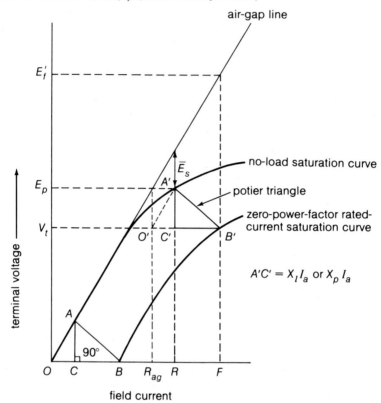

the vertical side (AC or $A'C' = X_p\,I_a$) of the Potier triangle replaces the leakage reactance X_l, and the voltage behind the Potier reactance $\overline{E}_p$ (obtained by the phasor combination of $\overline{V}_t$ and jX_p $\overline{I}_a$, as shown in Figure 8.7.2) is the air-gap voltage, of which saturation is assumed now to be a function. E_f of Figure 8.5.1 is proportional to the field current of an unsaturated machine. $\overline{E}_s$, the correction for saturation, is a function of $\overline{E}_p$, as shown in Figure 8.7.1, and would be parallel to $\overline{E}_p$ in the phasor diagram. $\overline{E}_f'$ is a phasor summation of the voltages $\overline{E}_f$ and $\overline{E}_s$. However, for convenience, $\overline{E}_s$ is sometimes taken parallel to $\overline{E}_f$ itself, as shown in Figure 8.7.2, rather than to $\overline{E}_p$; the theoretical justification of this is possibly greater with a salient-pole machine than with a non-salient-pole machine. E_f', a fictitious armature voltage, is now proportional to the field current of a saturated machine. Figure 8.7.2 shows the phasor diagram of a saturated salient-pole synchronous generator in the steady state, while that of a non-salient-pole synchronous generator is left to the reader's imagination. The per-unit Potier reactance X_p is of the order of 0.15 for the case of turboalternators and of 0.32 for water-wheel generators as well as synchronous condensers.

FIGURE 8.7.2 Steady-state phasor diagram of a salient-pole synchronous generator, while taking the effect of saturation into account approximately.

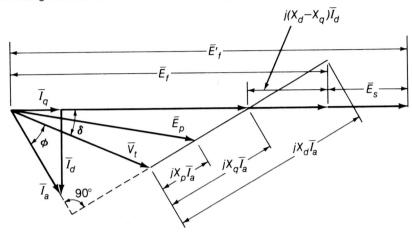

While X_p is assumed equal to X_l in a number of cases, the ratio (X_p/X_l) is about 1.3 for round-rotor machines and may be as much as 3 for some salient-pole machines with long slender poles and sizable field-leakage flux.

Referring to Figure 8.7.1, a *saturation factor* is defined by the ratio

$$k = R/R_{ag} \tag{8.7.1}$$

and the saturated synchronous reactance X_{ds} is obtained in terms of the unsaturated synchronous reactance X_{du} as follows:

$$X_{ds} = X_p + \frac{X_{du} - X_p}{k} \tag{8.7.2}$$

The calculation of voltage regulation with the use of saturated synchronous reactance is illustrated in Example 8.3.1.

Since it is now an almost universal practice to operate synchronous generators with automatic voltage regulators to hold the voltage constant under varying loads, a knowledge of the voltage regulation is of less importance than a knowledge of the change in the field excitation that is required to maintain constant voltage under varying loads. However, both involve the same factors for consideration.

8.8   PARALLEL OPERATION OF INTERCONNECTED SYNCHRONOUS GENERATORS

In order to assure continuity of the power supply within prescribed limits of frequency and voltage at all the load points scattered over the service area, it becomes necessary in any modern power system to operate several alternators in parallel, interconnected by various transmission lines, in a well-coordinated and optimized manner such that the operation is most economical. A generator can be paralleled with an infinite bus (or another generator running at rated voltage and frequency supplying the load) by driving it at synchronous speed corresponding to the system frequency and adjusting its field excitation so that its terminal voltage equals that of the bus. If the frequency of the incoming machine is not exactly equal to that of the system, the phase relation between its voltage and the bus voltage will vary at a frequency equal to the difference between the frequencies of the machine and bus voltages. In normal practice this difference can be made usually very small, to a fraction of a hertz; in polyphase systems, it is essential that the same phase sequence be maintained on either side of the synchronizing switch. Thus *synchronizing* requires the following conditions of the incoming machine: *(a)* correct phase sequence, *(b)* phase voltages in phase with those of the system, *(c)* frequency almost exactly equal to that of the system, and *(d)* machine terminal voltage approximately equal to the system voltage.

A *synchroscope* or a *bright-lamp synchronizing method* or a *dark-lamp synchronizing method*[4] is used for indicating the appropriate moment for synchronization. After the machine has been synchronized and is now a part of the system, it can be made to take its share of the active and reactive power by appropriate adjustments of its prime-mover throttle and field rheostat. The system frequency and the division of active power among the generators are controlled by means of prime-mover throttles regulated by governors and automatic frequency regulators, whereas the terminal voltage and reactive volt-ampere division among the generators are controlled by voltage regulators acting on the generator-field circuits and by transformers with automatic tap-changing devices.

EXAMPLE 8.8.1

Two three-phase, 6.6-kV, wye-connected synchronous generators operating in parallel supply a load of 3,000 kW at 0.8 power-factor lagging. The synchronous impedance per phase of machine A is $(0.5 + j10)\ \Omega$ and of machine B is $(0.4 + j12)\ \Omega$. The excitation of machine A is adjusted so that it delivers 150 A at a lagging power factor, and the governors are so set that the load is shared equally between the two machines. Determine the armature current, power factor, excitation voltage, and power angle of each machine.

[4]For details, see Lawrence, R. R., and H. E. Richards, *Principles of Alternating-Current Machinery*, 4th ed. (New York: McGraw-Hill, 1953), chap. 29; Langsdorf, A. S., *Theory of Alternating-Current Machinery*, 2d ed. (New York: McGraw-Hill, 1955), chap. 12.

SOLUTION

One phase of each generator and one phase of the equivalent wye of the load are shown in the figure below:

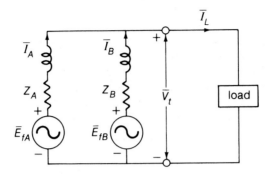

The load current $\bar{I}_L$ is calculated as follows:

$$\bar{I}_L = \frac{3{,}000}{\sqrt{3} \times 6.6 \times 0.8}\angle-\cos^{-1} 0.8 = 328(0.8 - j\,0.6)$$

$$= (262.4 - j196.8) \text{ A}$$

For machine A,

$$\cos \phi_A = \frac{1{,}500}{\sqrt{3} \times 6.6 \times 150} = 0.875 \text{ lagging; } \phi_A = 29°;$$

$$\sin \phi_A = 0.485$$

$$\bar{I}_A = 150(0.874 - j0.485) = (131.1 - j\,72.75) \text{ A}$$

$$\bar{I}_B = \bar{I}_L - \bar{I}_A = (131.3 - j124) = 180.6\angle-\cos^{-1}\left(\frac{131.3}{180.6}\right)$$

$$\cos \phi_B = \frac{131.3}{180.6} = 0.726 \text{ lagging}$$

With the terminal voltage $\overline{V}_t$ as reference, one has

$$\overline{E}_{fA} = \overline{V}_t + \overline{I}_A Z_A = (6.6/\sqrt{3})$$
$$+ (131.1 - j72.75)(0.5 + j10) \times 10^{-3}$$
$$= (4.6 + j1.27) \text{ kV per phase}$$

Power angle $\delta_A = \tan^{-1}(1.27/4.6) = 15.4°$

Line-to-line excitation
voltage for machine $A = \sqrt{3} \sqrt{(4.6)^2 + (1.27)^2} = 8.26$ kV

$$\overline{E}_{fB} = \overline{V}_t + \overline{I}_B Z_B = (6.6/\sqrt{3})$$
$$+ (131.3 - j124)(0.4 + j12) \times 10^{-3}$$
$$= (5.35 + j1.52) \text{ kV per phase}$$

Power angle $\delta_B = \tan^{-1}(1.52/5.35) = 15.9°$

Line-to-line excitation
voltage for machine $B = \sqrt{3} \sqrt{(5.35)^2 + (1.52)^2} = 9.6$ kV

The corresponding phasor diagram is sketched in the figure below:

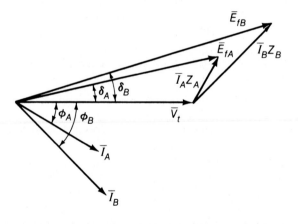

FIGURE 8.9.1   Generator connected to an infinite bus.

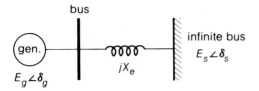

FIGURE 8.9.2   Positive-sequence reactance diagram corresponding to the system shown in Figure 8.9.1.

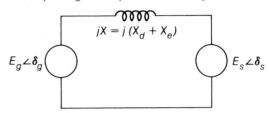

8.9   STEADY-STATE STABILITY

The property of a power system that ensures that it will remain in operating equilibrium under both normal and abnormal conditions is known as *power-system stability. Steady-state stability* is concerned with slow and gradual changes, while *transient stability* is concerned with severe disturbances such as sudden changes in load or fault conditions. The largest possible flow of power through a particular point without the loss of stability is known as the *steady-state stability limit* when the power is increased gradually, while it is referred to as the *transient-stability limit* when a sudden disturbance occurs.

When the power system is subjected to disturbances small enough and gradual enough so that the system can be regarded electrically as being in steady state, the steady-state equations of the machines can be utilized as an approximation for the steady-state stability analysis, while neglecting the effect of the excitation system. Such an approach is justified only in the case of generators equipped with non-continuously acting automatic voltage regulators that cannot act in time to keep the machine stable because of the inherent dead-bands and time delays. The ability of interconnected synchronous machines to remain in synchronism is analyzed under small and gradual disturbances affecting the system, without representing the excitation-system behavior.

The effect of resistance is usually neglected. When a machine is connected through appreciable impedance to other machines in the system, as is most often the case, the saliency effects can be neglected for all practical calculations. In terms of steady-state stability, the effect of saturation in machines is usually more significant than saliency effects.

For a generator connected to a very large system compared to its own size, the system in Figure 8.9.1 may be used and the corresponding positive-sequence reactance diagram is shown in Figure 8.9.2. Recall that, for the case of a single machine, the load angle δ is the angle between the terminal voltage and the q-axis. However, for a system in which several machines are interconnected, it is more convenient to measure an angle between the quadrature axis of each machine and a common reference axis that rotates at synchronous speed at all times. Let δ_g and δ_s, associated respectively with voltages $\overline{E}_g$ and $\overline{E}_s$ of Figure 8.9.1, be such angles measured from a common reference axis. The power-angle equation for the system under consideration becomes

$$P = \frac{E_g E_s}{X} \sin \delta_{gs} \qquad (8.9.1)$$

FIGURE 8.9.3   Power-angle curve of Equation 8.9.1.

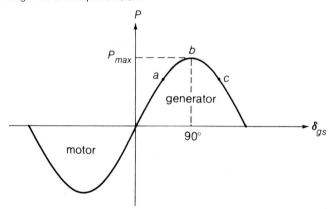

where δ_{gs} is the angular difference between δ_g and δ_s. The power-angle characteristic given by Equation 8.9.1 is plotted in Figure 8.9.3. Let us now examine the operating conditions at three points *a, b, c,* shown in Figure 8.9.3.

At the operating point *a,* the slope of the curve $dP/d\delta_{gs}$ is positive. If there is a small increase of the angle δ_{gs}, the electrical power can be seen to increase and will exceed the mechanical power input, if the mechanical power input corresponding to the initial power input at the operating point *a* remains constant. This results in deceleration, which will make the angle approach the initial operating point *a.* Then the machine is said to be stable. The peak of the power-angle curve given by P_{max} is known as the *steady-state power limit,* representing the maximum power that can be theoretically transmitted in a stable manner. A zero slope may be observed corresponding to such an operating condition shown at point *b* in Figure 8.9.3. A machine is usually operated at less than the power limit, thereby leaving a steady-state margin, as otherwise a slight increase in the angle δ_{gs} would lead to instability. The operation at point *c* shown in Figure 8.9.3 is clearly unstable, where the slope of the curve is negative. A slight increase in the angle δ_{gs} leads to a decrease in electrical power, which becomes smaller than the mechanical power. This in turn would accelerate the machine, thereby making the angle larger and losing synchronism eventually.

Steady-state stability analysis neglecting saturation leads to unduly pessimistic estimates regarding the ability of a machine in a system to remain in synchronism when small and gradual disturbances occur. The effect of saturation is usually taken into account by modifying the synchronous reactance X_d suitably, as discussed in Section 8.7. The saturated value usually lies within the range of 0.4 to 0.8 of the unsaturated synchronous reactance, smaller values being attributed to the newer and larger turbogenerators. It is also a common practice for overexcited operation to choose the saturated value as $(0.8X_d)$ for light loads and $(0.6X_d)$ for full loads.

FIGURE 8.10.1   Essential elements of an automatic feedback control system.

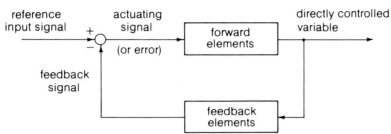

The slope $(dP/d\delta_{gs})$ of the power-angle curve gives a measure of the steady-state stability of a system, and this is most frequently used to check the steady-state stability of synchronous machines in a system. The test is made between pairs of machines that are electrically most remote from each other, since this is the situation most likely to give rise to instability. If the slopes of both machines are positive, the system is stable; if either slope is negative, the system is unstable. This criterion leads to conservative estimates, particularly when one of the tested machines is much larger than the other. Refined methods, presented in texts on stability,[5] may then be used.

The steady-state stability limits of a system may be increased either by an increase in the excitation of the generator or by a reduction in the reactance of the network, as seen from Equation 8.9.1. Installation of parallel transmission lines contributes to an increase in the stability limit and the dependability of the system. Series capacitors are sometimes used in lines either to improve the voltage regulation or to raise the stability limit by effectively decreasing the line reactance. Continuously acting automatic voltage regulators can raise the steady-state stability limits considerably. Such regulators may be represented mathematically by differential equations, and the overall performance—including the regulator, excitation system, generator, and system transient characteristics—may be evaluated by solving the equations for small changes.

8.10 EXCITATION SYSTEMS

The source of field current for the excitation of a synchronous machine is an excitation system that includes the exciter, regulator, and manual control. The modern excitation control system of a large synchronous machine is a *feedback control system,* whose essential elements are illustrated in Figure 8.10.1. Figure 8.10.2 shows the general layout of the components included in an excitation system. The use of solid-state components and the creation of new excitation systems prompted the requirement for new definitions.[6]

---

[5]See, for example, Kimbark, E. W., *Power-System Stability,* vols. 1, 2, and 3 (New York: John Wiley and Sons, 1948).

[6]IEEE Committee Report, "Proposed Excitation System Definitions for Synchronous Machines," *IEEE Transactions on Power Apparatus and Systems,* vol. PAS-88, no. 8, August 1969, pp. 1248–58. (Figures 8.10.1 and 8.10.2 are adapted from the above IEEE Committee Report.)

FIGURE 8.10.2 General layout of the components included in an excitation system.

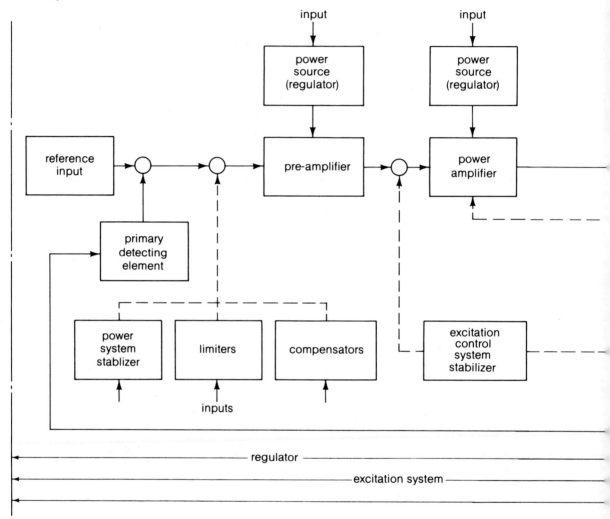

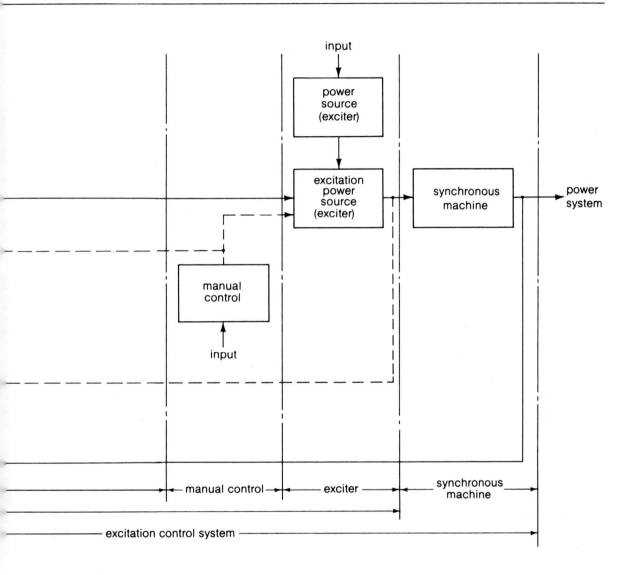

The main *exciter* could be one of the following:

a. DC generator-commutator exciter, whose energy is derived from a dc generator with its commutator and brushes. The exciter may be driven by a motor, prime mover, or the shaft of the synchronous machine.

b. Alternator-rectifier exciter, whose energy is derived from an alternator and converted to dc by rectifiers. The power rectifiers may be either controlled or noncontrolled; they may be either stationary or rotating with the alternator shaft. The alternator may be driven by a motor, prime mover, or the shaft of the synchronous machine.

c. Compound-rectifier exciter, whose energy is derived from the currents and potentials of the ac terminals of the synchronous machine and converted to dc by rectifiers. The exciter includes the power transformers (current and potential), power reactors, and power rectifiers, which may be either noncontrolled or controlled, including the gate circuitry.

d. Potential-source-rectifier exciter, whose energy is derived from a stationary ac potential source and converted to dc by rectifiers. The exciter includes the power potential transformers, if used, and power rectifiers, which may be controlled or noncontrolled, including the gate circuitry.

A synchronous-machine *regulator* couples the output variables of the synchronous machine to the input of the exciter through feedback and forward controlling elements for the purpose of regulating the synchronous machine output variables. The regulator generally consists of an error detector, preamplifier, stabilizers, auxiliary inputs, and limiters. A regulator may be continuously acting or noncontinuously acting; the former type initiates a corrective action for a sustained in-finitesimal change in the controlled variable, while the latter requires a sustained finite change in the controlled variable to initiate corrective action. Regulators in which the regulating function is accomplished by mechanically varying a resistance are known as rheostatic-type regulators.

Power-system stabilizers provide additional input to the regulator to improve power-system dynamic performance. Quantities such as shaft speed, frequency, and synchronous machine power may be used as input to the power-system stabilizer. Other signal modifiers include *limiters* and *compensators*. Limiters act to limit a variable by modifying or replacing the function of the primary detector element when predetermined conditions have been reached. Examples include maximum excitation, minimum excitation, and maximum volts per hertz. Compensators act to compensate for the effect of a variable by modifying the function of the primary detecting element. Examples include reactive-current compensators, active-current compensators, and line-drop compensators. Generator-voltage regulators are equipped with reactive compensators in order to share reactive current among generators operating in parallel. Line-drop compensators introduce within the regulator-input circuit a voltage equivalent to the impedance drop and thereby modify the generator voltage. The voltage drops of the resistive and reactive portions of the impedance are obtained respectively by active and reactive compensators.

The *excitation-control-system stabilizer* modifies the forward signal by either series or feedback compensation to improve the dynamic performance of the excitation-control system; undesired oscillations and overshoot of the regulated voltage are eliminated.

The *exciter-control system* provides the proper field voltage to maintain a desired system voltage. An important characteristic of such a control system is its ability to respond rapidly to voltage deviations during both normal and abnormal or emergency system operations. While many different types of exciter-control systems are employed in practice, the basic principle of operation may be described in the following way: The voltage deviation signal is amplified to produce the signal required to change the exciter field current; that change in turn produces a change in exciter-output voltage, thereby resulting in a new level of excitation for the synchronous generator.

BIBLIOGRAPHY

Del Toro, V. *Electromechanical Devices for Energy Conversion and Control Systems*. Englewood Cliffs, N.J.: Prentice-Hall, 1968.

Fitzgerald, A. E.; Kingsley, C., Jr.; and Kusko, A. *Electric Machinery*. 3d ed. New York: McGraw-Hill, 1971.

Hindmarsh, J. *Electrical Machines and Their Applications*. 2d ed. New York: Pergamon, 1970.

Institute of Electrical and Electronics Engineers Committee Report. "Proposed Excitation System Definitions for Synchronous Machines." *IEEE Transactions on Power Apparatus and Systems,* vol. PAS-88, no. 8, August 1969, pp. 1248–58.

Institute of Electrical and Electronics Engineers. *Standard No. 115–1965, Code SH00554, Test Procedures for Synchronous Machines*. New York, 1965.

Kimbark, E. W. *Power System Stability: Synchronous Machines*. New York: Dover, 1968.

Knowlton, A. E., ed. *Standard Handbook for Electrical Engineers*. 8th ed. New York: McGraw-Hill, 1949.

Matsch, L. W. *Electromagnetic and Electromechanical Machines*. New York: Intext, 1972.

Nasar, S. A., and Unnewehr, L. E. *Electromechanics and Electric Machines*. New York: John Wiley & Sons, 1979.

Puchstein, A. F.; Lloyd, T. C.; and Conrad, A.G. *Alternating-Current Machines*. 3d ed. New York: John Wiley & Sons, 1954.

Sarma, M. S. *Synchronous Machines (Their Theory, Stability and Excitation Systems)*. New York: Gordon and Breach, 1979.

Say, M. G. *Alternating-Current Machines*. New York: John Wiley & Sons, Halsted Press, 1976.

Slemon, G. R., and Straughen, A. *Electric Machines*. Reading, Mass.: Addison-Wesley, 1980.

Westinghouse Electric Corporation Central Station Engineers. *Electrical Transmission and Distribution Reference Book*. 4th ed. East Pittsburgh, Pa.: Westinghouse Electric Corporation, 1950.

PROBLEMS

8-1.

A three-phase, wye-connected, cylindrical-rotor synchronous generator rated at 10 kVA, 230 V has a synchronous reactance of 1.5 ohms per phase and an armature resistance of 0.5 ohm per phase.

   a. Determine the voltage regulation at full load
      (i) with 0.8 lagging power factor.
      (ii) with 0.8 leading power factor.
   b. Calculate the power factor for which the voltage regulation becomes zero on full load.

8-2.

The following readings are taken from the results of an open-circuit and a short-circuit test on a 10-MVA, 3-phase, wye-connected, 13.8-kV, two-pole turbine generator driven at synchronous speed:

Field current	170 A	200 A
Armature current, short-circuit test	418 A	460 A
Line-to-line voltage, open-circuit test	13,000 V	13,800 V
Line-to-line voltage, air-gap line	15,500 V	17,500 V

The armature resistance may be neglected.

   a. Determine the unsaturated value of the synchronous reactance in ohms per phase and also in per unit.

   b. Compute the saturated value of the synchronous reactance in ohms per phase and also in per unit.

   c. Find the short-circuit ratio.

   d. Calculate the field current required at rated voltage, rated kVA, and 0.8 lagging power factor, while accounting for saturation under load.

8-3.

The loss data for the synchronous generator of Problem 8–2 are given below:

Open-circuit core loss at 13.8 kV is 70 kW
Short-circuit load loss at 418 A, 75°C is 50 kW
Friction and windage loss is 80 kW
Field-winding resistance at 75°C is 0.3 ohms
Stray-load loss at full load is 20 kW

Determine the efficiency of the generator at rated load, rated voltage, and 0.8 power factor lagging.

8-4.

The following data are obtained from the open-circuit and short-circuit characteristics of a three-phase, wye-connected, four-pole, 150-MW, 0.85-pf, 12.6-kV, 60-Hz, hydrogen-cooled turbine generator with negligible armature resistance:

Open-circuit characteristic:

Field current, A	200	300	400	500	600	700	800	900
Line-to-line terminal voltage, kV	3.8	5.8	7.8	9.8	11.3	12.6	13.5	14.2

Short-circuit characteristic:

Armature current, A	4,043	8,086
Field current, A	350	700

Determine *(a)* the unsaturated synchronous reactance, *(b)* saturated synchronous reactance at rated voltage, *(c)* short-circuit ratio, *(d)* estimated field current and regulation for rated voltage, rated current at 0.85 power-factor lagging, and *(e)* estimated field current and regulation for rated voltage, rated current at 0.85 power-factor leading.

8–5.

A 4,000-V, 5,000-hp, 60-Hz, 12-pole synchronous motor with a synchronous reactance of 4 ohms per phase (based on cylindrical-rotor theory) is excited to produce unity power factor at rated load. Neglect all losses.

    a. Find the rated and maximum torques.

    b. What is the armature current corresponding to the maximum torque?

8–6.

A synchronous motor with $X_d = 1.0$ per unit has a rated power factor of 0.8 leading. Neglecting the effect of saliency and losses, determine the ratio of the pull-out torque when the field is excited for 0.8 leading power factor at rated mechanical load to that when the motor delivers its rated load at unity power factor.

8–7.

A three-phase, wye-connected, four-pole, 400-V, 60-Hz, 15-hp synchronous motor has a synchronous reactance of 3 ohms per phase and negligible armature resistance. The data for its no-load magnetization curve are given below:

Field current, A	2	3.5	4.4	6	8	10	12
Line-to-neutral voltage, V	100	175	200	232	260	280	295

    a. When the motor operates at full load at 0.8 leading power factor, determine the power angle and the field current. Neglect all the losses.

    b. Compute the minimum line current for the motor operating at full load, and the corresponding field current.

    c. When the motor runs with an excitation of 10 A while taking an armature current of 25 A, calculate the power developed and the power factor.

    d. If the excitation is adjusted such that the magnitudes of the excitation voltage and terminal voltage are equal, and if the motor is taking 20 A, find the torque developed.

8–8.

Equation 8.4.6 in the text is derived by neglecting the effect of armature resistance. Including the effect of armature resistance R_a, show that the power developed per phase by a round-rotor synchronous machine is given by

$$P = -\frac{E_f V_t}{Z_s}\cos(\delta + \theta) + E_f^2\,\frac{R_a}{Z_s^2}$$

where V_t is the terminal voltage per phase, E_f is the excitation voltage per phase, Z_s is the synchronous impedance per phase, θ is the impedance angle, and δ is the power angle.

8-9.

Consider the per-phase equivalent circuit shown below of a three-phase synchronous generator supplying a synchronous motor through reactors:

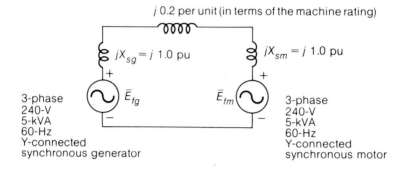

j 0.2 per unit (in terms of the machine rating)

$jX_{sg} = j$ 1.0 pu $jX_{sm} = j$ 1.0 pu

$\bar{E}_{fg}$ $\bar{E}_{fm}$

3-phase
240-V
5-kVA
60-Hz
Y-connected
synchronous generator

3-phase
240-V
5-kVA
60-Hz
Y-connected
synchronous motor

The motor drives a mechanical load, the torque of which is such that the mechanical power is 3.2 kW at rated speed.

   a. If the field currents of the two machines are adjusted so that rated voltage is applied to the motor terminals at rated frequency with a motor power factor of 0.8 lagging, determine the internally generated emfs E_{fg} and E_{fm}.
   b. With the generator field current as in part *(a)*, what is the minimum value of the motor-excitation voltage for which the machines will remain in synchronism?

8-10.

A turbogenerator rated at 200 MVA, 11 kV, 0.9 lagging power factor has a synchronous reactance of 1.7 per unit and a field current of 300 A for rated load. Draw an operating chart (circle diagram) for the machine when running on 11-kV infinite bus bars. Find from the diagram the field current required when the machine delivers 100 MW at 0.8 lagging power factor. Determine also the load angle and the armature current.

8-11.

A three-phase, wye-connected, 2,300-V, 60-Hz round-rotor synchronous motor has a synchronous reactance of 2 ohms per phase and negligible armature resistance.

   a. If the motor takes a line current of 350 A operating at 0.8 power factor leading, calculate the excitation voltage and the power angle.
   b. If the motor is operating on load with a power angle of $-20°$, and the excitation is so adjusted that the excitation voltage is equal in magnitude to the terminal voltage, determine the armature current and the power factor of the motor.

8-12.

An induction motor takes 350 kW at 0.8 power factor lagging while driving a load. When an overexcited synchronous motor taking 150 kW is connected in parallel with the induction motor, the overall power factor is improved to 0.95 lagging. Determine the kVA rating of the synchronous motor.

8-13.

An industrial plant consumes 500 kW at a lagging power factor of 0.6.

    a. Find the required kVA rating of a synchronous capacitor to improve the power factor to 0.9.

    b. If a 500-hp, 90 percent-efficient synchronous motor operating at full load and 0.8 leading power factor is added instead of a capacitor, calculate the resulting power factor.

8-14.

A three-phase, wye-connected, cylindrical-rotor synchronous motor with negligible armature resistance and a synchronous reactance of 1.27 ohms per phase is connected in parallel with a three-phase, wye-connected load taking 50 A at 0.707 lagging power factor from a three-phase, 220-V, 60-Hz source. If the power developed by the motor is 33 kW at a power angle of 30°, determine (a) the overall power factor of the motor and the load, and (b) the reactive power of the motor.

8-15.

A three-phase salient-pole synchronous generator has the reactances of $X_d = 1.0$ per unit and $X_q = 0.7$ per unit. When the machine is delivering full-load current at rated terminal voltage of 3.3 kV and a power factor of 0.8 lagging, determine the excitation voltage (internally generated emf). If the excitation is removed, what is the maximum per-unit power that can be generated?

8-16.

A salient-pole synchronous generator with $X_d = 1.0$ per unit and $X_q = 0.6$ per unit is connected to an infinite bus.

    a. If the field circuit is open, determine the maximum real power in per unit this generator can deliver without losing synchronism.

    b. Calculate the reactive power in per unit, current in per unit, and the power factor. Is the power factor leading or lagging?

8-17.

A three-phase, wye-connected, salient-pole synchronous generator rated at 20 kVA, 220 V has negligible armature resistance and the reactances per phase of $X_d = 4$ ohms, $X_q = 2$ ohms. When the machine is supplying rated load at 0.8 lagging power factor, determine (a) the power angle, (b) the voltage regulation, and (c) the power developed due to saliency.

8–18.

A salient-pole synchronous generator is operating under balanced steady-state conditions at rated voltage, delivering rated current at a lagging power-factor angle of 15°. The reactances in per-unit are given to be $X_d = 1.2$ and $X_q = 0.8$.

    a. Compute the per-unit excitation voltage and the power angle.

    b. Using the excitation voltage found in part *(a)*, to what value could the real power output be increased without loss of synchronism?

8–19.

Equation 8.5.10 gives the steady-state power-angle equation for a salient-pole synchronous machine with negligible armature resistance and fixed field excitation. Show that the condition for maximum power is given by

$$\cos \delta = -\frac{X_q E_f}{4(X_d - X_q) V_t} + \sqrt{\left[\frac{X_q E_f}{4(X_d - X_q) V_t}\right]^2 + \frac{1}{2}}$$

8–20.

Starting from Equation 8.6.2 of the text, taking the terminal bus voltage $\overline{V}_t$ as reference draw the steady-state phasor diagram of a synchronous machine, showing the three components of current making up the armature current $\overline{I}_a$. With reference to such a diagram, try to identify the four possible cases of operation of a synchronous machine.

8–21.

The following readings are obtained in a slip test on a 5-kVA, 240-V, 60-Hz synchronous machine:

Line-to-line volts:	200 maximum;	180 minimum
Line current in amps:	12 maximum;	8 minimum

Calculate X_d and X_q in ohms and in per unit, assuming the armature to be wye-connected. What would the per-phase and per-unit values be if the armature were delta-connected?

8–22.

A four-pole, 60-Hz, three-phase, wye-connected, 13.2-kV, 93.75-MVA turbogenerator is delivering rated load at 0.8 power-factor lagging. The Potier reactance of the generator is given to be 0.248 ohm per phase; effective armature resistance is 0.004 ohm per phase; unsaturated synchronous reactance is 2.13 ohms per phase; E_s, the correction for saturation, corresponding to E_p of 8,280 V per phase (see Figure 8.7.1) is 1,155 V per phase; a field current of 425 A corresponds to the rated voltage on the air-gap line of the open-circuit saturation curve of the machine. Compute the field current required for the given operating condition of the turboalternator.

8–23.

Compute the field current required for a power factor of 0.80-leading current when a 45-kVA, three-phase, wye-connected, 220-V, six-pole, 60-Hz synchronous machine is run as a synchronous motor at a terminal voltage of 230 V, with a power input to its armature of 45 kW. It is given that the unsaturated value of the synchronous reactance, based on cylindrical-rotor theory, is 0.92 per unit, Potier reactance of 0.227 per unit, effective armature resistance of 0.04 per unit, and E_s, the correction for saturation, corresponding to E_p of 1.2 per unit (see Figure 8.7.1) is 0.45 per unit. Further, it is given that 1.00 per-unit excitation is 2.40 field amperes, when unit excitation is defined as the value corresponding to unit voltage on the air-gap line of the open-circuit characteristic of the machine.

8–24.

Two identical three-phase, 33-kV, wye-connected synchronous generators operating in parallel share equally a total load of 12 MW at 0.8 lagging power factor. The synchronous reactance of each machine is 8 ohms per phase and the armature resistance is negligible.

   a. If one of the machines has its field excitation adjusted such that it delivers 125 A lagging current, determine the armature current, power factor, excitation voltage and power angle of each machine.
   b. If the power factor of one of the machines is 0.9 lagging, find the power factor and current of the other machine.

8–25.

Consider two identical three-phase, 13.8-kV, 100-MVA, 60-Hz turbine generators, with a synchronous reactance of each of 1.10 per unit, operating in parallel supplying a total load of 150 MVA at 0.8 lagging power factor, rated voltage, and rated frequency. Calculate for each generator (i) the real power, (ii) the reactive power, (iii) the armature current, (iv) the power factor, (v) the excitation voltage, and (vi) the torque angle,

   a. if the prime movers are adjusted so that the machines share the real power equally, and the field excitations are such that the reactive power output of each of the two machines is the same.
   b. if the machines deliver equal real power, but with the field excitation of one of the generators (say, generator 1) increased by 20 percent above its value in part *(a)*, while the field current of the other generator is adjusted so that the terminal voltage remains constant at rated value.
   c. if the excitation of generator 1 is kept constant at the value in part *(b)*, while the input to its prime mover is increased by 20 percent, and adjustments are made in the field current of generator 2 and in its prime mover such that the frequency and terminal voltage remain at their rated values.

8–26.

Repeat Problem 8–25 for two generators with the same voltage and frequency ratings as in Problem 8–25, but with generator 1 rated at 80 MVA with a synchronous reactance of 1.0 per unit while generator 2 is rated at 120 MVA with a synchronous reactance of 1.2 per unit.

9

DIRECT-CURRENT MACHINES

It can be stated without exaggeration that conventional dc generators are becoming obsolete and are being increasingly replaced by solid-state rectifiers in most applications where an ac supply is available. However, the same is not true with motors. It is the torque-speed characteristics of dc motors that make them extremely valuable in many industrial applications. Significant features of the direct-current drives include adjustable motor speed over wide ranges, constant mechanical power output or constant torque, rapid acceleration or deceleration, and responsiveness to feedback signals.

The dc commutator machine is built in a wide range of sizes, from small control devices in the one-watt power rating up to very large motors of 10,000 hp or more employed in rolling mill applications. The dc machines today are principally applied as industrial drive motors, particularly where high degrees of flexibility in the control of speed and torque are demanded. Such motors are used in steel and aluminum rolling mills, traction motors, overhead cranes, forklift trucks, electric trains, and golf carts. Commutator machines are also used in portable tools supplied from batteries, in automobiles as starter motors, blower motors, and in control applications as actuators and speed- or position-sensing devices.

In this chapter we shall examine the steady-state performance characteristics of dc machines to help us understand their applications. A study of the flux, mmf conditions, and switching action at the commutator brings out conditions limiting the machine capability and suggests methods for combating such conditions. The analysis, restricted to the steady state in this chapter, illustrates the versatility of the dc machine.

9.1 CONSTRUCTIONAL FEATURES OF DC MACHINES

The dc machines differ from ac machines in having a commutator and the armature on the rotor. They also have salient poles on the stator, and, except for a few small machines, they have what are known as *commutating poles* between the main poles. Medium and large machines are provided with large heat-dissipating surfaces and effectively placed ventilating ducts for passage of cooling air in order to prevent hot spots. Rotor punchings for small machines are mounted solidly on the shaft, whereas, on large machines, it is customary to employ a spider, consisting of a hub and projecting arms, so that the annular laminations may be rigidly fastened to the shaft, thereby permitting free flow of air to keep the armature ventilated and cooled. See Photo 9–1 for a cross-sectional view of a typical dc machine.

The armature-core punchings, slotted to receive the insulated armature winding, are usually of high-permeability electrical sheet steel, 0.4 to 0.65 mm thick, and have an insulating film between them. Small and medium units use doughnut-shaped circular punchings; but large units, above 1 m in diameter, utilize segmental punchings along with appropriately placed ventilating ducts. The main and commutating-pole punchings are generally thicker than rotor punchings because only the pole faces are subjected to high-frequency flux changes. The pole punchings range from 1.5 to 3 mm thick, and they are normally riveted. The frame yoke is usually made from rolled mild steel plate, but, on high-demand large generators for rapidly changing loads, laminations may be used. The electromagnets are bolted to the cylindrical yoke or frame. End bells with their bearings and brush-rigging become part of the stator when the machine is assembled. The yoke may have a base with feet or a supporting bracket upon which the entire structure rests. See Photo 9–2 for a view of a fully assembled dc machine with pedestal-type construction.

PHOTO 9.1   Cross-sectional view of a typical dc machine.   Photo courtesy of
Westinghouse Electric Corporation.

PHOTO 9.2   View of a fully-assembled dc machine with pedestal-type
construction.   Photo courtesy of Westinghouse Electric Corporation.

PHOTO 9.3   Cross-sectional view of a typical V-ring commutator of a dc machine.   Photo courtesy of Westinghouse Electric Corporation.

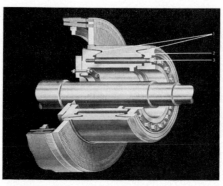

PHOTO 9.4   A completely wound armature along with the commutator of a dc machine.   Photo courtesy of Westinghouse Electric Corporation.

The commutator is truly the heart of the dc machine. It is made up of hard copper bars, drawn accurately in a wedge shape, separated from each other by mica plate segments 0.5 to 1.25 mm thick, depending on the size of the machine and on the maximum voltage that can be expected between bars during operation. The mica segments and bars are clamped between two metal V-rings and insulated from them by cones of mica. On very high-speed commutators, shrink rings of steel are used to hold the bars. Mica is used under the rings. The carbon brushes with rounded contact surfaces ride on the commutator bars and carry the load current from the rotor coils to the external circuit. The brush holders hold the brushes against the commutator surface by springs to maintain a fairly constant pressure and smooth riding. See Photos 9–3 and 9–4 for a cross-sectional view of a V-ring commutator and a completely wound armature along with the commutator of a dc machine.

FIGURE 9.1.1   Section of a dc machine illustrating the arrangement of various field windings.

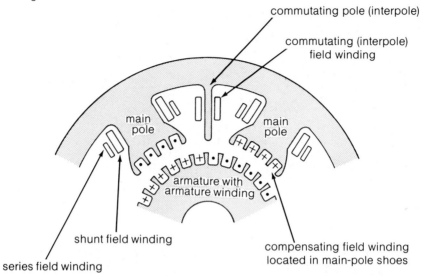

commutating pole (interpole)

commutating (interpole) field winding

main pole

main pole

armature with armature winding

shunt field winding

series field winding

compensating field winding located in main-pole shoes

FIGURE 9.1.2   Schematic connection diagram of a dc machine.

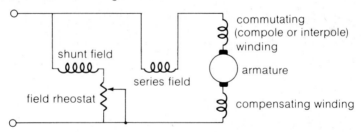

shunt field

field rheostat

series field

commutating (compole or interpole) winding

armature

compensating winding

A dc generator or motor may have as many as four field windings, depending upon the type and size of the machine and the kind of service the machine is to have. These consist of two normal exciting fields, the *shunt* and *series* windings, and two fields (that act in a corrective capacity to combat the detrimental effects of armature reaction), the *commutating (compole or interpole)* and *compensating* windings connected in series with the armature. The type of machine, whether shunt, series, or compound, is determined solely by the normal exciting-field circuit connections, as shown in Figure 5.3.14. Figure 9.1.1 illustrates how different field windings are arranged with respect to one another in part of a cross section of a dc machine, while Figure 9.1.2 shows the

PHOTO 9.5 Commutating poles and compensating windings of a dc machine. Photo courtesy of Westinghouse Electric Corporation.

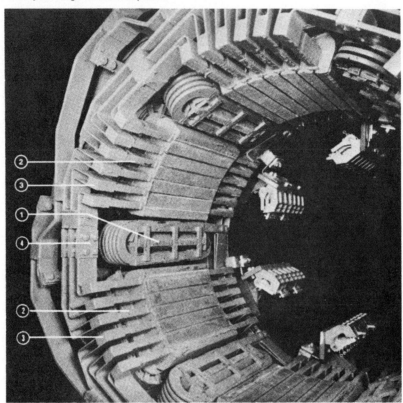

schematic connection diagram of a dc machine. The commutating and compensating windings, their purpose as well as circuit connections, are further discussed in detail in the sections to follow. See Photo 9–5 for a view of commutating poles and compensating windings of an assembled dc machine.

9.2 EQUIVALENT CIRCUIT OF A DC MACHINE

A review of the material presented under "Elementary Direct-Current Machines" in Section 5.3 is very helpful at this stage to recall the principle of operation of dc machines.

Various arrangements of armature windings have been discussed in Chapter 6, where it has been noted that the commutator brush is normally situated on the center line of a main pole, although it is connected to a coil in the interpolar gap. It is a common practice to use a schematic

FIGURE 9.2.1   Representations of a dc machine: (a) physical arrangement, (b) schematic diagram, (c) circuit representation of a dc generator under steady-state conditions $V_f = R_f I_f$; $V_t = E_a - I_a R_a$, (d) circuit representation of a dc motor under steady-state conditions $V_f = I_f R_f$; $V_t = E_a + I_a R_a$.

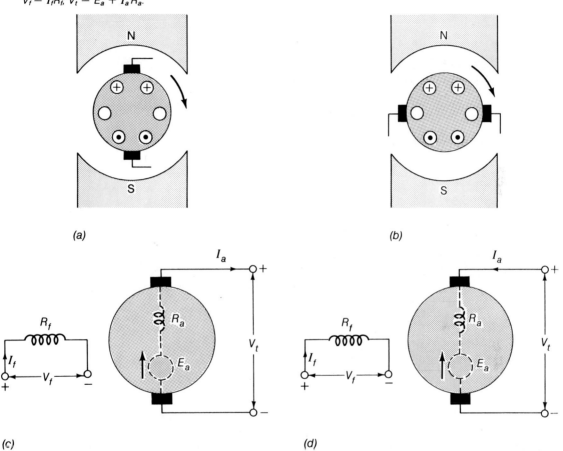

(a)   (b)

(c)   (d)

representation of a dc machine in which the brushes are shown in the position of the coil to which they are connected. The physical arrangement together with the schematic diagram and circuit representations are shown in Figure 9.2.1. Under *steady-state conditions*, the interrelations between the voltage and current are given by

$$V_f = I_f R_f \qquad (9.2.1)$$

and

$$V_t = E_a \pm I_a R_a \qquad (9.2.2)$$

where the plus sign is used for a motor and the minus sign for a generator. V_f is the voltage applied to the field circuit, I_f is the field current, and R_f is the field-winding resistance. V_t is the terminal voltage, E_a is the generated emf, I_a is the armature current, and R_a is the armature resistance. The generated emf E_a is given by Equation 5.3.24 as

$$E_a = K_a \, \phi \, \omega_m \qquad (9.2.3)$$

which is the speed (motional) voltage induced in the armature circuit due to the flux of the stator-field current. The electromagnetic torque T_e is given by Equation 5.3.26 as

$$T_e = K_a \, \phi \, I_a \qquad (9.2.4)$$

where K_a is the design constant. The product $E_a I_a$ is known as the electromagnetic power being converted and is related to the electromagnetic torque by the relation

$$P_{em} = E_a \, I_a = T_e \, \omega_m \qquad (9.2.5)$$

In the case of a motor, the terminal voltage is always greater than the generated emf and the electromagnetic torque produces rotation against a load. For a generator, the terminal voltage is less than the generated emf and the electromagnetic torque opposes that applied to the shaft by the prime mover. If the magnetic circuit of the machine is not saturated, note that the flux ϕ in Equations 9.2.3 and 9.2.4 is proportional to the field current I_f producing the flux.

9.3 COMMUTATOR ACTION

In the discussion on dc armature windings in Section 6.3, it has been shown why a rotating commutator winding is called a pseudostationary winding that produces a stationary flux when carrying a direct current. As a consequence of the arrangement of the commutator and brushes, the currents in all conductors under the north pole are in one direction and the currents in all conductors under the south pole are in the opposite direction. Thus, the magnetic field of the armature currents is stationary in space in spite of the rotation of the armature. As seen from Figures 6.3.2 and 6.3.3, during the time when the brushes are simultaneously in contact with two adjacent commutator segments, the coils connected to these segments are short-circuited by the brushes and are temporarily removed from the main circuits through the winding; the directions of the currents in the coils are then reversed. The coils are then said to be undergoing *commutation* during this interval of time, and the process of reversal of currents in the coils is known as commutation. If the width of the brush is infinitesimally small, then the current reversal will be instantaneous; in practice, however, there is some finite time Δt, depending upon the speed of the rotor and the width of the brushes, during which the current reversal takes place. The current changes from $+I$ to $-I$ in a time Δt. Ideally, the current in the coils being commutated should reverse linearly with time, as shown in Figure 9.3.1. Serious departure from *linear commutation* results in sparking at the brushes. Means for achieving sparkless commutation are discussed in Section 9.5. As shown in Figure 9.3.1, with linear commutation, the waveform of the current in any coil as a function of time is trapezoidal.

FIGURE 9.3.1 Waveform of current in an armature coil undergoing *linear* commutation.

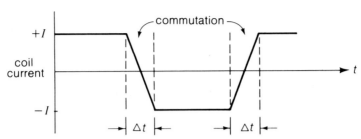

While the air-gap flux distribution produced by the field windings is symmetrical about the center line of the field poles, known as the *field (direct* or *d-) axis,* the axis of the armature-mmf wave is 90 electrical degrees away from the axis of the field poles, i.e., in the *interpolar (quadrature* or *q-) axis,* as shown in Figure 5.3.5. The brushes are so located that commutation occurs when the coil sides are in the *neutral zone,* midway between the field poles. Because of the position of the coils to which the brushes are connected, the brushes are shown in the quadrature axis in Figure 5.3.5, and the armature-mmf wave is then directed along the brush axis as shown. However, the geometrical (physical) position of the brushes in an actual machine is approximately 90 electrical degrees away from their position in the schematic diagram, i.e., along the center lines of the main poles, because of the shape of the end connections to the commutator, which can be seen from Figures 6.3.2 and 6.3.3. As stated earlier, the magnetic torque and the speed voltage appearing at the brushes are independent of the spatial waveform of the flux distribution.

9.4 ARMATURE REACTION

It is shown in Section 6.7 and by Figure 6.7.4 that the armature-mmf wave may be closely approximated by a triangle, corresponding to the wave produced by a finely distributed armature winding or current sheet. Figure 9.4.1*(a)* shows the flux-density distribution due to the main field alone, while Figure 9.4.1*(b)* shows the armature-mmf distribution for a machine with brushes in the geometric neutral axes as well as the flux-density distribution due to the armature mmf alone. Note that the armature-reaction flux is appreciably decreased in the interpolar space. Observe also that the axis of the armature mmf is fixed at 90 electrical degrees from the main-field axis by the brush position. Hence, armature reaction of this type is known as the *cross-magnetizing armature reaction,* which apparently causes a decrease in the resultant air-gap flux-density distribution under one half of the pole and an increase under the other half. The resultant air-gap flux-density distribution, when the armature and field windings are both excited, is shown in Figure 9.4.1*(c).* It can be seen from Figure 9.4.1*(c)* that the effect of armature reaction is to distort the flux wave and to shift the position of the magnetic neutral axis in the direction of rotation for the generating case and against the direction of rotation for the motoring case.

In general, the resultant flux-density distribution is not the algebraic sum of the flux distributions due to main field and armature mmf because of the nonlinearity of the iron-magnetic circuit. Because of the saturation of iron, the flux density is decreased by a greater amount under one pole tip than it is increased under the other. Hence, the resultant flux per pole is lower than that which

FIGURE 9.4.1 Main field, armature, and resultant flux-density distributions with brushes on the geometric neutral axes. *(a)* Flux-density distribution due to main field alone, *(b)* flux-density distribution due to armature mmf alone, *(c)* resultant flux-density distribution.

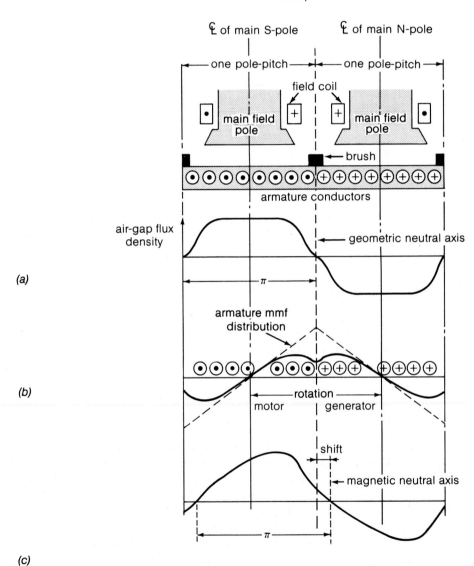

FIGURE 9.5.1 Undercommutation (delayed commutation).

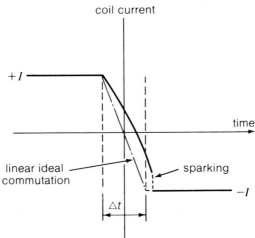

would be produced by the field winding alone; and this is known as the *demagnetizing effect of the cross-magnetizing armature reaction,* caused by saturation. At the flux densities usually employed for normal machine operation, the effect is usually significant, especially at heavy loads.

The distortion of the flux distribution caused by the cross-magnetizing armature reaction may have a detrimental influence on commutation, and may limit the short-time overload of a dc machine. However, its effect may be limited in the design and construction of the machine by increasing the reluctance of the cross-flux path; such techniques as using chamfered or eccentric pole faces, which increase the air gap at the pole tips, may be employed. The best but also the most expensive corrective measure is to compensate the armature mmf by means of a distributed *compensating winding* embedded in slots in the main-pole faces. Such a measure is presented in Section 9.5.

9.5 INTERPOLES AND COMPENSATING WINDINGS

When the commutator bar that is connected to a particular armature coil rotates past a fixed commutator brush, the coil current must reverse. In normal practice, a brush spans more than one commutator segment, so that, during commutation, several coils in series are short-circuited. The reversal of the coil current in a short time induces a voltage of self-inductance, the so-called *reactance voltage,* which opposes the change of current so that a spark could appear at the trailing edge of the brush. Not only the voltage of self-induction in the commutated coil but also the voltage of mutual induction from other coils (particularly those in the same slot, undergoing commutation at the same time) oppose changes in current in the commutated coil. The sum of these two voltages is often referred to as the reactance voltage. While the ideal process of commutation is linear as shown in Figure 9.3.1, the reactance voltage causes the condition known as *undercommutation (delayed commutation),* as illustrated in Figure 9.5.1.

The effect of a given reactance voltage in delaying commutation is minimized by the resistive brush-contact voltage drop. That is why carbon brushes with appreciable contact drop are usually employed. When good commutation is achieved by virtue of resistive drops, the process is known as the *resistance commutation,* which is used these days exclusively only in fractional horsepower machines.

It can be seen from Figure 9.4.1 that one of the reasons for poor commutation is the effect of armature-reaction flux in shifting the position of the magnetic neutral axis. The cross-magnetizing armature reaction causes distortion in the resultant air-gap flux distribution and creates flux in the interpolar region. The rotational voltage induced in the short-circuited coil becomes another important factor in the commutation process. The direction of the induced voltage being the same as the current under the immediately preceding pole face, this voltage then encourages the continuance of the current in the old direction and, like the reactance voltage, opposes the reversal of the current in the commutated coil. The general principle of producing a rotational voltage that approximately compensates for the reactance voltage in the coil undergoing commutation is known as *voltage commutation.* Voltage commutation is used in almost all modern commutating machines. The appropriate flux density is introduced in the commutating zone by means of small, narrow, auxiliary poles located between the main poles and centered on the interpolar gap; these auxiliary poles are known as *interpoles* or *commutating poles* or simply as *compoles,* which produce the so-called *commutation voltage* in order to counteract the bad effects of flux distortion and the reactance voltage on commutation. In the case of small machines, such a voltage can be introduced by shifting the brush position. *Brush shift* in the direction of rotation in a generator or against rotation in a motor produces a direct demagnetizing mmf that may result in unstable operation of a motor or excessive drop in the voltage of a generator.

The most general method for aiding commutation is by providing the machine with interpoles, and the commutating (interpole) winding must be connected in series with the armature, since both the armature mmf and the reactance voltage are proportional to the armature current. The interpole mmf must be sufficient not only to neutralize the cross-magnetizing armature mmf in the interpolar region but also to furnish the flux density required for the rotational voltage in the short-circuited armature coil to counteract the reactance voltage. Since the compole must produce flux proportional to armature current, to preserve the desired linearity, it must operate with an unsaturated magnetic circuit; hence, the air gap between the armature surface and compole is normally greater than that between the armature and main pole. The arrangement of a two-pole generator with interpoles is shown in Figure 9.5.2 and the compole has the same polarity as the main pole ahead in the direction of rotation. For the case of a motor, the arrangement is shown in Figure 9.5.3 and the interpole has the same polarity as the main pole behind, relative to the direction of rotation. Under these circumstances, with appropriate design of interpoles, a dc machine can be operated as either a motor or a generator with either direction of rotation.

Even with interpoles present, for machines subjected to heavy overloads, rapidly changing loads, or operation with a weak main field, the armature reaction under the poles may badly distort the flux wave and cause induced coil voltages high enough to result in flashover or arcing between

FIGURE 9.5.2 DC generator with interpoles.

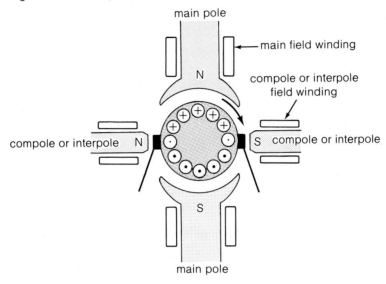

FIGURE 9.5.3 DC motor with interpoles.

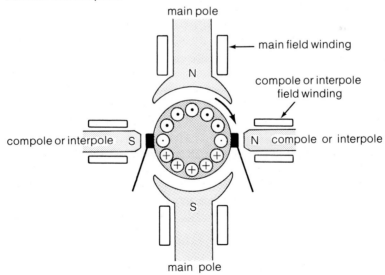

FIGURE 9.5.4 Component and total air-gap flux-density distributions for a dc machine with interpoles and compensating winding.

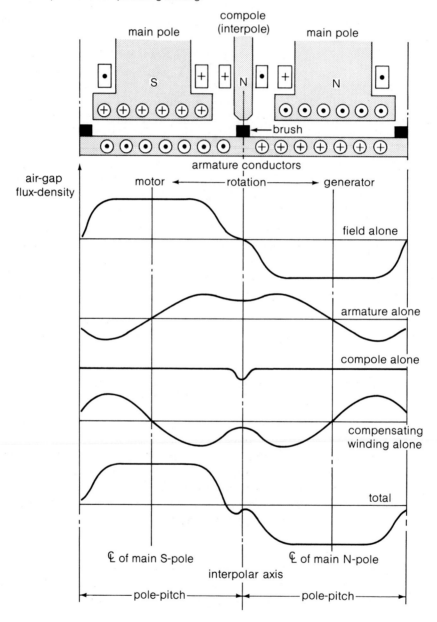

commutator segments. This process may quickly spread around the entire commutator, leading to a direct short circuit on the line, besides its destructive effects on the commutator itself. The demagnetizing effect of the armature mmf under pole faces may be compensated or neutralized by providing a *compensating (pole-face) winding,* arranged in slots in the pole face, having a polarity opposite to that of the adjoining armature winding. Given the proper number of turns and connected in series with the armature in order to carry a proportional current, the compensating winding has the same axis as that of the armature, and will almost completely neutralize the armature reaction of the armature inductors under the pole faces. The net effect of the main field, armature, commutating winding, and compensating winding on the air-gap flux distribution is that, except for the commutation zone in the interpolar region, the resultant flux-density distribution is substantially the same as that produced by the main field alone, as sketched in Figure 9.5.4.

Furthermore, with the addition of a compensating winding, the armature-circuit time constant will be reduced; hence, the speed of response will be improved. Because of their expense, pole-face windings are usually employed only in machines designed for heavy overloads or rapidly changing loads, such as steel-mill motors subjected to reverse duty cycles or in motors intended to operate over wide speed ranges by shunt-field control. The schematic diagram of Figure 9.1.2 shows the relative positions of various windings, indicating that the commutating and compensating fields act along the armature axis (i.e., the quadrature axis), and the shunt as well as series fields act along the axis of the main poles (i.e., the direct axis). It is thus possible to achieve rather complete control of the air-gap flux around the entire armature periphery, along with smooth sparkless commutation.

EXAMPLE 9.5.1

A six-pole, 600-V, 600-kW dc generator has a simplex lap winding with 720 armature conductors, and is equipped with six commutating poles.

a. If the mmf of the commutating poles is 1.2 times that of the armature, calculate the number of turns in the interpole winding.
b. If the generator is also provided with a compensating winding, and the pole-face covers 70 percent of the pole-span, compute the number of conductors for the compensating winding in each pole-face.
c. When the pole-face winding of part *(b)* is in the circuit, what should be the number of turns per pole in the interpole winding?

SOLUTION

a. The number of armature conductors being 720, the number of turns is 720/2 = 360; the number of armature turns per pole is 360/6 = 60; and the number of parallel paths in the armature is 6, the same as the number of poles, the winding being simplex lap.

Since the interpole winding is in series with the armature carrying the armature current I_a, the mmf per interpole must be 1.2 × 60I_a/6, and the number of turns in the interpole winding per pole is then given by 12.

b. The armature conductors per pole being 720/6 = 120, the number of armature conductors under each pole face is given by 120 × 0.7 = 84. Since the compensating winding carries the entire armature current, i.e., 6 times that of the armature conductors, the number of conductors per pole for the compensating winding is then 84/6 = 14.

c. Since good commutation requires 12 turns per pole, of which 14/2 = 7 are provided by the compensating winding, the interpole winding should make up the difference, i.e., 12 − 7 = 5 turns per pole.

9.6 DC GENERATOR CHARACTERISTICS

Figure 5.3.18 shows schematic diagrams of field-circuit connections for dc machines without including commutating-pole or compensating windings. Shunt generators may be either *separately excited* or *self-excited* as shown in Figures 5.3.18(a) and (b), respectively. Compound machines may be connected *long-shunt,* as in Figure 5.3.18(d), or *short-shunt,* in which the shunt-field circuit is connected directly across the armature, without including the series field.

In general, the three characteristics that specify the *steady-state* performance of a dc generator are:

a. The *open-circuit characteristic,* also known as the *no-load magnetization curve,* which gives the relationship between the generated emf and field current at constant speed.

b. The *external characteristic,* which gives the relationship between the terminal voltage and load current at constant speed.

c. The *load characteristic,* which gives the relationship between the terminal voltage and field current, with constant armature current and speed.

All other characteristics depend on the form of the open-circuit characteristic, the load, and the method of field connection. Under steady-state conditions, the currents being constant or, at most, slowly varying, voltage drops due to inductive effects are negligible. While the self-inductance of armature coils undergoing commutation and the mutual inductance between these coils and the rest of the armature winding influence commutation, their effect on load characteristics of conventional dc machines is negligible.

As stated earlier and shown in Figure 9.2.1, the terminal voltage V_t of a dc generator is related to the armature current I_a and the generated emf E_a by

$$V_t = E_a - I_a R_a \qquad (9.6.1)$$

where R_a is the total internal armature resistance, including that of interpole and compensating windings as well as that of the brushes. The value of the generated emf E_a, given by Equation 9.2.3, is governed by the direct-axis field flux (which is a function of the field current and armature reaction) and the angular velocity ω_m of the rotor.

The open-circuit and load characteristics of a separately excited dc generator, along with its schematic diagram of connections, are shown in Figure 9.6.1. It can be seen from the form of the external volt-ampere characteristic, shown in Figure 9.6.1(c), that the terminal voltage falls slightly

FIGURE 9.6.1 Open-circuit and load characteristics of a separately excited dc generator. *(a)* Schematic diagram of connections.

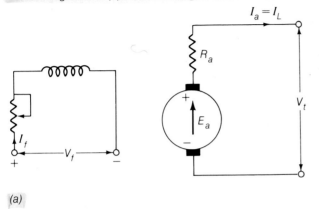

(a)

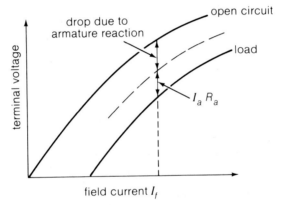

(b)

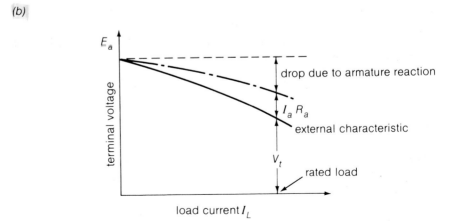

(c)

FIGURE 9.6.2 Self-excited dc shunt generator: *(a)* schematic diagram of connections, *(b)* open-circuit characteristic.

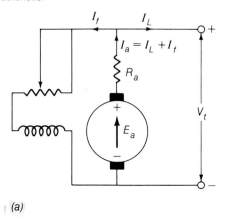

(a)

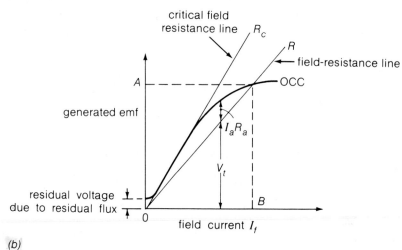

(b)

as the load current increases. *Voltage regulation* is defined as the percentage change in terminal voltage when full load is removed, so that, from Figure 9.6.1*(c)*, it follows that

$$\text{Voltage regulation} = \frac{E_a - V_t}{V_t} \times 100\% \qquad (9.6.2)$$

Since the separately excited generator requires a separate dc field supply, its use is limited to applications where a wide range of controlled voltage is essential.

Let us next consider the case of a self-excited dc shunt generator for which the schematic diagram of connections and the open-circuit characteristic are given in Figure 9.6.2*(a)* and *(b)*, respectively. With a field resistance, given by a line *OR* of slope *OA/OB*, shown in Figure 9.6.2*(b)*,

FIGURE 9.6.3 Graphical technique to compute the V_t-I_a curve for a self-excited dc shunt generator.

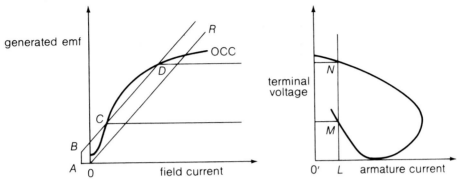

the open-circuit voltage will build up to a value of *OA*. It follows that there must be a *critical-field resistance,* when the corresponding critical-field resistance line OR_C is coincident with the linear part of the characteristic. That is the maximum permissible value of field resistance if the armature voltage is to build up. Reasons for failure of voltage buildup can be summarized as follows:

    a. Insufficient residual flux.
    b. Incorrect polarity of the field winding due to which the field current is in a direction as to reduce the residual flux.
    c. Field resistance above critical value if the speed is normal.
    d. Speed below critical value if the resistance of the field circuit is normal.

When the form of the open-circuit characteristic, the field resistance and turns, the armature resistance, and the demagnetizing ampere-turns of armature reaction are known, various graphical techniques are available to determine the variation of terminal voltage with armature current. One such technique is illustrated in Figure 9.6.3. For a value of armature current *O'L,* knowing the effective reduction in total ampere-turns caused by demagnetization, the equivalent reduction in field current *OA* is given by the demagnetization ampere-turns divided by the field turns. The armature resistance drop for the same armature current is given by *AB.* A line through *B* parallel to the field-resistance line *OR* is now drawn to meet the OCC at the points *C* and *D.* Horizontal lines *CM* and *DN* are then drawn to meet the vertical from *L* at points *M* and *N,* respectively. *M* and *N* are then the points on the terminal-voltage versus armature-current characteristic; by repeating this process for several different values of armature current, the complete V_t - I_a characteristic can be obtained, as in Figure 9.6.3. By subtracting the field current I_f from the armature current I_a for each point on the V_t - I_a characteristic, the external V_t - I_L characteristic can be obtained. It can be observed from Figure 9.6.3 that there is a maximum value of armature current beyond which a decrease in load resistance will cause a decrease in armature current. Under short-

FIGURE 9.6.4   DC series generator: *(a)* schematic diagram of connections, *(b)* volt-ampere characteristic.

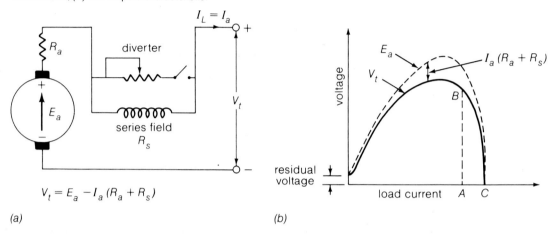

$$V_t = E_a - I_a (R_a + R_s)$$

(a)   *(b)*

circuit conditions, the armature current is given by the ratio of the residual voltage to the internal armature resistance. If the demagnetizing effect is neglected, the distance OA becomes zero and the construction for the V_t - I_a or V_t - I_L characteristic is simplified.

A shunt generator maintains approximately constant voltage on load, and finds wide application as an exciter for the field circuit of large ac generators. The shunt generator is also used sometimes as a tachogenerator when a signal proportional to motor speed is required for control or display purposes.

The schematic diagram and the volt-ampere characteristic of a dc series generator at constant speed are shown in Figure 9.6.4. The resistance of the series-field winding must be low for good efficiency as well as for low voltage drop. The series generator was employed in early constant-current systems by operating in a range *BC,* where the terminal voltage fell off very rapidly with increasing current.

The volt-ampere characteristics of dc compound generators at constant speed are shown in Figure 9.6.5. *Cumulatively compounded* generators, in which the series and shunt-field-winding mmfs are aiding, may be *overcompounded, flat-compounded,* or *undercompounded,* depending upon the strength of the series field. Overcompounding may be used to counteract the effect of a decrease in the prime-mover speed with increasing load, or to compensate for the line drop when the load is at a considerable distance from the generator. *Differentially compounded* generators, in which the series-winding mmf opposes that of the shunt-field winding, are used in applications where a wide variation in load voltage can be tolerated and where the generator may be exposed to load conditions approaching those of short circuit.

FIGURE 9.6.5 Volt-ampere characteristics of dc compound generators at constant speed.

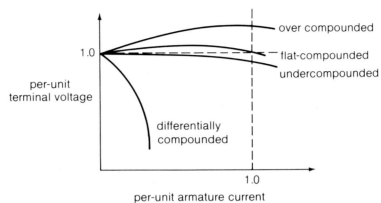

per-unit armature current

EXAMPLE 9.6.1

The open-circuit and short-circuit test data of a four-pole dc generator rated at 200 V, 100 A, and 1,000 rpm are given in Table 9.6.1. The armature is wave-wound with 704 conductors. The armature resistance is 0.15 ohm; the series-field resistance is 0.01 ohm; and the separately excited field resistance is 100 ohms with 2,500 turns per pole. The brush drop is 2 volts.

TABLE 9.6.1

Open-Circuit Characteristic								
field current (amps)	0.1	0.4	0.6	1.0	1.3	1.5	2.1	2.3
terminal voltage (volts)	17	80	120	170	190	200	217	220
Short-Circuit Characteristic								
field current (amps)	0.1	0.2	0.3	0.4				
armature current (amps)	50	80	100	120				

Calculate the following:

    a. The demagnetizing mmf due to full-load armature reaction, expressed as a percentage of full-load armature mmf.
    b. The no-load voltage when the separately excited field current is adjusted to give 200 V on full load.
    c. The new field current and the no-load voltage when a series winding having 25 turns per pole is cumulatively connected in circuit and the voltage again adjusted to 200 V on full load.
    d. The value of the series-diverter resistance to give a flat-compounded curve at 200 V.

SOLUTION

a. Full-load armature mmf $F_a = [(100/2)/4] \times 704/2 = 4{,}400$ At/pole. (Note that the number of parallel paths for a wave winding is 2.) Total excitation for 100 A on short circuit $= 0.3$ A $= F_f/2{,}500$. The internal voltage drop at full load $= (100 \times 0.15) + 2 = 17$ V. Excitation for 17 V from the OCC $= 0.1$ A $= F_r/2{,}500$.

Armature-demagnetizing mmf in terms of the equivalent field amperes is given by $0.3 - 0.1 = 0.2$ A $= -F_a'/2{,}500$. (Note that the effective or resultant mmf is $F_r = F_f + F_a'$, where F_a' is usually negative because of the demagnetizing effect.) As a percentage, $F_a'/F_a = (0.2 \times 2{,}500/4{,}400)100 = 11.36$ percent.

b. The full-load voltage plus the voltage drop at full load $= 200 + 17 = 217$ V. The required number of resultant ampere-turns, F_r, as read from the OCC corresponding to 217 V, is $2.1 \times 2{,}500 = 5{,}250$. The demagnetizing mmf due to full-load armature reaction, $-F_a' = 0.2 \times 2{,}500 = 500$. Since F_a' is negative, the total excitation required, F_f, is 5,750 At; the total field currrent required is then $5{,}750/2{,}500 = 2.3$ A, yielding 220 V on no load from the OCC.

c. Neglecting the small, additional series-field resistance drop, the total excitation must still be 5,750 At; the series-field winding contributes $25 \times 100 = 2{,}500$ At; hence, the separate-field winding must provide $5{,}750 - 2{,}500 = 3{,}250$ At or a field current of $3{,}250/2{,}500 = 1.3$ A, corresponding to which the no-load voltage from the OCC is 190 V.

d. For 200 V on no load, the separate-field excitation will have to be 1.5 (as read from the OCC) $\times 2{,}500 = 3{,}750$ At; for the total excitation at full load to be 5,750 At, the series-field contribution must be reduced to $5{,}750 - 3{,}750 = 2{,}000$ At; the corresponding series-field current must be $2{,}000/25 = 80$ A. Therefore, the required series-diverter resistance is $(80/20) \times 0.01 = 0.04$ ohm.

9.7 DC MOTOR CHARACTERISTICS

An understanding of the speed-torque characteristics of a dc motor can be gained from Equations 9.2.2 to 9.2.4. In shunt motors, the field current can be simply controlled by the use of a variable resistance in series with the field winding; the load current influences the flux only through armature reaction, and its effect is therefore relatively small. In series motors, the flux is largely determined by the armature current, which is also the field current; it is somewhat difficult to control the armature and field currents independently. In the compound motor, the effect of the armature current on the flux depends on the degree of compounding. Most motors are designed to develop a given horsepower at a specified speed, and it follows from Equations 9.2.2 and 9.2.3 that the angular velocity ω_m can be expressed as

$$\omega_m = \frac{V_t - I_a R_a}{K_a \phi} \tag{9.7.1}$$

FIGURE 9.7.1 Schematic diagrams of dc motors: *(a)* shunt motor, *(b)* series motor, *(c)* cumulatively compounded motor.

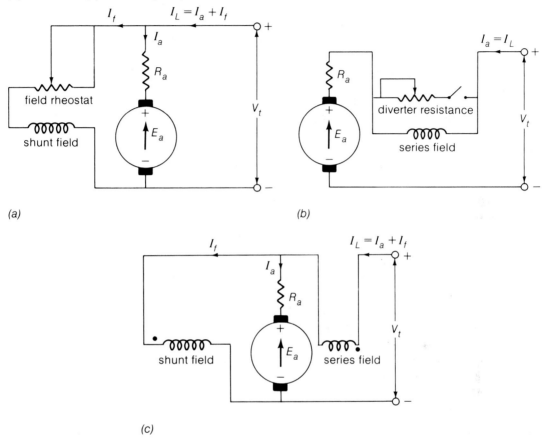

(a)

(b)

(c)

Thus the speed of a dc motor depends on the values of the applied voltage V_t, the armature current I_a, the resistance R_a, and the field flux per pole ϕ.

The schematic arrangement of a shunt motor is shown in Figure 9.7.1(a). For a given applied voltage and a field current, Equations 9.2.4 and 9.2.3 can be rewritten as

$$T_e = (K_a \phi) I_a = K_m I_a \tag{9.7.2}$$

$$E_a = (K_a \phi) \omega_m = K_m \omega_m \tag{9.7.3}$$

Since

$$V_t = E_a + I_a R_a, \text{ or } I_a = (V_t - E_a)/R_a,$$

FIGURE 9.7.2 Characteristic curves for dc motors.

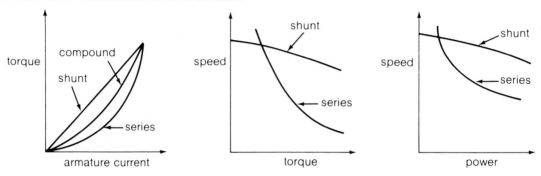

it follows that

$$T_e = \frac{K_m V_t}{R_a} - \frac{K_m^2 \, \omega_m}{R_a} \tag{9.7.4}$$

$$P_{em} = E_a I_a = T_e \, \omega_m = \frac{V_t \, K_m \, \omega_m}{R_a} - \frac{K_m^2 \, \omega_m^2}{R_a} \tag{9.7.5}$$

The forms of the torque-armature current, speed-torque, and speed-power characteristics for a shunt-connected dc motor are illustrated in Figure 9.7.2.

The shunt motor is essentially a constant speed machine with a low speed regulation. As seen from Equation 9.7.1, the speed is inversely proportional to the field flux, so that it can be varied by controlling the field flux. However, when the motor operates at very low values of field flux, the speed will be high, and if the field becomes open-circuited, the speed will rise rapidly beyond the permissible limit governed by the mechanical structure. In order to limit the speed to a safe value, if a shunt motor is to be designed to operate with a low value of shunt-field flux, it is usually fitted with a small cumulative series winding, known as a *stabilizing winding*.

The schematic diagram of a series motor is shown in Figure 9.7.1*(b)*. The field flux is directly determined by the armature current, and if there is no saturation, the flux is proportional to the armature current so that

$$T_e = K_a \, \phi \, I_a = K \, I_a^2 \tag{9.7.6}$$

and with negligible armature resistance,

$$V_t = E_a = K_a \, \phi \, \omega_m = K \, I_a \, \omega_m \tag{9.7.7}$$

$$T_e = \frac{V_t^2}{K \, \omega_m^2} \tag{9.7.8}$$

$$P_{em} = \omega_m \, T_e = \frac{V_t^2}{K \, \omega_m} \tag{9.7.9}$$

FIGURE 9.7.3 Typical speed-torque characteristics of various electric motors.

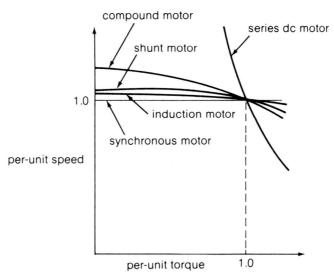

and the speed-power curve is a rectangular hyperbola. The forms of the torque-armature current, speed-torque, and speed-power characteristics for a series-connected dc motor are illustrated in Figure 9.7.2. Note that the no-load speed is very high and care must be taken to ensure that the machine always operates on load. However, in practice, the series machine normally has a small shunt-field winding to limit the no-load speed. The assumption that the flux is proportional to the armature current is valid only on light load in the linear region of magnetization; in general, performance characteristics of the series motor have to be obtained through the use of the magnetization curve. The series motor is ideally suited to traction where large torques are required at low speeds and relatively low torques are needed at high speeds.

A schematic diagram for a cumulatively compounded dc motor is shown in Figure 9.7.1(c). The operating characteristics of such a machine lie between those of the shunt and series motors shown in Figure 9.7.2. The differentially compounded motor has little application, since it is inherently unstable, particularly at high loads. Figure 9.7.3 shows a comparison of the speed-torque characteristics for various types of electric motors in which 1.0 per unit represents rated values.

EXAMPLE 9.7.1

A 200-V dc shunt motor has a field resistance of 200 Ω and an armature resistance of 0.5 Ω. On no-load, the machine operates with full field flux at a speed of 1,000 rpm with an armature current of 4 A.

   a. If the motor drives a load requiring a torque of 100 N·m, find the armature current and speed of the motor.

b. If the motor is required to develop 10 hp at 1,200 rpm, compute the required value of external series resistance in the field circuit.

Magnetic saturation and armature reaction can be neglected.

SOLUTION

a. Full field current $I_f = 200/200 = 1$ A. On no load, $E_a = V_t - I_a R_a$
$= 200 - (4 \times 0.5) = 198$ V. Since $E_a = k_1 I_f \omega_m$, where k_1 is a constant,

$$k_1 = \frac{198}{1\left(\dfrac{2\pi}{60} \times 1,000\right)} = 1.89$$

On load, $T_e = k_1 I_f I_a$, or $100 = 1.89 \times 1.0 \times I_a$; therefore the armature current $I_a = 100/1.89 = 52.9$ A. Now, $V_t = E_a + I_a R_a$, or $E_a = 200 - (52.9 \times 0.5) = 173.55$ V. Since $E_a = k_1 I_f \omega_m$, it follows that

$$\omega_m = \frac{173.55}{1.89 \times 1.0} = 91.8 \text{ rad/s}$$

i.e., the load speed $= (91.8 \times 60)/2\pi = 876$ rpm

b. For 10 hp at 1,200 rpm

$$T_e = \frac{10 \times 746}{\dfrac{2\pi}{60} \times 1,200} = 59.34 \text{ N·m}$$

Then $59.34 = 1.89 I_f I_a$, or $I_f I_a = 31.4$. Since $V_t = E_a + I_a R_a$, it follows that

$$200 = 1.89\left(\frac{2\pi}{60} \times 1,200\right) I_f + 0.5 I_a$$

or

$$200 = 237.6 I_f + 0.5 I_a = 237.6 I_f + \frac{0.5 \times 31.4}{I_f}$$

$I_f = 0.754$ A or 0.088 A; and $I_a = 31.4/I_f = 41.6$ A or 356.8 A
Since the value of $I_f = 0.088$ A will produce very high armature currents, it will not be considered.
   Thus, with $I_f = 0.754$ A, $R_f = 200/0.754 = 265.25 \ \Omega$
External resistance required $= 265.25 - 200 = 65.25 \ \Omega$

FIGURE 9.8.1   Shunted-armature method of speed control: (a) as applied
to a series motor, (b) as applied to a shunt motor.

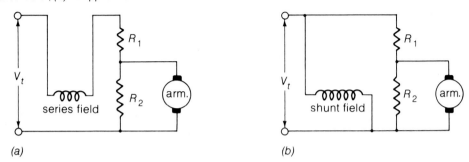

(a)

(b)

9.8   CONTROL OF DC MOTORS

Equation 9.7.1 shows that the speed of a dc motor can be varied by control of the field flux, the armature resistance, and the armature applied voltage. The three most common speed-control methods are *shunt-field-rheostat control, armature-circuit-resistance control,* and *armature-terminal-voltage control.* The *base speed* of the machine is defined as the speed with rated armature voltage and normal armature resistance and field flux. Speed control above the base value can be obtained by varying the field flux. By inserting a series resistance in the shunt-field circuit of a dc shunt motor (or a compound motor), speed control over a wide range above the base speed can be achieved. However, it is important to note that a reduction in the field flux causes a corresponding increase in speed, so that the generated emf does not change appreciably while the speed is increased, but the machine torque is reduced as the field flux is reduced. The dc motor with shunt-field-rheostat speed control is accordingly referred to as a *constant-horsepower drive.* This method of speed control is suited to applications where the load torque falls as the speed increases. In the case of a machine with a series field, speed control above the base value can be achieved by placing a diverter-resistance in parallel with the series winding so that the field current is less than the armature current.

When speed control below the base speed is required, the effective armature resistance can be increased by inserting external resistance in series with the armature, and this method can be applied to shunt, series, or compound motors. However, it has the disadvantage that the series resistance, carrying full armature current, will cause significant power loss with an associated reduction in overall efficiency. The speed of the machine will be governed by the value of the voltage drop in the series resistor, and will therefore be a function of the load on the machine. The application of this method of control is limited; because of its low initial cost, however, the series-resistance method or a variation of it is often attractive economically for short-time or intermittent slowdowns. Unlike shunt-field control, armature-resistance control offers a *constant-torque drive* because both flux and, to a first approximation, allowable armature current remain constant as speed varies. The *shunted-armature method* is a variation of this control scheme, as shown in Figure 9.8.1(a) as applied to a series motor and Figure 9.8.1(b) as applied to a shunt motor. Resistors R_1 and R_2 act as a voltage divider applying a reduced voltage to the armature, and offer greater flexibility in their adjustment to provide the desired performance.

FIGURE 9.8.2   Ward-Leonard method of speed control.

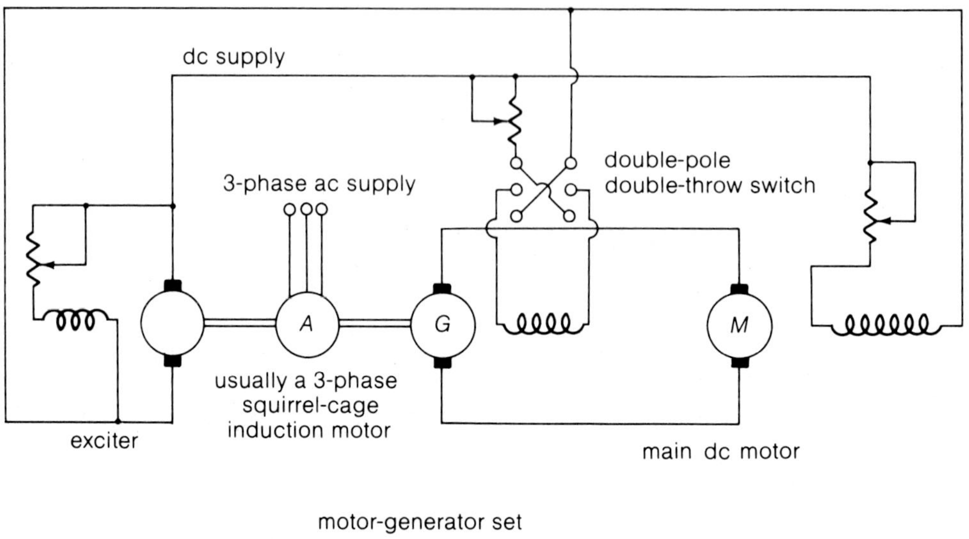

motor-generator set

The *Ward-Leonard system* (or several variations of it) is a conventional scheme that utilizes the armature-terminal-voltage control. While this system is one of the most versatile methods of speed control available, the recent development of solid-state controlled rectifiers capable of handling many kilowatts has opened up a whole new field of solid-state dc motor drives with precise control of motor speed. However, these are not presented here, and only the basic elements of the Ward-Leonard method of speed control are discussed. Since the power available is usually of constant-voltage alternating current, auxiliary equipment in the form of a rectifier or a motor-generator set is required to provide the controlled armature voltage for the motor whose speed is to be controlled.

The schematic diagram of a basic Ward-Leonard system is shown in Figure 9.8.2. An individual motor-generator set supplies power to the armature of the main motor M whose speed is to be controlled; the field-rheostat adjustment in the separately excited generator G controls the armature voltage of the main dc motor M, thereby permitting close control of speed over a wide range. The motor field control (effected by means of the rheostat in the field of motor M) combined with the control of generator voltage makes it possible to achieve the widest possible speed range. With such dual control, base speed being defined as the normal-armature-voltage, full-field speed of the motor, speeds above base speed are obtained by motor-field control at approximately constant horsepower, and speeds below base speed are obtained by armature-voltage control at approximately constant torque. The overall output limitations are as shown in Figure 9.8.3; continuous speed control over very wide ranges is available, with the major advantage that large torques are available at low speeds.

The Ward-Leonard system also has the advantage that, when the armature voltage of the main motor M is reduced to produce a decrease in speed, the motor M will momentarily operate as a generator and the generator G will operate as a motor driving the ac motor A as a generator. Thus,

FIGURE 9.8.3 Output limitations of the Ward-Leonard system combining the armature-voltage and field-rheostat methods of speed control.

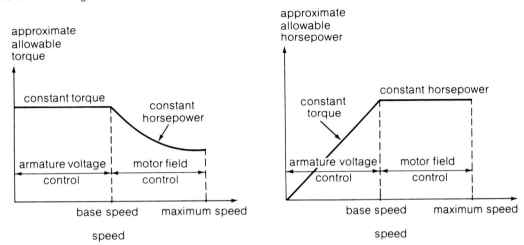

the change in kinetic energy of the main motor during retardation is converted to electrical energy and returned to the power system; such a process is known as *regeneration*. The obvious disadvantage of the Ward-Leonard method of speed control is the requirement of a separate motor-generator set and its associated initial investment.

Next let us consider *dc motor starting*. When voltage is applied to the armature of a dc motor with the rotor stationary, there is no generated emf and the armature current is limited only by the internal armature resistance of the machine. So, in order to limit the starting current to the value that the motor can commutate successfully, all except very small dc motors are started with variable external resistance in series with their armatures. This starting resistance is cut out either manually or automatically as the motor comes up to speed.

Typical forms of manual starters for dc motors are shown in Figure 9.8.4. The *three-point starter* shown in Figure 9.8.4(a) has the holding coil connected in series with the field circuit so that the starter arm, which is spring loaded, drops back to the off position if the field circuit is open-circuited. With sufficient resistance in the field circuit, the holding-coil current may no longer be able to create an electromagnetic pull strong enough to overcome the spring tension, and the starter arm may fall back to the off position. To overcome this objection, in the *four-point starter* shown in Figure 9.8.4(b), the holding coil is connected directly across the supply along with a current-limiting resistor in series; with this arrangement in parallel with the armature and shunt field, the holding-coil current is independent of any field-rheostat changes. The starter arm drops back to the off position if the voltage supply to the machine is removed. In both cases, the field circuit is completed when the moving arm is connected to the first stud of the variable resistor, and it is common to include a relay in the armature circuit to give overload protection. The starting series resistance is cut out in steps with values of resistance between steps such as to keep the initial current at each step within the proper limits as the motor accelerates. Automatic starters are generally used for larger motors, with relays to short out the resistance in series with the armature in successive steps.

FIGURE 9.8.4   Typical forms of manual starters for dc motors: *(a)* three-point starter connected to a shunt motor, *(b)* four-point starter connected to a shunt motor.

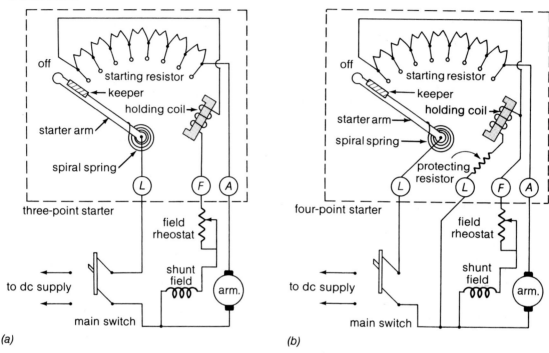

(a)   (b)

When facilities for changing the field current are incorporated within the starter, it is known as a *controller*. It is common to have mechanical interlocks in the controller to ensure that the machine can be started only with full field flux. The *drum controller* for a series motor, with the external series resistor rated to carry full-load current continuously, will provide facilities for starting, speed control, and reversal of the direction of rotation.

EXAMPLE 9.8.1

A 10-hp, 230-V, 500-rpm shunt motor having a full-load armature current of 37 A is started with a four-point starter. The resistance of the armature circuit, including the interpole winding is 0.39 ohm, and the resistances of the steps in the starting resistor are 1.56, 0.78, and 0.39 ohm in the order in which they are cut out successively. When the armature current has dropped to its rated value, the starting box is switched to the next point, thus eliminating a step at a time in the starting resistance. Neglecting field-current changes, armature reaction and armature inductance, find the initial and final values of the armature current and speed corresponding to each step.

SOLUTION

First step: At this point the entire resistance of the starting resistor is in series with the armature circuit; thus,

$$R_{T1} = 1.56 + 0.78 + 0.39 + 0.39 = 3.12 \ \Omega$$

At starting, the counter emf is zero; the armature starting current is then

$$I_{st} = \frac{V_t}{R_{T1}} = \frac{230}{3.12} = 73.7 \ A$$

By the time the armature current drops to its rated value of 37 A, the counter emf is

$$E = 230 - (3.12)(37) = 230 - 115.44 = 114.56 \ V$$

The counter emf, when the motor is delivering rated load at rated speed of 500 rpm with series-starting resistor completely cut out, is

$$230 - (0.39)(37) = 230 - 14.43 = 215.57 \ V$$

The speed corresponding to the counter emf of 114.56 V is then given by

$$N = \frac{114.56}{215.57} \times 500 = 265.7 \ rpm$$

Second step: The 1.56-ohm step is cut out, leaving a total resistance of $3.12 - 1.56 = 1.56$ ohms in the armature circuit. The initial motor speed at this step is 265.7 rpm, which means that the counter emf is still 114.56 V, if the effect of armature reaction is neglected. Accordingly, the resistance drop in the armature circuit is still 115.44 V, so that

$$I_a R_{T2} = 115.44 \quad \text{or} \quad I_a = \frac{115.44}{1.56} = 74 \ A$$

That is, if the inductance of the armature is neglected, the initial current is 74 A. With the final current at 37 A, the counter emf is

$$E = 230 - (37)(1.56) = 230 - 57.72 = 172.28 \ V$$

corresponding to which the speed is

$$N = \frac{172.28}{215.57} \times 500 = 399.6 \ rpm$$

Third step: The total resistance included in the armature circuit is 0.78 ohm at the beginning of this step. The initial motor speed is 399.6 rpm, and the counter emf is 172.28 V. The initial armature current is then $(230 - 172.28)/0.78 = 57.72/0.78 = 74$ A. With the final current at 37 A, the counter emf is

$$E = 230 - (37)(0.78) = 230 - 28.86 = 201.14 \text{ V}$$

corresponding to which the speed is

$$N = \frac{201.14}{215.57} \times 500 = 466.5 \text{ rpm}$$

Thus we have the following results:

Step Number	Current (amps)		Speed (rpm)	
	Initial	Final	Initial	Final
1	74	37	0	266
2	74	37	266	400
3	74	37	400	467
4	74	37	467	500

9.9   TESTING AND EFFICIENCY

As for any other machine, the efficiency of a dc machine can be expressed as

$$\text{Efficiency} = \frac{\text{Output}}{\text{Input}} = \frac{\text{Input} - \text{Losses}}{\text{Input}}$$

$$= 1 - \frac{\text{Losses}}{\text{Input}} \qquad (9.9.1)$$

The losses are made up of rotational losses (3 to 15 percent), armature-circuit copper losses (3 to 6 percent), and shunt-field copper loss (1 to 5 percent). Figure 9.9.1 shows the schematic diagram of a dc machine along with the power division in a generator and a motor. The resistance voltage drop, also known as the *arc drop,* between the brushes and commutator is generally assumed constant at 2 V, and the brush-contact loss is therefore calculated as $(2I_a)$; in such a case, the resistance of the armature circuit should not include the resistance between brushes and commutator.

FIGURE 9.9.1 *(a)* Schematic diagram of a dc machine, *(b)* power division in a dc generator, *(c)* power division in a dc motor.

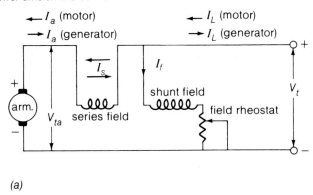

(a)

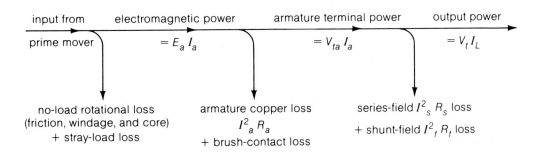

(b)

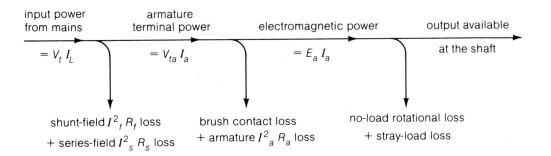

(c)

EXAMPLE 9.9.1

The following data apply to a 100-kW, 250-V, six-pole, 1,000-rpm long-shunt compound generator:

> No-load rotational losses = 4,000 watts
> Armature resistance at 75°C = 0.015 ohm
> Series-field resistance at 75°C = 0.005 ohm
> Interpole field resistance at 75°C = 0.005 ohm
> Shunt-field current = 2.5 A

Assuming a stray-load loss of 1 percent of the output and a brush-contact-resistance drop of 2 V, compute the rated-load effficiency.

SOLUTION

The total armature-circuit resistance (not including that of brushes) is

$$R_a = 0.015 + 0.005 + 0.005 = 0.025 \text{ ohm}$$

$$I_a = I_L + I_f = \frac{100,000}{250} + 2.5 = 402.5 \text{ A}$$

The losses are then computed as follows:

No-load rotational loss =	4,000
Armature-circuit copper loss = $(402.5)^2(0.025)$ =	4,050
Brush-contact loss = $2I_a$ = 2 × 402.5 =	805
Shunt-field circuit copper loss = 250 × 2.5 =	625
Stray-load loss = 0.01 × 100,000 =	1,000
Total losses	10,480 watts

The efficiency at rated load is then given by

$$\eta = 1 - \frac{10,480}{100,000 + 10,480} = 1 - 0.095$$
$$= 0.905 \text{ or } 90.5 \text{ percent}$$

It is usual to determine the efficiency by some method based on the measurement of losses. See the bibliography given at the end of the chapter for test codes and standards.

FIGURE 9.9.2 Schematic diagram of connections for the Kapp-Hopkinson test.

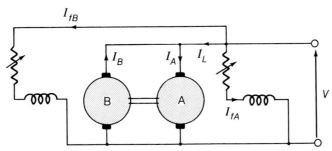

In a method known as the *Swinburne's Test,* the machine is run on no load at rated speed with rated applied voltage. The armature resistance R_a is measured and the armature copper loss is calculated for an assumed armature current I_a. The remainder of the losses is assumed to be constant and independent of the armature current. For a no-load applied voltage V, if the no-load armature current is I_0 and the field current is I_f, the efficiency for a shunt motor is given by

$$\eta_M = 1 - \frac{V(I_f + I_0) + I_a^2 R_a}{V(I_f + I_a)} \tag{9.9.2}$$

The corresponding efficiency as a generator is given by

$$\eta_G = 1 - \frac{V(I_f + I_0) + I_a^2 R_a}{V(I_a + I_0) + I_a^2 R_a} \tag{9.9.3}$$

The value of efficiency obtained by this method is usually greater than the actual efficiency.

A more accurate method of predicting the efficiency is the *Kapp-Hopkinson Test* in which two similar machines are mechanically coupled and electrically connected *back-to-back* as shown in Figure 9.9.2.

Machine A is run up to normal speed as a motor with machine B unexcited. Excitation is then applied to machine B and increased until the open-circuit voltage of this machine is equal to the system voltage. Machine B is then connected to the system; by increasing the excitation of machine B while decreasing that of machine A so that the speed is unchanged, machine A can be made to motor while machine B generates. In this situation, the supply system provides total losses. Assuming η to be the efficiency of each machine, it follows that

$$\text{Generator input} = \text{Motor output}$$
$$= \eta \, (\text{Motor input}) = \eta \, V \, I_A \tag{9.9.4}$$

$$\text{Generator output} = V \, I_B$$
$$= \eta \, (\text{Generator input}) = \eta^2 \, V \, I_A \tag{9.9.5}$$

Then

$$\eta = \sqrt{I_B/I_A} \tag{9.9.6}$$

When the efficiencies of the two machines are not the same, the following analysis applies. The total fixed loss of the system, as seen from Figure 9.9.2, is given by

$$P = V I_L - (I_A^2 R_A + I_B^2 R_B) \tag{9.9.7}$$

so that the fixed loss of each machine can be assumed to be $(P/2)$.

With the input power to the motor being $(V I_A + V I_{fA})$ and the output power from the generator being $V I_B$, the efficiency is given by

$$\text{Motor efficiency} = 1 - \frac{(P/2) + I_A^2 R_A + V I_{fA}}{V(I_A + I_{fA})} \tag{9.9.8}$$

and

$$\text{Generator efficiency}$$
$$= 1 - \frac{(P/2) + I_B^2 R_B + V I_{fB}}{V(I_B + I_{fB}) + (P/2) + I_B^2 R_B} \tag{9.9.9}$$

The values of efficiency thus found are close to the actual efficiency.

BIBLIOGRAPHY

American National Standards Institute. *ANSI Std. No. C.50.4–1965, Rotating Electrical Machinery.* New York, 1965.

Clayton, A. E. *Performance and Design of DC Machines.* London: Pitman, 1938.

Daniels, A. R. *Introduction to Electric Machines.* London: Macmillan, 1976.

Del Toro, V. *Electromechanical Devices for Energy Conversion and Control Systems.* Englewood Cliffs, N.J.: Prentice-Hall, 1968.

Fitzgerald, A. E., et al. *Electric Machinery.* 3d ed. New York: McGraw-Hill, 1971.

Hindmarsh, J. *Electrical Machines and Their Applications.* 2d ed. New York: Pergamon, 1970.

Institute of Electrical and Electronics Engineers. *Standard No. 113–1962, Test Code for Direct-Current Machines.* New York, 1962.

Kloeffler, R. G., et al. *Direct-Current Machinery.* New York: Macmillan, 1950.

Knowlton, A. E., ed. *Standard Handbook for Electrical Engineers.* 8th ed. New York: McGraw-Hill, 1949.

Langsdorf, A. S. *Principles of Direct-Current Machines.* New York: McGraw-Hill, 1959.

Matsch, L. W. *Electromagnetic and Electromechanical Machines.* New York: Intext, 1972.

Say, M. G., and Taylor, E. O. *Direct Current Machines.* New York: John Wiley & Sons, Halsted Press, 1980.

Siskind, C. S. *Direct-Current Machinery.* New York: McGraw-Hill, 1952.

Thaler, G. J., and Wilcox, M. L. *Electric Machines: Dynamic and Steady-State.* New York: John Wiley & Sons, 1966.

PROBLEMS

9-1.

A 100-kW, 250-V shunt generator has an armature-circuit resistance of 0.05 ohm and field-circuit resistance of 60 ohms. With the generator operating at rated voltage, determine the induced voltage at *(a)* full load, and *(b)* half-full load, while neglecting brush-contact drop.

9-2.

A 10-hp, 250-V shunt motor has an armature-circuit resistance of 0.5 ohm and a field resistance of 200 ohms. At no load, rated voltage, and 1,200 rpm, the armature current is 3 A. At full load and rated voltage, the line current is 40 A, and the flux is 5 percent less than its no-load value because of armature reaction. Compute the full-load speed.

9-3.

A 10-kW, 230-V self-excited shunt generator, when delivering rated load, has an armature-circuit voltage drop that is 6 percent of the terminal voltage and a shunt-field current equal to 4 percent of the rated load current. Calculate the resistance of the armature circuit and that of the field circuit.

9-4.

A 20-hp, 250-V shunt motor has a total armature-circuit resistance of 0.25 ohm and a field-circuit resistance of 200 ohms. At no load and rated voltage, the speed is 1,200 rpm and the line current is 4.5 A. At full load and rated voltage, the line current is 65 A; assume the field flux to be reduced by 6 percent from its value at no load due to the demagnetizing effect of armature reaction. Compute the full-load speed.

9-5.

A dc series motor is connected to a load, the torque of which varies as the square of the speed. With the diverter-circuit open, the motor takes 20 A and runs at 500 rpm. Determine the motor current and speed when the diverter-circuit resistance is made equal to the series-field resistance. You may neglect saturation and the voltage drops across the series-field resistance as well as the armature resistance.

9-6.

A 50-kW, 230-V compound generator has the following data:

> Armature-circuit resistance = 0.5 Ω
> Series-field resistance = 0.05 Ω
> Shunt-field circuit resistance = 125 Ω

Assuming the total brush-contact drop to be 2 V, find the induced armature voltage at rated load and rated terminal voltage, for *(a)* short-shunt, and *(b)* long-shunt compound connection.

9–7.

A 20-hp, 440-V, 500-rpm dc shunt motor has rotational losses of 800 W at rated speed. The armature-circuit resistance is 0.5 ohm and the field-circuit resistance is 220 ohms.

    a. Calculate the armature current when the motor is delivering rated load at rated voltage and rated rpm.
    b. Determine the electromagnetic torque, shaft-output torque, and efficiency.
    c. Let the load on the shaft be removed. Find the no-load speed of the motor for the given field current and terminal voltage, neglecting saturation.

9–8.

A 10-kW, 230-V shunt generator, with an armature-circuit resistance of 0.1 ohm and a field-circuit resistance of 230 ohms, delivers full load at rated voltage and 1,000 rpm. If the machine is run as a motor while absorbing 10 kW from 230-V mains, find the speed of the motor. You may neglect the brush-contact drop.

9–9.

The magnetization curve taken at 1,000 rpm on a 200-V dc series motor has the following data:

Field current (A)	5	10	15	20	25	30
Voltage (V)	80	160	202	222	236	244

The armature-circuit resistance is 0.25 ohm and the series-field resistance is 0.25 ohm.
    Calculate the speed of the motor when *(a)* the armature current is 25 A, and *(b)* the electromagnetic torque is 36 N·m. Neglect armature reaction.

9–10.

The open-circuit characteristic data of a shunt generator at 1,200 rpm are given below:

Field current (A)	1.0	1.5	2.0	2.5	3.0	3.5	4.0	4.5	5.0	6.0
Terminal voltage (V)	67	100	134	160	180	200	210	220	230	242

    a. Determine the critical field resistance for self-excitation at 1,200 rpm.
    b. Find the total field-circuit resistance if the induced voltage is 230 V.

9–11.

A dc series motor operates at 750 rpm with a line current of 100 A from the 250-V mains. Its armature-circuit resistance is 0.15 ohm and its series-field resistance is 0.1 ohm. Assuming that the flux corresponding to a current of 25 A is 40 percent of that corresponding to a current of 100 A, determine the motor speed at a line current of 25 A at 250 V.

9–12.

The data for the magnetization curve of a 250-V dc series motor taken at 300 rpm are given below:

Field current (A)	20	30	40	50
Terminal voltage (V)	162	215	250	274

The armature-circuit resistance is 0.3 ohm and the series-field resistance is 0.1 ohm. Determine the speed-torque curves when operating on a 250-V supply for the following cases:

a. Without any external resistance.
b. With an external series resistance of 2 ohms.
c. With a series-winding diverter of 0.1 ohm.
d. With an external series resistance of 2 ohms and an armature diverter of 10 ohms connected across the armature.

9–13.

A 7.5-hp, 250-V, 1,800-rpm shunt motor having a full-load line current of 26 A is started with a four-point starter. The resistance of the armature circuit including the interpole winding is 0.48 ohm and the resistance of the shunt-field circuit including the field rheostat is 350 ohms. The resistances of the steps in the starting resistor are 2.24, 1.47, 0.95, 0.62, 0.40, and 0.26 ohms, in the order in which they are cut out successively. When the armature current is dropped to its rated value, the starting box is switched to the next point, thus eliminating a step in the starting resistance. Neglecting field-current changes, armature reaction, and armature inductance, find the initial and final values of the armature current and speed corresponding to each step.

9–14.

Two shunt generators operate in parallel to supply a total load current of 3,000 A. Each machine has an armature resistance of 0.05 ohm and a field resistance of 100 ohms. If the generated emfs are 200 V and 210 V respectively, determine the load voltage and the armature current of each machine.

9–15.

Three dc generators are operating in parallel with excitations such that their external characteristics are almost straight lines over the working range with the following pairs of data points:

Load Current (A)	Terminal Voltage (V)		
	Generator I	Generator II	Generator III
0	492.5	510	525
2,000	482.5	470	475

Compute the terminal voltage and current of each generator when

a. the total load current is 4,350 A.
b. the load is completely removed without change of excitation currents.

9–16.

The external-characteristic data of two shunt generators in parallel are given below:

Load current (A)	0	5	10	15	20	25	30
Terminal voltage (V) I:	270	263	254	240	222	200	175
Terminal voltage (V) II:	280	277	270	263	253	243	228

Calculate the load current and terminal voltage of each machine when

    a. the generators supply a load resistance of 6 ohms.
    b. the generators supply a battery of emf 200 V and resistance 1.5 ohms.

9–17.

Two similar shunt machines connected to a 240-V dc supply are tested by the Hopkinson method. The generator is delivering an output of 42 kW, and the current taken from the supply is 40 A. The shunt-field circuit resistances of the motor and generator are 300 and 240 ohms respectively. The armature-circuit resistance of each machine is 0.03 ohm. Calculate the efficiency of the generator.

9–18.

A 25-hp, 500-V, 500-rpm dc shunt motor has a field circuit resistance of 500 ohms and an armature-circuit resistance of 0.5 ohm. The full-load armature current is 42 A. The magnetization-characteristic data taken at 400 rpm are given below:

Field current (A)	0.4	0.6	0.8	1.0	1.2
Generated emf (V)	236	300	356	400	432

The difference between the electromagnetic torque and useful-output torque may be assumed to vary directly in proportion with speed.
Calculate the following:

    a. The field current required to operate at full-load torque when running at rated speed. With this field current, find also the no-load speed.
    b. The speed, with this field current, at which the machine must be driven in order to regenerate with full-load armature current.
    c. The extra field-circuit resistance required to run at 600 rpm on no load and on full-load torque.
    d. The external armature-circuit resistance needed to operate at 300 rpm with full-load torque.

9-19.

The data for the magnetization curve of a 150-hp, 250-V, 500-A dc series motor taken at 1,000 rpm are given below:

Field current (A)	100	200	300	400	500	600
Voltage (V)	93	163	194	212	220	227

The resistance of the armature-circuit (including that of the interpole winding and brushes) is 0.015 ohm and that of the series-field winding is 0.01 ohm. The series-field winding has 10 turns per pole. The demagnetizing mmf of armature reaction is 250 At per pole at rated current and may be assumed to vary linearly with current.

    a. Determine (i) the speed, (ii) the electromagnetic power, and (iii) the electromagnetic torque when the armature current is 500 A.

    b. Repeat part *(a)* for an armature current of 250 A.

9-20.

The field of the series motor of Problem 9-19 is replaced by a shunt field of 600 turns and a series field of four turns per pole connected cumulatively for a long-shunt compound motor operation at 250 V. The armature winding is unchanged; the resistance of the shunt field winding is 50 ohms and that of the series-field winding is 0.005 ohm.

Calculate *(a)* the no-load speed and *(b)* the following corresponding to an armature current of 500 A: (i) the speed, (ii) the electromagnetic power, and (iii) the electromagnetic torque.

9-21.

A 100-kW, 250-V long-shunt cumulatively-compound generator has an armature-circuit resistance of 0.025 ohm and a series-field resistance of 0.005 ohm. The open-circuit test data taken at 1,200 rpm are given below:

Shunt-field current (A)	0	1	2	3	4	5	6	7	8	9
Generated voltage (V)	6	57	114	171	218	250	275	293	308	320

There are 1,000 shunt-field turns per pole and 4 series-field turns per pole. Determine the terminal voltage at rated-current output when the shunt-field current is 5 A and the speed is 1,000 rpm. Neglect armature reaction.

9-22.

The magnetization curve, including armature-reaction effects, at an armature-current of 400 A for the generator of Problem 9-21 is given below at 1,200 rpm:

Shunt-field current (A)	0	1	2	3	4	5	6	7	8	9
Generated voltage (V)	6	57	114	171	210	240	262	282	299	313

If the generator is to be flat-compounded so that the full-load voltage at 1,200 rpm is 250 V when the shunt-field rheostat is adjusted to yield a no-load voltage of 250 V at 1,200 rpm, calculate the diverter resistance to be placed across the series field.

9–23.

A 100-hp, 250-V dc shunt motor with 1,000 shunt-field turns per pole has an armature-circuit resistance of 0.025 ohm. The data at 1,200 rpm for the no-load saturation curve and for the magnetization curve including armature-reaction effects at an armature current of 400 A are the same as given in Problems 9–21 and 9–22. The field rheostat is adjusted for a no-load speed of 1,100 rpm.

a. Determine the speed corresponding to an armature current of 400 A.
b. If a stabilizing winding consisting of two cumulative series turns per pole is added, and the resistance of this winding is negligible, find the speed corresponding to an armature current of 400 A.

9–24.

The open-circuit-characteristic data of a 500-kW, 500-V, shunt, six-pole, dc generator obtained at 1,000 rpm are given below:

Field current (A)	0	1	2	3	4	5	6	6.5	7	7.5	8
Terminal voltage (V)	10	100	200	300	390	450	490	500	510	520	525

The armature-circuit resistance including that of brushes is 0.02 ohm.

a. Calculate the resistance of the shunt-field circuit when the no-load voltage is the rated voltage at a speed of 1,000 rpm.
b. With the resistance of the shunt-field circuit as in part *(a)*, neglecting the effect of armature reaction, determine (by graphical means) the terminal voltage when the armature current is 1,000 A and the speed is 1,000 rpm.
c. If the demagnetizing mmf of armature reaction corresponds to a field current of 0.3 A, how would the answer in part *(b)* be affected?

9–25.

The shunt-field winding of the 500-kW, 500-V generator of Problem 9–24 has 2,000 turns per pole. The armature characteristic or field-compounding curve of the shunt generator obtained at a speed of 1,000 rpm is almost a straight line with the following two data points:

Armature current (A)	0	1,000
Shunt-field current (A)	6.5	7.8

Determine the number of turns per pole of a series-field winding for flat-compounding at 1,000 rpm. Neglect the resistance of the series-field winding, and check whether a diverter resistance is necessary if the number of turns in the series-field winding is to be an integer.

9–26.

A separately excited dc generator with an armature-circuit resistance R_a is operating at a terminal voltage V_t while delivering an armature current I_a, and has a constant loss P_c. Find the value of I_a for which the generator efficiency is a maximum.

9–27.

The open-circuit-characteristic data of a 25-hp, 250-V, 85-A shunt motor taken at 1,800 rpm are given below:

Shunt-field current (A)	0.6	0.8	1.0	1.2	1.3	1.5	1.7	2.0	2.4
Terminal voltage (V)	150	185	214	237	245	260	270	280	290

The resistance of the shunt-field circuit, including the field rheostat, is 185 ohms, and the resistance of the armature circuit (including that of the interpole winding and brushes) is 0.08 ohm. The field winding has 3,000 turns per pole. The demagnetizing mmf of armature reaction, at rated armature current, is 0.08 A in terms of the shunt-field current.

The friction, windage, and core losses (known as the no-load rotational losses) are 1,500 W, and the stray-load loss is to be assumed as 1 percent of the output. The field-circuit resistance may be assumed constant at 185 ohms. Calculate the following when the motor draws 85 A from the line: *(a)* speed of the motor, *(b)* electromechanical power, *(c)* mechanical-power output, *(d)* output torque, and *(e)* efficiency.

PART 3

TRANSIENTS AND DYNAMICS

10

TRANSIENTS AND DYNAMICS OF AC MACHINES

Energy storage in magnetic and electric fields is the key to energy conversion by electromagnetic methods. However, as pointed out in Chapter 5, almost all industrial electric machines are magnetic-field devices. Changes in stored magnetic energy, in response to changes in operating conditions, cannot occur instantaneously: a period of accommodation intervenes between the initial and final operating conditions. Not only the electric transients but also the mechanical transients may need to be considered.

An ac machine constitutes a multitude of windings characterized by time-varying self and mutual inductances. Also, saturation and magnetic nonlinearity tend to complicate matters considerably. Physical behavior may often become obscure in the light of mathematical analysis unless one is very careful. The main objective of this chapter is to present a unified development and an adequate background of the fundamental coupled-circuit theory of the transient performance of ac machines, out of which steady-state theory can be deduced as a special case. Emphasis is placed on a more or less rigorous mathematical development that is sound enough from a practical engineering point of view and on presenting a fundamental physical understanding of the machine so that the reader is well equipped with the concepts required to extend the theory as he or she needs it.

10.1 MATHEMATICAL DESCRIPTION OF A THREE-PHASE SYNCHRONOUS MACHINE

The assumptions on which the model and the present analysis are based are listed below:

1. The stator windings are sinusoidally distributed around the periphery along the air gap, as far as all mutual effects with the rotor are concerned.
2. The effect of stator slots on the variation of any of the rotor inductances with rotor angle is neglected.
3. Saturation is neglected, at least for the present.

Justification for any assumption comes from the comparison of performance calculated on that basis with actual performance obtained by test. It will be found that the first two assumptions are quite reasonable, while the third one is not, as saturation does significantly affect certain aspects of machine performance.

The electrical performance equations of a synchronous machine are now developed from a coupled-circuit viewpoint. The generator convention for polarities is adopted so that positive current corresponds to generator action. One can write the voltage relations for the armature (stator) circuits as follows:

$$e_a = p\lambda_a - ri_a \qquad\qquad (10.1.1a)$$

$$e_b = p\lambda_b - ri_b \qquad\qquad (10.1.1b)$$

$$e_c = p\lambda_c - ri_c \qquad\qquad (10.1.1c)$$

FIGURE 10.1.1   Diagram showing conventions.

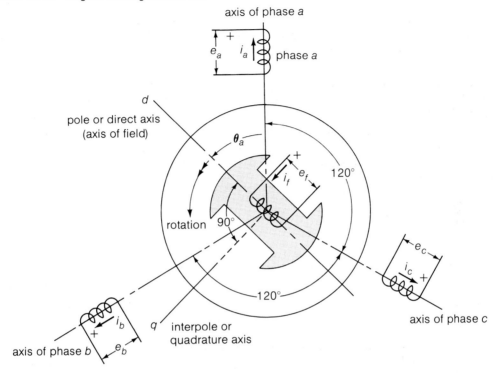

where e_a is the terminal voltage of phase a, λ is the total flux linkage of phase a, and i_a is the current in phase a; a,b,c are the three phases, and r is the resistance of each armature winding, assumed to be the same for the three phases; p is the derivative operator d/dt; and t denotes time.

As for the rotor field circuit, whenever the flux linkage is subject to change, a voltage drop in the direction of positive current occurs in the coil, and the resistance of the field circuit also contributes to a voltage drop in the same direction, so that the voltage equation can be written as

$$e_{fd} = p\lambda_{fd} + r_{fd}\, i_{fd} \tag{10.1.2}$$

The symbols e, λ, and i have the same meaning indicated earlier, and the subscript fd denotes the circuit in question, noting further that the field axis aligns itself with the pole (direct) axis.

From the physical description of a synchronous machine, the field winding has its axis in line with the pole axis, and assuming that the rotor magnetic paths and electric circuits are symmetrical about both the pole and interpolar axes for a salient-pole machine, one may choose conveniently the pole axis as direct axis and the interpolar axis as quadrature axis, the angle between the two axes being 90 electrical degrees. The quadrature axis is located 90 electrical degrees ahead of the direct axis in the direction of rotor rotation. Figure 10.1.1 shows the schematic diagram of

FIGURE 10.1.2   Flattened layout of damper circuits.

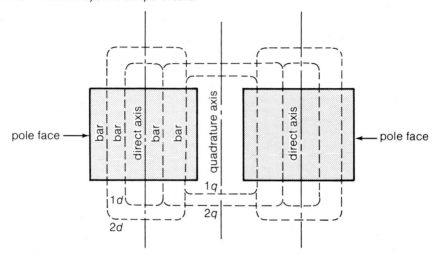

phase windings and field winding along with the notations adopted for voltages and currents, as well as the location of the direct and quadrature axes. A *salient two-pole three-phase machine* has been chosen for simplicity and convenience.

A flattened layout of damper circuits is given in Figure 10.1.2, showing the direct-axis circuits (numbered as $1d$ and $2d$) and quadrature-axis circuits (numbered as $1q$ and $2q$). The symmetrical choice of the rotor circuits has the virtue of making all mutual inductances and resistances between direct- and quadrature-axis rotor circuits equal to zero. Considering only one circuit in each axis, the voltage equations of direct-axis and quadrature-axis amortisseur circuits are given by:

$$e_{1d} = 0 = p\lambda_{1d} + r_{1d} i_{1d} \tag{10.1.3a}$$

$$e_{1q} = 0 = p\lambda_{1q} + r_{1q} i_{1q} \tag{10.1.3b}$$

INDUCTANCES   The self-inductance of any stator winding varies periodically from a maximum (when the direct axis coincides with the phase axis) to a minimum (when the quadrature axis is in line with the phase axis). The inductance has a period of 180 electrical degrees and is expressible by a series of cosines of even harmonics of angle, in view of the rotor symmetry. The variation may approximately be represented as

$$l_{aa} = L_{aa0} + L_{aa2} \cos 2\theta \tag{10.1.4a}$$

$$l_{bb} = L_{aa0} + L_{aa2} \cos 2(\theta - 120°) \tag{10.1.4b}$$

$$l_{cc} = L_{aa0} + L_{aa2} \cos 2(\theta + 120°) \tag{10.1.4c}$$

where $\theta = \theta_a$ is the angle between the direct axis and the axis of phase *a*, as shown in Figure 10.1.1.

The mutual inductances between any two stator phases are also periodic functions of rotor angular position because of the rotor saliency. One may conclude from symmetry considerations that the mutual inductance between phases *a* and *b* should have a negative maximum when the pole axis is lined up 30° behind phase *a* or 30° ahead of phase *b*, and a negative minimum when it is midway between the two phases. It can be shown[1] that the variable part of the mutual inductance is of exactly the same magnitude as that of the variable part of the self-inductance, and that the constant part has a magnitude of very nearly half that of the constant part of the self-inductance. Thus, the variations of stator mutual inductances may be represented as:

$$l_{ab} = l_{ba} = -[L_{ab0} + L_{aa2} \cos 2(\theta + 30°)] \qquad (10.1.5a)$$

$$l_{bc} = l_{cb} = -[L_{ab0} + L_{aa2} \cos 2(\theta - 90°)] \qquad (10.1.5b)$$

$$l_{ca} = l_{ac} = -[L_{ab0} + L_{aa2} \cos 2(\theta + 150°)] \qquad (10.1.5c)$$

All the rotor self-inductances, l_{fdfd}, l_{1d1d}, and l_{1q1q}, are constants since the effects of stator slots and of saturation are neglected. The mutual inductance between any rotor direct- and any rotor quadrature-axis circuit vanishes, i.e., $l_{fd1q} = l_{1qfd} = 0$ and $l_{1d1q} = l_{1q1d} = 0$. The mutual inductance between any two circuits both in direct axis (or both in quadrature axis) is constant, i.e., $l_{fd1d} = l_{1dfd} = $ constant.

Next let us consider the mutual inductances between stator and rotor circuits, which are periodic functions of rotor angular position. Since only the space-fundamental component of the produced flux links the sinusoidally distributed stator, all stator-rotor mutual inductances vary sinusoidally, reaching a maximum when the two coils in question align. Thus their variations may be written as:

$$l_{afd} = l_{fda} = L_{afd} \cos \theta \qquad (10.1.6a)$$

$$l_{bfd} = l_{fdb} = L_{afd} \cos (\theta - 120°) \qquad (10.1.6b)$$

$$l_{cfd} = l_{fdc} = L_{afd} \cos (\theta + 120°) \qquad (10.1.6c)$$

$$l_{a1d} = l_{1da} = L_{a1d} \cos \theta \qquad (10.1.7a)$$

$$l_{b1d} = l_{1db} = L_{a1d} \cos (\theta - 120°) \qquad (10.1.7b)$$

$$l_{c1d} = l_{1dc} = L_{a1d} \cos (\theta + 120°) \qquad (10.1.7c)$$

$$l_{a1q} = l_{1qa} = L_{a1q} \cos (\theta + 90°) = -L_{a1q} \sin \theta \qquad (10.1.8a)$$

$$l_{b1q} = l_{1qb} = -L_{a1q} \sin (\theta - 120°) \qquad (10.1.8b)$$

$$l_{c1q} = l_{1qc} = -L_{a1q} \sin (\theta + 120°) \qquad (10.1.8c)$$

[1]See Concordia, C., *Synchronous Machines* (New York: John Wiley & Sons, 1951), Chap. 2.

FLUX-LINKAGE RELATIONS   The rotor flux-linkage equations are given below:

$$\lambda_{fd} = l_{fdfd}\, i_{fd} + l_{fd1d}\, i_{1d} + l_{fd1q}\, i_{1q}$$
$$- l_{fda}\, i_a - l_{fdb}\, i_b - l_{fdc}\, i_c \tag{10.1.9a}$$

$$\lambda_{1d} = l_{1dfd}\, i_{fd} + l_{1d1d}\, i_{1d} + l_{1d1q}\, i_{1q}$$
$$- l_{1da}\, i_a - l_{1db}\, i_b - l_{1dc}\, i_c \tag{10.1.9b}$$

$$\lambda_{1q} = l_{1qfd}\, i_{fd} + l_{1q1d}\, i_{1d} + l_{1q1q}\, i_{1q}$$
$$- l_{1qa}\, i_a - l_{1qb}\, i_b - l_{1qc}\, i_c \tag{10.1.9c}$$

The above equations may be rewritten as follows after substituting the inductance variations that have been obtained earlier:

$$\lambda_{fd} = - L_{afd}[i_a \cos \theta + i_b \cos (\theta - 120°)$$
$$+ i_c \cos (\theta + 120°)] + L_{fdfd}\, i_{fd} + L_{fd1d}\, i_{1d} \tag{10.1.10a}$$

$$\lambda_{1d} = -L_{a1d}[i_a \cos \theta + i_b \cos (\theta - 120°)$$
$$+ i_c \cos (\theta + 120°)] + L_{1dfd}\, i_{fd} + L_{1d1d}\, i_{1d} \tag{10.1.10b}$$

$$\lambda_{1q} = +L_{a1q}[i_a \sin \theta + i_b \sin (\theta - 120°)$$
$$+ i_c \sin (\theta + 120°)] + L_{1q1q}\, i_{1q} \tag{10.1.10c}$$

The form of the above equations suggests the introduction of new variables for simplification:

$$i_d = K[i_a \cos \theta + i_b \cos (\theta - 120°)$$
$$+ i_c \cos (\theta + 120°)] \tag{10.1.11a}$$

$$i_q = -K[i_a \sin \theta + i_b \sin(\theta - 120°)$$
$$+ i_c \sin (\theta + 120°)] \tag{10.1.11b}$$

where K is a constant that can be conveniently chosen. The currents i_d and i_q are proportional to the components of mmf in the direct and quadrature axes respectively, produced by the resultant of all three armature currents i_a, i_b, and i_c. For balanced phase currents of a given maximum magnitude, the maximum values of i_d and i_q can be of the same magnitude. Under balanced conditions, the maximum magnitude of any one of the phase currents is then given by $\sqrt{i_d^2 + i_q^2}$. In order to achieve this, a value of $2/3$ is assigned to the constant K in the above equations.

The next logical step is to eliminate the old variables i_a, i_b, and i_c in favor of the newly introduced variables. If three currents are to be eliminated, three substitute variables are required in general. Hence, we need to introduce another new variable i_0, which is the conventional zero-phase sequence current of the symmetrical component theory.

$$i_0 = + \frac{1}{3} (i_a + i_b + i_c) \tag{10.1.11c}$$

We have now established the *transformation* from the *"abc"* phase variables to the *"dq*0*"* Blondel variables, which is expressed as follows in matrix notation:

$$
\begin{bmatrix} i_d \\ i_q \\ i_0 \end{bmatrix} = \begin{bmatrix} \dfrac{2}{3}\cos\theta & \dfrac{2}{3}\cos(\theta - 120°) & \dfrac{2}{3}\cos(\theta + 120°) \\ -\dfrac{2}{3}\sin\theta & -\dfrac{2}{3}\sin(\theta - 120°) & -\dfrac{2}{3}\sin(\theta + 120°) \\ \dfrac{1}{3} & \dfrac{1}{3} & \dfrac{1}{3} \end{bmatrix} \begin{bmatrix} i_a \\ i_b \\ i_c \end{bmatrix} \qquad (10.1.12)
$$

The above is rewritten below for the sake of simplicity:

$$[i_B] = [A][i_p] \qquad (10.1.12a)$$

where $[i_B]$ is the *Blondel-component column matrix*, $[i_p]$ is the *phase-component column matrix*, and $[A]$ is the *Blondel-transformation matrix*. The same transformation matrix can be applied to the flux linkages as well as the voltages, just as it has been applied to the currents.

One can find the phase components from the Blondel components through inverse Blondel transformation.

$$[i_p] = [A]^{-1}[i_B] \qquad (10.1.13)$$

where $[A]^{-1}$ can be seen to be

$$
[A]^{-1} = \begin{bmatrix} \cos\theta & -\sin\theta & 1 \\ \cos(\theta - 120°) & -\sin(\theta - 120°) & 1 \\ \cos(\theta + 120°) & -\sin(\theta + 120°) & 1 \end{bmatrix} \qquad (10.1.13a)
$$

Going back to the rotor flux-linkage equations 10.1.10 and substituting the new variable currents, one obtains:

$$\lambda_{fd} = -\frac{3}{2} L_{afd}\, i_d + L_{fdfd}\, i_{fd} + L_{fd1d} \qquad (10.1.14a)$$

$$\lambda_{1d} = -\frac{3}{2} L_{a1d}\, i_d + L_{1dfd}\, i_{fd} + L_{1d1d}\, i_{1d} \qquad (10.1.14b)$$

$$\lambda_{1q} = -\frac{3}{2} L_{a1q}\, i_q + L_{1q1q}\, i_{1q} \qquad (10.1.14c)$$

The above equations contain inductances that are all *constant*, as indicated by the capital L s. Next, the armature flux-linkage relations can be written as:

$$\lambda_a = -l_{aa} i_a - l_{ab} i_b - l_{ac} i_c + l_{afd} i_{fd} + l_{a1d} i_{1d}$$
$$+ l_{a1q} i_{1q} \tag{10.1.15a}$$

$$\lambda_b = -l_{ba} i_a - l_{bb} i_b - l_{bc} i_c + l_{bfd} i_{fd} + l_{b1d} i_{1d}$$
$$+ l_{b1q} i_{1q} \tag{10.1.15b}$$

$$\lambda_c = -l_{ca} i_{ca} - l_{cb} i_b - l_{cc} i_c + l_{cfd} i_{fd} + l_{c1d} i_{1d}$$
$$+ l_{c1q} i_{1q} \tag{10.1.15c}$$

The inductance variations that have been obtained earlier are now substituted in the above equations. The new variables of flux linkages (λ_d, λ_q, and λ_0) are now introduced in terms of the phase variables (λ_a, λ_b, and λ_c), and relatively simple relations given below are obtained after considerable simplification, the details of which are left to the student as a desirable exercise:

$$\lambda_d = -(L_{aa0} + L_{ab0} + \frac{3}{2} L_{aa2}) i_d$$
$$+ L_{afd} i_{fd} + L_{a1d} i_{1d} \tag{10.1.16a}$$
$$= -L_d i_d + L_{afd} i_{fd} + L_{a1d} i_{1d}$$

$$\lambda_q = -(L_{aa0} + L_{ab0} - \frac{3}{2} L_{aa2}) i_q + L_{a1q} i_{1q}$$
$$= -L_q i_q + L_{a1q} i_{1q} \tag{10.1.16b}$$

$$\lambda_0 = -(L_{aa0} - 2L_{ab0}) i_0 = -L_0 i_0 \tag{10.1.16c}$$

λ_d and λ_q may be interpreted as the flux linkages in coils moving with the rotor and centered over the direct and quadrature axes respectively. The equivalent direct-axis moving armature circuit is seen to have the self-inductance

$$L_d = L_{aa0} + L_{ab0} + \frac{3}{2} L_{aa2} \tag{10.1.17a}$$

which is known as the *direct-axis synchronous inductance*. The equivalent quadrature-axis moving armature circuit has the self-inductance

$$L_q = L_{aa0} + L_{ab0} - \frac{3}{2} L_{aa2} \tag{10.1.17b}$$

which is known as the *quadrature-axis synchronous inductance*. Also in the zero-sequence axis, there is an equivalent coil, which is completely separated magnetically from all the other coils and which has the self-inductance

$$L_0 = L_{aa0} - 2L_{ab0} \qquad (10.1.17c)$$

known as the *zero-sequence inductance*.

ARMATURE VOLTAGE EQUATIONS   The new voltages e_d, e_q, and e_0 are now defined in the same manner as the currents and flux linkages through Blondel transformation. The equations 10.1.1 are then substituted for e_a, e_b, and e_c; the new currents i_d, i_q, and i_0 are introduced in order to eliminate the phase quantities i_a, i_b, and i_c. Then the armature voltage relations can be rewritten as

$$e_d = p\lambda_d - \lambda_q p\theta - ri_d \qquad (10.1.18a)$$

$$e_q = p\lambda_q + \lambda_d p\theta - ri_q \qquad (10.1.18b)$$

$$e_0 = p\lambda_0 - ri_0 \qquad (10.1.18c)$$

From a physical standpoint the manipulations through the transformation correspond to the specification of the armature quantities along axes fixed to the rotor and rotating with speed $p\theta$ with respect to the stator. One can therefore naturally expect to find the generated (speed) voltages as well as the induced voltages produced by the rotating flux linkages. The representation of a synchronous machine in terms of the direct-axis, quadrature-axis, and zero-sequence windings is given in Figure 10.1.3.

The complete set of machine-performance equations, consisting of flux-linkage relations and circuit-voltage equations, have now been obtained. These are known as *Park's equations*, which are linear differential equations with constant coefficients, if the rotor speed is constant.

Looking at the flux-linkage equations given by Equations 10.1.14 and 10.1.16, the *reciprocity* of mutual-inductance coefficients, which is an essential condition for the existence of a static equivalent circuit, is not fulfilled. This difficulty occurs because the same transformation has been used for both currents and flux linkages, with the choice of $K = 2/3$. It could have been avoided by other choices of transformation. However, the difficulty is circumvented by suitably defining a per-unit system, which is presented later in this section.

It is often desirable to rewrite the armature flux-linkage relations in a more suitable form. This can be done by substituting the rotor flux-linkage relations into the rotor-circuit voltage equations, solving those for the rotor currents in terms of the field voltage e_{fd} and the armature currents i_d, i_q, and plugging the resultant relations into the armature flux-linkage equations. During these

FIGURE 10.1.3   Representation of a synchronous machine in terms of
direct-axis, quadrature-axis, and zero-sequence windings.

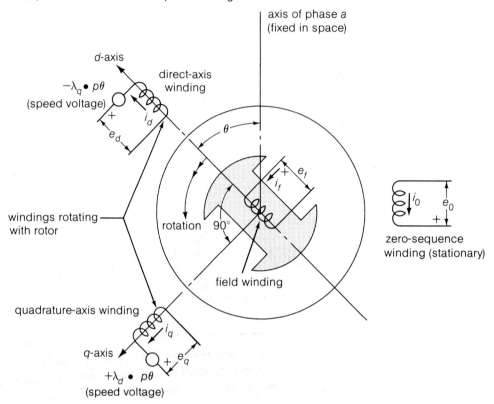

manipulations, it is easier to treat the derivative operator $p(= d/dt)$ algebraically, which is legit-
imate for many problems since all the flux-linkage equations and all the rotor-circuit voltage relations are linear. As a result, one arrives at the equations of the following form:

$$\lambda_d = G(p) \, e_{fd} - L_d(p) \, i_d \tag{10.1.19a}$$

$$\lambda_q = -L_q(p) \, i_q \tag{10.1.19b}$$

$$\lambda_0 = -L_0 i_0 \tag{10.1.19c}$$

where $G(p)$, $L_d(p)$ and $L_q(p)$ are operators expressed as functions of the derivative operator p. $L_d(p)$,
$L_q(p)$, and L_0 are known as the *operational inductances*.

POWER OUTPUT   The instantaneous power output of a three-phase synchronous generator is given by

$$P = [e_a i_a + e_b i_b + e_c i_c]$$ (10.1.20)

In terms of the dq_0-components, it can be shown to be

$$P = \frac{3}{2} [e_d i_d + e_q i_q + 2e_0 i_0]$$ (10.1.21)

For balanced operation, under which zero-sequence quantities vanish, the power output is expressed as

$$P = \frac{3}{2} [e_d i_d + e_q i_q]$$ (10.1.21a)

Equation 10.1.21 may be rewritten by substituting the armature voltage relations given by Equations 10.1.18 for e_d, e_q, and e_0:

$$P = \frac{3}{2} [i_d p\lambda_d + i_q p\lambda_q + 2i_0 p\lambda_0]$$
$$+ \frac{3}{2} [i_q \lambda_d - i_d \lambda_q] p\theta$$ (10.1.22)
$$- \frac{3}{2} r [i_d^2 + i_q^2 + 2i_0^2]$$

The above expression for the power output may be interpreted further: the first term on the right-hand side shows the rate of decrease of armature magnetic energy; the second term indicates the power transferred across the air gap electromagnetically; and the third term is the armature resistance loss.

ELECTROMAGNETIC TORQUE   The electromagnetic torque is obtained by dividing the power transferred across the air gap by the rotor speed, $p\theta$. Thus one has

$$T = \frac{3}{2} [\lambda_d i_q - \lambda_q i_d]$$ (10.1.23)

We have so far developed the volt-ampere relations, flux-linkage equations, and expressions for power output and electromagnetic torque produced for the case of a two-pole synchronous machine. For the general case of a P-pole machine, the expression for torque in Equation 10.1.23 is to be multiplied by $(P/2)$.

L_{ad}-BASE PER-UNIT REPRESENTATION   The rated volt-amperes of the machine and the rated frequency are chosen as the stator three-phase base volt-amperes (or power) and base frequency, respectively. The stator base voltage is selected as the peak value of the rated line-to-neutral voltage, and the stator base current is the peak value of the rated phase current.

$$e_{s_{base}} = \sqrt{2} \, V_{ph} \; ; \, i_{s_{base}} = \sqrt{2} \, I_{ph} \qquad \qquad (10.1.24a,b)$$

where V_{ph} is the rated rms line-to-neutral voltage, and I_{ph} is the rated rms phase current given by

$$I_{ph} = \frac{\text{Rated volt-amperes}}{3 V_{ph}} \qquad \qquad (10.1.25)$$

The stator base impedance, base inductance, and base flux linkage are given by

$$Z_{s_{base}} = \frac{e_{s_{base}}}{i_{s_{base}}} \qquad \qquad (10.1.26a)$$

$$L_{s_{base}} = \frac{Z_{s_{base}}}{\omega_{base}} = \frac{Z_{s_{base}}}{2\pi f_{base}} \qquad \qquad (10.1.26b)$$

$$\lambda_{s_{base}} = \frac{e_{s_{base}}}{\omega_{base}} \qquad \qquad (10.1.26c)$$

and also, the volt-ampere rating of the machine is expressed as

$$\text{Stator 3-phase base volt-amperes (or power)}$$
$$= \frac{3}{2} e_{s_{base}} i_{s_{base}} \qquad \qquad (10.1.27)$$

The time t (seconds) can be made dimensionless by multiplying it by ω_{base}, and the rotor speed $p\theta$ for steady-state synchronous-speed operation becomes unity in per-unit notation. The per-unit reactance can be seen to be the same as the per-unit inductance, and as such the terms can be used interchangeably.

Having chosen the stator-base quantities, the armature voltage equations may now be expressed in per-unit notation. It turns out that Equations 10.1.18 are unchanged when all quantities involved are expressed in the per-unit system described above.

Next the rotor-base quantities are to be specified. In order to maintain the *reciprocity* of mutual-inductance coefficients in the flux-linkage equations expressed in per-unit (i.e., to have in the per-unit system $L_{afd} = L_{fda}; L_{a1d} = L_{1da}; L_{a1q} = L_{1qa}$) it can be shown[2] that we have to choose the volt-ampere base of each rotor circuit equal to the three-phase stator volt-ampere base:

$$e_{fd_{base}} i_{fd_{base}} = e_{1d_{base}} i_{1d_{base}} = e_{1q_{base}} i_{1q_{base}}$$
$$= \text{Rated 3-phase volt-amperes} \qquad (10.1.28)$$

It is desirable to make all per-unit mutual inductances equal between the stator and rotor circuits in each axis. Toward this end, let us consider each of the inductances L_d and L_q to be made up of two parts:

$$L_d = L_{ad} + L_l \; ; L_q = L_{aq} + L_l \qquad (10.1.29a,b)$$

where L_{ad} is the mutual (magnetizing) inductance between the stator and rotor in the *d*-axis, L_{aq} is the mutual (magnetizing) inductance in the *q*-axis between the stator and rotor, and L_l is the leakage inductance (assumed to be the same in both the axes) due to the flux (such as air-gap leakage, slot leakage, and end-turn leakage) that does not link any rotor circuit. The per-unit L_{ad}, L_{aq}, and L_l are obtained by dividing the respective inductances (in henries) by the $L_{s_{base}}$ (in henries).

In order to have in per-unit representation $L_{ad} = L_{afd} = L_{a1d}$, and $L_{aq} = L_{a1q}$, it can be shown that the base quantities are to be chosen in the following way:

$$i_{fd_{base}} = \frac{L_{ad}}{L_{afd}} i_{s_{base}} \qquad (10.1.30a)$$

$$i_{1d_{base}} = \frac{L_{ad}}{L_{a1d}} i_{s_{base}} \qquad (10.1.30b)$$

$$i_{1q_{base}} = \frac{L_{aq}}{L_{a1q}} i_{s_{base}} \qquad (10.1.30c)$$

From the base voltage and base current, the base impedance and the base inductance can be calculated for each of the rotor circuits. Also it can be shown that

$$L_{fd1d_{base}} = L_{fd_{base}} i_{fd_{base}}/i_{1d_{base}} \qquad (10.1.31)$$

---

[2]See Sarma, M. S., *Synchronous Machines (Their Theory, Stability, and Excitation Systems)* (New York: Gorden and Breach, 1979), Chap. 1.

The performance equations of a synchronous machine (known as *Park's equations*) in *per-unit* notation as per the adopted per-unit system, known as the L_{ad}-*base system,* are then given by the following:

$$e_d = p\lambda_d - \lambda_q\, p\theta - ri_d \tag{10.1.32a}$$

$$e_q = p\lambda_q + \lambda_d\, p\theta - ri_q \tag{10.1.32b}$$

$$e_0 = p\lambda_0 - ri_0 \tag{10.1.32c}$$

$$e_{fd} = p\lambda_{fd} + r_{fd}\, i_{fd} \tag{10.1.32d}$$

$$e_{1d} = 0 = p\lambda_{1d} + r_{1d}\, i_{1d} \tag{10.1.32e}$$

$$e_{1q} = 0 = p\lambda_{1q} + r_{1q}\, i_{1q} \tag{10.1.32f}$$

where

$$
\begin{aligned}
\lambda_d &= -L_d\, i_d + L_{ad}\, i_{fd} + L_{ad}\, i_{1d} \\
&= L_{ad}(-i_d + i_{fd} + i_{1d}) - L_l\, i_d \\
&= -L_d(p)\, i_d + G(p)\, e_{fd} + L_{ad}\, i_{1d}
\end{aligned} \tag{10.1.33a}
$$

$$\lambda_q = -L_q(p)\, i_q + L_{aq}\, i_{1q} \tag{10.1.33b}$$

$$\lambda_0 = -L_0\, i_0 \tag{10.1.33c}$$

$$\lambda_{fd} = -L_{ad}\, i_d + L_{fdfd}\, i_{fd} + L_{fd1d}\, i_{1d} \tag{10.1.33d}$$

$$\lambda_{1d} = -L_{ad}\, i_d + L_{1dfd}\, i_{fd} + L_{1d1d}\, i_{1d} \tag{10.1.33e}$$

$$\lambda_{1q} = -L_{aq}\, i_q + L_{1q1q}\, i_{1q} \tag{10.1.33f}$$

Note that the *Heaviside operator p* can be replaced by the *Laplace operator s*. The L_{ad}-base per-unit system assumes a sinusoidal mutual flux-density distribution in the air gap. The base field current is that which establishes the same space fundamental air-gap flux as the unit-peak three-phase armature currents. The significant features of the L_{ad}-base system are that it permits the performance equations to be represented by simple equivalent circuits in the main axes, and that the equivalent-circuit reactances correspond to those normally calculated by the designer. In the *per-unit* notation, the power output and the electromagnetic torque are given by

$$p = e_d\, i_d + e_q\, i_q + 2e_0\, i_0 \tag{10.1.34}$$

and

$$T = \lambda_d\, i_q - \lambda_q\, i_d \tag{10.1.35}$$

FIGURE 10.1.4 Direct-axis equivalent circuit.

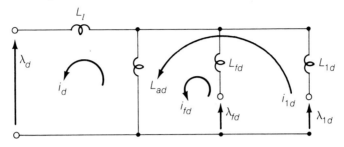

FIGURE 10.1.5 Quadrature-axis equivalent circuit.

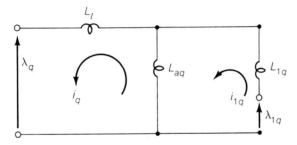

For quantities expressed in their normal physical units and in the per-unit system, two distinctly different notations have not been adopted since the only difference is the factor of 3/2 occurring in the rotor flux-linkage equations and the expressions for power output and electromagnetic torque. Further, we shall use only the per-unit performance equations hereafter.

We introduce the following new inductances:
the field-leakage inductance,

$$L_{fd} = L_{fdfd} - L_{ad}$$

the direct-axis amortisseur-leakage inductance,

$$L_{1d} = L_{1d1d} - L_{ad}$$

and the quadrature-axis amortisseur-leakage inductance,

$$L_{1q} = L_{1q1q} - L_{aq}$$

and neglecting the peripheral leakage $(L_{fd1d} - L_{ad})$, i.e., $L_{fd1d} \simeq L_{ad}$, the equivalent direct-axis and quadrature-axis circuits based on the per-unit flux-linkage equations are shown in Figures 10.1.4 and 10.1.5.

FIGURE 10.2.1   Direct-axis equivalent circuit.

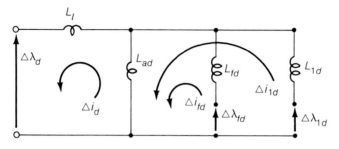

FIGURE 10.2.2   Quadrature-axis equivalent circuit.

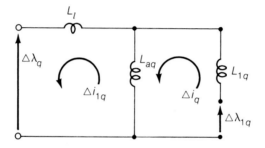

10.2   SYNCHRONOUS-MACHINE TRANSIENT REACTANCES AND TIME CONSTANTS

During transients, one is interested in the changes that occur in the values of variables just after $(t = 0_+)$ and just before $(t = 0_-)$ a disturbance. By defining such differences as

$$\Delta\lambda_d = \lambda_d\big|_{t=0_+} - \lambda_d\big|_{t=0_-} \;\; ; \Delta i_d = i_d\big|_{t=0_+} - i_d\big|_{t=0_-} \;\; ; \text{etc.,}$$

one could verify that the flux-linkage relations in terms of the change-variables are of the same form as the original equations, and hence the equivalent circuits of the same form as before can be drawn for the change-variables as shown in Figures 10.2.1 and 10.2.2.

From the *theorem of constant flux linkage,* the flux linkage of any closed circuit of finite resistance and emf cannot change instantly, and the flux linkage of any closed circuit having no resistance or emf remains constant. If the changes occur in a time that is short compared with the time constants of the circuit, the flux linkages will remain substantially constant during the change. Then the equivalent circuits, which hold good for the short period of disturbance, can be redrawn as in Figures 10.2.3 and 10.2.4.

When the damper circuits are considered along with the main field winding, the armature flux linkage per armature ampere is defined as the *direct-axis subtransient inductance L_d''.* The decrement of the currents in the damper circuits is very rapid compared to that of the field current,

FIGURE 10.2.3   Direct-axis equivalent circuit for the subtransient period.

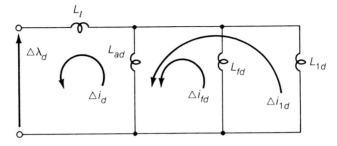

FIGURE 10.2.4   Quadrature-axis equivalent circuit for the subtransient period.

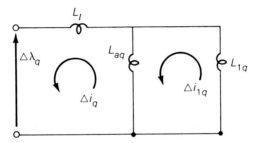

so that all rotor currents except the field current become negligibly small after a few cycles. The effective armature reactance will then increase from the *subtransient* to the *transient* value. The idea of a transient of very short time duration is conveyed through the term "subtransient."

The *subtransient direct-axis reactance* and the *subtransient quadrature-axis reactance* are then defined in per-unit notation as

$$X_d'' = \frac{\Delta\lambda_d}{\Delta i_d} \text{ and } X_q'' = \frac{\Delta\lambda_q}{\Delta i_q} \qquad (10.2.1a,b)$$

It follows from the equivalent circuits of Figures 10.2.3 and 10.2.4 that

$$X_d'' = L_l + \frac{1}{\dfrac{1}{L_{ad}} + \dfrac{1}{L_{fd}} + \dfrac{1}{L_{1d}}} \qquad (10.2.2a)$$

$$X_q'' = L_l + \frac{1}{\dfrac{1}{L_{aq}} + \dfrac{1}{L_{1q}}} \qquad (10.2.2b)$$

Neglecting the amortisseur winding circuits, one can go through a similar procedure and define the *transient direct-axis* and *quadrature-axis reactances:*

$$X_d' = L_l + \cfrac{1}{\cfrac{1}{L_{ad}} + \cfrac{1}{L_{fd}}};$$

$$X_q' = L_l + \cfrac{1}{\cfrac{1}{L_{aq}}} = L_l + L_{aq} = X_q \qquad (10.2.3a,b)$$

NEGATIVE SEQUENCE REACTANCE X_2   If the unexcited field structure is rotated forward at synchronous speed with all rotor circuits closed and the negative-sequence currents are applied to the armature terminals, a backward mmf rotating at synchronous speed with respect to the armature is set up. This mmf, rotating backward at twice the synchronous speed with respect to the rotor, induces currents of twice the rated frequency in the rotor circuits, while keeping the flux linkages of those circuits almost constant at zero value. The armature flux linkage per armature ampere under these conditions is defined as the negative-sequence inductance L_2. The value of X_2 lies between X_d'' and X_q'', and is usually taken as the average value given by

$$X_2 = \frac{X_d'' + X_q''}{2} \qquad (10.2.4)$$

SYNCHRONOUS-MACHINE RESISTANCES   The positive-sequence resistance of a synchronous machine is its armature ac resistance, which is somewhat greater than the dc armature resistance. It is mostly neglected in power-system studies under normal operation and also in system-stability studies. The negative-sequence resistance of a synchronous machine is greater than the positive-sequence resistance, since currents are induced in all the rotor circuits under negative-sequence conditions. Its value depends greatly on the resistance of the damper windings. The zero-sequence resistance of a synchronous machine is usually neglected.

SYNCHRONOUS-MACHINE TIME CONSTANTS   The *direct-axis transient open-circuit time constant T_{d0}'*, also known as the field open-circuit time constant, in per unit is given by

$$T_{d0}' = \frac{X_{fdfd}}{r_{fd}} = \frac{1}{r_{fd}} (X_{ad} + X_{fd}) \qquad (10.2.5)$$

When the armature is open-circuited and there is no amortisseur winding, the change of field current in response to the sudden application, removal, or change of emf in the field circuit is governed by this time constant. The open-circuit ac terminal voltage of an unsaturated machine is directly proportional to the field current, and therefore changes with the same time constant. This time constant is of the order of 750 to 4,000 radians (2 to 11 seconds), and is greater than any of the other time constants discussed below.

The *direct-axis transient short-circuit time constant* T_d' is related to T_{d0}' as

$$T_d' = \frac{X_d'}{X_d} T_{d0}'$$

(10.2.6)

because the ratio of the short-circuit inductance of the field winding (when the armature is short-circuited) to its open-circuit inductance (when the armature is open-circuited) is X_d'/X_d and its resistance is unchanged. This time constant is about one-fourth as large as the open-circuit time constant.

For a machine with damper windings, the *direct-axis subtransient open-circuit time constant* T_{d0}'' and the *short-circuit time constant* T_d'' may be introduced. The subtransient time constant is shorter than the transient time constant, and the short-circuit value is related to the open-circuit value as

$$T_d'' = \frac{X_d''}{X_d'} T_{d0}''$$

(10.2.7)

The open-circuit time constant with one amortisseur circuit is given by

$$T_{d0}'' = \frac{1}{r_{1d}} \left[X_{1d1d} - \frac{X^2_{fd1d}}{X_{fdfd}} \right]$$

$$= \frac{1}{r_{1d}} \left[\frac{X_{ad} X_{fd}}{X_{ad} + X_{fd}} + X_{1d} \right]$$

(10.2.8)

noting that the amortisseur winding and field winding are coupled, and a mutual reactance exists between the two.

The *quadrature-axis subtransient open-circuit time constant* T_{q0}'' may now be introduced for the machine with one additional rotor circuit as the ratio given by

$$T_{q0}'' = \frac{X_{1q1q}}{r_{1q}} = \frac{X_{aq} + X_{1q}}{r_{1q}}$$

(10.2.9)

and the *short-circuit time constant* T_q'' is related to T_{q0}'' as

$$T_q'' = \frac{X_q''}{X_q} T_{q0}''$$

(10.2.10)

Usually T_q'' is very nearly equal to T_d''.

Next, the *quadrature-axis transient open-circuit time constant* T_{q0}' and the *short-circuit time constant* T_q' may be introduced. For the case of a salient-pole machine, these time constants are meaningless; however, for a machine with a solid round rotor, these constants may be applicable; the value of T_q' is about one-half that of T_d'.

TABLE 10.2.1   TYPICAL AVERAGE VALUES OF SYNCHRONOUS-MACHINE CONSTANTS

Machine Constant	Turbogenerator (solid rotor)	Waterwheel Generator (with dampers)	Synchronous Condenser	Synchronous Motor
X_d	1.1	1.15	1.80	1.20
X_q	1.08	0.75	1.15	0.90
X_d'	0.23	0.37	0.40	0.35
X_q'	0.23	0.75	1.15	0.90
X_d''	0.12	0.24	0.25	0.30
X_q''	0.15	0.34	0.30	0.40
X_2	0.13	0.29	0.27	0.35
X_0	0.05	0.11	0.09	0.16
r (dc)	0.003	0.012	0.008	0.01
r (ac)	0.005	0.012	0.008	0.01
r_2	0.035	0.10	0.05	0.06
T_{d0}'	5.6×377	5.6×377	9.0×377	6.0×377
T_d'	1.1×377	1.8×377	2.0×377	1.4×377
$T_d'' = T_q''$	0.035×377	0.035×377	0.035×377	0.035×377
T_a	0.16×377	0.15×377	0.17×377	0.15×377

Adapted from E. W. Kimbark, *Power-System Stability: Synchronous Machines* (New York: Dover Publications, 1956/1968), Chap. 12.

The *armature short-circuit time constant* T_a applies to the direct current in the armature windings and to the induced alternating currents in the field and damper windings, when both are closed. It is equal to the ratio of the average armature short-circuit inductance to the armature resistance under the stated conditions. It is approximately given by

$$T_a = \frac{2X_d'' X_q''}{r(X_d'' + X_q'')} \qquad (10.2.11a)$$

The armature time constant is also sometimes taken to be

$$T_a = \frac{X_2}{r} = \frac{(X_d'' + X_q'')}{2r} \qquad (10.2.11b)$$

for simplicity and convenience.

The currents and voltages of a synchronous machine under transient conditions have components whose magnitudes change in accordance with one or more of the above time constants discussed. Investigation of this aspect is carried out in the following section. A proper understanding of the nature of short-circuit currents of synchronous machines (particularly of large ones) is necessary to evaluate short-circuit forces, stresses, and torques, and to design proper protective relaying and switchgear. In practice, however, one needs to consider an integrated power system as a whole, of which the synchronous machine is just one of the components.

Table 10.2.1 gives typical average values of synchronous-machine constants for various types of machines in per-unit notation, while Table 10.2.2 gives a summary of reactances and time-constants.

TABLE 10.2.2   SUMMARY OF REACTANCES AND TIME CONSTANTS

Reactances		
Synchronous:	d-axis	$X_d = X_{ad} + X_l$
	q-axis	$X_q = X_{aq} + X_l$
Transient:	d-axis	$X_d' = \dfrac{X_{ad} X_{fd}}{X_{ad} + X_{fd}} + X_l$
Subtransient:	d-axis	$X_d'' = \dfrac{X_{ad} X_{fd} X_{1d}}{X_{ad} X_{fd} + X_{fd} X_{1d} + X_{1d} X_{ad}} + X_l$
	q-axis	$X_q'' = \dfrac{X_{aq} X_{1q}}{X_{aq} + X_{1q}} + X_l$
Time-Constants		
Open-circuit transient:	d-axis	$T_{d0}' = \dfrac{1}{r_{fd}}[X_{ad} + X_{fd}]$
Open-circuit subtransient:	d-axis	$T_{d0}'' = \dfrac{1}{r_{1d}}\left[\dfrac{X_{ad} X_{fd}}{X_{ad} + X_{fd}} + X_{1d}\right]$
	q-axis	$T_{q0}'' = \dfrac{1}{r_{1q}}[X_{aq} + X_{1q}]$
Short-circuit transient:	d-axis	$T_d' = \dfrac{1}{r_{fd}}\left[\dfrac{X_{ad} X_l}{X_{ad} + X_l} + X_{fd}\right] = \dfrac{X_d'}{X_d} T_{d0}'$
Short-circuit subtransient:	d-axis	$T_d'' = \dfrac{1}{r_{1d}}\left[\dfrac{X_{ad} X_{fd} X_l}{X_{ad} X_{fd} + X_{fd} X_l + X_{ad} X_l} + X_{ld}\right] = \dfrac{X_d''}{X_d'} T_{d0}''$
	q-axis	$T_q'' = \dfrac{1}{r_{1q}}\left[\dfrac{X_{aq} X_l}{X_{aq} + X_l} + X_{1q}\right] = \dfrac{X_q''}{X_q} T_{q0}''$
Short-circuit armature (dc):		$T_a = \dfrac{1}{r}\left[\dfrac{2X_d'' X_q''}{X_d'' + X_q''}\right]$

10.3   SYNCHRONOUS-MACHINE TRANSIENT PARAMETERS FROM SUDDEN THREE-PHASE SHORT-CIRCUIT TEST DATA

Reactances and time constants of a synchronous machine are of great assistance for predicting short-circuit currents. Conversely, a short-circuit current oscillogram may be utilized to evaluate some of the reactances and time constants. Before we go into details of computing the constants from test data, let us try to get a physical picture of what happens under short-circuit conditions.

Consider a three-phase, initially unloaded, synchronous generator operating at synchronous speed with constant excitation. Let a three-phase short-circuit be suddenly applied at the armature terminals. We shall now attempt to explore the nature of the three armature-phase currents and the field current.

The flux that is produced by the field circuit links the armature circuits. When the three-phase short-circuit fault occurs at $t = 0$, the trapped armature-flux linkages are given by

$$\lambda_a \propto \cos \alpha \qquad (10.3.1a)$$

$$\lambda_b \propto \cos (\alpha - 2\pi/3) \qquad (10.3.1b)$$

$$\lambda_c \propto \cos (\alpha + 2\pi/3) \qquad (10.3.1c)$$

where α is the angle at $t = 0$ between the phase-a axis and the d-axis. Thus the flux linking each phase is different. As the field moves away after $t = 0$, since the flux cannot change immediately, the dc current of appropriate magnitude appears in each phase to preserve the flux. Since the flux is different for all the phases depending on the angle α, the dc currents which appear are also of different magnitudes, depending on α. These currents damp out eventually with the armature time constant T_a.

If the magnitudes of all the dc currents appearing in the armature phases were to be the same, there would be no net resultant flux. But as they happen to be all unequal, there is a resultant flux, which produces a damped fundamental current in the field circuit because of the relative motion between the field and armature circuits, the damping being dictated by the armature time constant T_a.

To an observer on the rotor, the above-mentioned fundamental component produces a uniaxial pulsating flux, which can be resolved into two rotating flux waves of the same magnitude traveling in opposite directions at synchronous speed. The one traveling in the positive direction with respect to the rotor travels at twice the synchronous speed with respect to the stator, and the other one is stationary with respect to the stator. So the latter does not induce anything; but the former produces double-frequency component currents of the same magnitude in the armature phases. These second-harmonic currents give rise to zero net resultant flux. Hence the reflection across the air gap ceases. These second-harmonic currents in the armature phases damp out eventually with the armature time constant T_a.

Thus we see that the unequal dc currents in the armature phases give rise to the fundamental component in the field circuit, which in turn is responsible for producing equal second-harmonic components in the armature phases. Since the process has started in the armature, all these damp out with the armature time constant T_a.

The constant flux from the sustained dc current in the field circuit induces sustained fundamental three-phase armature currents. These currents produce an mmf rotating forward at synchronous speed with respect to the stator, but stationary with respect to the field and centered on the direct axis of the field. This armature mmf opposes the field mmf and tends to reduce the field flux linkages and damper-winding flux linkages. In order to prevent such flux linkage changes, increased field current as well as amortisseur currents are induced. Thus there exist the transient and subtransient dc components in the rotor windings, damped by the direct-axis transient short-circuit time constant T_d', and the direct-axis subtransient short-circuit time constant T_d'', respectively. Correspondingly, the fundamental ac components appear in the armature windings; the transient and subtransient fundamental components damp out with time constants T_d' and T_d'' respectively.

Thus we have the following components in the field current: *(i)* sustained dc, *(ii)* damped dc with time constant T_d', *(iii)* damped dc with time constant T_d'', and *(iv)* damped fundamental ac component with time constant T_a.

The armature phase windings have the following component currents: *(i)* sustained fundamental ac, *(ii)* damped fundamental ac with time constant T_d', *(iii)* damped fundamental ac with time constant T_d'', *(iv)* damped dc components with time constant T_a, which depend on the instant the fault occurs, and *(v)* damped second-harmonic ac with time constant T_a. Thus each current wave in general consists of two kinds of components: *(a)* alternating-current components, and *(b)* direct-

FIGURE 10.3.1 Alternating component of a symmetrical short-circuit armature current of a synchronous machine.

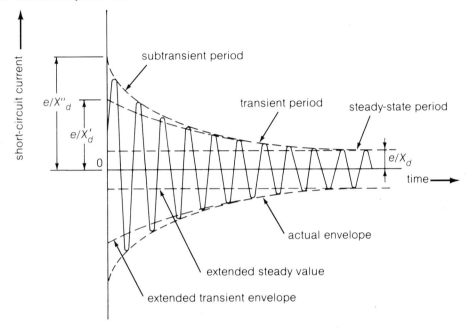

current components, the former of which are equal in all the three phases and the latter of which are dependent upon the particular point on the cycle at which the short circuit occurs.

Note that the short-circuit currents would not decay if the flux linkages were to remain absolutely constant. Actually they have decrements controlled by different time constants of the synchronous machine, as explained above. The induced second-harmonic armature currents are rather small if the machine has damper windings.

Figure 10.3.1 shows an enlarged view of the alternating components of a symmetrical short-circuit armature current of a synchronous machine. The envelope of the current for the subtransient, transient, and steady-state periods may be observed. The subtransient period lasts only for the first few cycles, during which the decrement of the current is very rapid; the transient period covers a relatively longer time, during which the decrement of the current is more moderate; and finally the steady state is attained, during which the current has a sustained value. From the open-circuit prefault armature voltage and the initial values of different armature-current components, the direct-axis reactances X_d, X_d', and X_d'' can be computed. The direct-axis short-circuit time constants T_d' and T_d'' can be evaluated from logarithmic plots of the transient and subtransient components.

For a prolonged short circuit the armature current finally attains a sustained value (during the steady-state period), the magnitude of which is given by e/X_d, where e is the open-circuit voltage of an unsaturated machine, or the voltage read from the air-gap line of a saturated machine. The initial value of the alternating component of the armature current at the beginning of the transient period is given by e/X_d', which is seen by extrapolating the transient envelope in Figure 10.3.1.

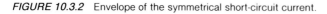

FIGURE 10.3.2   Envelope of the symmetrical short-circuit current.

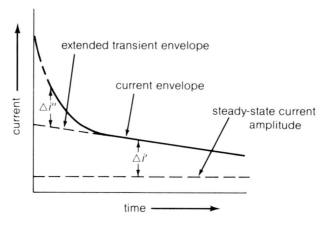

FIGURE 10.3.3   Semilog plot of $\Delta i'$ as a function of time.

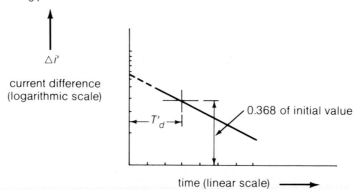

The initial value of the alternating component of the armature current at the commencement of the subtransient period is given by e/X_d''.

Figure 10.3.2 shows the envelope of the symmetrical short-circuit current. The difference $\Delta i'$ indicated in Figure 10.3.2 between the transient envelope (including the extended part) and the steady-state amplitude is plotted to a logarithmic scale as a function of time in Figure 10.3.3. Similarly, the difference $\Delta i''$ between the subtransient and extrapolated transient envelope is plotted to a logarithmic scale as a function of time in Figure 10.3.4. It is seen that both the plots are approximately straight lines, thereby illustrating the essentially exponential nature of the current decrement. From Figures 10.3.3 and 10.3.4, one can evaluate the time constants T_d' and T_d'', by reading the times corresponding to 0.368 of their initial current values.

A typical short-circuit oscillogram in Figure 10.3.5 shows the three-phase armature-current waves as well as the field current. The traces of the armature-phase currents are not symmetrical about the zero-current axis, and definitely exhibit the dc components responsible for offset waves.

FIGURE 10.3.4   Semilog plot of $\Delta i''$ as a function of time.

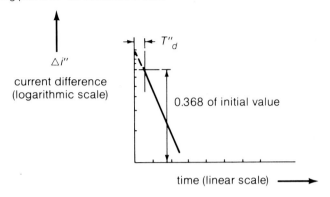

FIGURE 10.3.5   Short-circuit three-phase armature current and field current waves.   Source: Adapted, by permission, from E. W. Kimbark. *Power System Stability: Synchronous Machines* (New York: Dover Publications, 1968), 41.

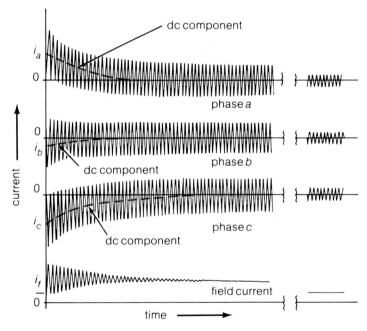

FIGURE 10.4.1   Simple model of a synchronous machine for steady-state stability analysis.

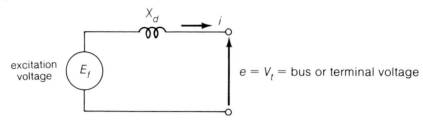

A symmetrical wave, such as that of Figure 10.3.1, is easily obtained by replotting the offset wave with the dc component subtracted. The armature time constant T_a controls the decay of the dc component and is equal to the time required for the dc component to decay to 0.368 of its initial value. Semilog plots of the dc components of the armature currents, shown in Figure 10.3.5, and the ac component of the field current can be utilized for determining the time constant T_a. From the semilog plot of the excess dc component of the field current over sustained value, time constants T_d' and T_d'' can also be evaluated.

For the experimental determination of the quadrature-axis transient and subtransient reactances, the negative-sequence resistance and reactance, the reader is referred to the other published literature.[3]

10.4 SYNCHRONOUS-MACHINE DYNAMICS

Transient stability is concerned with the operation of an interconnected power system subjected to severe disturbances such as faults, switching of circuits, and sudden changes in load. Both electrical and mechanical systems will obviously be in a transient state. Transient stability analysis involves some of both electrical and mechanical properties of the machines of the system, because the machines must adjust the relative angles of their rotors after every disturbance to meet the conditions of power transfer imposed. The duration of the fault depends on the relay and circuit-breaker characteristics, and it is usually of the order of 10 cycles or less for systems with high-speed fault clearing.

The simplest model that can be used for a synchronous machine for steady-state stability studies is given in Figure 10.4.1, while neglecting the armature resistance and saliency. For transient-stability studies, neglecting transient saliency and armature resistance and assuming *constant field-*

---

[3]Institute of Electrical and Electronics Engineers, *Standard No. 115-1965, Code SH0054, Test Procedures for Synchronous Machines* (New York, 1965). Wagner, C. F., and R. D. Evans, *Symmetrical Components* (New York: McGraw-Hill, 1933). Wright, S. H., "Determination of Synchronous Machine Constants by Test," *AIEE Transactions,* vol. 50, pp. 1331–50, December 1931.

FIGURE 10.4.2 Simple model of a synchronous machine for transient-stability studies.

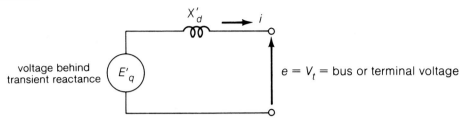

voltage behind transient reactance E'_q

$e = V_t$ = bus or terminal voltage

flux linkages, the simple equivalent circuit of Figure 10.4.2 is often used. Neglecting damper-circuit currents, one has in per-unit notation

$$\lambda_d = -X_d\, i_d + X_{ad}\, i_{fd} \tag{10.4.1}$$

and

$$\lambda_{fd} = -X_{ad}\, i_d + X_{fdfd}\, i_{fd} \tag{10.4.2}$$

Eliminating the field current i_{fd}, one obtains

$$\lambda_{fd} = \frac{X_{fdfd}}{X_{ad}}\left[\lambda_d + \left(X_d - \frac{X_{ad}^2}{X_{fdfd}}\right) i_d\right] \tag{10.4.3}$$

The quantity $[X_d - (X_{ad}^2/X_{fdfd})]$ is a short-circuit reactance of the armature direct-axis circuit, indicating the demagnetizing effect of the armature current, and can be measured at the direct-axis armature terminals with zero field resistance. One may define here for convenience

$$\left(X_d - \frac{X_{ad}^2}{X_{fdfd}}\right) = X_d' \tag{10.4.4}$$

which is called the transient reactance, and

$$\frac{X_{ad}}{X_{fdfd}}\lambda_{fd} = E_q' \tag{10.4.5}$$

which is the *voltage behind transient reactance* and is a quantity proportional to the field-flux linkage. Using the above new definitions, one gets from Equation 10.4.3

$$E_q' = \lambda_d + X_d'\, i_d \tag{10.4.6}$$

FIGURE 10.4.3 Phasor diagram of a salient-pole synchronous generator
$(X'_q = X_q)$ operating at a lagging power factor in the transient state.

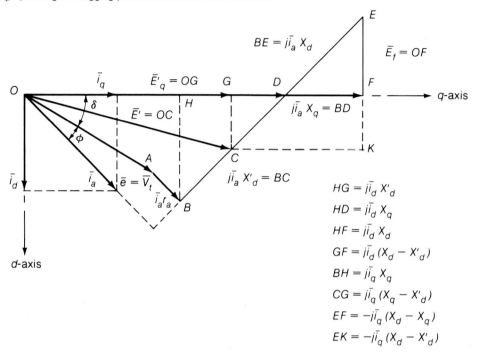

$$HG = j\bar{i}_d X'_d$$

$$HD = j\bar{i}_d X_q$$

$$HF = j\bar{i}_d X_d$$

$$GF = j\bar{i}_d (X_d - X'_d)$$

$$BH = j\bar{i}_q X_q$$

$$CG = j\bar{i}_q (X_q - X'_d)$$

$$EF = -j\bar{i}_q (X_d - X_q)$$

$$EK = -j\bar{i}_q (X_d - X'_d)$$

and since

$$e_q = \lambda_d - ri_q \tag{10.4.7}$$

it follows then

$$e_q = E_q' - X_d' i_d - ri_q \tag{10.4.8}$$

From Equations 10.4.5, 10.4.2, and 10.4.4, one obtains

$$E_q' = -\frac{X_{ad}^2}{X_{fdfd}} i_d + E_f = E_f - (X_d - X_d') i_d \tag{10.4.9}$$

where $E_f = X_{ad} i_{fd}$ is the excitation voltage.

A phasor diagram *similar to that of Figure 8.5.1* can now be developed as shown in Figure
10.4.3 for the case of a salient-pole synchronous generator (with $X_q' = X_q$) operating at a lagging
power factor in the transient state. A quantity corresponding to the field-flux linkage, E_q' or E'
neglecting transient saliency, is now identified on the phasor diagram of Figure 10.4.3.

FIGURE 10.4.4 Transient power-angle characteristic of a salient-pole synchronous machine (with negligible armature resistance).

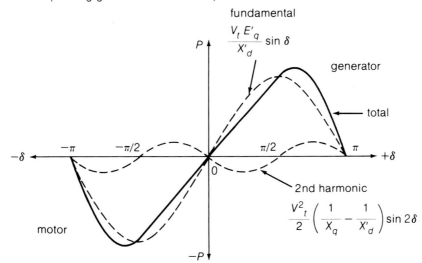

Assuming the field-flux linkage to remain constant, one may derive a *transient power-angle characteristic* in terms of the voltage E_q' in a similar manner to that of a steady-state power-angle characteristic of a salient-pole synchronous machine developed in Section 8.5. The procedure for derivation is the same as before, except that E_q' replaces E_f, and X_d' replaces X_d. Then one obtains the following with negligible armature resistance similar to Equation 8.5.10:

$$P = \frac{V_t E_q'}{X_d'} \sin \delta + \frac{1}{2} V_t^2 \left(\frac{1}{X_q} - \frac{1}{X_d'} \right) \sin 2\delta \qquad (10.4.10)$$

Noting that $(X_d' - X_q)$ is negative, the second-harmonic term, which represents the *transient-reluctance power*, reaches its positive maximum at $\delta = 135°$, corresponding to generator action, so that the maximum transient power occurs at an angle between $\pi/2$ and $3\pi/4$. Equation 10.4.10 is sketched in Figure 10.4.4.

The usual assumptions regarding the representation of a synchronous machine for transient-stability analysis are as follows: *(a)* The effects of speed variation are neglected and the speed is treated as a constant by setting $(p\theta)$ equal to unity in per-unit notation. *(b)* The electromagnetic torque T_e is taken to be equal to the electric power developed or the air-gap power P_e for a generator, and the mechanical-torque input is assumed to be a constant. *(c)* Only the fundamental-frequency voltages and currents are considered.

The effects of saturation are either neglected or approximately taken care of. Amortisseur effects may either be neglected or approximately considered in transient-stability studies. The transient saliency is usually neglected. Assuming the field-flux linkage to remain constant, the transient power-angle or torque-angle relationship is given in per-unit notation by

$$P_e = T_e = \frac{V_t E'}{X_d'} \sin \delta \qquad (10.4.11)$$

in which transient saliency is neglected, V_t is the terminal voltage, and δ is now the angle between $\overline{E'}$ and $\overline{V_t}$.

Variation of field-flux linkage, excitation response, speed changes, and saturation effects can be adequately considered in the step-by-step solution procedure[4] and other digital-computer methods[5] for transient-stability studies.

THE SWING EQUATION   The per-unit mechanical acceleration is given by

$$p^2\theta = \frac{1}{H'}(T_m - T_e) \qquad (10.4.12)$$

where p is the operator d/dt, with t in radians; θ is the angle in electrical radians between the d-axis and the centerline of the phase-a axis; H' is the per-unit inertia constant in $(\text{kW} - \text{rad})/\text{kVA}$; T_m represents the per-unit mechanical-torque input; and T_e stands for the per-unit electromagnetic torque developed. Since θ is continuously changing with time, it is more convenient to measure the angular position with respect to a synchronously rotating reference axis. θ is then expressed as

$$\theta = t + \delta \qquad (10.4.13)$$

where δ is the angle between the reference axis and the quadrature axis of the machine. Equation 10.4.12 is then rewritten as

$$p^2\delta = \frac{1}{H'}(T_m - T_e) \qquad (10.4.14)$$

As it is more common to express the angle in degrees and time in seconds, the acceleration equation is also given by

$$p^2\delta = \frac{180f_b}{H}(T_m - T_e) \qquad (10.4.15)$$

[4]Stevenson, Jr., W. D., *Elements of Power System Analysis*, 3d ed. (New York: McGraw-Hill, 1975), Chap. 14.

[5]Stagg, G. W., and A. H. El-Abiad, *Computer Methods in Power System Analysis* (New York: McGraw-Hill, 1968), Chaps. 7–10.

where p is the operator d/dt, with t in seconds; δ is in electrical degrees; f_b is the base rated frequency; H is the per-unit inertia constant in $(kW - s)/kVA$; and T_m as well as T_e are the per-unit values. The torque caused by friction, windage, and core loss in a machine is usually disregarded. Since the per-unit electrical torque is equal to the per-unit air-gap power developed, one may rewrite Equation 10.4.15 as

$$p^2\delta = \frac{180 f_b}{H} (P_m - P_e) \qquad (10.4.16)$$

where P_m is the per-unit shaft-power input (minus rotational losses, if they are to be considered); and P_e represents the per-unit electrical power developed (electrical power output of the machine plus the armature copper losses). While the equations are developed here with a generator in mind, one can easily modify them appropriately for the case of a motor. Quite often in the literature, the quantities involved are expressed in their natural units without the use of per-unit notation, in which case Equation 10.4.16 becomes

$$p^2\delta = \frac{1}{M} (P_m - P_e) \qquad (10.4.17)$$

where p is the operator d/dt, with t in seconds; δ is in electrical degrees; M is the angular momentum given by $(GH/180 f_b)$ kilojoule-seconds per electrical degree; G is the rating of the machine in kilovolt-amperes; H is in $(kW - s)/kVA$; f_b is the base rated frequency in hertz; P_m and P_e are expressed in kilowatts. Equation 10.4.16 or 10.4.17 is known as the *swing equation*. Note that the inertia constant, defined as the angular momentum at synchronous speed, is truly a constant, while the angular momentum of a machine is not a constant if the angular velocity is changing. However, M is treated as a constant since the speed of the machine does not differ much from the synchronous speed unless the stability limit is exceeded. The shaft-power input P_m is usually taken as a constant for simplifying the calculations, although it is possible in digital-computer programs to take account of the governor action and hence the change in input from the prime mover. The electrical power developed P_e is of the form given by Equation 10.4.11. The solution of Equation 10.4.17 yields the swing curve, which is a graph of δ as a function of time t. The swing equation in the form of Equation 10.4.17 is used for further discussion in this section.

Formal analytical solution of the swing equations is almost impossible for a multimachine system. Elliptic integrals result even for the simple case of one machine connected to an infinite bus, when the resistance is neglected and P_m is taken to be zero. Transient stability studies are usually carried out with the aid of digital computers, utilizing numerical integration schemes, and even then a point-by-point solution is attempted. If the swing curve indicates that the angle δ starts to decrease after reaching a maximum value, it is usually assumed that the system will not lose stability and that the oscillations of δ around the equilibrium point will become successively smaller and eventually damp out. It is possible in certain cases to make use of the equal-area criterion of stability in order to gain an understanding of the transient-stability conditions without formally solving the swing equation. The equal-area criterion cannot be used directly in systems where three or more machines exist and the assumption of an infinite bus is not valid.

EQUAL-AREA CRITERION FOR TRANSIENT STABILITY   For each of the two interconnected machines, the acceleration equations are given by

$$p^2\delta_1 = \frac{P_{a1}}{M_1} \; ; p^2\delta_2 = \frac{P_{a2}}{M_2} \tag{10.4.18}$$

in which subscripts 1 and 2 refer to the interconnected machines, and P_a is the accelerating power
given by $(P_m - P_e)$. The above is rearranged as

$$p^2\delta_{12} = p^2(\delta_1 - \delta_2) = \frac{P_{a1}}{M_1} - \frac{P_{a2}}{M_2} \tag{10.4.19}$$

Multiplying the above by $(2p\delta_{12})$ and integrating both sides, one obtains

$$(p\delta_{12})^2 = 2\int\left(\frac{P_{a1}}{M_1} - \frac{P_{a2}}{M_2}\right) d\delta_{12} \tag{10.4.20}$$

The relative speed between the two machines becomes zero when $(p\delta_{12})$ equals zero, which then
forms the basis of the equal-area criterion. The machines will not remain at rest with respect to
each other the first time $(p\delta_{12})$ becomes zero; but the fact that δ_{12} has momentarily stopped changing may be taken to indicate stability. This is equivalent to the assumption that the swing curve
indicates stability when the angle δ_{12} reaches a maximum and starts to decrease.
    For the case of one machine connected to an infinite bus, the equal-area criterion is given by

$$\int_{\delta_0}^{\delta_s} \frac{2P_{a1}}{M_1} d\delta_{12} = 0 \text{ or } \int_{\delta_0}^{\delta_s} P_{a1} d\delta_{12} = 0 \tag{10.4.21}$$

in which δ_0 is the angle prior to the disturbance when the machine is operating at synchronous
speed, and δ_s is the angle after the disturbance when the machine is again operating at synchronous
speed and the angle ceases to change. M_2 corresponding to the infinite bus is infinite in view of
the infinite inertia of that system. Figure 10.4.5 shows a generator connected to an infinite bus,
as well as the equivalent circuit of the system for transient-stability study. The accelerating power
P_{a1} is given by

$$P_{a1} = P_m - P_e \tag{10.4.22}$$

where P_m is usually assumed to be a constant equal to the value P_{e0} prior to the disturbance, and
P_e is given by the corresponding power-angle equation as

$$P_e = \frac{E_g' E_s}{(X_d' + X_e)} \sin\delta_{12} = P_{max} \sin\delta_{12} \tag{10.4.23}$$

which is shown in Figure 10.4.6.

FIGURE 10.4.5 *(a)* Generator connected to an infinite bus, *(b)* equivalent circuit of the system for transient stability study. (Note: The angles δ_1 and δ_2 are measured from a common reference axis. E'_g is the voltage behind transient reactance of the generator. The resistance is neglected here.)

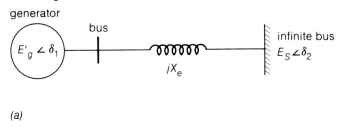

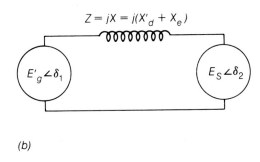

FIGURE 10.4.6 Equal-area criterion for transient stability of a generator connected to an infinite bus, for critical clearing time.

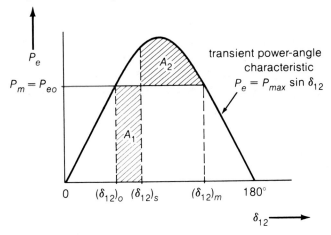

FIGURE 10.4.7   Equal-area criterion for determining the power limit.

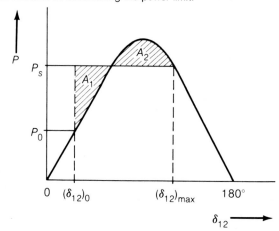

Let us now consider a solid three-phase fault at the generator terminals occurring at $t = 0$, with the fault cleared at a time $t = t_s$ without any alteration in the reactance X_e. The generator is isolated from the system during the fault period. The shaded area A_1 of Figure 10.4.6 represents the energy that tends to increase the rotor speed and cause loss of synchronism between the generator and the infinite bus, while the shaded area A_2 represents the energy that tends to stabilize and restore synchronism. The algebraic sum of the areas producing stability is given by

$$A_2 - A_1 = \int P_{a1} \, d\delta_{12} \qquad \qquad (10.4.24)$$

When the area A_2 is greater than the area A_1, the machine is transiently stable; when A_2 is less than A_1, the machine is transiently unstable. The limiting case between being stable or unstable arises when A_2 is just equal to A_1, in which case the corresponding $(\delta_{12})_s = \delta_c$ is the *critical-clearing angle* and the corresponding clearing time is the *critical-clearing time*.

The equal-area criterion is also easily applied to find the value to which the input power could be suddenly increased without loss of synchronism, for the case of a synchronous generator (connected to an infinite bus) initially operating under steady-state conditions while delivering power P_0, shown in Figure 10.4.7. When areas A_1 and A_2 are equal in Figure 10.4.7, P_s represents such a value.

It is possible to transmit some power even during the fault, to an extent that depends on the nature of the fault. Such a case is easily analyzed by means of the equal-area criterion as shown in Figure 10.4.8. Let P_m be the mechanical input from the prime mover; $(P_{max} \sin\delta_{12})$ be the electric

FIGURE 10.4.8 Equal-area criterion applied to fault-clearing, when power is transmitted during the fault. (Note: Areas A_1 and A_2 are equal.)

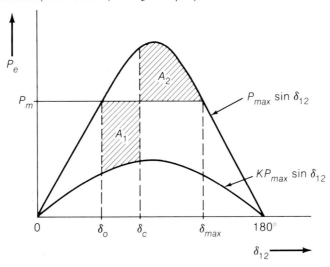

power that can be transmitted by the generator before the fault and also after clearance of the fault; and ($K\,P_{max}\sin\delta_{12}$) be the power that can be transmitted during the fault, where K is a constant depending on the nature of the fault. The largest possible value of δ_{12} for clearing to occur without exceeding the transient-stability limit is called the critical-clearing (or switching) angle, and it is given by δ_c when the areas A_1 and A_2 are equal. The critical-clearing time corresponding to the critical-clearing angle is the time taken for the machine to swing from its original position to its critical-clearing angle. If the fault-clearing should occur at a value of δ_{12} greater than δ_c, the area A_2 will be less than the area A_1 and δ_{12} will continue to increase beyond δ_{max}; the speed will increase further and the generator cannot regain synchronism. The severity of the fault affects the value of K and hence the critical-clearing angle. An increase in the severity of the fault results in a decrease of the value of K. In terms of decreasing severity, the fault conditions are given below:

1. Three-phase fault.
2. Double line-to-ground fault.
3. Line-to-line fault.
4. Single line-to-ground fault.

The effect of grounding impedance in system-neutrals is to decrease the severity. While the three-phase fault is the most severe one, it is the one least likely to occur. For complete reliability a system should be designed for transient stability for three-phase faults at the worst locations. However, reliability is sometimes sacrificed from an economic standpoint.

FACTORS AFFECTING TRANSIENT STABILITY   It is important for the power-system designer and operating engineer to understand the causes of instability in a system, so that appropriate means for its prevention—from economic and technical viewpoints—can be reliably employed. Some of the factors that affect stability considerations are more remote plant siting, larger plant ratings, larger unit sizes, fewer transmission circuits, more heavily loaded transmission, increased pool interchanges, increased emphasis on reliability, and increased severity of various disturbance criteria as well as contingencies. Most of these are just manifestations of the problems of environmental considerations or economic pressures or both.

A summary of the methods used for improving transient stability is given below:

1. Increasing the system voltage.
2. Reducing the overall reactance by appropriately choosing individual system components, providing series compensation, and adding more parallel transmission lines.
3. Providing a strong transmission network with various interconnections and additional transmission lines.
4. Controlling the high-response and optimally stabilized excitation systems.
5. Using high-speed circuit breakers, including reclosing breakers.
6. Introducing a breaking resistor in the case of a sudden loss of a large load.
7. Independent-pole switching of the circuit breakers.
8. Single-pole switching to trip only the faulted phase.
9. Load shedding during the periods of insufficient generation.
10. Generation shedding in response to the line faults.
11. Applying fast turbine-valve control.
12. Providing a dc-transmission line.

Countermeasures taken against sudden, undesired changes in the network should match the disturbance with respect to the location, amount, duration, and dependence on angular difference. A proper combination of measures accomplished rapidly under suitable control can improve system stability to a large extent. Various means of improving power-system stability should be considered in terms of a coordinated system and not as individual measures applied locally. Methods such as the resistor breaking, load shedding, generation shedding, and switching of series capacitors on the electrical side have to be coordinated with fast valving and bypassing of steam or water on the mechanical side. Coordination and cooperation between different ownerships of interconnected power systems are required for economical design and reliable operation. Although a central control by a digital computer may be conceived for the coordination of various measures in the future, much can be accomplished for the time being by local analog-type or digital-type devices.

10.5 MATHEMATICAL DESCRIPTION OF A THREE-PHASE INDUCTION MACHINE

A three-phase induction machine has three distributed stator windings and three distributed rotor windings, as shown schematically in Figure 10.5.1. By assuming the machine to be cylindrical, the self inductances and mutual inductances between stator phases or rotor phases are constant. However, mutual inductances between the stator and rotor coils are functions of the rotor position.

FIGURE 10.5.1 Schematic representation of an idealized three-phase two-pole induction machine.

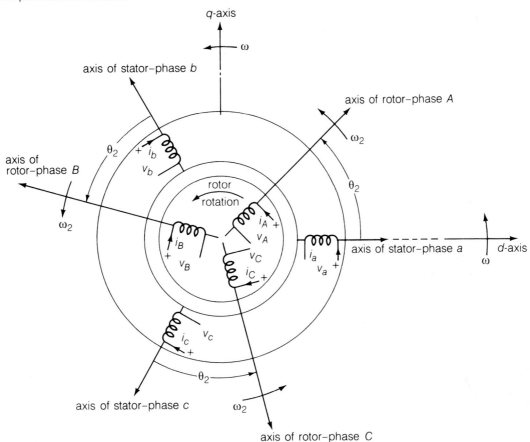

Space mmf and flux waves are considered to be sinusoidally distributed, thereby neglecting the effects of teeth and slots. The machine is regarded as a group of linear coupled circuits, permitting superpositions to be applied, while neglecting saturation, hysteresis, and eddy currents. The stator and rotor phases are balanced and may be connected in wye or delta. The rotor may be a squirrel-cage structure short-circuited on itself by end rings, or a wound rotor with the terminals brought to slip rings on the shaft for connecting to an external circuit. With the rotor terminals short-circuited, the rotor-terminal voltages V_A, V_B, and V_C become zero; however, these are retained in Figure 10.5.1 for generality in our study.

The magnetic axes of each of the stator and rotor phases are shown in Figure 10.5.1, in which the axis of a rotor phase is displaced by an angle θ_2 from the axis of the corresponding stator phase. The angle θ_2 varies with time, as the rotor turns; with a constant rotor angular velocity ω_2, or for a constant slip **S**,

$$\theta_2 = \omega_2 t = (1 - \textbf{S})\, \omega t \tag{10.5.1}$$

where ω is the stator electrical angular velocity.

The voltage relations for the stator and rotor windings from a coupled-circuit viewpoint are given by the following:

$$v_1 = p\lambda_1 + r_1 i_1 \tag{10.5.2}$$

$$v_2 = p\lambda_2 + r_2 i_2 \tag{10.5.3}$$

where the subscript 1 denotes a stator phase *a, b,* or *c;* subscript 2 denotes a rotor phase *A, B,* or *C; r* is the resistance per phase; λ is the flux linkage; and *p* is the differential operator d/dt. As shown in Figure 10.5.1, the motor conventions are used here.

Let the maximum value of mutual inductance between any stator phase and any rotor phase be *M.* Let the inductance per phase and mutual inductance between phases be L_1 and M_1 for the stator winding, and L_2 and M_2 for the rotor winding. The flux-linkage equations are then given by the following:

Stator:

$$
\begin{aligned}
\lambda_a = {}& L_1 i_a + M_1(i_b + i_c) \\
& + M\,[i_A \cos \theta_2 + i_B \cos (\theta_2 + 120°) \\
& + i_C \cos (\theta_2 - 120°)]
\end{aligned}
\tag{10.5.4a}
$$

$$
\begin{aligned}
\lambda_b = {}& L_1 i_b + M_1 (i_a + i_c) \\
& + M\,[i_A \cos (\theta_2 - 120°) + i_B \cos \theta_2 \\
& + i_C \cos (\theta_2 + 120°)]
\end{aligned}
\tag{10.5.4b}
$$

$$
\begin{aligned}
\lambda_c = {}& L_1 i_c + M_1 (i_a + i_b) \\
& + M\,[i_A \cos (\theta_2 + 120°) \\
& + i_B \cos (\theta_2 - 120°) + i_C \cos \theta_2]
\end{aligned}
\tag{10.5.4c}
$$

Rotor:

$$
\begin{aligned}
\lambda_A = {}& L_2 i_A + M_2 (i_B + i_C) \\
& + M\,[i_a \cos \theta_2 + i_b \cos (\theta_2 - 120°) \\
& + i_c \cos (\theta_2 + 120°)]
\end{aligned}
\tag{10.5.5a}
$$

$$\lambda_B = L_2 \, i_B + M_2 \, (i_A + i_C) \\ \quad + M \, [i_a \cos (\theta_2 + 120°) + i_b \cos \theta_2 \\ \quad + i_c \cos (\theta_2 - 120°)]$$

(10.5.5b)

$$\lambda_C = L_2 \, i_C + M_2 \, (i_A + i_B) \\ \quad + M \, [i_a \cos(\theta_2 - 120°) \\ \quad + i_b \cos (\theta_2 + 120°) + i_c \cos \theta_2]$$

(10.5.5c)

In general one may assume

$$i_a + i_b + i_c = 0$$

(10.5.6a)

and

$$i_A + i_B + i_C = 0$$

(10.5.6b)

since there is no neutral connection in the great majority of induction machines, or the external circuitry is such that there are no zero-sequence currents.

Let the self inductance of the equivalent two-phase stator coils and the corresponding quantity for the rotor coils be introduced as

$$L_{11} = L_1 - M_1$$

(10.5.7a)

and

$$L_{22} = L_2 - M_2$$

(10.5.7b)

By making use of Equations 10.5.6 and 10.5.7, one may simplify the flux-linkage equations as given below:

Stator:

$$\lambda_a = L_{11} \, i_a + M \, [i_A \cos \theta_2 + i_B \cos(\theta_2 + 120°) \\ \quad + i_C \cos(\theta_2 - 120°)]$$

(10.5.8a)

$$\lambda_b = L_{11} \, i_b + M \, [i_A \cos(\theta_2 - 120°) + i_B \cos \theta_2 \\ \quad + i_C \cos(\theta_2 + 120°)]$$

(10.5.8b)

$$\lambda_c = L_{11} \, i_c + M \, [i_A \cos(\theta_2 + 120°) \\ \quad + i_B \cos(\theta_2 - 120°) + i_C \cos \theta_2]$$

(10.5.8c)

Rotor:

$$\lambda_A = L_{22} \, i_A + M \, [i_a \cos \theta_2 + i_b \cos(\theta_2 - 120°) \\ + i_c \cos(\theta_2 + 120°)] \tag{10.5.9a}$$

$$\lambda_B = L_{22} \, i_B + M \, [i_a \cos(\theta_2 + 120°) + i_b \cos \theta_2 \\ + i_c \cos(\theta_2 - 120°)] \tag{10.5.9b}$$

$$\lambda_C = L_{22} \, i_C + M \, [i_a \cos(\theta_2 - 120°) \\ + i_b \cos(\theta_2 + 120°) + i_c \cos \theta_2] \tag{10.5.9c}$$

If we now substitute the flux-linkage expressions in the voltage relations of Equations 10.5.2 and 10.5.3, we obtain an algebraically complicated set of nonlinear differential equations; the nonlinearity is introduced by the trigonometric terms of Equations 10.5.8 and 10.5.9 because θ_2 is a function of time. However, the algebra can be greatly simplified by an appropriate transformation and the resulting equations made linear for all constant-speed cases. One can try by splitting the air-gap mmfs into components along the two perpendicular axes (the direct and quadrature axes) as is done for synchronous machines, even though there is now no obvious geometric feature of the machine, such as salient poles, to dictate the specific choice for the location of the axes.

We shall choose the d-axis to coincide with the phase-a axis at $t = 0$, so that its displacement from the phase-a axis at any time t is given by ωt; we shall choose *synchronously rotating axes*[6] rotating at synchronous speed as determined by the electrical angular velocity ω of the impressed stator voltages, with the q-axis located 90 electrical degrees ahead of the d-axis in the direction of rotation. The new stator-current variables i_{1d} and i_{1q} are now defined as follows:

$$i_{1d} = k_d[i_a \cos \omega t + i_b \cos(\omega t - 120°) \\ + i_c \cos(\omega t + 120°)] \tag{10.5.10a}$$

$$i_{1q} = -k_q[i_a \sin \omega t + i_b \sin(\omega t - 120°) \\ + i_c \sin(\omega t + 120°)] \tag{10.5.10b}$$

where each of the constants k_d and k_q is taken to be 2/3. Expressed in matrix form, we have

$$\begin{bmatrix} i_{1d} \\ i_{1q} \\ i_{10} \end{bmatrix} = \frac{2}{3} \begin{bmatrix} \cos \omega t & \cos(\omega t - 120°) & \cos(\omega t + 120°) \\ -\sin \omega t & -\sin(\omega t - 120°) & -\sin(\omega t + 120°) \\ 1/2 & 1/2 & 1/2 \end{bmatrix} \begin{bmatrix} i_a \\ i_b \\ i_c \end{bmatrix} \tag{10.5.11}$$

[6]Other choices of axes are contained in the problems at the end of the chapter.

in which the zero-sequence component i_{10} is zero because of Equation 10.5.6a and the factor 1/3 is chosen for the zero-sequence component. The phase components can then be expressed in terms of the d, q, 0 components as follows:

$$\begin{bmatrix} i_a \\ i_b \\ i_c \end{bmatrix} = \begin{bmatrix} \cos \omega t & -\sin \omega t & 1 \\ \cos(\omega t - 120°) & -\sin(\omega t - 120°) & 1 \\ \cos(\omega t + 120°) & -\sin(\omega t + 120°) & 1 \end{bmatrix} \begin{bmatrix} i_{1d} \\ i_{1q} \\ i_{10} \end{bmatrix} \qquad (10.5.12)$$

where i_{10} is zero, as stated already.

The component currents i_{1d} and i_{1q} produce the same magnetic fields as the actual phase currents. The flux-linkage and voltage transformations are performed using the same transformation matrix as for the currents. Equations 10.5.11 and 10.5.12 can accordingly be written twice more, once with λ replacing i and once with v replacing i.

Corresponding transformations for rotor quantities need now be made in relation to the same synchronously rotating dq-axes. Let θ_s be the angle from the rotor phase-A axis to the d-axis. If the rotor is rotating at a slip $\mathbf{S}$, the d-axis is advancing continuously with respect to a point on the rotor at the rate

$$\frac{d\theta_s}{dt} = p\,\theta_s = \mathbf{S}\omega \qquad (10.5.13)$$

The rotor dq-component currents are then defined as

$$\begin{bmatrix} i_{2d} \\ i_{2q} \\ i_{20} \end{bmatrix} = \frac{2}{3} \begin{bmatrix} \cos \theta_s & \cos(\theta_s - 120°) & \cos(\theta_s + 120°) \\ -\sin \theta_s & -\sin(\theta_s - 120°) & -\sin(\theta_s + 120°) \\ 1/2 & 1/2 & 1/2 \end{bmatrix} \begin{bmatrix} i_A \\ i_B \\ i_C \end{bmatrix} \qquad (10.5.14)$$

where i_{20} is zero because of Equation 10.5.6b. The phase components are then expressed in terms of the dq-components as

$$\begin{bmatrix} i_A \\ i_B \\ i_C \end{bmatrix} = \begin{bmatrix} \cos \theta_s & -\sin \theta_s & 1 \\ \cos(\theta_s - 120°) & -\sin(\theta_s - 120°) & 1 \\ \cos(\theta_s + 120°) & -\sin(\theta_s + 120°) & 1 \end{bmatrix} \begin{bmatrix} i_{2d} \\ i_{2q} \\ i_{20} \end{bmatrix} \qquad (10.5.15)$$

where i_{20} is zero, as stated already. Once more, exactly the same transformations hold good for the rotor flux-linkages λ_A, and λ_B, and λ_C in terms of λ_{2d} and λ_{2q}, and for the rotor voltages v_A, v_B, and v_C in terms of v_{2d} and v_{2q}.

Recognizing that the angle θ_2 of Equation 10.5.1 can be replaced by

$$\theta_2 = \omega t - \theta_s \tag{10.5.16}$$

the flux-linkage equations are obtained as given below after simplifications by the use of trigonometric reduction formulas:

Stator:

$$\lambda_{1d} = L_{11}\, i_{1d} + L_{12}\, i_{2d} \tag{10.5.17a}$$

$$\lambda_{1q} = L_{11}\, i_{1q} + L_{12}\, i_{2q} \tag{10.5.17b}$$

Rotor:

$$\lambda_{2d} = L_{22}\, i_{2d} + L_{12}\, i_{1d} \tag{10.5.18a}$$

$$\lambda_{2q} = L_{22}\, i_{2q} + L_{12}\, i_{1q} \tag{10.5.18b}$$

where $L_{12} = (3/2)\, M$. Next, by making use of the transformation matrix of Equation 10.5.11 and the voltage relations of Equation 10.5.2, the expressions for the dq-component stator voltages v_{1d} and v_{1q} can be obtained in terms of λ_{1d} and λ_{1q}. A similar process is used for the rotor voltages v_{2d} and v_{2q}. The final results are given below:

Stator:

$$v_{1d} = r_1\, i_{1d} + p\lambda_{1d} - \omega\, \lambda_{1q} \tag{10.5.19a}$$

$$v_{1q} = r_1\, i_{1q} + p\lambda_{1q} + \omega\, \lambda_{1d} \tag{10.5.19b}$$

Rotor:

$$v_{2d} = r_2\, i_{2d} + p\lambda_{2d} - \lambda_{2q}(p\theta_s) \tag{10.5.20a}$$

$$v_{2q} = r_2\, i_{2q} + p\lambda_{2q} + \lambda_{2d}(p\theta_s) \tag{10.5.20b}$$

Equations 10.5.17 to 10.5.20 constitute the basic idealized induction machine relations for analysis in dq-variables. The quantity $(p\theta_s)$ in the speed-voltage terms of Equations 10.5.20 is the slip angular velocity given by Equation 10.5.13, which is the relative angular velocity of the synchronously rotating dq-axes with respect to the rotor. A positive value of the slip **S** corresponds to motor action, when $(p\theta_s)$ is positive; negative **S** corresponds to generator action, when $(p\theta_s)$ is negative. If the rotor speed is constant, $(p\theta_s)$ is constant and Equations 10.5.19 and 10.5.20 become linear differential equations with constant coefficients.

The $p\lambda_{1d}$ and $p\lambda_{1q}$ terms in Equations 10.5.19 are often neglected because these are usually small compared with the terms $\omega\,\lambda_{1q}$ and $\omega\,\lambda_{1d}$. This approximation, as in the case of a synchronous machine, corresponds to ignoring the dc-component of the stator short-circuit current following a transient. The r_1-terms are frequently neglected in transient analysis, while they are usually included in steady-state analysis.

The instantaneous power input to the three-phase stator is given by

$$p_1 = v_a\,i_a + v_b\,i_b + v_c\,i_c \qquad\qquad (10.5.21)$$

or, in terms of the *dq*-variables, it can be shown to be

$$p_1 = \frac{3}{2}\,(v_{1d}\,i_{1d} + v_{1q}\,i_{1q}) \qquad\qquad (10.5.22)$$

The electromagnetic torque can be obtained as the power associated with the speed voltages divided by the corresponding speed in mechanical radians per second. The power associated with the speed-voltage terms of Equations 10.5.20 is given by

$$\frac{3}{2}\,[-\lambda_{2q}\,i_{2d}(p\theta_s) + \lambda_{2d}\,i_{2q}(p\theta_s)]$$

$(p\theta_s)$ is positive for motor action and the rotor goes backward with respect to the synchronously rotating *dq*-axes; the corresponding speed is therefore $[-(p\,\theta_s)\,2/\text{poles}]$, expressed in mechanical radians per second. The electromagnetic torque, positive for motor action, is then given by

$$T = \frac{3}{2} \cdot \frac{\text{poles}}{2}\,(\lambda_{2q}\,i_{2d} - \lambda_{2d}\,i_{2q}) \qquad\qquad (10.5.23)$$

The above result is compatible with the concept of torque production from interacting magnetic fields: the *d*-axis magnetic field with the *q*-axis mmf, and the *q*-axis magnetic field with the *d*-axis mmf. The sine of the space-displacement angle has a magnitude of unity for both interactions. This torque is available to furnish the rotational losses, drive the load, and provide acceleration. The inertia torque required to accelerate the rotating mass is given by

$$T_{inertia} = J\,\frac{d\omega_0}{dt} = J\,\frac{d^2\theta_0}{dt^2} \qquad\qquad (10.5.24)$$

where J is the moment of inertia of the rotor and the mechanical equipment coupled to it, θ_0 is the shaft-position angle and ω_0 is the shaft angular velocity, the angular measurements being in mechanical radians.

10.6 INDUCTION-MACHINE ELECTRICAL TRANSIENTS

Let us consider an induction machine operating as a motor or a generator, and let a *three-phase short-circuit* occur at its terminals. Because of the "trapped" flux linkages with the rotor circuits, in either case, the machine will feed current into the fault; this fault current will, of course, decay to zero in time. Let us determine its initial magnitude, while neglecting the stator-phase resistance r_1 as well as the $p\lambda_{1d}$ and $p\lambda_{1q}$ terms in the stator-voltage equation 10.5.19; that is to say, we shall consider the fundamental component alone, ignoring the dc component in the short-circuit current. Further, let us consider the machine to be operating at small values of slip so that the speed-voltage terms $\lambda_{2q}(p\theta_s)$ and $\lambda_{2d}(p\theta_s)$ in the rotor-voltage equations 10.5.20 could be ignored. Such an assumption is justified because an induction machine usually operates at small values of slip, and also the short-circuit current decays so rapidly that the speed is not changed appreciably.

The voltage equations 10.5.19 and 10.5.20 are then simplified as follows:
Stator:

$$v_{1d} = -\omega\lambda_{1q} \qquad\qquad (10.6.1a)$$

$$v_{1q} = \omega\lambda_{1d} \qquad\qquad (10.6.1b)$$

Rotor:

$$v_{2d} = r_2 i_{2d} + p\lambda_{2d} = 0 \qquad\qquad (10.6.2a)$$

$$v_{2q} = r_2 i_{2q} + p\lambda_{2q} = 0 \qquad\qquad (10.6.2b)$$

Note that the rotor voltages in Equations 10.6.2 are set equal to zero to indicate either a squirrel-cage rotor or a wound rotor with short-circuited slip rings. The rotor external impedance, if any, for the case of a wound rotor can be included along with the rotor parameters.

Solving for i_{2d} from Equation 10.5.18a, we get

$$i_{2d} = \frac{\lambda_{2d} - L_{12} i_{1d}}{L_{22}} \qquad\qquad (10.6.3)$$

and, upon substitution into Equation 10.5.17a

$$\lambda_{1d} = \left(L_{11} - \frac{L_{12}^2}{L_{22}}\right) i_{1d} + \frac{L_{12}}{L_{22}} \lambda_{2d} \qquad\qquad (10.6.4)$$

which may be written as

$$\lambda_{1d} = L' i_{1d} + \frac{L_{12}}{L_{22}} \lambda_{2d} \qquad\qquad (10.6.5)$$

where

$$L' = L_{11} - \frac{L_{12}{}^2}{L_{22}} \qquad (10.6.6)$$

Similarly, from Equations 10.5.17b and 10.5.18b, one obtains

$$\lambda_{1q} = L' i_{1q} + \frac{L_{12}}{L_{22}} \lambda_{2q} \qquad (10.6.7)$$

The quantity L' is known as the *transient inductance* of an induction machine, which is analogous to the direct-axis transient inductance L_d' of a synchronous machine.

The voltages v_{1d} and v_{1q} go to zero when a three-phase short-circuit occurs at the terminals; it follows from Equations 10.6.1 that the stator-flux linkages λ_{1d} and λ_{1q} must also go to zero. Note that the dc component of the stator current, which physically prevents λ_{1d} and λ_{1q} from changing suddenly, is ignored here by neglecting the $p\lambda_{1d}$ and $p\lambda_{1q}$ terms. With λ_{1d} and λ_{1q} each equal to zero under these circumstances, in order to maintain the rotor-flux linkages constant at their initial prefault values λ_{2d0} and λ_{2q0}, one requires that the stator component currents be, from Equations 10.6.5 and 10.6.7,

$$i_{1d} = - \frac{(L_{12}/L_{22})}{L'} \lambda_{2d0} \; ; i_{1q} = - \frac{(L_{12}/L_{22})}{L'} \lambda_{2q0} \qquad (10.6.8a,b)$$

The above equations yield *initial* values immediately after the short-circuit takes place, i.e., at $t = 0_+$; these would continue to exist if there were no rotor resistance to cause their decay. The instantaneous phase currents can be obtained from the transformation equation 10.5.12. The corresponding rms stator current is given by

$$\begin{aligned}
\bar{I}_1 = I_{1d} + jI_{1q} &= \frac{i_{1d}}{\sqrt{2}} + j \frac{i_{1q}}{\sqrt{2}} \\
&= - \frac{1}{X'} \omega \frac{L_{12}}{L_{22}} \left(\frac{\lambda_{2d0}}{\sqrt{2}} + j \frac{\lambda_{2q0}}{\sqrt{2}} \right) \qquad (10.6.9)
\end{aligned}$$

where $X' = \omega L'$ is known as the *transient reactance*. The rms magnitude of $\bar{I}_1$ is given by

$$I_1 = \frac{1}{X'} \omega \frac{L_{12}}{L_{22}} \frac{1}{\sqrt{2}} \sqrt{\lambda_{2d0}^2 + \lambda_{2q0}^2} \qquad (10.6.10)$$

which is known as the *initial symmetrical short-circuit current.*

Next we shall examine how this short-circuit current decays, and how we can evaluate it from prefault conditions.

DECAY OF THE SHORT-CIRCUIT CURRENT   Because of the rotor resistance r_2, the short-circuit current will decay to zero in time. The decrement can be found by examining the decay of rotor currents i_{2d} and i_{2q}. Substituting the value of λ_{2d} from Equation 10.5.18a into Equation 10.6.2a, one has

$$r_2 \, i_{2d} + L_{22} \, pi_{2d} + L_{12} \, pi_{1d} = 0 \qquad\qquad (10.6.11)$$

With $\lambda_{1d} = 0$ from Equation 10.5.17a

$$i_{1d} = -\frac{L_{12}}{L_{11}} i_{2d} \qquad\qquad (10.6.12)$$

Using the above in Equation 10.6.11, one gets

$$r_2 \, i_{2d} + \frac{1}{L_{11}} \left(L_{11} - \frac{L_{12}^2}{L_{22}} \right) L_{22} \, pi_{2d} = 0$$

or

$$i_{2d} + \frac{L'}{L_{11}} \frac{L_{22}}{r_2} \frac{di_{2d}}{dt} = 0 \qquad\qquad (10.6.13)$$

in which the quantity $(L'/L_{11})(L_{22}/r_2)$ is known as the *short-circuit time constant T'*. The ratio (L_{22}/r_2) is the *time constant of the rotor circuit alone*, T_0', of the induction machine that is responsible for the decay of rotor transients when the stator is open-circuited. Because of the coupling between the rotor and stator windings, with the stator short-circuited, the apparent inductance is lower. T' can be expressed in terms of T_0' as

$$T' = \frac{L'}{L_{11}} \cdot \frac{L_{22}}{r_2} = T_0' \frac{L'}{L_{11}} = T_0' \frac{X'}{X_{11}} \qquad\qquad (10.6.14)$$

Identical results are obtained by considering the decay of the rotor component i_{2q}. The exponential decay of the transient current whose initial rms magnitude is given by Equation 10.6.10 is characterized by the short-circuit time constant T', which is similar to the direct-axis short-circuit transient time constant T_d' of a synchronous machine. The results are about the same as in a synchronous machine except that the short-circuit current in an induction machine decays to zero. Note again that the dc component of the short-circuit current has been ignored in the present discussion.

TRANSIENT EQUIVALENT CIRCUIT   We shall now attempt to evaluate the flux linkages λ_{2d0} and λ_{2q0} from the prefault operating conditions of the machine, and also, in the process, develop a transient equivalent circuit of the induction machine.

Let the rms value of each of the stator-phase currents prior to the short-circuit be I_{10} with its dq-components I_{1d0} and I_{1q0} given by $i_{1d0}/\sqrt{2}$ and $i_{1q0}/\sqrt{2}$, respectively. Rewriting Equation 10.6.5 for prefault conditions and multiplying by ω, one obtains

$$\omega \frac{L_{12}}{L_{22}} \lambda_{2d0} = \omega \lambda_{1d0} - X' \, i_{1d0} \qquad (10.6.15)$$

Replacing $\omega \lambda_{1d0}$ by v_{1q0} as per Equation 10.6.1b and expressing Equation 10.6.15 in phasor form, we get

$$\omega \frac{L_{12}}{L_{22}} \frac{\overline{\lambda}_{2d0}}{\sqrt{2}} = \overline{V}_{1q0} - X' \, \overline{I}_{1d0} \qquad (10.6.16)$$

where V_{1q0} is given by $v_{1q0}/\sqrt{2}$. Similarly, Equation 10.6.7 yields

$$\omega \frac{L_{12}}{L_{22}} \frac{\overline{\lambda}_{2q0}}{\sqrt{2}} = -\overline{V}_{1d0} - X' \, \overline{I}_{1q0} \qquad (10.6.17)$$

Equations 10.6.16 and 10.6.17 can be combined to give

$$\begin{aligned} \omega \frac{L_{12}}{L_{22}} & \left(\frac{\overline{\lambda}_{2d0}}{\sqrt{2}} + j \frac{\overline{\lambda}_{2q0}}{\sqrt{2}} \right) \\ &= \frac{1}{j} \left[(\overline{V}_{1d0} + j\overline{V}_{1q0}) - jX'(\overline{I}_{1d0} + j\overline{I}_{1q0}) \right] \\ &= \frac{1}{j} (\overline{V}_{10} - jX' \, \overline{I}_{10}) = \frac{1}{j} \overline{E}_1' \end{aligned} \qquad (10.6.18)$$

in which $\overline{V}_{10}$ is the prefault rms terminal voltage, and $\overline{E}_1'$ is given by

$$\overline{E}_1' = \overline{V}_{10} - jX' \, \overline{I}_{10} \qquad (10.6.19)$$

which is known as the initial *voltage behind the transient reactance* of an induction machine; it is proportional to the rotor-flux linkages. The rms magnitude of the initial symmetrical short-circuit current expressed earlier by Equation 10.6.10 is now given by

$$I_1 = E_1'/X' \qquad (10.6.20)$$

The rms magnitude of the symmetrical short-circuit current at any time t after the fault may then be expressed as

$$I_1' = \frac{E_1'}{X'} e^{-t/T'} \qquad (10.6.21)$$

FIGURE 10.6.1 Simple transient equivalent circuit of an induction machine.

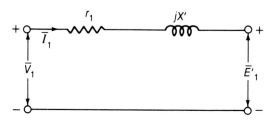

in which the decrement characterized by the short-circuit time constant T' of Equation 10.6.14 is applied. In case there is an external reactance between the machine terminals and the fault, the time constant can be modified by adding that reactance to the numerator and denominator of the fraction in Equation 10.6.14.

From the foregoing discussion, it follows that an induction machine can be represented by a simple transient equivalent circuit of Figure 10.6.1, in which X' is the transient reactance and r_1, the stator-phase resistance, may be added as shown for somewhat greater precision. The results are once again very similar to those obtained for a synchronous machine. However, in the case of an induction machine, while the initial short-circuit current is relatively high compared with its normal current, the electrical transient usually subsides quite rapidly compared with the duration of the motional transient. Hence the electrical transients in an induction machine are ignored in many problems. We shall illustrate this statement through Example 10.6.1. It should be noted here that large 3,600-rpm induction motors are among the principal exceptions to this statement.

EXAMPLE 10.6.1

A 500-hp, 440-volt, 60-Hz, three-phase, star-connected, six-pole squirrel-cage induction motor has a full-load efficiency of 0.95 and a lagging power factor of 0.9. Determine the motor rms short-circuit current if a three-phase short-circuit occurs on its supply lines at the motor terminals while the motor is operating in the steady state under rated full-load conditions.

The motor parameters in ohms per phase referred to the stator are given below:

$$X_{l1} = 0.075; \quad X_{l2} = 0.075; \quad X_m = 3.00$$
$$r_1 = 0.009; \quad r_2 = 0.008$$

SOLUTION

Since $X_{l1} = \omega(L_{11} - L_{12})$, $X_{l2} = \omega(L_{22} - L_{12})$, $X_m = \omega L_{12}$, and $X' = \omega L' = \omega[L_{11} - (L_{12}^2/L_{22})]$, the motor transient reactance can be worked out as

$$X' = X_{l1} + X_m - \frac{X_m^2}{X_{l2} + X_m}$$

or

$$X' = 0.075 + 3.00 - \frac{(3.00)^2}{0.075 + 3.00} = 0.148 \text{ ohm per phase}$$

The prefault stator current with the motor operating in the steady state under rated full-load conditions is given by

$$I_1 = \frac{500 \times 746}{0.90 \times 0.95 \times 440 \times \sqrt{3}} = 572.46 \text{ A}$$

Choosing the terminal voltage as the reference phasor, the prefault voltage behind transient reactance is then

$$\overline{E_1}' = \frac{440}{\sqrt{3}} - (0.009 + j0.148)(572.46\angle - \cos^{-1} 0.90)$$

or

$$\overline{E_1}' = 225\angle - 19.2° \text{ V}$$

The initial rms short-circuit current, from Equation 10.6.20, is

$$\frac{225}{0.148} = 1,520 \text{ A}$$

The open-circuit time constant T_0' given by $\dfrac{L_{22}}{r_2}$ is

$$T_0' = \frac{3.00 + 0.075}{2\pi \times 60 \times 0.008} = 1.02 \text{ sec.}$$

and the short-circuit time constant T', given by Equation 10.6.14, is

$$T' = 1.02 \times \frac{0.148}{3.075} = 0.05 \text{ sec.}$$

The rms short-circuit current is then given by Equation 10.6.21 as

$$I_1 = 1,520e^{-t/0.05} \text{ A}$$

The short-circuit time constant of 0.05 sec. corresponds to three cycles on a 60-Hz base; that is to say, the short-circuit current decreases to 36.8% of its initial value in three cycles. It will substantially disappear in about 10 cycles. Thus one can see that the electrical transient usually decays rapidly, even though the initial short-circuit current of an induction machine may be relatively high compared with its normal current.

10.7 INDUCTION-MACHINE DYNAMICS

For the study of electrical transients in ac machines, one has to make a compromise between the detailed analysis that may become necessary when only conditions within a single machine are of concern and the more approximate models that must be adopted for a simplified representation of the machine as a single element of a large dynamic system. From the systems point of view, one has to decide on reasonable approximations that can be allowed from an engineering viewpoint and come up with simple equivalents for the analysis of dynamic problems.

In many power-system transient-stability studies, the system loads are represented by shunt impedances, which are treated as static impedances. However, not all loads behave as static impedances, and a notable exception is the case of large induction-motor loads. The load inertia has to be considered to account for mechanical transients, since these transients may have a significant effect on generator swings. Several representations may be used for induction-motor loads, depending upon the degree of refinement desired:

(i) Simple impedance.
(ii) Steady-state equivalent circuit.
(iii) Steady-state behavior including only the motor's mechanical transients.
(iv) Transient equivalent circuit including both mechanical and rotor electrical transients.
(v) Transient behavior including the motor's mechanical, rotor electrical, and stator electrical transients.

Let us further consider the model given by case *(iv)* for the analysis of induction-machine dynamics. The mechanical transients are identified by the differential equation for the slip:

$$\frac{d\mathbf{S}}{dt} = \frac{1}{2H} (T_m - T_e) \qquad (10.7.1)$$

where H is the inertia constant in $kW \cdot s/kVA$, T_m is the mechanical load torque in per unit, T_e is the electromagnetic torque in per unit, and $\mathbf{S}$ is the motor slip in per unit. The differential equation describing the rate of change of voltage E_1' behind the transient reactance X' of Figure 10.6.1 is given by[7]

$$\frac{d\overline{E_1'}}{dt} = -j\omega\mathbf{S}\overline{E'} - \frac{1}{T_0'}[\overline{E_1'} - j(X-X')\overline{I_1}] \qquad (10.7.2)$$

where T_0' is the rotor open-circuit time constant in seconds, X is the reactance in per unit with open-circuited rotor, X' is the reactance in per unit with blocked rotor, and $\overline{I_1}$ in per unit (from Figure 10.6.1) is given by

$$\overline{I_1} = \frac{\overline{V_1} - \overline{E_1'}}{r_1 + jX'} \qquad (10.7.3)$$

Equation 10.7.2, along with the swing equation 10.7.1 representing the rotor acceleration and the equations for rotor electrical and mechanical torques, is solved to account for the transient behavior of an induction machine including both mechanical and rotor electrical transients. A step-by-step method of solution is indicated below:

1. Let the disturbance be initiated on the system at a time t_n. The electrical torque of the induction motor is computed corresponding to the time t_n.
2. The change in motor speed in a certain time interval Δt is calculated from Equation 10.7.1, based on the difference between the electrical and mechanical torques of the motor. The motor speed at $(t_n + \Delta t)$ is then found.
3. The change in $\overline{E_1'}$, proportional to the change in rotor flux linkage, during Δt is computed from Equation 10.7.2 based on the conditions at t_n.
4. The procedure is then repeated at each time interval.

Besides power-system stability studies involving large concentrations of induction-motor loads, common induction-machine dynamic problems are associated with starting and stopping and with the ability of the motor to continue its operation during serious disturbances of the supply system. A typical industrial problem may be to analyze the starting situation of a large motor, the associated voltage reduction caused by the heavy inrush starting current, and its effect on the operation of other parallel motors in the system. The period of heavy current inrush during starting of an induction motor usually lasts for about the time that is required to reach the slip $\mathbf{S}_{max\ T}$ corresponding to maximum torque, after which it comes down to the normal running value.

[7]Brereton, D. S., D. G. Lewis, and C. C. Young, "Representation of Induction Motor Loads during Power System Stability Studies," *AIEE Transactions*, vol. 76, pt. III, pp. 451–61, August 1957.

Piecewise linear representations of the torque-speed characteristic are often used in analytical solutions. Since the curve is practically linear in normal operating region of small slips, the relations

$$T_e = k\mathbf{S} \tag{10.7.4}$$

where k is a constant, may be used. Alternatively, when a wider region of the torque-speed curve is involved, a number of straight lines may be needed; or sometimes, a reasonably good representation, given by Equation 7.4.12, is used. When the terminal voltage is varied, to include the effect that the electromagnetic torque varies as the square of the voltage, Equation 10.7.4 is modified as

$$T_e = k' V_1^2 \mathbf{S} \tag{10.7.5}$$

in which k' is a constant and V_1 is the applied voltage. Typical analytical expressions such as

$$T_m = [A (1 - \mathbf{S})^2 + B (1 - \mathbf{S}) + C] P \tag{10.7.6}$$

where P is the per-unit net power transferred to the rotor through the air gap, and A, B, and C are constants of the mechanical load speed-torque characteristic, are used to represent the mechanical load torque as a function of slip.

BIBLIOGRAPHY

Adkins, B. *The General Theory of Electrical Machines*. New York: John Wiley & Sons, 1957.

Alger, P. L. *Induction Machines—Their Behavior and Uses*. 2d ed. New York: Gordon and Breach, 1970.

Concordia, C. *Synchronous Machines*. New York: John Wiley & Sons, 1951.

Fitzgerald, A. E., and Kingsley, C., Jr. *Electric Machinery*. 2d ed. New York: McGraw-Hill, 1961.

Kimbark, E. W. *Power System Stability: Synchronous Machines*. New York: Dover, 1968.

Knowlton, A. E., ed. *Standard Handbook for Electrical Engineers*. 8th ed. New York: McGraw-Hill, 1949.

Meisel, J. *Principles of Electromechanical Energy Conversion*. New York: McGraw-Hill, 1966.

Sarma, M. S. *Synchronous Machines (Their Theory, Stability and Excitation Systems)*. New York: Gordon and Breach, 1979.

Westinghouse Electric Corporation Central Station Engineers. *Electrical Transmission and Distribution Reference Book*. 4th ed. East Pittsburgh, Pa.: Westinghouse Electric Corporation, 1964.

White, D. C., and Woodson, H. H. *Electromechanical Energy Conversion*. New York: John Wiley & Sons, 1959.

Woodson, H. H., and Melcher, J. R. *Electromechanical Dynamics—Part I: Discrete Systems*. New York: John Wiley & Sons, 1968.

PROBLEMS

10-1.

Following the procedure indicated in the text, obtain the flux-linkage relations given by Equations 10.1.16 starting from Equations 10.1.15.

10-2.

Develop the armature voltage relations given by Equations 10.1.18 starting from Equations 10.1.1, indicating clearly the transformations you carry out.

10-3.

A three-phase, 13,800-volt (rms line-to-line voltage), 60-Hz, two-pole, wye-connected synchronous machine operates at its rated speed. The maximum value of the mutual inductance between the field winding and any one of the armature phase windings is 0.04 henry. Calculate the required field current for the machine to develop normal rated voltage on open circuit.

10-4.

Suppose the negative-sequence stator currents from a generator have the form given below:

$$i_a = 1.0 \cos (t + \alpha)$$

$$i_b = 1.0 \cos (t + \alpha + 120°)$$

$$i_c = 1.0 \cos (t + \alpha - 120°)$$

Let the machine be operating at synchronous speed under steady-state conditions. Evaluate the corresponding i_d, i_q, and i_0.

10-5.

Consider a machine operating at synchronous speed under steady-state conditions, and let dc stator currents flow from the generator having the following magnitudes:

$$i_a = 1.0; \; i_b = i_c = -1/2$$

Find the corresponding i_d, i_q, and i_0.

10-6.

Consider the flux-linkage and armature-voltage relations of a synchronous machine given by Equations 10.1.14, 10.1.16, and 10.1.18.

    a. For the case of the machine operating at steady-state synchronous speed, rewrite these equations making all possible simplifications.

    b. Repeat part *(a)* with the additional constraint that the armature is open-circuited. If the rated frequency is 60 Hz and $L_{afd} = 0.05$ henry, calculate the line-to-line voltage of the machine corresponding to a field current of 1,500 amperes.

    c. Repeat part *(a)* with the additional constraint of the armature being short-circuited. Neglect armature resistance. If the rated frequency is 60 Hz, $L_{afd} = 0.05$ henry, and $L_d = 0.025$ henry, calculate the line current for a wye-connected generator corresponding to a field current of 1,500 amperes.

10–7.

A three-phase, 60-Hz, 12.1-kV, 20-MVA, four-pole synchronous generator has the following inductances and resistances:

$$L_l = 0.00214 \text{ H}$$

$$L_{ad} = 0.02145 \text{ H}$$

$$L_{aq} = 0.0191 \text{ H}$$

$$L_{afd} = 0.045 \text{ H}$$

$$L_{fdfd} = 0.1538 \text{ H}$$

$$r_{fd} = 0.0208 \text{ ohm}$$

$$L_{a1d} = 0.017 \text{ H}$$

$$L_{1d1d} = 0.0226 \text{ H}$$

$$L_{fd1d} = 0.0541 \text{ H}$$

$$r_{1d} = 0.0315 \text{ ohm}$$

$$L_{a1q} = 0.007 \text{ H}$$

$$r_{1q} = 0.0611 \text{ ohm}$$

$$r_a = 0.0147 \text{ ohm}$$

$$L_{1q1q} = 0.00446 \text{ H}$$

Convert all of the above to per-unit quantities.

10–8.

A three-phase, 60-Hz, 13-kV, 25-MVA, wye-connected synchronous generator is given to have $X_d = 1.2$ per unit and a time constant of 1,000 radians. Find X_d in ohms and the time constant in seconds.

10–9.

Starting from the fundamental flux-linkage equations, justify the equivalent circuits shown in Figures 10.1.4 and 10.1.5.

10–10.

A four-pole, three-phase, 10-MVA, 11.8-kV, 60-Hz turbogenerator is subjected to a sudden three-phase short circuit from an unloaded condition when the open-circuit voltage is 5.9 kV. An oscillogram of one of the armature currents is taken and the measurements made on the ordinates of the envelope of the oscillogram are given below in a table. The scale factor of the oscillogram based on the instantaneous value is 1,930 amperes per centimeter. Compute the direct-axis reactances X_d, X_d', X_d'' in per unit, and the time constants T_d', T_d'', and T_a in seconds.

TABLE FOR PROBLEM 10–10

Time (cycles)	Ordinates of Envelopes (cm)		Time (cycles)	Ordinates of Envelopes (cm)	
	Upper	Lower		Upper	Lower
0	2.63	−0.63	15	0.86	−0.62
1	2.24	−0.50	20	0.75	−0.63
2	1.96	−0.44	25	0.67	−0.61
3	1.73	−0.41	30	0.63	−0.59
4	1.56	−0.42	40	0.54	−0.54
5	1.44	−0.44	50	0.48	−0.48
6	1.33	−0.45	60	0.43	−0.43
7	1.24	−0.48	90	0.32	−0.32
8	1.16	−0.50	120	0.25	−0.25
10	1.05	−0.55	∞	0.15	−0.15

10–11.

Consider a three-phase short circuit of a synchronous machine occurring at the armature terminals of the machine initially unloaded and normally excited. Which of the following are affected by the time at which the short-circuiting switch is closed?

    a. The dc component in phase *a*.
    b. The dc component in direct axis.
    c. The dc component in field.
    d. The torque.

10–12.

If $i_a(s)$ is given to be

$$i_a(s) = \frac{1}{X_d'} \frac{s + 1/T_{d0}'}{s(s + 1/T_d')(s^2 + 2as + 1)}$$

where *a* is a constant, find the sustained (steady-state) value of $i_a(t)$.

10-13.

A synchronous machine is initially operating at unity power factor, rated kVA, and rated terminal voltage. The machine parameters are given below:

$$X_d = X_q = 1.0; \ X_d' = 0.15; \ X_d'' = 0.10; \ X_q'' = 0.12; \ r = 0$$

Corresponding to a three-phase terminal short circuit, determine:

a. the initial subtransient fundamental frequency fault current.
b. the maximum dc offset that can occur in any phase current. Comment on the base of which this is expressed in per unit.
c. the initial transient fundamental frequency fault current (using the voltage behind transient reactance) and the dc field current corresponding to this value (using E_q').
d. the steady-state fault current.

10-14.

Consider a synchronous machine with no amortisseur windings and negligible armature resistance. Let a sudden three-phase short circuit occur at the armature terminals of the machine, initially unloaded, normally excited, and run at synchronous speed. That is to say, in per-unit notation,
at $t = 0_-$ (just prior to short circuit),

$$e_d = 0; \ e_q = 1, \ e_{fd} = \frac{r_{fd}}{X_{ad}}$$

at $t = 0_+$ (just after short circuit),

$$e_d = 0, \ e_q = 0, \ e_{fd} = \frac{r_{fd}}{X_{ad}}$$

Apply the performance equations of the synchronous machine and show that

$$i_d(t) = \frac{1}{X_d} + \left(\frac{1}{X_d'} - \frac{1}{X_d}\right) e^{-t/T_d'} - \frac{1}{X_d'} \cos t \ e^{-t/T_a}$$

where T_a is the armature time constant given by $(1/a)$, in which

$$a = \frac{r}{2} \frac{X_d' + X_q}{X_d' X_q}$$

10–15.

Let

$$
i_d(t) = \left(\frac{1}{X_d''} - \frac{1}{X_d'}\right) e^{-t/T_d''} + \left(\frac{1}{X_d'} - \frac{1}{X_d}\right) e^{-t/T_d'}
$$
$$
+ \frac{1}{X_d} - \frac{1}{X_d''} \cos t\, e^{-t/T_a}
$$

$$
i_q(t) = \frac{1}{X_q''} \sin t\, e^{-t/T_a}
$$

and

$$
i_0(t) = 0
$$

Show that

$$
i_a(t) = \left[\left(\frac{1}{X_d''} - \frac{1}{X_d'}\right) e^{-t/T_d''} + \left(\frac{1}{X_d'} - \frac{1}{X_d}\right) e^{-t/T_d'} + \frac{1}{X_d}\right]
$$

$$
\cos(t+\alpha) - \left[\frac{1}{2}\left(\frac{1}{X_d''} + \frac{1}{X_q''}\right) \cos \alpha\right.
$$

$$
\left. + \frac{1}{2}\left(\frac{1}{X_d''} - \frac{1}{X_q''}\right) \cos(2t + \alpha)\right] e^{-t/T_a}
$$

10–16.

Consider the conditions of Problem 10–14 and show that

$$
i_{fd}(t) = \frac{1}{X_{ad}} + \frac{X_{ad}}{X_{fdfd} X_d'} e^{-t/T_d'} - \frac{X_{ad}}{X_{fdfd} X_d'} \cos t\, e^{-t/T_a}
$$

10–17.

Consider the conditions of Problem 10–14 in order to calculate the short-circuit torque as in Equation 10.1.35. Neglect damping for the first half-cycle and show that

$$
T = \frac{1}{X_d'} \sin t + \left(\frac{1}{2X_q} - \frac{1}{2X_d'}\right) \sin 2t
$$

10-18.

The short-circuit torque of a typical synchronous machine is given by $T = 3.44 \sin \phi - 1.095 \sin 2\phi$ for the first half-cycle.

    a. Sketch the alternating components as well as the total torque as functions of the angle ϕ for the first half-cycle.

    b. Find the angle ϕ corresponding to the maximum torque and calculate the maximum torque in terms of the full-load torque.

10-19.

Equation 10.4.10 gives the transient power-angle equation for a salient-pole synchronous machine with negligible armature resistance. Show that the condition for maximum transient power is given by

$$\cos \delta = \frac{X_q E_q'}{4(X_q - X_d') V_t} - \sqrt{\left[\frac{X_q E_q'}{4(X_q - X_d') V_t}\right]^2 + \frac{1}{2}}$$

10-20.

A waterwheel generator with negligible armature resistance has the following per-unit parameters:

$$X_d = 1.0, \ X_q = 0.6 \ X_d' = 0.3$$

Let the machine operate as a generator connected to an infinite bus having a voltage of 1.0 per unit. Plot the steady-state and transient power-angle characteristics for the following cases:

    a. The no-load excitation and initial field flux.

    b. The full-load excitation with unit armature current at 0.8 lagging power factor and initial field flux.

10-21.

The power-angle characteristic of a synchronous machine is given by $P = (630 \sin \delta)$ kW. Compute the maximum shaft load in hp that may be suddenly applied to the motor operating initially unloaded. Neglect damping.

10-22.

By carrying out the manipulations needed, show that Equations 10.5.17 and 10.5.18 are correct.

10-23.

Carry out the manipulations described in the text to arrive at the Equations 10.5.19 and 10.5.20.

10–24.

The steady-state analysis of a polyphase induction machine (presented in Chapter 7) can also be accomplished from the mathematical model set up for the machine in Section 10.5. Let the three stator-phase currents be given by

$$i_a = \sqrt{2}\, I_1 \cos(\omega t + \alpha); \; i_b = \sqrt{2}\, I_1 \cos(\omega t + \alpha - 120°);$$

$$i_c = \sqrt{2}\, I_1 \cos(\omega t + \alpha + 120°)$$

where α is an arbitrary phase angle.

a. Show that the currents, viewed from the synchronously rotating dq-axes, appear as steady dc quantities.

b. In steady-state analysis, one is usually interested in relations involving rms currents and voltages rather than instantaneous values. So, express the stator-phase current and voltage as well as the rotor-phase current in phasor form; obtain the voltage equations for the stator and rotor by making use of the flux-linkage equations; and show that they can be arranged in the following form:

$$\overline{V}_1 = r_1 \overline{I}_1 + j X_{l1} \overline{I}_1 + j X_m(\overline{I}_1 + \overline{I}_2)$$
$$0 = \frac{r_2}{S} \overline{I}_2 + j X_{l2} \overline{I}_2 + j X_m(\overline{I}_1 + \overline{I}_2)$$

where X_{l1}, the stator leakage reactance per phase, is $\omega(L_{11} - L_{12})$; X_{l2}, the rotor leakage reactance per phase, is $\omega(L_{22} - L_{12})$; and X_m, the magnetizing reactance per phase, is ωL_{12}.

c. With all quantities referred to the stator, obtain the per-phase equivalent circuit of a polyphase induction machine satisfying the volt-ampere equations of part *(b)*, and show that it is the same as Figure 7.2.5*(c)* developed from a generalized transformer point of view.

d. Through the application of Equation 10.5.23, show that the expression for the internal electromagnetic torque is the same as Equation 7.4.1 obtained in Chapter 7.

10–25.

The dq-transformation presented in the text makes use of synchronously rotating axes. As an alternative, some induction machine analysts have used axes that are fixed with respect to the stator structure and are labeled as α- and β-axes. The α-axis coincides with that of phase *a* and the β-axis is located 90° ahead of the α-axis in the direction of rotation.

Let the axis of each rotor phase be displaced by an angle θ from the axis of the correspondingly labeled stator phase.

    a. By replacing ωt by zero in Equations 10.5.11 and 10.5.12 and simplifying, obtain the transformation equations relating abc-phase currents and $\alpha\beta0$-currents for the stator.

    b. By replacing θ_s by $(-\theta)$ in Equations 10.5.14 and 10.5.15, obtain the transformation equations for the rotor quantities.

    c. Show that the component flux-linkage equations are the same as Equations 10.5.17 and 10.5.18, if one replaces d and q by α and β, respectively.

    d. Show that the voltage equations are then
Stator:

$$e_{1\alpha} = r_1 i_{1\alpha} + p\lambda_{1\alpha}$$

$$e_{1\beta} = r_1 i_{1\beta} + p\lambda_{1\beta}$$

Rotor:

$$e_{2\alpha} = r_2 i_{2\alpha} + p\lambda_{2\alpha} + \lambda_{2\beta}(p\theta)$$

$$e_{2\beta} = r_2 i_{2\beta} + p\lambda_{2\beta} - \lambda_{2\alpha}(p\theta)$$

    e. Show that, in the steady state, the component currents and voltages in the $\alpha\beta$-system are stator-frequency quantities.

10–26.

As an alternative to the use of synchronously rotating dq-axes for the induction-machine analysis, some analysts have used dq-axes fixed with respect to the rotor. Let the axis of each rotor phase be displaced by an angle θ from the axis of the correspondingly labeled stator phase.

    a. By replacing ωt by θ in Equations 10.5.11 and 10.5.12, obtain the transformation equations relating abc-phase currents and the new $dq0$-currents for the stator.

    b. By replacing θ_s by zero in Equations 10.5.14 and 10.5.15, obtain the transformation equations for the rotor quantities.

    c. Show that the component flux-linkage equations are the same as Equations 10.5.17 to 10.5.18.

    d. Show that the voltage equations are then
Stator:

$$e_{1d} = r_1 i_{1d} + p\lambda_{1d} - \lambda_{1q}(p\theta)$$

$$e_{1q} = r_1 i_{1q} + p\lambda_{1q} + \lambda_{1d}(p\theta)$$

Rotor:

$$e_{2d} = r_2 i_{2d} + p \lambda_{2d}$$

$$e_{2q} = r_2 i_{2q} + p \lambda_{2q}$$

e. Show that, in the steady state, the component currents and voltages in the new dq-system are slip-frequency quantities.

10–27.

This problem is concerned with the analysis of a two-phase induction machine instead of the three-phase machine presented in the text, with the use of synchronously rotating dq-axes. Let the magnetic axis of phase b be 90° ahead of the phase a axis in the direction of rotation. Let there be two corresponding windings A and B on the rotor with their magnetic axes displaced by 90°. At any instant of time, let the axis of a rotor phase be displaced by an angle θ_2 from the axis of the correspondingly labeled stator phase. Following the general procedure adopted in the text,

a. develop the flux-linkage equations corresponding to Equations 10.5.8 and 10.5.9.

b. show that the appropriate dq-transformation of variables is typically given by the following:.
Stator:

$$i_{1d} = i_a \cos \omega t + i_b \sin \omega t$$

$$i_{1q} = -i_a \sin \omega t + i_b \cos \omega t$$

Rotor:

$$i_{2d} = i_A \cos \theta_s + i_B \sin \theta_s$$

$$i_{2q} = -i_A \sin \theta_s + i_B \cos \theta_s$$

c. obtain the relations in terms of dq-variables for phase variables.

d. show that the flux-linkage equations in terms of dq-variables are still given by Equations 10.5.17 and 10.5.18; find L_{11}, L_{22}, and L_{12} in terms of L_1, L_2, and M.

e. show that the component voltage equations 10.5.19 and 10.5.20 hold good in this case as well.

f. show that the instantaneous power input to the two-phase stator is given by

$$p_1 = v_{1d} i_{1d} + v_{1q} i_{1q}$$

g. show that the motor torque is given by

$$T = \frac{\text{poles}}{2} (\lambda_{2q} i_{2d} - \lambda_{2d} i_{2q})$$

10-28.

When unbalanced voltages are applied, the method of three-phase symmetrical components is often used for steady-state analysis. The *abc*-phasors are related to the symmetrical-component phasors by the following:

$$\bar{I}_a = \bar{I}_a{}^+ + \bar{I}_a{}^- + \bar{I}_a{}^0$$

$$\bar{I}_b = \bar{I}_a{}^+ e^{-j120°} + \bar{I}_a{}^- e^{j120°} + \bar{I}_a{}^0$$

$$\bar{I}_c = \bar{I}_a{}^+ e^{j120°} + \bar{I}_a{}^- e^{-j120°} + \bar{I}_a{}^0$$

where the $\bar{I}_a{}^+$-set is a balanced positive-sequence set of components with an *abc*-phase sequence, and the $\bar{I}_a{}^-$-set is a balanced negative-sequence set of components with an *acb*-phase sequence.

a. Obtain the transformation relating the symmetrical-component phasors and the *abc*-phasors.
b. Assuming that zero-sequence-component currents and voltages do not exist for purposes of this problem, and assuming that the principle of superposition holds good, outline a method for computing the torque-slip characteristic of an induction motor with unbalanced applied voltages.

10-29.

A 400-hp, 440-volt, 60-Hz, three-phase, wye-connected, four-pole wound-rotor induction motor has the following parameters in ohms per phase referred to the stator:

$$X_{l1} = X_{l2} = 0.05 \; ; X_m = 2.5 \; ; r_1 = 0.005 \; ; r_2 = 0.006$$

The motor is supplied at the rated terminal voltage through a step-down transformer that can be represented by a series reactance of 0.03 ohm per phase. The induction motor is operating at its full load with an efficiency and power factor of 90 percent each, with its slip rings short-circuited. If a three-phase short-circuit occurs at the high-voltage terminals of the transformer bank, determine the initial symmetrical short-circuit current in the motor and show how it is decremented.

10–30.

For an induction motor started at full voltage, develop an expression for the time t required to reach the speed corresponding to $S_{max\ T}$. The motor is unloaded at starting and rotational losses may be neglected. Let J be the combined inertia of the rotor and the connected mechanical equipment, and let the torque-slip curve be given by the relation

$$\frac{T}{T_{max}} = \frac{2}{\dfrac{S}{S_{max\ T}} + \dfrac{S_{max\ T}}{S}}$$

where T_{max} is the maximum torque and $S_{max\ T}$ is the slip at maximum torque.

11

DIRECT-CURRENT MACHINE DYNAMICS

The steady-state operation and performance of direct-current machines are discussed in Chapters 5 and 9. In this chapter a study is made of the dynamic behavior of direct-current machines to examine the response to sudden changes. The effects of inductance and inertia, which are negligible during the steady-state operation, come into play during a rapid transition from one operating condition to another and must therefore be considered when analyzing the dynamic behavior of a dc machine. As loads change and shift, the dynamic characteristics of the individual machines taken together will determine whether the system can react in a stable manner to such disturbances. Because of the versatility of dc machines and the ease with which they can be controlled, dc motors with solid-state controls are often used in applications requiring a wide range of motor speeds or precise control of motor output. Our primary objective in this chapter is to develop mathematical models of dc machines and analyze their dynamic characteristics.

11.1 DYNAMIC MODELS

Because of the complexity of dynamic-system problems, the usual simplifying assumptions made in many problems involving the behavior of a dc machine as a system component are the following:

- a. The air-gap flux distribution produced by field windings is symmetrical about the center line of the field poles, which is known as the field axis or the direct axis.
- b. The axis of the armature-mmf wave is fixed in space along the quadrature-axis; the brushes are narrow, and commutation is linear.
- c. Because the armature-mmf axis is perpendicular to the field axis, the armature-mmf has no effect on the total direct-axis flux; that is to say, the demagnetizing effect of armature reaction is neglected.
- d. The effects of magnetic saturation will be neglected, at least for the time being, thereby allowing superposition and considering inductances to be independent of the currents.
- e. A two-pole machine is considered for the model, while the results can be applied to a P-pole machine.

The schematic representation of the model of a dc machine is shown in Figure 11.1.1 with motor and generator conventions. The arrow representing the reference direction for a current also represents that of its associated magnetic field. Based on Equations 5.3.24 and 5.3.26, one can write

$$e_a = K \phi_d \omega_m = k i_f \omega_m \qquad (11.1.1)$$

$$T_e = K \phi_d i_a = k i_f i_a \qquad (11.1.2)$$

where K and k are constants, ϕ_d is the direct-axis air-gap flux that is linearly proportional to the field current i_f, ω_m is the angular velocity corresponding to the speed of rotation, i_a is the armature current, e_a is the generated speed voltage, and T_e is the electromagnetic torque developed by the

FIGURE 11.1.1 Schematic representation of the model of a dc machine: *(a)* motor conventions, *(b)* generator conventions.

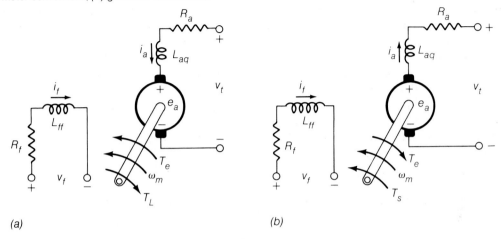

(a)   (b)

machine. The system performance is described by these equations together with the differential equation of motion of the mechanical system, the volt-ampere equations for the armature and field circuits, as well as the magnetization curve, which is taken here to be linear.

The voltage equation for the field circuit is given by

$$v_f = L_{ff} \frac{di_f}{dt} + R_f i_f = L_{ff} p i_f + R_f i_f \qquad (11.1.3)$$

where p is the derivative operator d/dt, and v_f, i_f, R_f, and L_{ff} are the terminal voltage, current, resistance, and self-inductance of the field circuit, respectively.

Note that the axis of the field mmf lies along the direct axis, and the mutual inductance between the field and armature circuits is zero because their axes are perpendicular to each other. The voltage equation for the armature circuit of a motor of Figure 11.1.1(a) is given by

$$\begin{aligned} v_t &= e_a + L_{aq} p i_a + R_a i_a \\ &= k\, i_f\, \omega_m + L_{aq} p i_a + R_a i_a \end{aligned} \qquad (11.1.4)$$

where v_t, R_a, and L_{aq} are the terminal voltage, resistance, and self-inductance of the armature circuit, respectively. The subscript q is used with the inductance L_{aq} to indicate the axis of the armature-mmf to be along the quadrature axis. The inductance L_{aq} includes the effect of any quadrature-axis stator windings in series with the armature, such as interpoles and compensating windings discussed in Section 9.5. The dynamic equation for the mechanical system of a motor is given by

$$T_e = k \, i_f \, i_a = J \, p\omega_m + B\omega_m + T_L \qquad (11.1.5)$$

or

$$T_e - T_L = J \, p\omega_m + B\omega_m \qquad (11.1.6)$$

where J is the combined polar moment of inertia of the load and the rotor of the motor, B is the equivalent viscous friction constant of the load and the motor (in the latter case of which it is used for approximating the rotational losses of the motor), and T_L is the mechanical load torque (delivered to the load) opposing rotation.

For the case of a dc generator shown in Figure 11.1.1(b), the armature voltage and torque equations are given by

$$v_t = e_a - L_{aq} \, pi_a - R_a \, i_a$$
$$= k \, i_f \, \omega_m - L_{aq} \, pi_a - R_a \, i_a \qquad (11.1.7)$$

$$T_S - T_e = J \, p\omega_m + B\omega_m \qquad (11.1.8)$$

where

$$T_e = k \, i_f \, i_a \qquad (11.1.9)$$

and T_S is the mechanical driving torque applied to the shaft in the direction of rotation.

The field and armature currents and the speed can be considered as *state variables* because the state of a physical system can be described in terms of its stored energy, and the energy storage is associated with the magnetic fields produced by the field and armature currents and with the kinetic energy of the rotating parts. Equations 11.1.3 to 11.1.9 are first-order differential equations containing product nonlinearities, such as $i_f \, i_a$ and $i_f \, \omega_m$, of these state variables. These equations, together with the torque-speed characteristics of the mechanical system connected to the shaft and the Kirchhoff-law equations for the circuits connected to the armature and field terminals, describe the system performance.

11.2 DYNAMIC ANALYSIS

The *transfer function* is a means by which the dynamic characteristics of electromechanical devices are described. The transfer function is a mathematical formulation that relates the output variable of a device to the input variable. For linear devices, the transfer function is independent of the input quantity and solely dependent upon the parameters of the device together with any operations of time, such as differentiation and integration, that it may possess. In order to obtain

FIGURE 11.2.1 Basic building block of a block diagram.

$$R(s) \qquad\qquad\qquad C(s)$$
$$\text{input} \qquad G(s) \qquad \text{output}$$

the transfer function, one usually goes through the following three steps: *(i)* determining the governing equation for the device expressed in terms of the output and input variables; *(ii)* Laplace transforming the governing equation, assuming all initial conditions to be zero; and *(iii)* rearranging the equation to yield the ratio of the output to input variable. The properties of transfer functions are summarized as follows:

  a. A transfer function is defined only for a linear time-invariant system.
  b. The transfer function between an input variable and an output variable of a system is defined as the Laplace transform of the output to the Laplace transform of the input, or as the Laplace transform of the impulse response; the impulse response of a linear system is defined as the output response of the system when the input is a unit impulse function.
  c. All initial conditions of the system are assumed to be zero.
  d. The transfer function is independent of the input.

The *block diagram* is a pictorial representation of the equations of the system. Each block represents a mathematical operation, and the blocks are interconnected to satisfy the governing equations of the system. The block diagram thus provides a chart of the procedure to be followed in combining the simultaneous equations, from which useful information can often be obtained without finding a complete analytical solution. The block-diagram technique has been highly developed in connection with studies of feedback control systems,[1] often leading to programming a problem for solution on an analog computer.

The simple configuration shown in Figure 11.2.1 is actually the basic building block of a complex block diagram. The arrows on the diagram imply that the block diagram has a unilateral property; or in other words, signal can pass only in the direction of the arrows. A box is the symbol for multiplication; the input quantity is multiplied by the function in the box to obtain the output. With circles indicating summing points (in an algebraic sense), and with boxes or blocks denoting multiplication, any linear mathematical expression may be represented by block-diagram notation as in Figure 11.2.2 for the case of an *elementary feedback control system.*

[1]Kuo, B. C., *Automatic Control Systems,* 4th ed. (Englewood Cliffs, N.J.: Prentice-Hall, 1982).

FIGURE 11.2.2   Block diagram of an elementary feedback control system.

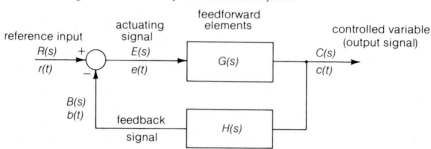

R(s): reference input
C(s): output signal (controlled variable)
B(s) = H(s)C(s): feedback signal
E(s) = [R(s) - B(s)]: actuating signal (error)
G(s) = C(s)/E(s): forward path transfer function or open-loop transfer function
M(s) = C(s)/R(s) = [G(s)/1+G(s)H(s)]: closed loop transfer function
H(s): feedback path transfer function
G(s)H(s): loop gain

The block diagrams of complex feedback control systems usually contain several feedback loops, and they may have to be simplified in order to evaluate an overall transfer function for the system. A few of the block diagram reduction manipulations are given in Table 11.2.1; no attempt is made here to cover all the possibilities.

SEPARATELY EXCITED DC GENERATOR   Let us consider the electrical transients in a separately excited dc generator resulting from changes in excitation. The analysis will be made on a linear basis, while neglecting the effects of saturation. Further, let the generator speed be a constant so that the dynamics of the mechanical drive do not enter into the problem. Considering the dc generator of Figure 11.1.1(b), the voltage equation for the field circuit is given by

$$v_f = R_f i_f + L_{ff}\, p i_f = R_f (1 + \tau_f p)\, i_f \qquad (11.2.1)$$

where $\tau_f = L_{ff}/R_f$ is the time constant of the field circuit. The Laplace transformation of Equation 11.2.1 with zero initial conditions yields

$$V_f(s) = I_f(s)\, R_f + L_{ff}\, s\, I_f(s) = R_f\, (1 + \tau_f s)\, I_f(s) \qquad (11.2.2)$$

TABLE 11.2.1   SOME OF THE BLOCK DIAGRAM REDUCTION MANIPULATIONS

original block diagram	manipulation	modified block diagram
	cascaded elements	
	addition or subtraction (eliminating the auxiliary forward path)	
	shifting of the pick-off point ahead of the block	
	shifting of the pick-off point behind the block	
	shifting the summing point ahead of the block	
	shifting the summing point behind the block	
	removing H from feedback path	
	eliminating the feedback path	

FIGURE 11.2.3   Block diagram representing Equations 11.2.3 and 11.2.4.

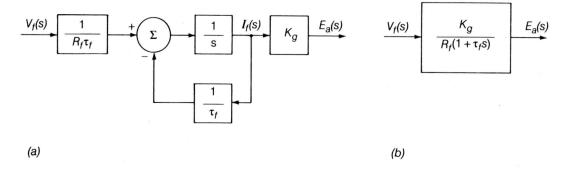

(a)   (b)

The above equation may be rearranged as follows:

$$sI_f(s) = \frac{1}{\tau_f}\left[\frac{V_f(s)}{R_f} - I_f(s)\right]$$

(11.2.3)

With an integrator $1/s$ in the forward path and $I_f(s)$ as the state variable, the block diagram corresponding to Equation 11.2.3 is given in Figure 11.2.3(a). Multiplication of the output $I_f(s)$ by a constant K_g gives the generated emf $E_a(s)$ at a speed ω_m; K_g is the slope of the air-gap line of the magnetization curve (taken at the speed ω_m) representing the relationship between e_a and i_f. This operation is also shown in the block diagram of Figure 11.2.3(a). The transfer function relating the armature induced voltage to the field winding voltage is then given by

$$\frac{E_a(s)}{V_f(s)} = \frac{K_g I_f(s)}{V_f(s)} = \frac{K_g}{R_f(1+\tau_f s)}$$

(11.2.4)

which is represented in Figure 11.2.3(b).

Let us next consider the generator dynamics including the effect of the load. Let R_L and L_L be the load resistance and load inductance, and R_a and L_{aq} be the armature winding resistance and leakage inductance, respectively. Then one can write

$$e_a = i_a R_a + L_{aq} pi_a + i_a R_L + L_L pi_a$$

or

$$e_a = i_a (R_a + R_L) + (L_{aq} + L_L) pi_a$$

(11.2.5)

FIGURE 11.2.4 Block diagram of Equation 11.2.8 representing the generator dynamics including the effect of the load.

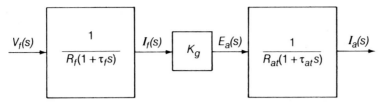

Laplace transformation gives

$$E_a(s) = I_a(s) (R_a + R_L) (1 + s \tau_{at})$$

where τ_{at} is the armature circuit time constant given by

$$\tau_{at} = \frac{L_{at}}{R_{at}} = \frac{L_{aq} + L_L}{R_a + R_L} \tag{11.2.6}$$

The transfer function then becomes

$$\frac{I_a(s)}{E_a(s)} = \frac{1}{R_{at} (1 + s \tau_{at})} \tag{11.2.7}$$

The total transfer function relating the armature current to the field voltage is given by

$$\frac{I_a(s)}{V_f(s)} = \frac{E_a(s)}{V_f(s)} \cdot \frac{I_a(s)}{E_a(s)} = \frac{K_g}{R_f (1 + \tau_f s)} \cdot \frac{1}{R_{at}(1 + \tau_{at} s)} \tag{11.2.8}$$

which is represented in the block diagram of Figure 11.2.4 with two state variables $I_f(s)$ and $I_a(s)$ for the second-order system. After finding the armature current i_a, the electromagnetic torque T_e can then be determined from Equation 11.1.2 or by

$$T_e = \frac{e_a}{\omega_m} i_a \tag{11.2.9}$$

The block diagram can thus be extended piece by piece to take into account as many additions as desired.

EXAMPLE 11.2.1

A separately excited dc generator is given to have the following parameters:

$$R_f = 100 \text{ ohms}; \ L_{ff} = 20 \text{ H}; \ R_a = 0.1 \text{ ohm}; \ L_{aq} = 0.1 \text{ H}$$

$$K_g = 100 \text{ V per field ampere at rated speed}$$

The load connected to the generator has a resistance $R_L = 4.5$ ohms and an inductance $L_L = 2.2$ H. Assume that the prime mover is rotating at rated speed, the load switch is closed, and the generator is initially unexcited. Determine the armature current as a function of time when a 230–V dc supply is suddenly applied to the field winding, assuming the generator speed to be essentially constant.

SOLUTION
The transfer function relating the armature current to the field voltage is given by Equation 11.2.8:

$$\frac{I_a(s)}{V_f(s)} = \frac{K_g}{R_f \, (1 + \tau_f \, s)} \cdot \frac{1}{R_{at} \, (1 + \tau_{at}s)}$$

With $R_{at} = R_a + R_L = 0.1 + 4.5 = 4.6$ ohms; $\tau_f = L_{ff}/R_f = 20/100 = 0.2$ and $\tau_{at} = L_{at}/R_{at} = (L_{aq} + L_L)/(R_a + R_L) = (0.1 + 2.2)/(0.1 + 4.5) = 0.5$,

$$\frac{I_a(s)}{V_f(s)} = \frac{100}{100 \, (1 + 0.2 \, s)} \cdot \frac{1}{4.6 \, (1 + 0.5 \, s)}$$

The Laplace transform of the field voltage for a step change of 230 V is

$$V_f(s) = 230/s$$

Hence

$$I_a(s) = \frac{230}{s} \, \frac{100}{100(1 + 0.2 \, s)} \, \frac{1}{4.6(1 + 0.5 \, s)}$$

or

$$I_a(s) = \frac{500}{s(s + 5)(s + 2)} = \frac{K_1}{s} + \frac{K_2}{s+5} + \frac{K_3}{s+2}$$

where

$$K_1 = \frac{500}{(s+5)(s+2)}\Bigg]_{s=0} = 50$$

$$K_2 = \frac{500}{s(s+2)}\Bigg]_{s=-5} = 33.3$$

$$K_3 = \frac{500}{s(s+5)}\Bigg]_{s=-2} = -83.3$$

Taking the inverse Laplace transform, one gets the current buildup as

$$i_a(t) = 50 + 33.3\, e^{-5t} - 83.3\, e^{-2t}$$

The effect of the smaller of the two time constraints, τ_f, on the current buildup may be ignored when its value is less than about one-quarter of the longer one.

SEPARATELY EXCITED DC MOTOR   Let us next consider the case of a separately excited dc motor with constant field excitation; we shall investigate how the speed of the motor responds to changes in the voltage applied to the armature terminals. The analysis will involve electrical transients in the armature circuit and the dynamics of the mechanical load driven by the motor. At a constant motor-field current I_f, the electromagnetic torque and the generated emf are given by

$$T_e = K_m\, i_a \tag{11.2.10}$$

$$e_a = K_m\, \omega_m \tag{11.2.11}$$

where $K_m = k\, I_f$ is a constant, which is also the ratio e_a/ω_m; in terms of the magnetization curve, e_a is the generated emf corresponding to the field current I_f at the speed ω_m. Let us now try to find the transfer function that relates $\Omega_m(s)$ to $V_t(s)$.

The differential equation for the motor armature current i_a is given by Equation 11.1.4, which may be rearranged as

$$R_a\,(1 + \tau_a p)\, i_a = v_t - e_a \tag{11.2.12}$$

where v_t is the terminal voltage applied to the motor, e_a is the back emf given by Equation 11.2.11, R_a and L_a include the series resistance and inductance of the armature circuit and electrical source put together, and $\tau_a = L_a/R_a$ is the *electrical time constant* of the armature circuit. The electro-

FIGURE 11.2.5   Block diagram representing Equations 11.2.15 and
11.2.16.

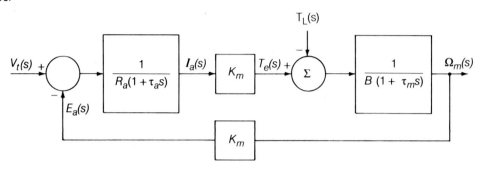

magnetic torque is given by Equation 11.2.10, and from the dynamic equation 11.1.6 for the mechanical system, the acceleration is given by

$$(B + Jp)\, \omega_m = T_e - T_L \tag{11.2.13}$$

or

$$B(1 + \tau_m P)\, \omega_m = T_e - T_L \tag{11.2.14}$$

where $\tau_m = J/B$ is the *mechanical time constant,* and the load torque T_L, in general, is a function of speed.

Laplace transforms of Equations 11.2.12 and 11.2.14 lead to the following:

$$I_a(s) = \frac{V_t(s) - E_a(s)}{R_a\,(1 + \tau_a s)} = \frac{V_t(s) - K_m\,\Omega_m(s)}{R_a\,(1 + \tau_a s)} \tag{11.2.15}$$

$$\Omega_m(s) = [T_e(s) - T_L(s)]\,\frac{1}{B}\,\frac{1}{(1 + \tau_m s)} \tag{11.2.16}$$

The corresponding block diagram representing the above operations is given in Figure 11.2.5 in terms of the state variables $I_a(s)$ and $\Omega_m(s)$ with $V_t(s)$ as input.

The application of the closed loop transfer function $M(s)$ shown in Figures 11.2.2 to 11.2.5 yields the following transfer function relating $\Omega_m(s)$ and $V_t(s)$ with $T_L = 0$:

$$\frac{\Omega_m(s)}{V_t(s)} = \frac{K_m/[R_a(1 + \tau_a s)\, B\,(1 + \tau_m s)]}{1 + [K_m^2/R_a(1 + \tau_a s)\, B(1 + \tau_m s)]} \tag{11.2.17}$$

FIGURE 11.2.6 Block diagram representing Equation 11.2.18.

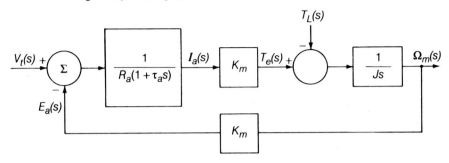

With mechanical damping B neglected, the above equation reduces to

$$\frac{\Omega_m(s)}{V_t(s)} = \frac{1}{K_m[\tau_i s(\tau_a s + 1) + 1]} \qquad (11.2.18)$$

where $\tau_i = JR_a/K_m^2$ is the *inertial time constant,* and the corresponding block diagram is shown in Figure 11.2.6. The transfer function relating speed to load torque with $V_t = 0$ can be obtained from Figure 11.2.6 by eliminating the feedback path as

$$\frac{\Omega_m(s)}{T_L(s)} = -\frac{(1/Js)}{1 + (1/Js)[K_m^2/R_a(1 + \tau_a s)]}$$

$$= -\frac{(\tau_a s + 1)}{Js(\tau_a s + 1) + (K_m^2/R_a)} \qquad (11.2.19)$$

Expressing the torque equation for the mechanical system as

$$T_e = K_m i_a = J p\omega_m + B\omega_m + T_L \qquad (11.2.20)$$

dividing by K_m, and substituting $\omega_m = e_a/K_m$, one gets

$$i_a = \frac{J}{K_m^2}\frac{de_a}{dt} + \frac{B}{K_m^2}e_a + \frac{T_L}{K_m} \qquad (11.2.21)$$

The above equation can be identified to be the node equation for a parallel C_{eq} - G_{eq} - Z_L circuit with

$$C_{eq} = J/K_m^2 \; ; \; G_{eq} = B/K_m^2 \; ; \text{ and } Z_L = K_m e_a/T_L \qquad (11.2.22)$$

FIGURE 11.2.7 Analog electrical circuit for a separately excited dc motor.

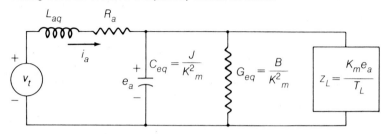

and a common voltage of e_a. An analog electrical circuit can then be drawn as in Figure 11.2.7 for a separately excited dc motor, in which the inertia is represented by a capacitance, the damping by a shunt conductance, and the load-torque component of current is shown flowing through an equivalent impedance Z_L. The time constants, τ_i associated with inertia and τ_m associated with the damping-type load torque that is proportional to speed, in terms of the analog circuit, are given by

$$\tau_i = R_a C_{eq} \text{ and } \tau_m = C_{eq}/G_{eq} \qquad (11.2.23)$$

Note in the above analysis that K_m is proportional to the constant motor field current I_f.

The self-inductance of the armature can often be neglected except in the case of motors driving a load that has rapid torque pulsations of appreciable magnitude.

EXAMPLE 11.2.2

A five-hp, 220-V, separately excited dc motor has the following parameters: $R_a = 0.5$ ohm; $k = 2$ H; $R_f = 220$ ohms; $L_{ff} = 110$ H. The armature winding inductance is negligible. The torque required by the load is proportional to the speed, and the combined constants of the motor armature and the load are $J = 3$ kg·m² and $B = 0.3$ kg·m²/s.

Consider the armature-controlled dc motor whose speed is made to respond to variations in the applied motor armature voltage v_t. Let the field current be maintained constant at 1 A.

    a. Develop a block diagram relating the motor speed and the motor applied voltage, and find the corresponding transfer function.

    b. Compute the steady-state speed corresponding to a step applied armature voltage of 220 V.

    c. How long does the motor take to reach 0.95 of the steady-state speed of part *(b)?*

    d. Determine the value of the total effective viscous damping coefficient of the motor-load configuration.

SOLUTION

a. Modifying Figure 11.2.5 for the negligible armature winding inductance, one has the following:

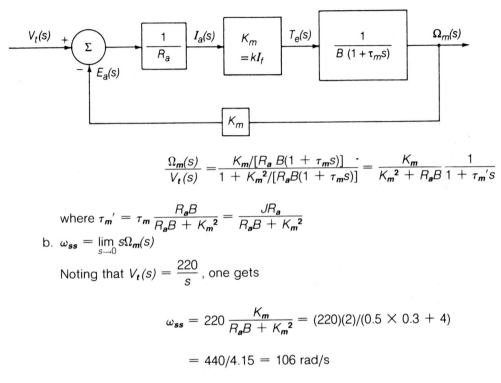

$$\frac{\Omega_m(s)}{V_t(s)} = \frac{K_m/[R_a B(1 + \tau_m s)]}{1 + K_m^2/[R_a B(1 + \tau_m s)]} = \frac{K_m}{K_m^2 + R_a B} \frac{1}{1 + \tau_m' s}$$

where $\tau_m' = \tau_m \dfrac{R_a B}{R_a B + K_m^2} = \dfrac{J R_a}{R_a B + K_m^2}$

b. $\omega_{ss} = \lim_{s \to 0} s\Omega_m(s)$

Noting that $V_t(s) = \dfrac{220}{s}$, one gets

$$\omega_{ss} = 220 \frac{K_m}{R_a B + K_m^2} = (220)(2)/(0.5 \times 0.3 + 4)$$

$$= 440/4.15 = 106 \text{ rad/s}$$

c. $\tau_m' = J R_a/(R_a B + K_m^2) = 3 \times 0.5/4.15 = 0.36$
For the motor to reach 0.95 of ω_{ss},

$$e^{-t/0.36} = 0.05 \text{ or } t = 0.36 \times 3 = 1.08 \text{ seconds}$$

d. Total effective viscous damping $= B + (K_m^2/R_a)$
$$= 0.3 + (4/0.5) = 8.3 \text{ kg} \cdot \text{m}^2/\text{s}$$

EFFECT OF SATURATION   The dynamic analysis of dc machines has so far been studied on a linear basis with magnetic saturation neglected. Such a treatment is justified if the transient response takes place in substantially the linear region. The response to small disturbances can be treated on an incrementally linear basis. For more comprehensive studies, however, transient investigations may require a nonlinear analysis because of saturation.

FIGURE 11.2.8 Block diagram of a separately excited dc generator including saturation.

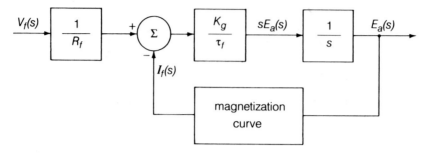

The linearized block diagram of a separately excited dc generator can be modified rather easily, as shown in Figure 11.2.8, to take care of magnetic saturation by feeding back the output e_a through the magnetization curve to obtain i_f. The equation to be satisfied is

$$v_f = R_f i_f + L_{ff} p\, i_f \tag{11.2.24}$$

or

$$v_f - R_f i_f = L_{ff} p\, i_f = \frac{L_{ff}}{K_g} p\, e_a \tag{11.2.25}$$

or

$$p\, e_a = \frac{K_g}{\tau_f}\left(\frac{v_f}{R_f} - i_f\right) \tag{11.2.26}$$

where K_g is the slope of the air-gap line in generated volts per field ampere at the speed ω_m, and $\tau_f = L_{ff}/R_f$ is the unsaturated value of the field-circuit time constant. Note that, when saturation is considered, the inductance of the field winding is no longer a constant. Laplace transformation of Equation 11.2.26 gives

$$sE_a(s) = \frac{K_g}{\tau_f}\left[\frac{V_f(s)}{R_f} - I_f(s)\right] \tag{11.2.27}$$

which is represented in the block diagram of Figure 11.2.8.

FIGURE 11.3.1 Cross-field machine (basic two-pole metadyne).

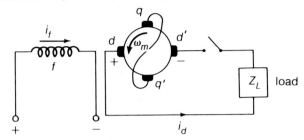

11.3 METADYNES AND AMPLIDYNES

We have so far considered the dc machines with brushes located only in the quadrature axis. The separately excited generator can be looked upon as a single-stage rotating power amplifier having a gain and one or more time constants. By the introduction of a shunt or series field, one can obtain performance control through feedback. The shunt generator can then be seen to be similar to a feedback amplifier. With the object of increasing the control possibilities, such as increasing the sensitivity or gain, decreasing the effective time constants or response time, and decreasing the sensitivity of the system to uncontrolled external disturbances, *cross-field machines* have been developed for control applications. Cross-field machines with more than two brush sets per pair of poles, called *metadynes,* can be used in a wide variety of applications requiring high speed response, or high power amplification, or some other special built-in characteristic. The purpose of this section is to examine the effects of additional brushes located in the direct axis; we shall concern ourselves with metadyne generators, with emphasis on the most commonly used form, the *amplidyne.*

The basic two-pole metadyne has a field winding as in the case of a dc machine and two pairs of commutator brushes, one pair in line with the field winding in the direct *d*-axis and the other pair at right angles to the field winding in the quadrature *q*-axis, as shown in Figure 11.3.1. With the generator driven at a constant speed and with magnetic saturation neglected, the voltage generated in the armature between the quadrature-axis brushes is proportional to the field current i_f:

$$e_{aq} = k_{qf} i_f \qquad (11.3.1)$$

where k_{qf} is a constant. If the quadrature-axis brushes are short-circuited, a weak control-field current will produce a relatively much larger quadrature-axis armature current because the impedance of the short-circuited armature is small. A corresponding flux-density wave will be centered on the quadrature axis and will produce an effect similar to that of a fictitious stator winding on the quadrature axis. With a pair of brushes *dd'* placed on the commutator in the

FIGURE 11.3.2 Basic two-pole amplidyne.

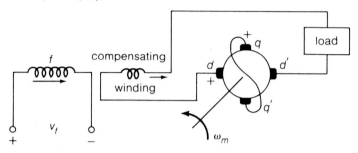

d-axis, the speed voltage generated in the armature by its rotation in the *q*-axis flux will appear across these brushes. Assuming constant speed and negligible saturation, this voltage will be proportional to the *q*-axis armature current i_q:

$$e_{ad} = k_{dq} \, i_q \qquad\qquad (11.3.2)$$

where k_{dq} is a constant. Now, if a load Z_L is connected to the *d*-axis brushes as shown in Figure 11.3.1, the *d*-axis armature current i_d will result, which in turn will produce an mmf *opposing* the control-field mmf. Under the assumed conditions of constant speed and negligible saturation, the *q*-axis generated emf is then given by

$$e_{aq} = k_{qf} \, i_f - k_{qd} \, i_d \qquad\qquad (11.3.3)$$

where k_{qd} is a constant. The metadyne generator is thus a two-stage power amplifier with rapid response and with strong negative current feedback from the final output stage to the input. Because of the effect of the negative feedback, the power amplification will be reduced. For a given value of the field current, the metadyne maintains nearly constant output current i_d over a wide range of load impedance, giving approximately a *constant-current* characteristic.

The most commonly used form of the metadyne is the amplidyne, which consists of the basic metadyne generator along with a *cumulative compensating winding* on the *d*-axis connected in series with the *d*-axis load current, as shown in Figure 11.3.2. If the compensating winding is designed to produce a flux as nearly as possible equal and opposite to the flux produced by the *d*-axis armature current, the negative-feedback effect of the load current is canceled and the control-field winding has almost complete control over the *d*-axis flux. Very little control-field power input is then required to produce a large current in the short-circuited *q*-axis of the armature. Power amplification of the order of 20,000:1 can easily be obtained. The fully compensated amplidyne gives approximately a *constant-voltage* characteristic. Any degree of compensation may, however, be used by varying the number of compensating turns.

FIGURE 11.3.3 Metadyne as a generalized machine: *(a)* schematic diagram of a two-pole metadyne with voltages impressed on both axes, *(b)* reference directions for speed voltages.

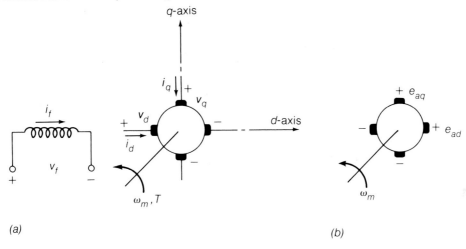

(a)

(b)

Assuming perfect compensation, constant speed, and negligible saturation, the transfer function relating the d-axis generated emf E_{ad} to the control-field applied voltage V_f can be obtained as

$$\frac{E_{ad}}{V_f} = \frac{k_{qf}/R_f}{\tau_f s + 1} \cdot \frac{k_{dq}/R_{aq}}{\tau_{aq} s + 1} \qquad (11.3.4)$$

where R_f and τ_f are the resistance and time constant of the control field, and R_{aq} and τ_{aq} are the resistance and time constant of the q-axis armature circuit. The d-axis armature-circuit resistance is considered along with the load, while the d-axis armature inductance is usually neglected. The time lag is essentially due to the q-axis time constant τ_{aq}. Various auxiliary or control-field windings may be added to either axis of an amplidyne either to improve performance characteristics or to obtain characteristics to meet performance specifications.

METADYNE AS A GENERALIZED MACHINE The two-axis metadyne may be regarded as a basic prototype or generalized machine. Let us consider the schematic diagram of a simple two-pole metadyne shown in Figure 11.3.3 with voltages v_d and v_q applied to the d-axis and q-axis brushes, respectively, and v_f applied to the d-axis stator field circuit. Let i_d, i_q, and i_f be the currents supplied to the respective circuits, as shown in Figure 11.3.3. For the *reference directions* shown in Figure 11.3.3, the flux linkages λ_f, λ_d, and λ_q associated with the field winding, the d-axis armature winding, and the q-axis armature winding can be expressed in a matrix form as follows:

$$\begin{bmatrix} \lambda_f \\ \lambda_d \\ \lambda_q \end{bmatrix} = \begin{bmatrix} L_{ff} & L_{fd} & 0 \\ L_{fd} & L_{ad} & 0 \\ 0 & 0 & L_{aq} \end{bmatrix} \begin{bmatrix} i_f \\ i_d \\ i_q \end{bmatrix} \qquad (11.3.5)$$

where L_{ff}, L_{ad}, and L_{aq} are the self-inductances of the field winding, the d-axis armature winding, and the q-axis armature winding, respectively; L_{fd} is the mutual inductance between the field winding and the d-axis armature winding, and it is a positive quantity for the reference directions chosen because the mmfs of the field current and of the d-axis armature current are in the same direction.

Noting the reference directions for speed voltages e_{aq} and e_{ad} as shown in Figure 11.3.3(b), the voltage equations can now be written as follows with the chosen reference directions:

$$v_f = R_f i_f + p \lambda_f \tag{11.3.6}$$

$$v_d = R_{ad} i_d + p \lambda_d - e_{ad} \tag{11.3.7}$$

$$v_q = R_{aq} i_q + p \lambda_q + e_{aq} \tag{11.3.8}$$

where R_f, R_{ad}, and R_{aq} are the resistances of the respective windings. For a two-pole machine with sinusoidally distributed windings, with saturation neglected, the speed voltages can be expressed as follows:

$$e_{ad} = \omega_m L_{aq} i_q \tag{11.3.9}$$

$$e_{aq} = \omega_m(L_{af} i_f + L_{ad} i_d) \tag{11.3.10}$$

where ω_m is the angular velocity corresponding to the speed of rotation, and L_{af} is the coefficient relating the no-load speed voltage generated in the q-axis to the field current and the speed. The voltage Equations 11.3.6 to 11.3.8 can then be expressed in matrix form as follows, after making use of Equations 11.3.5, 11.3.9, and 11.3.10:

$$\begin{bmatrix} v_f \\ v_d \\ v_q \end{bmatrix} = \begin{bmatrix} R_f + L_{ff} p & L_{fd} p & 0 \\ L_{fd} p & R_{ad} + L_{ad} p & -\omega_m L_{aq} \\ \omega_m L_{af} & \omega_m L_{ad} & R_{aq} + L_{aq} p \end{bmatrix} \begin{bmatrix} i_f \\ i_d \\ i_q \end{bmatrix} \tag{11.3.11}$$

The magnetic torque in the positive direction of rotation as shown in Figure 11.3.3(a) is given by the power input corresponding to the speed voltages divided by ω_m:

$$T = \frac{e_{aq} i_q - e_{ad} i_d}{\omega_m} = L_{af} i_f i_q + (L_{ad} - L_{aq}) i_d i_q \tag{11.3.12}$$

11.4 SOLID-STATE DRIVES FOR DC MOTORS

Direct-current motors are easily controllable and have dominated the adjustable-speed drive field. The torque-speed characteristics of a dc motor can be controlled by adjusting the armature voltage or the field current, or by inserting resistance into the armature circuit. Solid-state motor controls are designed to utilize each of these modes for particular purposes. The control resistors in which much energy is wasted are being eliminated through the development of power semiconductor

FIGURE 11.4.1 Basic principle of a chopper drive employing a dc-to-dc converter.

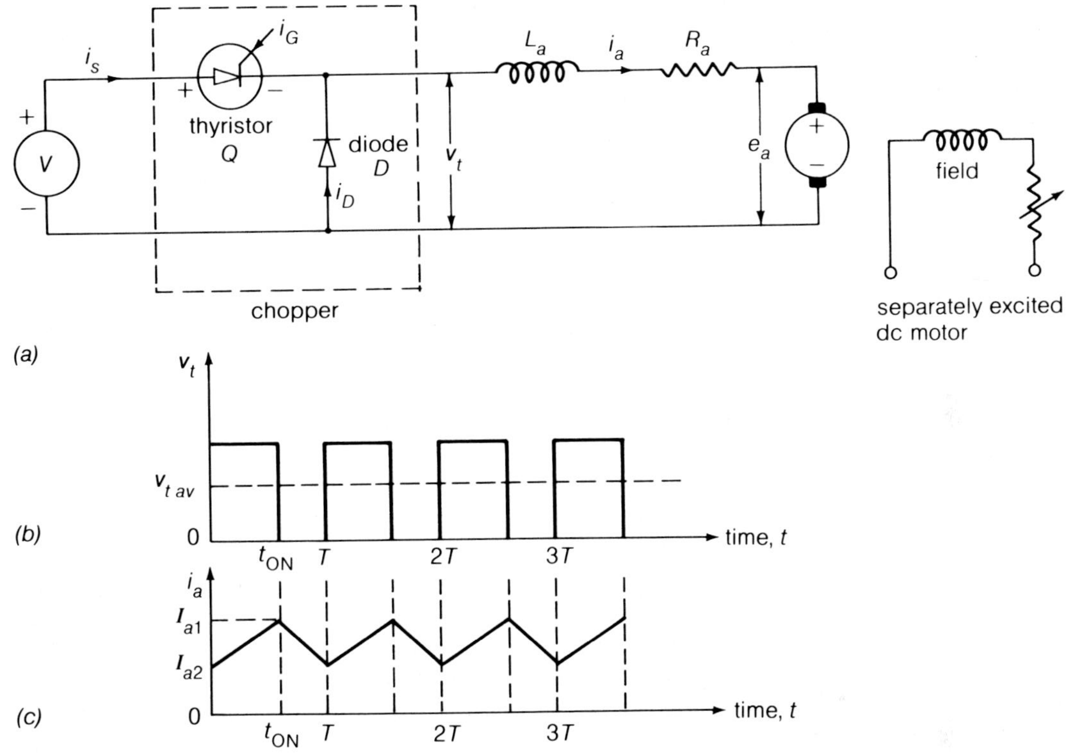

devices and the evolution of flexible and efficient converters. Thus, the inherently good controllability of a dc machine has been significantly increased in recent years.

The converters as applied to dc machines are of two principal types: *(a) dc-to-dc converters,* or *choppers,* which are employed when a dc source of suitable and constant voltage is already available; and *(b) phase-controlled rectifiers,* which are employed when only an ac source is available.

Figure 11.4.1 illustrates the basic principle of operation of a chopper drive employing a dc-to-dc converter for controlling a separately excited dc motor. The chopper can be considered as an ideal electronic switch that applies the source voltage V to the motor-armature terminals in a series of pulses, as shown in Figure 11.4.1(b). The average value $v_{t\ av}$ may be varied either by varying the pulse width or by varying the pulse frequency, or by a combination of both of these methods. The waveform of the armature current resulting from the pulsed v_t can be shown to be continuous, as in Figure 11.4.1(c), under conditions approaching full load, when the chopping frequency is sufficiently high to maintain the current and the motor torque is essentially constant.

In the chopper circuit of Figure 11.4.1, energy flow can take place only in one direction, from source to motor armature. Since i_a cannot reverse, the motor will be unable to regenerate. Regenerative braking is very desirable in high-inertia systems such as subway trains, because the

FIGURE 11.4.2   Two-quadrant chopper drive.

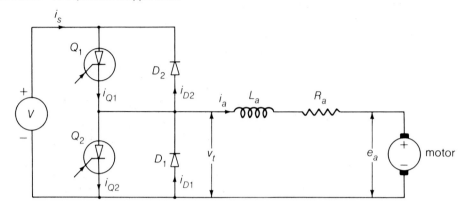

FIGURE 11.4.3   Three-phase, full-wave controlled bridge rectifier.

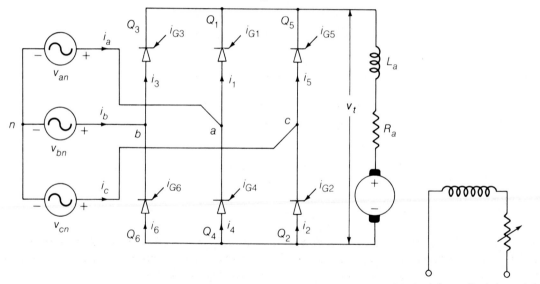

separately excited dc motor

FIGURE 11.4.4 Dual converter providing four-quadrant operation of a dc motor: *(a)* schematic diagram, *(b)* four-quadrant operation.

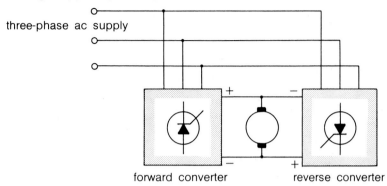

three-phase ac supply

forward converter   reverse converter

(a)

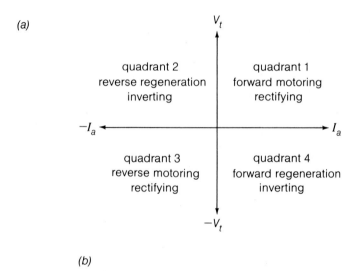

V_t

quadrant 2
reverse regeneration
inverting

quadrant 1
forward motoring
rectifying

$-I_a$ ←—————————————→ I_a

quadrant 3
reverse motoring
rectifying

quadrant 4
forward regeneration
inverting

$-V_t$

(b)

kinetic energy of the train would otherwise be dissipated as heat in the braking resistors or friction brakes. Hence slightly elaborated chopper circuits, such as the *two-quadrant chopper drive* shown in Figure 11.4.2, have been developed to permit regeneration to take place.

When the available power source is ac, a phase-controlled rectifier is employed to drive the dc machine. For motor-output power of 1 kW or less, a single-phase rectifier is satisfactory; for greater power, however, a three-phase source is necessary and the power circuit of such a drive with three-phase, full-wave controlled bridge rectifier is shown in Figure 11.4.3 as applied to a separately excited dc motor.

A controlled rectifier can transfer energy from the dc machine to the ac system. That is to say, it can invert. By employing a combination of two controlled rectifiers, as shown in Figure 11.4.4, a drive capable of operation in all four quadrants can be achieved. Such an arrangement of rectifiers is called a *dual converter.*

In spite of the variable-speed capabilities and the excellent speed-torque characteristics of conventional dc motors, the commutator limits the speed and voltage of dc machines, while posing a continual maintenance problem. In recent years considerable research has been devoted to the elimination of the commutator and brushes by electronic means, while at the same time retaining the desirable characteristics of the dc motor. The result is the *brushless dc motor*.

For further details on solid-state drives for dc motors and brushless dc motors, one is referred to books (some of which are cited at the end of the chapter) devoted to those special topics.

BIBLIOGRAPHY

Fitzgerald, A. E., and Kingsley, C., Jr. *Electric Machinery.* 2d ed. New York: McGraw-Hill, 1961.

Fitzgerald, A. E.; Kingsley, C., Jr.; and Kusko, A. *Electric Machinery.* 3d ed. New York: McGraw-Hill, 1971.

Kusko, A. *Solid-State DC Motor Drives.* Cambridge, Mass.: M.I.T. Press, 1969.

Matsch, L. W. *Electromagnetic and Electromechanical Machines.* New York: Intext, 1972.

Morgan, A. T. *General Theory of Electrical Machines.* London: Heyden & Son, 1979.

Pearman, R. A. *Power Electronics—Solid State Motor Control.* Reston, Va.: Reston, 1980.

Say, M. G., and Taylor, E. O. *Direct Current Machines.* New York: John Wiley & Sons, Halsted Press, 1980.

Slemon, G. R., and Straughen, A. *Electric Machines.* Reading, Mass.: Addison-Wesley, 1980.

White, D. C., and Woodson, H. H. *Electromechanical Energy Conversion.* New York: John Wiley & Sons, 1959.

Woodson, H. H., and Melcher, J. R. *Electromechanical Dynamics—Part I: Discrete Systems.* New York: John Wiley & Sons, 1968.

PROBLEMS

11–1.

A separately excited dc generator has the following parameters:

> Field winding resistance $R_f = 60$ ohms
> Field winding inductance $L_{ff} = 60$ H
> Armature resistance $R_a = 1$ ohm
> Armature inductance $L_{aq} = 0.4$ H
> Generated emf constant $K_g = 120$ V/field-amp, at rated speed

The armature terminals of the generator are connected to a low-pass filter with a series inductance of $L = 1.6$ H and a shunt resistance of $R = 1$ ohm. Determine the transfer function relating the output voltage $V_t(s)$ across the shunt resistance R and the input voltage $V_f(s)$ applied to the field winding.

11–2.

A separately excited dc generator, running at a constant speed, supplies a load having a 1-Ω resistance in series with a 1-H inductance. The armature resistance is 0.1 Ω and its inductance is negligible. The field, having a resistance of 50 Ω and an inductance of 5 H, is suddenly connected to a 100-V source.

Determine the armature current build-up as a function of time, if the generator voltage constant K_g is 50 V/field-amp at rated speed.

11–3.

The following test data are taken on a 20-hp, 250-V, 600-rpm dc shunt motor:

$$R_f = 150 \ \Omega; \ \tau_f = 0.5 \ \text{s}; \ R_a = 0.15 \ \Omega; \ \tau_a = 0.05 \ \text{s}$$

When the motor is driven at rated speed as a generator with no load, a field current of 2 A produces an armature emf of 250 V.

Determine the following: *(a)* L_{ff}, the self-inductance of the field circuit; *(b)* L_{aq}, the self-inductance of the armature circuit; *(c)* the coefficient k relating the speed voltage to the field current; and *(d)* the friction coefficient B_L of the load at rated load and rated speed, assuming that the torque required by the load T_L is proportional to the speed.

11–4.

A separately excited dc generator can be treated as a power amplifier when driven at constant angular velocity ω_m. If the armature circuit is connected to a load having a resistance R_L, obtain an expression for the voltage gain $V_L(s)/V_f(s)$.

With $R_a = 0.1 \ \Omega$, $R_f = 10 \ \Omega$, $R_L = 1 \ \Omega$, and $K_g = 100$ V/field-amp at rated speed, determine the voltage gain and the power gain if the generator is operating under steady state with 25 V applied across the field.

11–5.

Consider the motor of Problem 11–3 to be initially running at constant speed with an impressed armature voltage of 250 V, with the field separately excited by a constant field current of 2 A. Let the motor be driving a pure-inertia load with the combined polar moment of inertia of the armature and load of 3 kg·m². The rotational losses of the motor may be neglected.

    a. Determine the speed.
    b. Neglecting the self-inductance of the armature, obtain the expressions for the armature current and the speed as functions of time if the applied armature voltage is suddenly increased from 250 V to 260 V.
    c. Repeat part *(b)*, including the effect of the armature self-inductance.

11–6.

A separately excited dc motor carries a load of $(330 \ \dot{\omega}_m + \omega_m)$ N·m. The armature resistance is 1 Ω and its inductance is negligible. If 100 V is suddenly applied across the armature, while the field current is constant, obtain an expression for the motor speed build-up as a function of time, given the motor torque constant K_m to be 10 N·m/A.

11–7.

Consider the motor of Example 11.2.2 in the text to be operated as a separately excited dc generator at a constant speed of 900 rpm, with a constant field current of 1.5 A. Let the load current be initially zero.

Determine the armature current and the armature terminal voltage as functions of time for a suddenly applied load impedance consisting of a resistance of 11.5 ohms and an inductance of 0.1 H.

11–8.

Consider a motor supplied by a generator, each with a separate and constant field excitation. Assume the internal voltage E of the generator to be a constant, and neglect the armature reaction of both machines. With the motor running with no external load and the system being in steady state, let a load torque be suddenly increased from zero to T. The machine parameters are given below:

Armature inductance of motor $+$ generator, $L = 0.008$ H
Armature resistance of motor $+$ generator, $R = 0.04$ ohm
Internal voltage of the generator, $E = 400$ V
Moment of inertia of motor armature and load, $J = 42$ kg·m²
Motor constant, $K_m = 4.25$ N·m/A
No-load armature current, $i_0 = 35$ A
Suddenly applied torque, $T = 2,000$ N·m

Determine the following:

a. The undamped angular frequency of the transient speed oscillations, and the damping ratio of the system.
b. The initial speed, final speed, and the speed drop in rpm.

11–9.

Consider a separately excited dc motor having a constant field current, constant applied armature voltage, and accelerating a pure inertia load from rest. Neglecting the armature inductance and the rotational losses, show that by the time the motor reaches its final speed the energy dissipated in the armature resistance is equal to the energy stored in the rotating parts.

11–10.

Consider the shunt motor of Problem 9–13 to be coupled to a pure-inertia load. Let the polar moment of inertia of the load and the motor combined be 2 kg·m². Neglect the rotational losses of the motor and the self-inductance of the motor armature.

a. Assuming the magnetic circuit of the motor to be linear, calculate k of Equation 11.1.1 from the full-load data given in Problem 9–13.
b. While the motor accelerates on Step 3 of the starting box, determine the armature current and motor speed as functions of time. (Note that this problem deals with the transient response of a dc shunt motor when the initial values are other than zero.)

11–11.

This problem deals with an alternative derivation of the torque and voltage equations for a dc machine in a dynamic state from the coupled-circuit viewpoint. Consider a simple two-pole dc machine comprising a salient-pole stator with a field winding *f* and a cylindrical rotor with a finely distributed armature winding *a*. Neglect magnetic saturation. Let

$$\mathbf{L_{aa}} = \frac{L_{ad} + L_{aq}}{2} + \frac{L_{ad} - L_{aq}}{2} \cos 2\theta$$

$$\mathbf{L_{af}} = -L_{af} \cos \theta$$

$$\mathbf{L_{ff}} = L_{ff} = \text{constant}$$

where θ is the angle between the field and armature axes, measured from the direct axis or the centerline of the stator poles.

   a. First consider what would happen if connections were made to the armature winding at fixed points with respect to the rotor through slip rings. Then consider the effect of a commutator, noting that the brushes always make contact with the coils on the armature winding which are passing through the quadrature axis. By evaluating the electromechanical-coupling term $d\mathbf{L_{af}}/d\theta$ at $\theta = \pi/2$ in terms of armature conductors Z_a, the number of parallel paths a, and the direct-axis flux ϕ_d, show that the equations for e_a, v_t, and T_e are the same as obtained in the text.
   b. Considering the field circuit, with the assumptions of linear commutation and a narrow commutating zone, show that the rotational voltage induced in the field winding by the active part of the armature winding is canceled by the transformer voltage induced by the coils undergoing commutation. Then the field-voltage equation will be the same as Equation 11.1.3, just as if the armature winding were a stationary coil in the quadrature axis.

11–12.

Determine the parameters of the analog capacitive circuit (Figure 11.2.7 in the text) for the motor in Example 11.2.2 and its connected load. With the aid of the equivalent circuit, obtain the expression for the armature current under the conditions imposed and compare it with that obtained in part *(b)* of Example 11.2.2.

11–13.

Neglecting the self-inductance of the armature circuit, show that the time constant of the equivalent capacitive circuit for a separately excited dc motor with no load is $R_a R_{eq} C_{eq}/(R_a + R_{eq})$, where $R_{eq} = 1/G_{eq}$ in Figure 11.2.7 of the text.

11–14.

Consider the dc motor of Example 11.2.2 to be operating at rated voltage under steady state with a field current of 1 A and with zero load resistance in series with the armature.

    a. Obtain the equivalent capacitive circuit neglecting the armature self-inductance and calculate the steady armature current.

    b. If the field current is suddenly reduced to 0.8 A while the armature applied voltage is constant at 240 V, compute the initial armature current $i_a(0)$ on the basis that the kinetic energy stored in the rotating parts cannot change instantaneously.

    c. Determine the final armature current $i_a(\infty)$ for the condition of part *(b)*.

    d. Obtain the time constant τ'_{am} of the armature current for the condition of part *(b)* and express the armature current as a function of time on the basis that

$$i_a = i_a(\infty) + [i_a(0) - i_a(\infty)]\, e^{-t/\tau'am}$$

11–15.

The process of plugging a motor involves reversing the polarity of the supply to the armature of the machine. Plugging corresponds to applying a step voltage of $-Vu(t)$ to the armature of the machine, where V is the rated terminal voltage.

A separately excited 200-V dc motor operates at rated voltage with constant excitation on no-load. The torque constant of the motor is 2 N·m/A; its armature resistance is 0.5 Ω and the total moment of inertia of the rotating parts is 4 kg·m². Neglect the rotational losses and the armature inductance. Obtain an expression for the speed of the machine after plugging as a function of time, and calculate the time taken for the machine to stop.

11–16.

The Ward-Leonard system discussed in Chapter 9 is a highly flexible arrangement for effecting position and speed control of a separately excited dc motor. The schematic diagram of such a system including a separately excited dc generator, the armature of which is connected directly to the armature of the separately excited dc motor driving a mechanical load, is shown below:

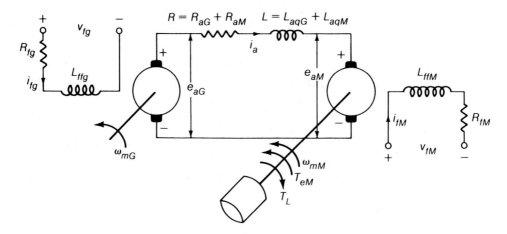

Let J be the combined polar moment of inertia of the load and motor, and B be the combined viscous friction constant of the load and motor.

Assuming that the mechanical angular speed of the generator ω_{mG} is a constant, develop the block diagram for the above system and obtain an expression for the transfer function $\Omega_{mM}(s)/V_{fG}(s)$.

11-17.

Consider an elementary motor-speed regulator scheme shown below for a separately excited dc motor, whose armature is supplied from a solid-state controlled rectifier. The motor speed is measured by means of a dc tachometer generator and its voltage e_t is compared with a reference voltage E_R. The error voltage $(E_R - e_t)$ is amplified and made to control the output voltage of the power-conversion equipment, so as to maintain substantially constant speed at the value set by the reference voltage.

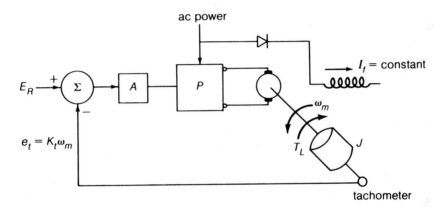

Let the armature-circuit parameters be R_a and L_a, and the speed-voltage constant of the motor be K_m $V - s/rad$.

Assume that the combination of A and P is equivalent to a linear controlled voltage source $v_S = K_A$ (error voltage), with negligible time lag and gain K_A. Assume also that the load torque T_L is independent of the speed, with zero damping. Neglect no-load rotational losses.

a. Develop the block diagram for the feedback speed-control system with E_R/K_t, the steady-state no-load speed setting, as input and Ω_m as output. K_t is the tachometer speed-voltage constant in volts/rpm.
b. With $T_L = 0$, evaluate the transfer function Ω_m/E_R.
c. With $E_R = 0$, obtain the transfer function Ω_m/T_L.
d. Find the expressions for the undamped natural frequency ω_n, the damping factor α, and the damping ratio $\xi = \alpha/\omega_n$.
e. For a step input ΔE_R, obtain the final steady-state response $\Delta\omega_m(\infty)$; i.e., evaluate $\Delta\omega_m(\infty)/\Delta E_R$.
f. Evaluate $\Delta\omega_m(\infty)/\Delta T_L$, for a step input ΔT_L of load torque.

11-18.

Consider the motor of Example 11.2.2 in the text to be used as a field-controlled dc machine. Let the armature be energized from a constant current source of 15 A. Assume no saturation.

    a. Develop a block diagram relating the motor speed and the applied field voltage.

    b. Determine the steady-state speed for a step applied field voltage of 220 V.

    c. How long does the motor take to reach 0.95 of the steady-state speed of part *(b)?*

11-19.

The output voltage of a 10-kW, 240-V dc generator is regulated by means of a closed-loop system shown on the following figure.

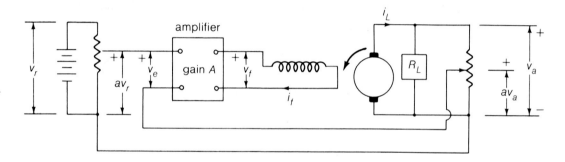

The generator parameters are given below:

$$R_f = 150\ \Omega;\ L_{ff} = 75\ H;\ R_a = 0.5\ \Omega;\ K_E = 150\ V/\text{field-amp}$$
at 1,200 rpm

The self-inductance of the armature is negligible. The amplifier has an amplification factor of $A = 10$ and the potentiometers are set such that a is unity and the reference voltage v_r is 250 V. The generator is driven by an induction motor the speed of which is almost 1,200 rpm when the generator output is zero and 1,140 rpm when the generator delivers an armature current of 42 A.

    a. Compute the steady-state armature terminal voltage at no-load and when the generator is delivering 42 A.

    b. Calculate the time constant for part *(a).*

    c. With the value of the field current as in part *(a),* for an armature current of 42 A, calculate the steady-state armature voltage.

11-20.

Perturbations in the voltage applied to the field or to the armature or in the torque of a dc motor may be regarded as small-signal inputs. The response of the motor to such inputs can be determined to a good degree of approximation by means of linearization techniques by considering the perturbations to be superimposed on quiescent operating points. Let the following equations with an additional subscript 0 represent the quiescent operating conditions for a dc motor:

$$T_{e0} = B\,\Omega_{m0} + T_{L0} = k\,I_{f0}\,I_{a0}$$

$$V_{f0} = R_f I_{f0}$$

$$V_{t0} = R_{a0}\,I_{a0} + k\,\Omega_{m0}\,I_{f0}$$

Let a small-signal input produce quantities with additional subscripts 1, superimposed on the quiescent points as shown below:

$$T_e = T_{e0} + T_{e1}$$

$$\omega_m = \Omega_{m0} + \omega_{m1}$$

$$i_f = I_{f0} + i_{f1}$$

$$v_t = V_{t0} + v_{t1}$$

$$i_a = I_{a0} + i_{a1}$$

When the above equations are substituted in the basic motor equations and the quiescent quantities are subtracted, show that the following equations result, when the products $(i_{f1}\,i_{a1})$ and $(i_{f1}\,\omega_{m1})$ are neglected:

$$T_{e1} = (J\,p + B)\,\omega_{m1} + T_{L1} = k(I_{f0}\,I_{a1} + I_{a0}\,i_{f1})$$

$$v_{f1} = (L_{ff}p + R_f)\,i_{f1}$$

$$v_{t1} = (L_{aq}p + R_a)\,i_{a1} + k(I_{f0}\,\omega_{m1} + \Omega_{m0}\,i_{f1})$$

Develop a linearized block diagram based on the above equations for a dc shunt motor.

11-21.

Pulsating loads produce oscillations in the torque that are superposed on the average value of torque and are accompanied by oscillations in the speed and armature current. If the torque oscillations are assumed to vary sinusoidally at an angular frequency of *r* rad/s, the torque and voltage can be expressed as phasors by substituting the imaginary quantity jr for p in the equations developed in Problem 11-20:

$$(B + jrJ)\,\overline{\Omega}_{m1} + \overline{T}_{L1} = k(I_{f0}\,\overline{I}_{a1} + I_{a0}\,\overline{I}_{f1})$$

$$\overline{V}_{f1} = (R_f + jr\,L_{ff})\,\overline{I}_{f1}$$

$$\overline{V}_{t1} = (R_a + jr\,L_{aq})\,\overline{I}_{a1} + k(I_{f0}\,\overline{\Omega}_{m1} + \Omega_{m0}\,\overline{I}_{f1})$$

where $\overline{\Omega}_{m1}$, $\overline{T}_{L1}$, $\overline{I}_{a1}$, $\overline{I}_{f1}$, $\overline{V}_{t1}$, and $\overline{V}_{f1}$ are phasors. Note that, if the voltage applied to the field is constant, $\overline{V}_{f1} = 0$ and $\overline{I}_{f1} = 0$ regardless of variations in the armature current, since there is no inductive coupling between the armature and the field. Let us apply the above analysis with phasor relationships for small oscillations in solving the following problem.

A 200-hp, 250-V, dc shunt motor has the following parameters:

$R_a = 0.025$ ohm; $L_{aq} = 0.001$ H; $R_f = 25$ ohms; $L_{ff} = 10$ H;
$k = 0.5$ H; the polar moment of inertia of the motor armature and the load together is 40 kg·m²; the load torque in N·m is

$$T_L = 2{,}500 + 600 \sin 20t$$

The no-load speed is 480 rpm. The rotational losses of the motor are negligible. If the motor is supplied from a 250-V dc source of negligible impedance, find the motor-armature current and the speed as functions of time.

11-22.

Consider the armature of the motor in Problem 11-21 to be energized at a constant terminal voltage of 250 V. Let the voltage applied to the field be $v_f = 150 + 60 \sin 3t$, and let the load be purely inertial with the combined polar moment of inertia of the motor armature and the load of 40 kg·m².

   a. Determine the quiescent values of the field current, armature current, and speed.
   b. Express the corresponding variable components in part *(a)* as phasors and as functions of time.
   c. Obtain the expressions for the corresponding total quantities in part *(a)* as functions of time.

11-23.

Repeat Problem 11-22 but with the voltage applied to the field to be constant at 250 V and the voltage applied to the armature to be

$$v_t = 150 + 60 \sin 3t$$

11–24.

Develop the block diagram of a simple metadyne generator, shown in Figure 11.3.1 of the text, with v_f as the input and i_d as the output, while operating at rated speed.

Also obtain the block diagram of the basic amplidyne corresponding to Equation 11.3.4.

11–25.

A 5-kW, 240-V amplidyne driven at its rated speed of 1,800 rpm has the following parameters:

$$R_f = 1,000 \text{ ohms}; \ L_{ff} = 200 \text{ H}; \ L_{af} = 1.5 \text{ H};$$
$$R_{aq} = 1 \text{ ohm}; \ L_{aq} = 0.3 \text{ H}$$

Assuming the machine to be completely compensated, express the open-circuit output voltage e_{ad} as a function of time after a constant voltage $v_f = 20$ V is applied to the field winding.

11–26.

Consider the steady-state operation of the basic metadyne generator and the amplidyne with all currents and voltages constant. Obtain the three voltage equations for the field, d-axis, and q-axis circuits. Express the output voltage in terms of the field current and the output current.

11–27.

A cross-field generator has the following parameters:

Inductance of d- or q-axis armature windings = 0.05 H
Resistance of d- or q-axis armature windings = 1 ohm
Resistance of compensating winding = 1 ohm
Mutual inductance between field winding and the d-axis armature winding = 1 H

If the machine rotates at a constant speed of 200 rad/s, plot curves of output current versus output voltage

a. when operating as a fully-compensated amplidyne with a field current of
(i) 0.1 A, (ii) 0.05 A.
b. when operating as a metadyne with a field current of (i) 1 A, (ii) 0.5 A.

11–28.

A two-pole, fully compensated amplidyne has the following parameters:

Field inductance = 25 H; Field resistance = 100 ohms;
Armature inductance = 0.05 H; Armature resistance = 1 ohm;
Compensating winding resistance = 1 ohm;
Field/armature mutual inductance = 1 H

If a step-function voltage of 100 V is applied to the field, determine the output current as a function of time if the machine is connected to a load of resistance 25 ohms and inductance 5 H. Assume the speed to remain constant at 200 rad/s.

APPENDIX A

UNITS, CONSTANTS,
AND CONVERSION FACTORS
FOR THE SI SYSTEM

TABLE A.1 PHYSICAL QUANTITIES

Physical Quantity	SI Unit	Symbol
length	meter	m
mass	kilogram	kg
time	second	s
current	ampere	A
admittance	siemen (A/V)	S
angle	radian	rad
angular acceleration	radian per second squared	rad/s^2
angular velocity	radian per second	rad/s
apparent power	voltampere (V·A)	VA
area	square meter	m^2
capacitance	farad (C/V)	F
charge	coulomb (A·s)	C
conductance	siemen (A/V)	S
electric field intensity	volt/meter	V/m
electric flux	coulomb (A·s)	C
electric flux density	coulomb/square meter	C/m^2
energy	joule (N·m)	J
force	newton (kg·m/s²)	N
frequency	hertz (1/s)	Hz
impedance	ohm (V/A)	Ω
inductance	henry (Wb/A)	H
linear acceleration	meter per second squared	m/s^2
linear velocity	meter per second	m/s
magnetic field intensity	ampere/meter	A/M
magnetic flux	weber (V·s)	Wb
magnetic flux density	tesla (Wb/m²)	T
magnetomotive force	ampere or ampere-turn	A or At
moment of inertia	kilogram-meter squared	$kg·m^2$
power	watt (J/s)	W
pressure	pascal (N/m²)	Pa
reactance	ohm (V/A)	Ω
reactive power	voltampere reactive	var
resistance	ohm (V/A)	Ω
resistivity	ohm-meter	$\Omega·m$
susceptance	siemen (A/V)	S
torque	newton-meter	N·m
voltage	volt (W/A)	V
volume	cubic meter	m^3

TABLE A.2   PREFIXES

Prefix	Symbol	Meaning
exa	E	10^{18}
peta	P	10^{15}
tera	T	10^{12}
giga	G	10^{9}
mega	M	10^{6}
kilo	k	10^{3}
hecto	h	10^{2}
deka	da	10^{1}
deci	d	10^{-1}
centi	c	10^{-2}
milli	m	10^{-3}
micro	μ	10^{-6}
nano	n	10^{-9}
pico	p	10^{-12}
femto	f	10^{-15}
alto	a	10^{-18}

TABLE A.3   PHYSICAL CONSTANTS

Quantity	Symbol	Value	Unit
Permeability constant	μ_0	1.257×10^{-6}	H/m
Permittivity constant	ϵ_0	8.854×10^{-12}	F/m
Gravitational acceleration constant	g_0	9.807	m/s^2
Speed of light in a vacuum	c	0.2998×10^{9}	m/s

TABLE A.4   CONVERSION FACTORS

Physical Quantity	SI Unit	Equivalents
Length	1 meter (m)	3.281 feet (ft) 39.37 inches (in.)
Angle	1 radian (rad)	57.30 degrees
Mass	1 kilogram (kg)	0.0685 slugs 2.205 pounds (lb) 35.27 ounces (oz)
Force	1 newton (N)	0.2248 pounds (lbf) 7.233 poundals 0.1×10^6 dynes 102 grams
Torque	1 newton-meter (N·m)	0.738 pound-feet (lbf-ft) 141.7 oz-in 10×10^6 dyne-centimeter 10.2×10^3 gram-centimeter
Moment of inertia	1 kilogram-meter^2 (kg·m^2)	0.738 slug-feet^2 23.7 pound-feet^2 (lb-ft^2) 54.6×10^3 ounce-inches^2 10×10^6 gram-centimeter^2 (g·cm^2)
Energy	1 joule (J)	1 watt-second 0.7376 foot-pounds (ft-lb) 0.2778×10^{-6} kilowatt-hours (kWh) 0.2388 calorie (cal) 0.948×10^{-3} British Thermal Units (BTU) 10×10^6 ergs
Power	1 watt (W)	0.7376 foot-pounds/second 1.341×10^{-3} horsepower (hp)
Resistivity	1 ohm-meter (Ω·m)	0.6015×10^9 ohm-circular mil/foot 0.1×10^9 micro-ohm-centimeter
Magnetic flux	1 weber (Wb)	0.1×10^9 maxwells or lines 0.1×10^6 kilolines
Magnetic flux density	1 tesla (T)	10×10^3 gauss 64.52 kilolines/in.^{2}
Magnetomotive force	1 ampere (A) or ampere-turn (At)	1.257 gilberts
Magnetic field intensity	1 ampere/meter (A/m)	25.4×10^{-3} ampere/in. 12.57×10^{-3} oersted

APPENDIX B

SPECIAL MACHINES

FIGURE B.1.1 Typical rotor punching for a four-pole reluctance-type
synchronous motor.

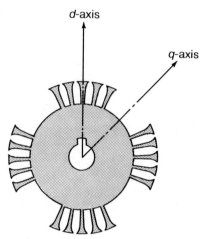

The basic operating principles of electrical machines have been discussed in the text and the performance analyses of the more common and conventional types of machines have been presented. Special machines are those that have features that distinguish them from the more conventional types. Because of their large variety, it would be unrealistic to attempt to cover all types of such machines. With an aim to familiarize the reader, this appendix is devoted to some special machines that are used to an appreciable extent in practice.

B.1 RELUCTANCE MOTORS

Reluctance motors depend for their operation on the difference between the reluctances in the direct and quadrature axes, and are in effect synchronous motors that operate without dc field excitation. Fractional horsepower motors are usually single-phase and find their application for such drives as electric clocks and other timing devices that require exact synchronous speed. In spite of their relatively large size and low power factor, polyphase reluctance motors in ratings up to 150 hp have been built because of their constructional simplicity and practically maintenance-free operation, owing to the lack of slip-rings, brushes, and dc field winding.

Reluctance motors are usually started as induction motors by making use of a squirrel-cage rotor from which teeth have been removed in locations so as to produce the desired number of salient poles, as for example the four-pole rotor shown in Figure B.1.1. In such rotors the teeth are removed, leaving all the rotor bars and the entire end-rings intact. Any of the single-phase starting arrangements discussed in the text may be used for single-phase reluctance motors.

B.2 HYSTERESIS MOTORS

The hysteresis loss in the rotor can be greatly magnified by constructing the rotor-core of an induction motor of hardened magnet steel, instead of the usual annealed low-loss silicon-steel laminations. Because of hysteresis, the magnetization of the rotor lags behind the inducing mmf wave; the synchronous-motor action is produced by an angular shift between the axis of rotating primary mmf and the axis of secondary magnetization. Such hysteresis motors, having smooth rotor surfaces without secondary teeth or windings, yield substantially the same torque from standstill all the way up to synchronous speed, while they are practically noiseless. Because of the inherently small torque obtainable from hysteresis losses, hysteresis motors are limited to small sizes, generally only a few watts to as high as 1/7 hp. These have found applications in clocks and phonographs.

B.3 PERMANENT-MAGNET MACHINES

A small synchronous machine (motor or generator) may have its excitation provided by a permanent magnet. A permanent-magnet synchronous motor is constructed by fitting the magnet inside the rotor cage, which is necessary for induction starting. A permanent-magnet generator, however, does not require a starting cage on its rotor. A major problem with such machines is to avoid demagnetization under service conditions; a generator may have its flux reduced to zero if a short-circuit occurs, and a motor is subjected to a reversal of flux every time it is started from standstill.

Permanent-magnet (PM) dc motors are built with power ratings ranging from a few watts to 100 kW or more. The PM motor is usually considered a linear device over its entire operating range of torque and speed, with a practically linear speed-torque characteristic. In the construction of small PM motors up to ratings of a few kilowatts, grain-oriented ferrite magnets are normally employed and magnetized during manufacture, before they are fitted into a stator. Typically, a 30 percent reduction in machine weight may be achieved by using permanent magnets instead of wound-field poles.

In large PM dc motors, it is not possible to assemble the large magnets, ferrite or alloy (such as Alnico), into the stator after they have been magnetized. It is therefore necessary to equip the pole pieces with magnetizing windings after the stator has been assembled with unmagnetized pole pieces. Their presence, as well as the interpoles and compensating windings that may be provided to improve commutation, result in a machine structure somewhat similar to that of a dc shunt motor.

A motor configuration that is radically different from that of a conventional machine is the *disk-armature* or *printed-circuit motor,* which has been widely used with permanent-magnet materials. The rotor is made of a disk of nonconducting and nonmagnetic material; on both sides of the disk, printed in copper, is the entire armature winding and commutator with appropriate connections made from side to side, as illustrated in Figure B.3.1(a). The stator consists essentially of two ferromagnetic end plates each carrying the required number of permanent-magnet pole pieces, as shown in Figure B.3.1(b). The low rotor inertia of such a configuration and consequent fast response of the motor make it suitable for control-system applications.

FIGURE B.3.1   Disk-armature or printed-circuit permanent-magnet dc motor: *(a)* one side of the rotor, *(b)* physical arrangement of the stator and rotor.

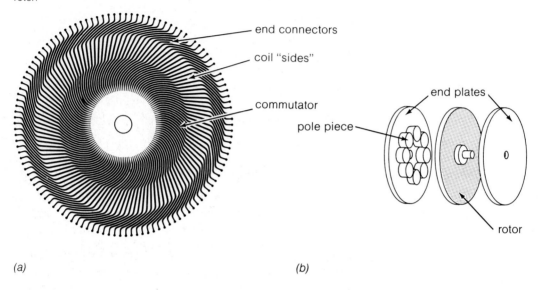

end connectors

coil "sides"

commutator

end plates

pole piece

rotor

(a)   *(b)*

The rare-earth/cobalt (such as samarium-cobalt) permanent magnets seem to permit the production of PM motors with improved performance and reduced size at a reasonable cost. Such motors of about 1-kW rating are already available, while higher ratings should be realizable in the near future. In order to employ minimum magnetic material, besides the printed-circuit motor configuration, the *stationary-armature* or the *"inside-out" configuration* has been developed, in which the armature winding and commutator are placed on the stator and the field poles are on the rotor. Two possible arrangements of the field poles on the rotor of the inside-out machine are shown in Figure B.3.2 with radial and circumferential magnetization.

B.4   STEPPING MOTORS

The stepping or stepper motor is a form of synchronous motor designed to rotate a specific number of degrees for each electrical pulse received by its control unit. Such a motor is used in digital control systems, where the motor receives open-loop commands as a train of pulses to turn a shaft or move a plate by a specific distance. A typical application is to position both the cutting tool and the workpiece very accurately during machine-tool operations in accordance with instructions on tape. With typical step angles of 15°, 7.5°, 5°, and 2° per pulse, the choice of which depends on the angular resolution required for the particular application, stepper motors are built to follow signals as rapid as 1,200 pulses per second and to have equivalent power ratings up to several horsepower.

FIGURE B.3.2 Rotor configurations for stationary-armature or inside-out PM machines: *(a)* radial magnetization, *(b)* circumferential magnetization.

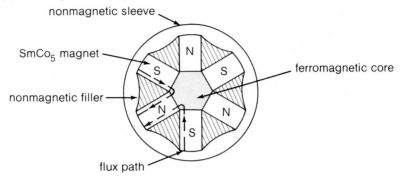

(a)

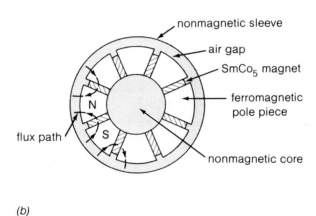

(b)

   Stepper motors are usually designed with a multipole, multiphase stator winding; typically, three- and four-phase windings are used with the number of poles determined by the required angular change per input pulse. The rotors are either of the *variable-reluctance type* or the *permanent-magnet type*. Stepper motors are operated with an external drive logic circuit. With a train of pulses applied to the input of the drive circuit, the circuit delivers appropriate currents to the motor-stator windings to make the axis of the air-gap field step around in coincidence with the input pulses. By virtue of the reluctance torque and/or the permanent-magnet torque, the rotor follows the axis of the air-gap magnetic field, depending upon the pulse rate and the load torque, including inertia effects.

FIGURE B.4.1 Variable-reluctance stepping motor having a rotor with eight poles and three separate eight-pole stators arranged along the rotor.

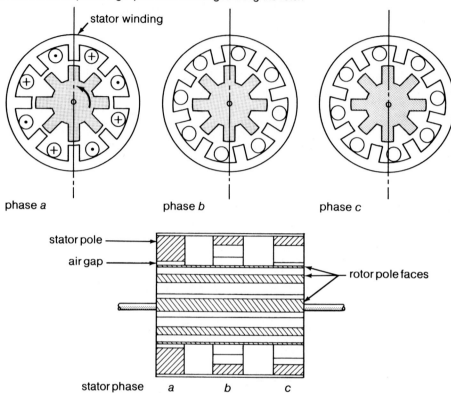

The *variable-reluctance stepping motor* shown in Figure B.4.1 has a rotor with eight poles and three separate eight-pole stators arranged along the rotor. When the poles of stator-phase-*a* are energized with alternate polarity by a set of series-connected coils carrying current i_a, the rotor poles tend to align with the poles of stator-phase-*a* as shown in Figure B.4.1. The stator-phase-*b* is identical with the phase-*a* stator, except that its poles are displaced by 15° in a counterclockwise direction. If the current i_a is now set equal to zero and the current i_b is established, the motor will develop a torque rotating the rotor counterclockwise by 15°. Stator-phase-*c* has its poles displaced a further 15° anticlockwise with respect to stator-phase-*b*. Setting the current i_b equal to zero and establishing current i_c will lead to a further 15° anticlockwise rotation of the rotor. Finally, interrupting the current i_c and reestablishing the current i_a will result in the completion of a 45° rotation of the rotor. Further current pulses in the *a-b-c* sequence will produce further anticlockwise stepping motions. Reversing the current-pulse sequence to *a-c-b* will produce reversed rotation. As seen from the preceding discussion, the step angle of a motor is determined by the number of poles.

FIGURE B.4.2   Stepping motor with a permanent-magnet rotor.

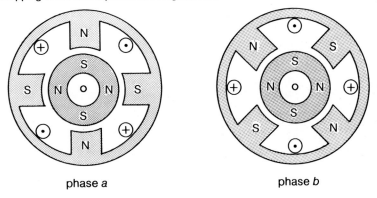

phase *a*   phase *b*

FIGURE B.4.3   Circuit with double coils in each phase for the motor of
Figure B.4.2.

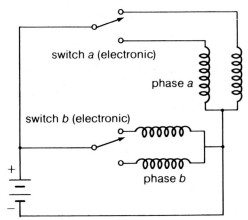

The torque developed by a stepping motor of a given size may be increased if a *permanent-magnet rotor* is used. Figure B.4.2 illustrates the stator arrangement for a four-pole, two-phase motor with a permanent-magnet rotor, typically made of ferrite. Energizing the phase-*a* winding only with the polarity shown will hold the rotor in the position shown. The additional energizing of phase-*b* winding with the polarity shown will result in a movement of 22.5° in an anticlockwise direction. De-energizing the phase-*a* winding will then result in a further 22.5° movement. Reversal of the phase-*a* winding current will now produce a further 22.5° movement, and so on. In order to simplify the switching operation required to reverse stator polarities, double coils are usually employed in each phase, as shown in Figure B.4.3 for the motor of Figure B.4.2.

FIGURE B.5.1 Homopolar-type inductor alternator.

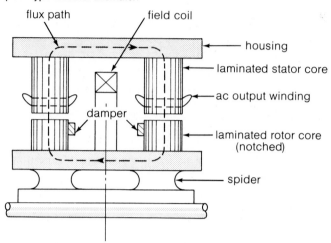

Typical variable-reluctance stepper motors operate at small steps, 15° or less, and at maximum position response rates up to 1,200 pps. Typical permanent-magnet types operate at larger steps, up to 90°, and at maximum response rates of 300 pps. Applications include table positioning for machine tools, tape drives, recorder-pen drives, and *X-Y* plotters.

B.5 INDUCTOR ALTERNATORS

The inductor alternator is a synchronous generator with stationary field and armature windings, and depends for its operation on a periodic variation in the reluctance of the air gap. Such machines can operate at high speeds to generate correspondingly high frequencies, usually from several hundred hertz up to 100,000 Hz, for use in such applications as steel and nonferrous melting by induction with high-frequency currents, for supplying power at radio frequency, and supplying military power demands including lighting, electric motors, and sophisticated electronic equipment requiring precise voltage and frequency regulation and control. They are suitable for direct connection to high-speed prime movers such as gas turbines.

The flux through an armature coil of an inductor alternator is unidirectional, and instead of periodically reversing its direction, it fluctuates between a maximum and a minimum value, with the design of the air gap being such that the induced voltage is practically sinusoidal. Inductor generators may generally be classified as *homopolar* or *heteropolar* types.

The *homopolar-type inductor alternator,* shown in Figure B.5.1, has two laminated stator cores and two laminated rotor cores. The ac output winding is arranged in the slots of the stator cores. The field winding, consisting of a coil that is wound concentric with the axis of the machine,

FIGURE B.5.2 Cross-sectional view of a four-section homopolar inductor alternator. Courtesy of U.S. Department of the Army.

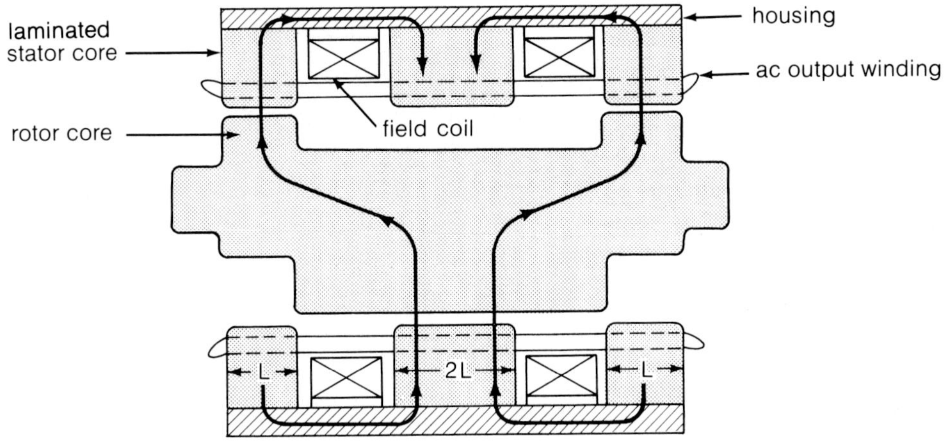

produces the unidirectional flux. The rotor cores are notched with open slots so as to produce the desired waveform of the flux-density distribution. One cycle of voltage is generated as the rotor advances through a distance equal to that between the centers of two adjacent rotor teeth, and the frequency is therefore given by

$$f = \frac{(\text{Number of rotor teeth})(\text{Synchronous speed in rpm})}{60} \qquad (B.5.1)$$

Figure B.5.2 shows a cross-sectional view of a four-section homopolar inductor alternator, which may be viewed as the equivalent of a combination of two two-section machines, such as the one shown in Figure B.5.1. Approximate flux paths of such an alternator are indicated in the figure.

The *heteropolar-type inductor alternator,* illustrated in Figure B.5.3, has a stator housing and stator cores, as well as a rotor, that are similar to those of the squirrel-cage induction motor. The stator punchings are notched to carry the ac output winding. The field winding, wound in several slots on the stator, produces a multipolar field, as indicated by the broken lines in Figure B.5.3*(b),* in contrast with the unipolar field produced by the field winding in the homopolar type. As in the homopolar type, the field fluctuates in the heteropolar machine due to variations in the air-gap length. The voltage response to changes in the field current is more rapid in the heteropolar machine than in the homopolar machine.

FIGURE B.5.3 Heteropolar-type inductor alternator.

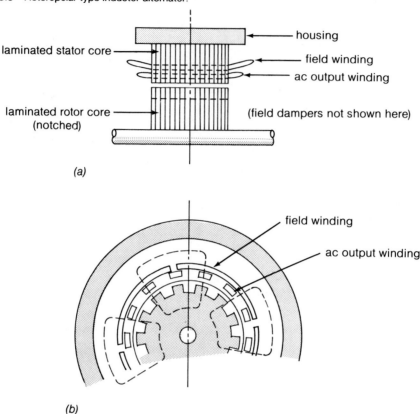

(a)

(b)

B.6 AC COMMUTATOR MOTORS

AC commutator motors have armatures that are similar to those in dc machines; their field structures, however, are laminated to reduce eddy currents. The two distinct advantages of the ac commutator motors over induction motors are a wide speed range and a high starting torque. Also, series motors may be operated at several times the induction-motor synchronous speed. AC commutator motors may be grouped into two classes: *(a) series motors,* which have similar characteristics to those of series-wound dc motors and whose speed at any given load may be varied by changing the applied voltage or, in some cases, by shifting the brushes; *(b) shunt motors,* whose speed is approximately constant for operation from a source of constant voltage and whose speed may be changed (independently of the load) by changing the voltage at the terminals of the motor, either by brush shifting or by providing suitably disposed and connected auxiliary coils.

The most common ac commutator machine is the single-phase series motor, although there are some three-phase series motors and three-phase shunt motors. Polyphase commutator motors have inherently better commutating ability. Single-phase motors are generally limited to sizes below

FIGURE B.6.1   AC single-phase series motor: (a) straight series motor,
(b) compensated series motor.

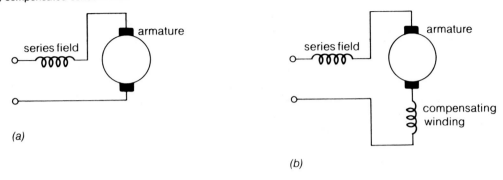

(a)

(b)

FIGURE B.6.2   Schematic diagram of a repulsion motor.

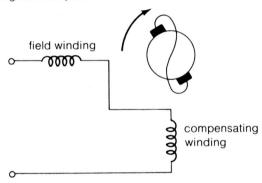

10 hp, except for railway application. With the advent of solid-state diodes and thyristors, the ac commutator motor is being displaced by the combination of rectifiers and dc motors, at less cost and superior performance.

A dc series motor with a laminated field structure can be made to operate on ac supply; in fact, small series motors up to about 1/2-hp rating, known as *universal motors,* are designed to operate on either dc or ac supply. Figure B.6.1 shows the schematic diagrams for the *straight series motor* and the *compensated series motor* with a compensating winding to overcome the armature mmf. Universal motors are generally used for small devices such as vacuum cleaners, food mixers, and portable tools operating at speeds in the range of 3,000 to 11,000 rpm.

If, instead of connecting the field and armature in series, the supply voltage is connected directly across the field and the armature brushes are short-circuited, the simple *repulsion motor* is obtained, as shown in Figure B.6.2. Actually, the field and compensating functions are performed by a single stator winding, and the brush axis is placed at an angle α with respect to the axis of this winding. If the brushes are in line with the stator-winding axis, i.e., $\alpha = 0$, the motor becomes

FIGURE B.6.3 Schematic diagram of a three-phase adjustable-speed brush-shifting Schrage-type motor.

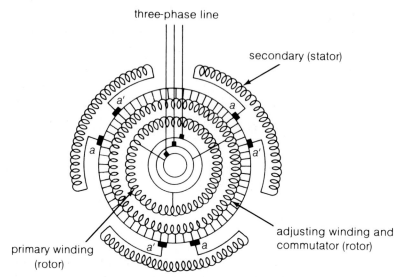

a short-circuited transformer, producing zero torque. With brushes located in a position corresponding to $\alpha = 90°$, the induced armature current is zero and hence the torque is also zero. The brushes must therefore occupy some intermediate position in order that the currents induced in the rotor by transformer action react on the stator flux and produce torque. When the brushes are shifted through a position of zero torque, the direction of rotation reverses.

The commutation of the repulsion motor is superior to that of the series motor up to synchronous speed and inferior at higher speeds because of the large short-circuit currents in the coils undergoing commutation. The armature, being isolated from the line, may be designed for any convenient low voltage. The repulsion motor has a torque-speed characteristic similar to that of the series motor but has the additional advantage that its speed can be adjusted by shifting the brushes. Prior to the widespread use of the capacitor-start single-phase induction motor, *repulsion-start induction-run motors* (see Section 7.8) were used in applications requiring high starting torque and practically constant running speed. There is also the *repulsion-induction motor* which has a squirrel cage in the bottom of the rotor slots, with the commutated winding occupying the top of the rotor slots; both repulsion and induction torques are combined in such a machine.

The *brush-shifting polyphase shunt motor,* also known as the *Schrage motor,* is basically an inside-out induction motor with its secondary winding on the stator and its primary winding on the rotor connected to the supply line through slip rings, as illustrated in Figure B.6.3. An adjusting winding is embedded in the same rotor slots and connected to a commutator. Line-frequency voltages are induced in the adjusting winding by transformer action from the primary, and slip-frequency voltages appear between brushes on the commutator. The two sets of brushes, indicated by *a* and *a'* in the figure, can be adjusted as to angular position and relative spacing

FIGURE B.7.1 Schematic diagram of a two-phase control motor.

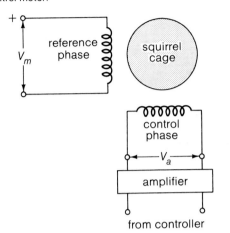

FIGURE B.7.2 Illustration of drag-cup rotor construction.

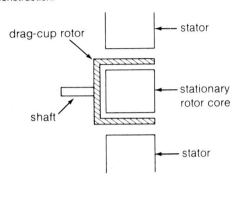

between them by means of a handwheel. The magnitude of the slip-frequency voltages inserted in series with the secondary windings depends on the spacing between the two sets of brushes, and the phase depends on their angular position. Thus, both the speed and power factor can be controlled. Machines of this type are mainly used in sizes up to 50 hp and in speed ranges of 6 to 1 or below, for a wide range of applications such as driving pumps, fans, paper mills and conveyors.

B.7 SERVO COMPONENTS

Two-phase control motors are two-phase squirrel-cage induction motors used in some control systems. Their output ranges from a fraction of a watt to several hundred watts. As shown in the schematic diagram of Figure B.7.1, the two stator windings, phase m (or the *reference phase*) and phase a (or the *control phase*), are identical and are displaced from each other by 90 electrical degrees. The reference phase is connected to a constant-voltage, constant-frequency source of V_m volts, while the control phase is supplied with the voltage V_a from the controller, usually with an amplifier intervening. The error voltage V_a, which is of the same frequency as V_m and is phase-displaced from V_m by 90 electrical degrees, is proportional to the required amount of correction. *High-resistance rotors* are used for such control motors in order to assure high torque near zero speed and to prevent the motor from running as a single-phase motor when the error signal is zero. When the maximum power output is below a few watts, a *drag-cup rotor* construction is used as illustrated in Figure B.7.2, in which the rotating member is like a thin metallic can with one end removed, thereby minimizing inertia. A stationary iron core, like a plug inside the cup, completes the magnetic circuit.

FIGURE B.7.3   Three-phase selsyn connections.

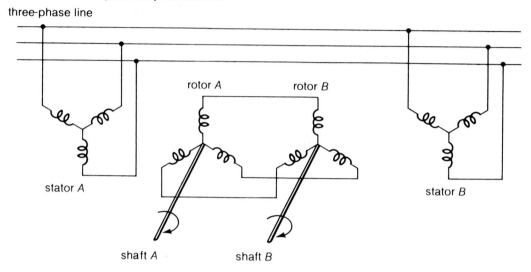

A small two-phase machine, often with drag-cup rotor construction, is used as an *ac tachometer* for measuring the angular velocity of a shaft. While the main winding, often referred to as the *fixed field* or *reference field*, is energized from a suitable alternating voltage of constant magnitude and frequency, a voltage of the same frequency is then generated in the auxiliary winding or *control field*; this voltage is applied to the high-impedance input circuit of an amplifier; thus, the auxiliary winding can be considered as open-circuited. Ideally, the electrical requirements are that the magnitude of the signal voltage generated in the auxiliary winding should be linearly proportional to the speed and that the phase of this voltage should be fixed with respect to the voltage applied to the main winding.

Self-synchronous devices, also known as *selsyns, synchros,* and *autosyns,* are used in many systems to synchronize the angular positions of two shafts at different locations where a mechanical interconnection of the shafts is rather impractical. Three-phase selsyns are used in systems where high torque transmission is required, while single-phase selsyns are used for low torque transmission. For transmitting shaft-position information, for indicating at some remote location the positions of some devices such as elevators and hoists, and for operating servomechanisms to control the position or motion of larger equipment, single-phase selsyns often find their applications.

The three-phase selsyn consists of two three-phase wound-rotor induction motors; as illustrated in Figure B.7.3, the stator windings are connected in parallel and energized from a three-phase source, while the rotors connected to their respective shafts are also electrically in parallel with each other. The single-phase selsyn, shown in Figure B.7.4, has a *selsyn-transmitter or generator* and a *selsyn-receiver or motor,* both of which have a single-phase winding (usually on the rotor) connected to a common ac voltage source; the other member of each (usually the stator) has a

FIGURE B.7.4   Single-phase synchrotransmitter-motor system.

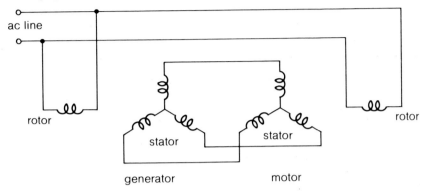

FIGURE B.7.5   Synchro generator-motor system with differential.

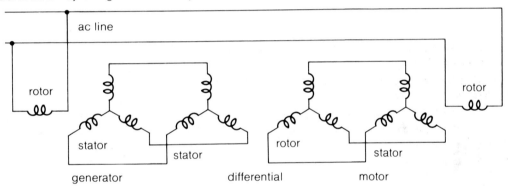

three-phase wye-connected winding, and the two windings are connected in parallel. The rotors are of salient-pole construction, which makes a simple winding arrangement possible. When the single-phase rotor windings are excited, voltages are induced by transformer action in the wye-connected stator windings. If the two rotors are in the same space position relative to their stator windings, the transmitter and receiver stator-winding voltages are equal, no current circulates in these windings, and hence no torque is transmitted. If, however, the two rotor space positions do not correspond, the stator-winding voltages are unequal; currents circulating in the stator windings, in conjunction with the air-gap magnetic fields, produce torques tending to align the two rotors. The receiver is generally provided with mechanical damping to reduce oscillations. The motor-torque at standstill or for slow rotation can be shown to depend closely upon the sine of the relative angular difference in position of the transmitter and receiver shafts.

The addition of a third selsyn known as a *differential selsyn,* as shown in Figure B.7.5, produces the rotation of its shaft as a function of the sum or difference of the rotation of two other shafts. The differential selsyn has a cylindrical rotor with a three-phase winding on the rotor as well as on the stator.

FIGURE B.7.6   Synchro generator-transformer system.

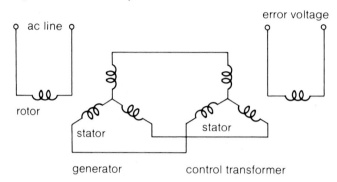

Synchro generator-transformer system with a *control transformer,* as illustrated in Figure B.7.6, is used to supply an error voltage, resulting from angular displacement between two shafts, to a corrective device such as a two-phase control motor, which has a greater torque capability than the selsyn system.

B.8   LINEAR INDUCTION MOTORS

Linear propulsion relates to a novel type of electric drive for ground-transportation systems. In such a scheme, the motor is split into two parts, of which one is carried by the vehicle and the other lies straight along the track. The force of interaction between these two structures is used directly as tractive effort, without the need for intermediate transmission or gear. The basic difference between rotating and linear motors is that in the former the air gap and the magnetic structures are *endless,* whereas in the latter the air gap and one of the magnetic structures (short side) are *open-ended.* All types of existing electric motors can be realized in the linear form. Further, the peculiar requirements of linear propulsion are giving rise to the development of novel configurations. As a result, a great variety of schemes is being considered, and only the test of time will decide on their practicality.

The *linear induction motor* (LIM) has reached the most advanced stage of development. In its simplest form, it consists of *(i)* a ferromagnetic structure located on the underside of the vehicle and carrying a winding energized by a polyphase power supply, and *(ii)* a passive *reaction rail* in the form of a bar of conducting material, usually aluminum, embedded in the guideway. In order to reduce the reluctance of the magnetic-flux path, the reaction rail is either sandwiched between two energized structures (double-sided LIM or DLIM) or backed by a passive ferromagnetic structure (single-sided LIM or SLIM), as illustrated in Figures B.8.1*(a)* and *(b)* respectively. The SLIM may require a more expensive rail system, but alleviates dynamic stability problems encountered with the DLIM.

FIGURE B.8.1   Linear induction motor configurations: *(a)* DLIM (double-sided), *(b)* SLIM (single-sided).

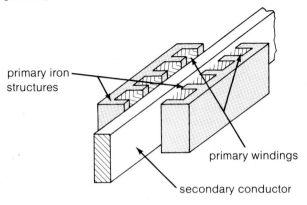

(a)

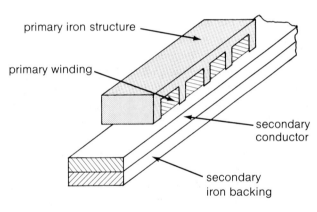

(b)

In its basic operation the LIM is similar to its rotating counterpart. The currents flowing in the *primary* energized winding set up an electromagnetic wave that travels backward with respect to the vehicle at synchronous speed. When the vehicle speed differs from the synchronous speed, the *secondary* reaction rail slips with respect to this traveling wave. The motionally induced voltage then generates a system of ac currents that interact with the impressed electromagnetic wave to produce a net thrust. There also develops a repulsive force that, however, is opposed by a force of attraction, whenever the secondary conductor is backed by ferromagnetic material. In addition to propulsion, the LIM can provide *levitation* and *guidance*. The translational or linear motion produced by a linear motor can proceed indefinitely, limited only by the length of its track.

The characteristic curve of rotational speed versus torque for a rotating induction motor is, for the case of a LIM, replaced by a curve of velocity versus thrust. The synchronous velocity of a LIM is given by

$$v_s = 2\tau f \ \text{m/s} \tag{B.8.1}$$

where τ is the pole pitch and f is the source frequency. The slip of the LIM is then given by

$$\mathsf{S} = \frac{v_s - v}{v_s} \tag{B.8.2}$$

In contrast with a rotating machine, the synchronous velocity of a LIM for a given frequency can be made to have any desired value and is not governed by the number of poles. Further, the number of poles on the primary need not be even, or indeed integral. The LIM operates at high slip with correspondingly low efficiency, and has a typical velocity-thrust curve that has the same general shape as the speed-torque curve of an induction motor with a rotor of reasonably high resistance. Also, the large gap between the ferromagnetic surfaces of the primary and secondary members in a SLIM or between the two primary members in a DLIM influences strongly the operating characteristics of a LIM.

A phenomenon present in the LIM, that has no parallel in the rotating induction motor, is known as the *end effect*. The LIM primary has an entry edge, at which the new secondary conductor continuously comes under the influence of the magnetic field, and an exit edge, at which the secondary conductor continuously leaves. By Lenz's law, the secondary current in the region of entry acts to prevent the build-up of gap-flux. As a result, when the motor is moving, the average gap-flux density for the first pair of poles near the entry edge tends to be significantly lower than that under later poles in the motor. Also, by Lenz's law, a current that tends to maintain the flux persists in the secondary conductor after it has left the exit edge. This current produces resistive loss without any corresponding development of thrust. While the general direction of the current beneath the primary is transverse, there are substantial regions where the orthogonal relationship between current, flux, and motion is not satisfied; in such regions, little or no thrust in the direction of motion is developed. The end effect, which is naturally most pronounced at high speed, reduces the maximum thrust that the motor can produce.

Large LIMs usually have three-phase primary windings and are employed for transportation, materials handling, extrusion presses, and the pumping of liquid metal. Small LIMs are used as curtain pullers, sliding-door closers, etc.

B.9   LINEAR SYNCHRONOUS MOTORS

The *linear synchronous machine* (LSM), although not so widely used at present, has such potential, particularly in the field of transportation, that a brief description of the mode of operation of two principal types is in order. The LSM develops a thrust when it operates at the synchronous velocity given by Equation B.8.1. The required system of interacting currents must be excited by a separate dc supply. The power involved, however, is small, because it needs to cover only the

FIGURE B.9.1 Schematic arrangement of windings in an air-cored linear synchronous motor (LSM).

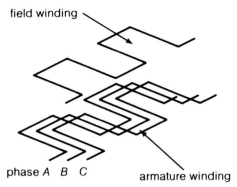

field winding

phase *A   B   C*   armature winding

ohmic loss; it can be drastically reduced by cryogenic cooling or totally eliminated by employing superconducting, or permanent magnets. Even with no excitation, a small amount of thrust can be obtained from the variable-reluctance effect of the nonenergized structure having salient poles. By varying the frequency of the supply through a power conditioner, such as a thyristor inverter, operation at variable speed is achieved.

In its simplest form, an air-cored LSM consists of two arrays of conductors, as illustrated in Figure B.9.1. The armature winding carries the polyphase system of currents to set up the traveling wave, and the field winding carries the dc excitation current. Such a scheme is practical only when the field is superconducting and is located on the vehicle.

The thrust obtained from a given system of conductor-currents can be increased by at least one order of magnitude by embedding both field and armature windings in iron structures. This leads to *iron-cored LSMs*. By placing the field and armature windings on board the vehicle, significant savings in the cost of the track can be achieved. Of particular interest are those machine types in which the track does not carry an alternating flux and, hence, need not be thinly laminated: the *claw-pole* or *Nadyne* (Figure B.9.2) and the *homopolar inductor* (Figure B.9.3).

In the intermediate speed range (150–350 km/h), iron-cored LSMs appear to have the edge over the LIMs, because they are less affected by end effects and can operate with larger clearances (3 cm), leading power factor, and better efficiency. Also, their strong attractive forces can be controlled independently of the thrust. At speeds of 350 km/h and over, the air-cored LSMs seem to have no competitors; they can operate with clearances up to 30 cm and require no power collectors. The practical realization, however, depends on major technological advances in superconducting magnets and power electronics.

FIGURE B.9.2   Sketch of a claw-pole motor (Nadyne).

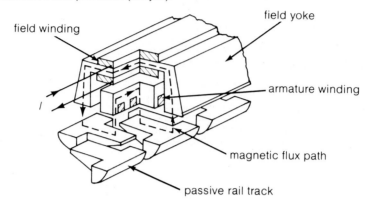

FIGURE B.9.3   Sketch of a homopolar linear synchronous motor.

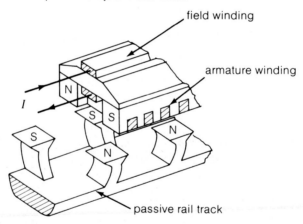

B.10   CRYOGENERATORS

The present development of technological bases for the production of future *cryogenerators* (also known as *superconductive generators*) indicates that large turboalternators with liquid helium-cooled, superconducting field windings are feasible, and that the projected advantages should be realizable through further intensive research and development. However, only the future will show at what rating the economic transition from the present conventional fluid cooling to cryo-cooling lies. Superconductivity does permit the present flux density limit of 2 to 2.5 T in the rotor iron to be exceeded considerably; however, one has to forgo the ferromagnetic properties of steel. Current loadings and power density can be increased significantly with practically no change in subtransient and transient reactances, and no deterioration in electrical stability.

FIGURE B.10.1   A typical basic layout of a cryogenerator with super-conducting rotor winding.

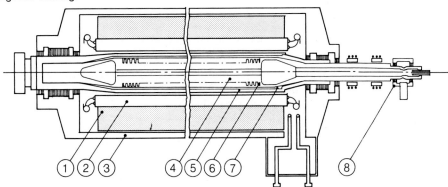

1. laminated iron shield
2. armature winding and supporting structure
3. stator housing
4. inner rotor
5. outer rotor shield (damper cylinder)
6. superconducting winding
7. rotor thermal insulation (vacuum)
8. rotor vacuum seals

Figure B.10.1 shows a typical basic layout of a superconducting generator.[1] The rotor is a rotating Dewar vessel containing a superconducting field winding of niobium-titanium (NbTi) alloy wire that is kept superconductive by liquid helium at 4 to 5 K. The cold rotor interior is thermally insulated in the radial direction by a vacuum space and a radiation shield, and in the axial direction by mechanical connections to the shaft with a high thermal resistance. The rotor outer cylinder is an electrical damping screen and a shield protecting the rotor winding from transient influences of the armature. The stator winding is directly water cooled, and its support structure must be of a nonmagnetic material and must resist high short-circuit forces and steady-state fatigue loads. Immediate surroundings must be shielded from the strong magnetic fields; the best rating per unit volume is achieved by using a shield of laminated magnetic iron, while minimum mass per unit volume results by using a conducting copper screen with a rating per unit volume reduced to about two-thirds.

[1]Abegg, K., "The Growth of Turbogenerators," *Philosophical Transactions of the Royal Society of London* 275, no. 1248, August 1973, pp. 51–67.

Potential advantages of superconductive generation appear to be the following as compared to the conventional fluid-cooling technology: reduced size and weight, higher efficiency, higher voltage and lower current levels, improved system performance, higher rotor critical speeds, and reduced cost.

Some of the key problem areas presenting serious design challenges are the following: electromagnetic rotor shield, overspeed control requirements, optimized superconducting rotor, and optimized stator winding and its structure. In the final analysis, relative reliability, availability, and cost become overriding considerations in a practical sense before the new class of cryogenerators can be put into commercial operation on a large scale.

SOLUTIONS TO ODD-NUMBERED PROBLEMS

CHAPTER 1

1-1

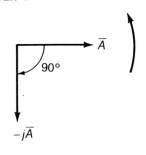

(a)

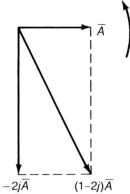

(c) $(1 - 2j)\,\overline{A} = \overline{A} - 2j\overline{A}$

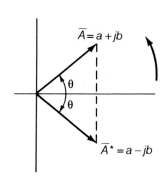

(b)

(d)

1-3.

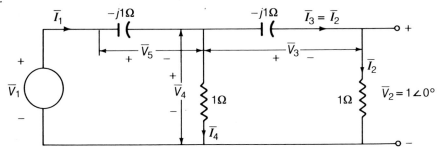

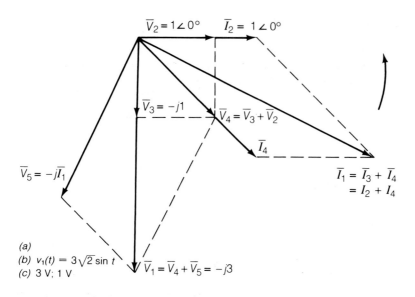

(a)

(b) $v_1(t) = 3\sqrt{2} \sin t$

(c) 3 V; 1 V

1–5. (a) 2,328.44 V

(b) 2,327.4 V

(c) 2,363.2 V

CHAPTER 2

2–1.

$$\overline{V}_{AN} = \frac{100}{\sqrt{3}} \angle -30°$$

$$\overline{V}_{BN} = \frac{100}{\sqrt{3}} \angle -150°$$

$$\overline{V}_{CN} = \frac{100}{\sqrt{3}} \angle 90°$$

(a)

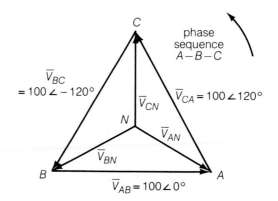

phase
sequence
$A - B - C$

$\overline{V}_{BC} = 100 \angle -120°$

$\overline{V}_{CA} = 100 \angle 120°$

$\overline{V}_{AB} = 100 \angle 0°$

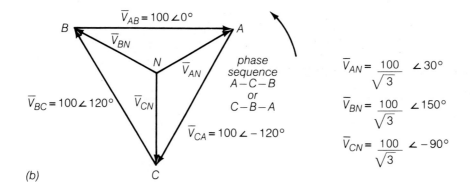

(b)

phase
sequence
$A - C - B$
or
$C - B - A$

$$\overline{V}_{AN} = \frac{100}{\sqrt{3}} \angle 30°$$

$$\overline{V}_{BN} = \frac{100}{\sqrt{3}} \angle 150°$$

$$\overline{V}_{CN} = \frac{100}{\sqrt{3}} \angle -90°$$

2-3. (a) $I_L = 32.8$ A $= I_{ph}$

    (b) $I_L = 32.8$ A; $I_{ph} = 18.94$ A

2-5. (a) 20 A; 0.8 lagging; 6 kVA; 4.8 kW; 3.6 kVAR

    (b) Fig. 2.2.2(b) of the text applies except that the diagram as a whole must be rotated such that $\overline{V}_{AB}$ becomes horizontal (i.e., a reference) with 0° angle.

    (c) Zero

2-7. (a) 21.2 kW

    (b) 45.56 kW

2-9. (a) 360.54 kW

    (b) 94.08 kW

    (c) 21.6 kW

2-11. 162 V

2-13. 751.34 V; 784.2 V; 16.325 kW; 12.824 kVAR

2-15. 274.2 $\angle \pm$ 40.89° Ω/ph.; No.

2-17. (a) 13.5 $\angle$75.6° Ω/ph.

    (b) 13.5 $\angle-$75.6° Ω/ph.

CHAPTER 3

3-1. $\nabla \cdot \overline{D} = \rho; \nabla \cdot \overline{B} = 0; \nabla \times \overline{E} = -\dfrac{\partial \overline{B}}{\partial t}; \nabla \times \overline{H} = \overline{J} + \dfrac{\partial \overline{D}}{\partial t}$

In low-frequency problems, the dielectric effects or displacement currents can be neglected. The material regions can be considered as void of the volume-charge density.

$\nabla \cdot \overline{J} = -\dfrac{\partial \rho}{\partial t}$

3-3. $\overline{F}_{21} = -\dfrac{\mu_o}{2\pi} \dfrac{I_1 I_2}{r} \overline{a}_r$ N/m, where $\overline{a}_r$ is the unit vector from $d\overline{l}_1$ to $d\overline{l}_2$. If I_2 and I_1 are in the same direction, the force pulls the wires together; when the wires carry current in opposite directions, the wires are repelled by the magnetic force.

3-5. (a) 11.43×10^{-3} Wb/m²

    (b) 11.43×10^{-3} Wb-turns

    (c) 0.652%

    (d) 0.146 Ω

3-7. 0.0006 Wb

3-9. 0.6×10^{-3} Wb; 0.96 T; 0.764×10^6 At/m.

3-11. 3.5 J; 6.5 J

3-13. (a) $v_1 = 0$; $v_2 = 0$; $W_m = 5$ μJ

    (b) $v_1 = 10 \cos 100t$ mV; $v_2 = 0.1 \cos 100t$ V; $W_m = 0$

    (c) $v_1 = [-10 \sin t + 3 \cos (t + 30°)]$ mV; $v_2 = [- \sin t + 30 \cos (t + 30°)]$ mV; $W_m = 1.775 \times 10^{-3}$ mJ

3-15.

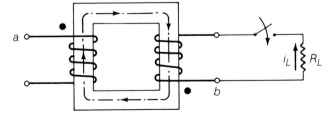

3-17. (a) $\phi = \dfrac{\mu_0 I l}{4\pi}$ Wb.

(b) $w_m = \dfrac{\mu_0 I^2 \rho^2}{8\pi^2 R^4}$ J/m³

(c) $W_m = \dfrac{\mu_0 I^2 l}{16\pi}$ J

(d) $L = \dfrac{\mu_0 l}{8\pi}$ H

3-19. 0.5 T

3-21. (a) 56.84 cm³; $l_m = 4.55$ cm; $A_m = 12.5$ cm².

(b) 170.5 cm³; $l_m = 5.46$ cm; $A_m = 31.23$ cm².

(c) 1.044 T

(d) 81.85 cm³; $l_m = 1.36$ cm; $A_m = 60.2$ cm².

CHAPTER 4

4-1. (a) 75.4 cos 377t V

(b) 53.32 V

4-3. (a) $a = 2$; $I_1 = 45.45$ A; $I_2 = 90.91$ A

(b) $I_{2(LV)} = 55$ A; $I_{1(HV)} = 27.5$ A

(c) 8 Ω

4-5. (a) $\lambda_1 = L_{11} i_1 - L_{12} i_2$; $\lambda_2 = -L_{22} i_2 + L_{12} i_1$

$$v_1 = R_1 i_1 + p\lambda_1 = R_1 i_1 + L_{11} \frac{di_1}{dt} - L_{12} \frac{di_2}{dt}$$

$$v_2 = -R_2 i_2 + p\lambda_2 = -R_2 i_2 - L_{22} \frac{di_2}{dt} + L_{12} \frac{di_1}{dt}$$

(b) $L_{m1} = \dfrac{N_1^2}{R_m}$, where R_m is the reluctance of the core

$L_{m2} = \dfrac{N_2^2}{R_m}$, which is the magnetizing inductance referred to the $N_2 -$ winding

$L_{12} = \dfrac{N_1 N_2}{R_m} = L_{21}$

(c) All inductances become infinite and the voltage equations become indeterminate.

4-7. (a)

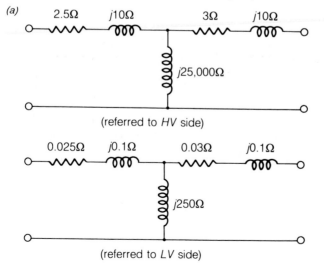

(referred to HV side)

(referred to LV side)

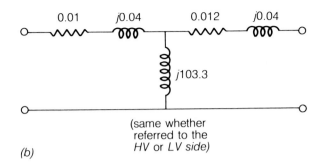

(b)

(same whether
referred to the
HV or *LV side*)

(c) 2,306.65 V

4-9. *(a)* 0.375 A

(b) 2.47 A

(c) 0.15

4-11. 0.964

4-13. *(a)* 0.977; 0.983; 0.982; 0.98; 0.978

(b) 0.971; 0.962 for 0.25 pu load

(c) $k = 0.577$ pu of full-load; 0.983; 0.979; 0.972

4-15. *(a)* 0.076

(b) 0.95

(c) 0.948

4-17.

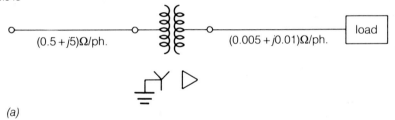

(a)

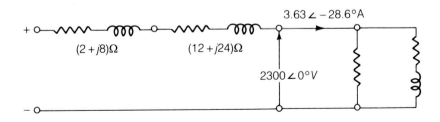

(b)

(c) 3.63 A; 4,159.6 V

4-19. $V_{L\text{-}L\ HV} = V_{ph\ HV} = 2{,}400$ V; $V_{L\text{-}L\ LV} = 208$ V; $V_{ph\ LV} = 120$ V;
  $I_{L\ LV} = 83.27$ A $= I_{ph\ LV}$; $I_{ph\ HV} = 4.16$ A; $I_{L\ LV} = 7.21$ A;
  $\dfrac{V_{L\text{-}L\ HV}}{V_{L\text{-}L\ LV}} = 11.54$; $\dfrac{I_{L\ HV}}{I_{L\ LV}} = 0.087$

4-21. $R_{eq\ HV} = 10.24$ Ω/ph.; $X_{eq\ HV} = 54.55$ Ω/ph.
  $R_{eq\ LV} = 4.87 \times 10^{-3}$ Ω/ph.; $X_{eq\ LV} = 26 \times 10^{-3}$ Ω/ph.
  $R_{eq} = 0.011$ pu; $X_{eq} = 0.06$ pu

4-23.   0.982; 0.0416

4-25.

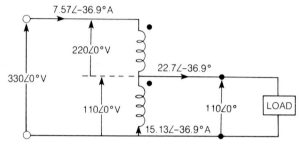

4-27. $j\,0.033$; $j\,0.054$; $j\,0.071$

4-29.   (a) $\epsilon_1 = 0.019$; $\epsilon_2 = 0.0198$; $\epsilon_3 = -0.005$
  (b) $\epsilon_{12} = 0.039$; $\epsilon_{13} = 0.014$; $\epsilon_{23} = 0.025$; $\epsilon_{32} = -0.025$

4-31.   (a) $(370 - j277)$; $(231 - j173)$; The 250-kVA transformer is overloaded because of its smaller per-unit leakage reactance.
  (b) $(359 - j305)$; $(241 - j305)$; The 250-kVA transformer operates with a 12.5% overload instead of 15.4% overload in Part (a).

CHAPTER 5

5-1. $F_e = -\dfrac{\partial W_e(v,\,x)}{\partial x} + v\,\dfrac{\partial q(v,\,x)}{\partial x}$; $F_e = +\dfrac{\partial W_e{}'(v,\,x)}{\partial x}$

  $F_e = -\dfrac{\partial W_e(q,\,x)}{\partial x}$; $F_e = +\dfrac{\partial W_e{}'(q,\,x)}{\partial x} - q\,\dfrac{\partial v(q,\,x)}{\partial x}$

5-3. $L = \dfrac{N^2 A\,\mu_0}{x + t}$; $F_e = -\dfrac{1}{2}\,i^2\,\dfrac{N^2 A \mu_0}{(x + t)^2}$

5-5.   333.3 N

5-7.   (a) $\pm\,377$ rad/s; $\pm\,188.5$ rad/s
  (b) 0.5 N·m; 188.5 W; 0.75 N·m; 141.4 W

5-9.   (a) 314 rad/s
  (b) 0.2 N·m; 62.8 W
  (c) 1.2 A

5-11.   (a) $T_e = -\,i_a\,i_f\,L\,\sin\theta + i_b\,i_f\,L\,\cos\theta$
  (b) $T_e = -\,I_a\,I_f\,L\,\sin\delta$; a steady torque

  (c) $v_{ta} = L_{aa}\dfrac{di_a}{dt} + e_{af}$; $e_{af} = -\,\omega\,L\,I_f\,\sin(\omega t + \delta)$

  $v_{tb} = L_{aa}\dfrac{di_b}{dt} + e_{bf}$; $e_{bf} = +\,\omega\,L\,I_f\,\cos(\omega t + \delta)$

5-13. (a) $T_e = -2 I_a^2 L_2 \sin 2\delta - \sqrt{2}\, I_a I_f L \sin \delta$; a steady torque

(b) An additional term proportional to $\sin 2\delta$ that does not depend on I_f.

(c) With a negative value of δ and a positive value of T_e, the machine will run as a synchronous motor; it will also run as a generator, if driven mechanically, with a positive value of δ and a negative value of T_e.

(d) Will run as a two-phase reluctance motor or generator.

5-15. $T_e = 0.98 \cos^2 314\, t \left[\dfrac{\sin \theta + 4 \sin 2\theta}{(6 + \cos \theta + 2 \cos 2\theta)^2} \right] \text{N·m}$

5-17. No for Parts (c), (d), (e), and (f).

Yes for Parts (a), (b), (g), and (h).

5-19. 5,100 or 2,100 rpm in either direction

5-21. (a) 160 V

(b) 277 V

(c) $e_{an} = 160 \sqrt{2} \sin 377t$; $e_{bn} = 160 \sqrt{2} \sin (377t - 120°)$; $e_{cn} = 160 \sqrt{2} \sin (377t + 120°)$

$e_{ab} = 277 \sqrt{2} \sin (377t + 30°)$; $e_{bc} = 277 \sqrt{2} \sin (377t - 90°)$; $e_{ca} = 277 \sqrt{2} \sin (377t + 150°)$

5-23. 1,200 rpm; 1,187.3 N·m

5-25. (a) 50 A; 159.15 N·m

(b) 56 A; 1,200 rpm; 178 N·m

(c) 1,320 rpm; 0 A

5-27. (a) 8

(b) 0.067; 3 1/3 Hz

(c) 50 rpm; 750 rpm

5-29. (i) Characteristic moves to the left with synchronous speed halved.

(ii) Torque is 1/4 of its previous value.

5-31. 180 Hz

5-33. 45 Ω; 48.23 μF

5-35. (a) 0.85

(b) 0.82

CHAPTER 6

6-1.

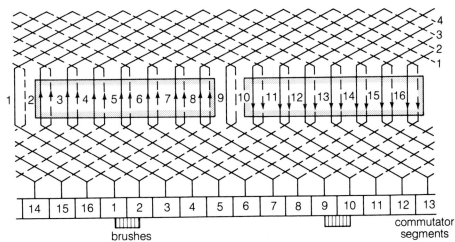

6-3.

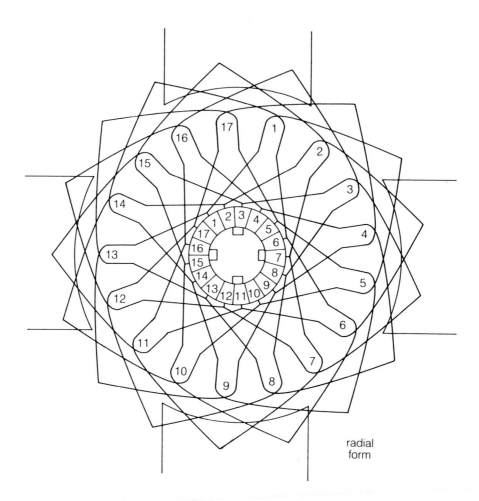

radial
form

6-5. (a) 100 A
(b) 50 A

6-7.

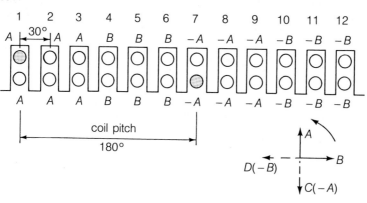

6–9.

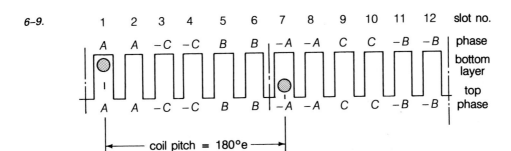

(a) 60 electrial degree phase spreading; full-pitch coils; $k_{d1} = 0.966$; $k_{p1} = 1$

1 2 3 4 5 6 7 8 9 10 11 12

A A −C −C B B −A −A C C −B −B

A −C −C B B −A −A C C −B −B A

$\dfrac{5}{6}$ pole pitch = 150°e

(b)   5/6 full-pitch coils; $k_{d1} = 0.966$; $k_{p1} = 0.966$

6–11.

phase belt

1 2 3 4 5 6 7 8 9 10 11 12 13 14 15 16 17 18 slot no.

A A A −C −C −C B B B −A −A −A C C C −B −B −B phase

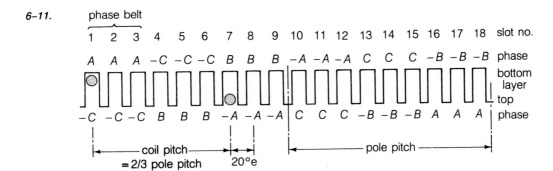

−C −C −C B B B −A −A −A C C C −B −B −B A A A phase

coil pitch
= 2/3 pole pitch   20°e   pole pitch

(a) 60°e spread; $k_{d1} = 0.96$; $k_{d5} = 0.2176$; $k_{d7} = -0.177$

$k_{p1} = 0.866$; $k_{p5} = -0.866$; $k_{p7} = -0.866$

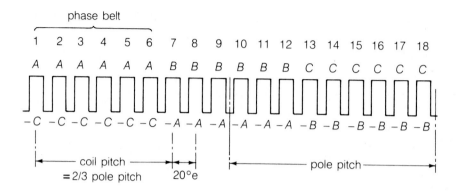

(b)   120°e spread; $k_{d1} = 0.8314$; $k_{d5} = -0.188$; $k_{d7} = 0.1535$

$$k_{p1} = 0.866; \ k_{p5} = -0.866; \ k_{p7} = -0.866$$

6-13. $k_{d1} = 0.958$; $k_{p1} = 0.966$; $E_{ph} = 1,972.27$ V; $E_{L-L} = 3,416$ V

6-15.   (a) 140 V
(b) 420 V

6-17.   6,890; 204; 63

CHAPTER 7

7-1.   (a) 3,600 rpm
(b) 180 rpm
(c) 3,600 rpm
(d) 0
(e) 3 Hz
(f) 50 V; 2.5 V

7-3. $R_1 = 0.1 \ \Omega$; $X_{l1} = 0.141 \ \Omega = X_{l2}'$; $R_2' = 0.105 \ \Omega$; $X_M = 6.159 \ \Omega$; 880 W

7-5.   62.5 N·m

7-7.   (b) $\dfrac{I_2'}{I_{2 \ max \ T}'} = \dfrac{\sqrt{2}\,(S/S_{maxT})}{\sqrt{(S/S_{maxT})^2 + 1}}$

7-9. $S_{maxT} = 0.5$; $S_{FL} = 0.134$

7-11.   (a) 0.1
(b) 38,060 W
(c) 459.2 hp
(d) 0.3
(e) $3 \ I_{2 \ FL}'$
(f) $1.2 \ T_{FL}$
(g) $4 \ I_{2FL}'$

7-13. $1.25 \ T_{FL}$; $2.63 \ T_{FL}$; 0.253

7-15.   1,150 W; 1,600 rpm; 0.818 Ω per phase of wye.

7-17.   (a) 1.52; 2,705 N·m
(b) 231 A; 2,083 N·m

7-19.   243.36 A; 867 kW; −465 kVAR

7–21.   *(i)* *(a)* 1/2
        (b) 4
        (c) 1
       (ii) *(a)* 1/2
        (b) 2
        (c) 2

7–23.   3.57 A; 0.645 lagging; 0.21 hp; 1,710 rpm; 0.87 N·m; 0.616

7–25.   174.3 μF

CHAPTER 8

8–1.   *(a)* 0.256
       (b) −0.053
       (c) 0.894 leading

8–3.   0.969

8–5.   *(a)* 59.35 × 10³ N·m; 87.05 × 10³ N·m
       (b) 978 A

8–7.   *(a)* −10.25°; 9.1 A
       (b) 16.15 A; 6.1 A
       (c) 14,350 W; 0.83 leading
       (d) 72.9 N·m

8–9.   *(a)* 1.753; 0.825
       (b) 0.8

8–11. *(a)* 3,179 V line-to-line; −17.76°
       (b) 230.5 A; 0.985 lagging

8–13. *(a)* 424.67 kVAR
       (b) 0.932 lagging

8–15. 5.87 kV line-to-line; 0.21

8–17. *(a)* 23.85°
       (b) 1.357
       (c) 4,473 W

8–21. 1.25; 0.75; 43.29 Ω or 1.25 pu; 25.98 Ω or 0.75 pu

8–23. 5.6 A

8–25. *(i)* *(a)* 60 MW
         (b) 45 MVAR
         (c) 3,138 A
         (d) 0.8 lagging
         (e) 22.55 kV line-to-line
         (f) 23.82°
       (ii) *(a)* $P_1 = P_2 = 60$ MW
         (b) $Q_1 = 76.84$ MVAR; $Q_2 = 13.16$ MVAR
         (c) $I_1 = 4,079$ A; $I_2 = 2,569$ A
         (d) $pf_1 = 0.6154$ lagging; $pf_2 = 0.977$ lagging
         (e) $E_{f1} = 1.96$ or 27.05 kV line-to-line; $E_{f2} = 1.32$ or 18.2 kV line-to-line
         (f) $\delta_1 = 19.7°$; $\delta_2 = 30°$
       (iii) *(a)* $P_1 = 72$ MW; $P_2 = 48$ MW
         (b) $Q_1 = 72.12$ MVAR; $Q_2 = 17.88$ MVAR
         (c) $I_1 = 4,267.5$ A; $I_2 = 2,143$ A
         (d) $pf_1 = 0.706$ lagging; $pf_2 = 0.937$ lagging
         (e) $E_{f1} = 1.96$ or 27.05 kV line-to-line; $E_{f2} = 1.31$ or 18.1 kV line-to-line
         (f) $\delta_1 = 23.8°$; $\delta_2 = 23.8°$

CHAPTER 9

9-1. (a) 270.2 V
(b) 260.2 V

9-3. 0.305 Ω; 132.2 Ω

9-5. 33.63 A; 594.6 rpm

9-7. (a) 38 A
(b) 300 N·m; 284.8 N·m; 0.848
(c) 522.6 rpm

9-9. (a) 794.5 rpm
(b) 888 rpm

9-11. 2,031 rpm

9-13.

Step No.	Current (A)			Speed (rpm)	
	Initial	Final		Initial	Final
1	39.0	25.3		0	667
2	39.0	25.3		667	1093
3	39.0	25.3		1093	1370
4	39.0	25.3		1370	1557
5	39.0	25.3		1557	1675
6	39.0	25.3		1675	1755
7	39.0	25.3		1755	1800

9-15. (a) $V_t = 485$ V; $I_{L1} = 1,500$ A; $I_{L2} = 1,250$ A; $I_{L3} = 1,600$ A
(b) $V_t = 500$ V; $I_{L1} = -1,500$ A; $I_{L2} = 500$ A; $I_{L3} = 1,000$ A

9-17. 0.9

9-19. (i) (a) 1,082 rpm
(b) 159.2 hp
(c) 1,048 N·m
(ii) (a) 1,393 rpm
(b) 81.7 hp
(c) 418 N·m

9-21. 227 V

9-23. (a) 1,099 rpm
(b) 1,043.5 rpm

9-25. 2.6 turns/pole; with $N_s = 3$, $R_{diverter} = 6.5 \, R_{se}$

9-27. (a) 1,806 rpm
(b) 20,352 W
(c) 25 hp
(d) 98.7 N·m
(e) 0.88

CHAPTER 10

10-3. 747 A

10-5. $\cos t$; $-\sin t$; 0

10-7. 0.11; 1.1; 0.9845; 1.1; 1.1996; 0.00043; 1.1; 1.235; 1.1178; 0.0046; 0.9845; 0.0414; 0.002; 1.1436

10-11. Only (a)

10-13. (a) 10.05
(b) 10.05; base value is the peak value of the rated armature current
(c) 6.73; 6.53
(d) 1.414

10-21. 610 hp

10-29. $1,778 \, e^{-t/0.057}$ A

CHAPTER 11

11–1. $\dfrac{V_t(s)}{V_f(s)} = \dfrac{1}{(1+s)^2}$

11–3.   (a) 75 H
   (b) 0.0075 H
   (c) 1.99 H
   (d) 3.78

11–5.   (a) 62.8 rad/s
   (b) $66.7\,e^{-35.2t}$ A; $(65.3 - 2.5\,e^{-35.2t})$ rad/s
   (c) $(54.2\,e^{-10t}\sin 24.6t)$ A; $[65.31 - 2.7\,e^{-10t}\cos(24.6t - 22.2°)]$ rad/s.

11–7. $23.57\,(1 - e^{-109.1t})$ A; $(271.04 - 13.9\,e^{-109.1t})$ V

11–9. $\dfrac{1}{2}\,J\,\omega_m^2$

11–15. $(200\,e^{-2t} - 100)$ rad/s; 0.278 s

11–17.

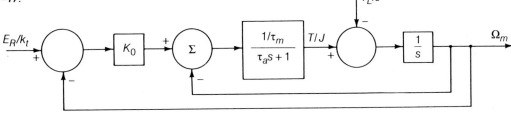

$$K_0 = \frac{K_t K_A}{K_m}\,;\; \tau_m = \frac{JR_a}{K_m^2}\,;\; \tau_a = \frac{L_a}{R_a}$$

(a)

(b) $\dfrac{K_0}{K_t}\,\dfrac{1}{\tau_m S\,(\tau_a S + 1) + 1 + K_0}$

(c) $-\dfrac{R_a}{K_m^2}\,\dfrac{\tau_a S + 1}{\tau_m S\,(\tau_a S + 1) + 1 + K_0}$

(d) $\sqrt{\dfrac{1+K_0}{\tau_a\,\tau_m}}\,;\; \dfrac{1}{2\tau_a}\,;\; \dfrac{1}{2}\sqrt{\dfrac{\tau_m}{\tau_a}\cdot\dfrac{1}{1+K_0}}$

(e) $\dfrac{1}{K_t}\cdot\dfrac{K_0}{1+K_0}$

(f) $-\dfrac{R_a}{K_m^2}\,\dfrac{1}{1+K_0}$

11–19.   (a) 227.3 V; 224.2 V
   (b) 0.045 s; 0.05 s
   (c) 194.2 V

11–21. $[500 + 136.8\sin(20t - 65.75°)]$ A; $[47.5 - 0.88\sin(20t - 27°)]$ rad/s.

11–23.   (a) 10 A; 0; 30 rad/s
   (b) 0; $285.7\angle 83.11°$A; $12\angle -6.89°$ rad/s.
   (c) 10 A; $[285.7\sin(3t + 83.11°)]$ A; $[30 + 12\sin(3t - 6.89°)]$ rad/s.

11–25. $(320.4 + 638.9\,e^{-5t} - 959.3\,e^{-3.33t})$ V

11–27.

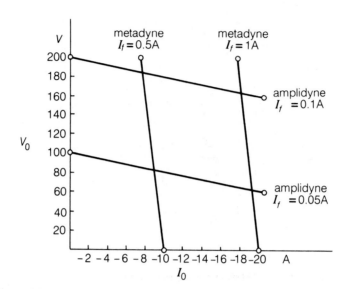

INDEX